The World of
MATHEMATICS
Volume 3

Edited by
James R. Newman

DOVER PUBLICATIONS, INC.
Mineola, New York

Bibliographical Note

This Dover edition, first published in 2000, is an unabridged republication of the 4-volume work originally published by Simon & Schuster, New York, in 1956. Page iii constitutes a continuation of this copyright page.

Library of Congress Cataloging-in-Publication Data

The world of mathematics / edited by James R. Newman.
     p. cm.
   Originally published: New York : Simon and Schuster, 1956.
   Includes bibliographical references and indexes.
   ISBN 0-486-41153-2 (pbk. : v. 1)—ISBN 0-486-41150-8 (pbk. : v. 2)—ISBN 0-486-41151-6 (pbk. : v. 3)—ISBN 0-486-41152-4 (pbk. : v. 4)
     1. Mathematics. I. Newman, James Roy, 1907–1966.

QA7 .W67 2000
510—dc21

                                                                    00-027006

Manufactured in the United States by Courier Corporation
41151605   2013
www.doverpublications.com

THE EDITOR wishes to express his gratitude for permision to reprint material from the following sources:

Appleton-Century-Crofts, Inc., for "History of Symbolic Logic," from *Symbolic Logic*. © 1932 by C. I. Lewis and C. H. Langford.

Blackwell & Mott Ltd. for "Mathematics and the World," by Douglas Gasking, from *Logic and Language*, 2nd series, edited by A. G. N. Flew.

Cambridge University Press for "The Theory of Groups," from *New Pathways in Science*, by Sir Arthur Stanley Eddington.

Dover Publications, Inc., for "Measurement" and "Numerical Laws and the Use of Mathematics in Science," from *What Is Science?*, by Norman Campbell.

Ginn and Company for "A Mathematical Science," from *Projective Geometry*, Vol. 1, by Oswald Veblen and John Wesley Young. © 1910, 1938.

Frau Doctor Lilly Hahn for "The Crisis in Intuition" and "Infinity," by Hans Hahn.

Harvard University Press for "The Essence of Mathematics," from *Collected Papers of Charles Sanders Peirce*, Vol. IV, edited by Charles Hartshorne and Paul Weiss, © 1933 by The President and Fellows of Harvard College; and for "Mathematical Postulates and Human Understanding," from *Positivism, A Study in Human Understanding*, by Richard von Mises, © 1951 by The President and Fellows of Harvard College.

Mrs. Sarah Y. Keyser for "The Group Concept," from *Mathematical Philosophy*, by Cassius Jackson Keyser.

Mathematical Association of America for "Geometry and Empirical Science" and "On the Nature of Mathematical Truth," by Carl G. Hempel, from *American Mathematical Monthly*.

W. W. Norton & Company for "Mathematics and the Metaphysicians," from *Mysticism and Logic*, by Bertrand Russell. © 1929 by W. W. Norton & Company, Inc.

The Public Trustee and the Society of Authors for "The Vice of Gambling and the Virtue of Insurance," from *Everybody's Political What's What*, by George Bernard Shaw.

Oliver and Boyd Ltd. for "Mathematics of a Lady Tasting Tea," from *Design of Experiments*, by Sir Ronald Fisher.

Oxford University Press for "Sampling and Standard Error," from *Statistics*, by L. H. C. Tippett; and for "Symbolic Logic," from *Introduction to Logic*, by Alfred Tarski, © 1941, 1946, by Oxford University Press, Inc.

Penguin Books Ltd. for "On the Average and Scatter," from *Facts from Figures*, by M. J. Moroney.

Messrs A. D. Peters for "Mathematics as an Art," from *Aspects of Science*, by John William Navin Sullivan.

Princeton University Press for excerpts from *How to Solve It*, by G. Polya, © 1945 by Princeton University Press.

Science for "The Mathematical Way of Thinking," by Hermann Weyl, from November 15, 1940, issue.

Simon and Schuster, Inc., for "New Names for Old" and "Paradox Lost and Paradox Regained," from *Mathematics and the Imagination*, © 1940 by Edward Kasner and James R. Newman.

John Wiley & Sons for "The Axiomatic Method," from *Introduction to the Foundations of Mathematics*. © 1952 by R. L. Wilder.

All four volumes of *The World of Mathematics* are available from Dover Publications. In addition to this volume, they are:

The World of Mathematics, edited by James R. Newman
Volume 1 (768 pp.)   ISBN 0-486-41153-2

Parts I–IV:
General Survey
Historical and Biographical
Arithmetic, Numbers and the Art of Counting
Mathematics of Space and Motion

The World of Mathematics, edited by James R. Newman
Volume 2 (720 pp.)   ISBN 0-486-41150-8

Parts V–VII:
Mathematics and the Physical World
Mathematics and Social Science
The Laws of Chance

The World of Mathematics, edited by James R. Newman
Volume 4 (464 pp.)   ISBN 0-486-41152-4

Parts XVIII–XXVI:
The Mathematician
Mathematical Machines: Can a Machine Think?
Mathematics in Warfare
A Mathematical Theory of Art
Mathematics of the Good
Mathematics in Literature
Mathematics and Music
Mathematics as a Culture Clue
Amusements, Puzzles, Fancies

Table of Contents

VOLUME THREE

PART X: Mathematics of Infinity

PART XI: Mathematical Truth and the Structure of Mathematics

PART XII: The Mathematical Way of Thinking

PART XIII: Mathematics and Logic

PART XIV: The Unreasonableness of Mathematics

PART XV: How to Solve It

PART XVI: The Vocabulary of Mathematics

PART XVII: Mathematics as an Art

COMMENTARY ON

An Ingenious Army Captain and on
a Generous and Many-sided Man

STATISTICS was founded by John Graunt of London, a "haberdasher of small-wares," in a tiny book called *Natural and Political Observations made upon the Bills of Mortality*.[1] It was the first attempt to interpret mass biological phenomena and social behavior from numerical data—in this case, fairly crude figures of births and deaths in London from 1604 to 1661. Graunt's tract appeared in 1662. Thirty years later, the Royal Society published in its "Philosophical Transactions" a paper on mortality rates written by the eminent astronomer Edmund Halley. This famous article was entitled "An Estimate of the Degrees of the Mortality of Mankind, drawn from curious Tables of the Births and Funerals at the City of Breslaw; with an Attempt to ascertain the Prices of Annuities upon Lives." It was followed by "Some further Considerations on the Breslaw Bills of Mortality." Together, the papers are the foundation for all later work on life expectancy, indispensable of course to the solvency of life-insurance companies.[2]

John Graunt was born in 1620 in Birchin Lane, London, "at the Sign of the Seven Stars," where his father kept a shop and home. He was early apprenticed to a merchant in small wares—buttons, needles and the like—and prospered in the trade. Success gave him the leisure to indulge interests somewhat broader than those of the notions counter. Aubrey describes him as "a very ingenious and studious person . . . [who] rose early·in the morning to his Study before shoptime."[3] He became a friend of Sir William Petty, later the author of a well-known book on the new study of political arithmetic, and probably discussed with him the ideas to be expressed in the *Observations*. The Bills of Mortality which attracted Graunt's attention were issued weekly by the company of parish clerks and listed the number of deaths in each parish, the causes, and also an "Accompt of all the Burials and Christnings,

---

[1] The full title is, *Natural and Political Observations Mentioned in a following Index, and made upon the Bills of Mortality*.

[2] "He not only gave a sound analysis of this problem (the calculation of annuity prices), but he put his results in such a convenient form that this first table of mortality has remained the pattern for all subsequent tables, as to its fundamental form of expression."—Lowell J. Reed in the introduction to *Degrees of Mortality of Mankind* by Edmund Halley, a reprint of the papers noted, issued by the Johns Hopkins Press, Baltimore, 1942; p. iv. The selection by Halley is based on this reprint.

[3] *Aubrey's Brief Lives*, edited by Oliver Lawson Dick; London, 1950, p. 114.

hapning that Week." They are described fully in the material selected from Graunt's book.

Charles II was so favorably impressed by the *Observations* that he specially proposed Graunt as an original member of the newly incorporated Royal Society. To forestall any possible objections on the ground that Graunt was a shopkeeper, "his Majesty gave this particular charge to his Society, that if they found any more such Tradesmen, they should be sure to admit them all, without any more ado." [4] He was elected F. R. S. in 1662.

The merit of the *Observations* was immediately recognized and encouraged the gathering and study of vital statistics on the Continent—particularly in France—as well as in England. The book went through several editions, the fifth of which, published after Graunt's death, was enlarged by Petty. Historians have long been vexed to decide how much Petty contributed to the original work. Aubrey, who delighted in retailing small malices, says only that Graunt had his "Hint" from Petty, but he implies much more. There seems no doubt that the book was a joint production. Graunt wrote by far the greater part, including the most valuable scientific portions; Petty, it may be supposed, added what Thomas Browne would have called "elegancy" and thereby increased the popularity of the book. Sir William was a bumptious and somewhat inflated man, unable to decide whether to patronize Graunt or to claim credit for his work. There is no evidence that he even understood the importance and originality of what his friend had done.[5] The last sentence of the preface is unmistakably Graunt's: "For herein I have, like a silly Scholeboy, coming to say my Lesson to the World (that Peevish, and Tetchie Master) brought a bundle of Rods wherewith to be whipt, for every mistake I have committed."

Graunt served as a member of the city common council and in other offices, but on turning Catholic—he was raised a Puritan—"layd down trade and all other publique Employment." Aubrey tells us that he was a man generally beloved, "a faythfull friend," prudent and just. "He had an excellent working head, and was very facetious and fluent in his conversation." He was accused of having had "some hand" in the great fire of London, and the fact that he was a Catholic gave impetus to the charge. It was said that, as an officer of a water company, he had given orders stopping the water supply just before the fire started. A diligent eighteenth-century historian proved this false by showing that Graunt had

---

[4] Tho. Sprat, *The History of the Royal Society of London, for the improving of Natural Knowledge*; 3rd Edition, London, 1722, p. 67.

[5] For a meticulous sifting of the evidence as to Graunt vs. Petty see the introduction to a reprint of the *Observations* (Baltimore, The Johns Hopkins Press, 1939), by Walter F. Willcox. As to Petty, no inconsiderable person even if he was inflated and bumptious, see E. Strauss, *Sir William Petty, Portrait of a Genius*, Glencoe (Ill.), 1954.

had no connection with the company until a month after the fire. Graunt died of jaundice on Easter-eve 1674, and was buried "under the piewes" in St. Dunstan's church. "What pitty 'tis," wrote Aubrey, "so great an Ornament of the Citty should be buryed so obscurely!"

*　　*　　*　　*　　*

Unlike poor Graunt, whom my edition of the *Britannica* does not deign even to notice, Edmund Halley has been amply celebrated. I shall dispose of him as briefly as possible. He was born in London in 1658, the son of a wealthy "Soape-boyler," and he enjoyed every advantage, including an excellent education, that rich and indulgent parents could confer. His passion for mathematics and astronomy showed itself in his youth: when he arrived at Queen's College, Oxford, he brought with him a large assortment of astronomical instruments, including a 24-foot telescope, whose use he had already mastered. His reputation as a theoretician and observer was established by the time he was twenty. He left the college before finishing his course, to make southern hemisphere observations at St. Helena. On his return, and by the King's command, he was awarded a Master of Arts degree; a few days later he was elected a Fellow of the Royal Society. He was then twenty-two. The next few years were spent on various astronomical labors which required him to travel widely on the Continent. Becoming deeply interested in the problem of gravity, he visited Newton at Cambridge in August 1684. It was a momentous meeting, for it resulted in the *Principia*, a work which might never have appeared except for Halley's extraordinary exertions. He suggested the project in the first place; he averted suppression of the third book; he bore all the expenses of printing and binding, corrected the proofs, and laid his own work entirely aside to see Newton's masterpiece through the press. The expense was assumed at a time when Halley could ill afford it. His father had suffered serious reverses before he died and had left an encumbered and almost worthless estate.

Halley's long life was crowded with literary and scientific activity. He was a classical scholar, hydrographer, mathematician, physicist, and astronomer. His writings include, besides a vast output in his specialty, such diverse items as "An Account of the Circulation of the Watery Vapours of the Sea, and of the Cause of Springs"; "Discourse tending to prove at what Time and Place Julius Caesar made his first Descent upon Britain"; "New and General Method of finding the Roots of Equations"; a translation from the Arabic—which language he learned for this purpose—of Apollonius' treatise *De sectione rationis* and a brilliant restoration of his two lost books *De sectione spatii*; an admirable edition of the same author's *Conics*; and more than eighty miscellaneous papers published by the Royal Society, which he served as secretary. In 1698 he commanded the war-sloop Paramour Pink in an expedition to the South Atlantic to study variations of the compass and to find new lands, if possible. On this journey he "fell in with great islands of ice, of so incredible a height and magnitude that I scarce dare write my thoughts of it." He was made Savilian professor of geometry at Oxford in 1703 and astronomer royal in 1721. One of his greatest achievements was a study of the orbits of comets, of which he described no less than twenty-four. Three of these were so much alike that he was convinced that the comets of 1531, 1607, and 1682 were one body. Assuming its period to be seventy-six years, he predicted its return in 1758. On Christmas Day of that year his conjecture was verified, and Halley's comet has since appeared in 1835 and 1910.

Halley died at the age of eighty-six. He was a generous, easygoing person, "free from rancor or jealousy," who spoke and acted with an "uncommon degree of sprightliness and vivacity." He enjoyed his work, had excellent health and owned a large circle of friends, among them Peter the Great of Russia to whose table he always had access. Bishop Berkeley thought Halley an "infidel," and it is true that in 1691 he was refused the Savilian professorship of astronomy at Oxford because of his alleged "materialistic views." The evidence is that he was a sensible man who spoke his mind freely—a dangerous practice in any age.

Halley's concern with the "curious tables" of Breslaw was one of his lesser diversions. This Silesian city had, for more than a century before his entry into the problem, kept regular records of its births and deaths. Dr. Caspar Neumann, a scientist and clergyman of Breslaw had analyzed some of these data, "disproving certain current superstitions with regard to the effect of the phases of the moon and the so-called 'climacteric' years, on health." [6] His results were submitted to Leibniz who sent them to the Royal Society. It was at about this time that the Society resumed publication of the "Transactions" after a lapse of several years. Halley promised to furnish five sheets in twenty of the forthcoming issues. He was never hard up for ideas, nor for the energy and ingenuity to express them. His Breslaw papers may therefore be regarded as a kind of filler for the "Transactions," to keep his word until something better came along. Nevertheless, the analysis reflects the exceptional power of his mind.

[6] Lowell J. Reed, *op. cit.*

Natural and *Political*

OBSERVATIONS

Mentioned in a following INDEX,

and made upon the

Bills of Mortality.

By *JOHN GRAUNT*,

Citizen of

LONDON.

With reference to the *Government, Religion, Trade, Growth, Ayre, Diseases*, and the several Changes of the said CITY.

—— *Non, me ut miretur Turba, laboro.*
Contentus paucis Lectoribus ——

LONDON,

Printed by *Tho: Roycroft*, for *John Martin*, *James Allestry*, and *Tho: Dicas*, at the Sign of the *Bell* in St. *Paul's* Church-yard, MDCLXII.

Our days on earth are as a shadow, and there is none abiding.
—I CHRONICLES XXIX

And so from hour to hour we ripe and ripe,
And then, from hour to hour, we rot and rot;
And thereby hangs a tale.   —SHAKESPEARE (As You Like It)

Let nature and let art do what they please,
When all is done, life's an incurable disease.   —ABRAHAM COWLEY

1 Foundations of Vital Statistics

By JOHN GRAUNT

TO THE RIGHT HONOURABLE JOHN LORD ROBERTS,
BARON OF TRURO, LORD PRIVIE-SEAL,
AND ONE OF HIS MAJESTIE'S MOST HONOURABLE PRIVIE COUNCIL.

My Lord,

AS the favours I have received from your Lordship oblige me to present you with some token of my *gratitude*: so the especial Honour I have for your Lordship hath made me *sollicitous* in the choice of the *Present*. For, if I could have given your Lordship any choice *Excerptions* out of the *Greek*, or Latine Learning, I should (according to our *English Proverb*) thereby but carry *Coals to Newcastle*, and but give your Lordship *Puddle-water*, who, by your own eminent *Knowledge* in those learned *Languages*, can drink out of the very *Fountains* your self.

Moreover, to present your Lordship with tedious *Narrations*, were but to speak my own *Ignorance* of the *Value*, which his Majesty, and the Publick have of your Lordship's Time. And in brief, to offer any thing like what is already in other Books, were but to derogate from your Lordship's learning, which the World knows to be universal, and unacquainted with few useful things contained in any of them.

Now having (I know not by what accident) engaged my thoughts upon the *Bills of Mortality*, and so far succeeded therein, as to have reduced several great confused *Volumes* into a few perspicuous *Tables*, and abridged such *Observations* as naturally flowed from them, into a few succinct *Paragraphs*, without any long Series of *multiloquious Deductions*, I have presumed to sacrifice these my small, but first publish'd, *Labours* unto your Lordship, as unto whose benigne acceptance of some other of my *Papers*, even the Birth of these is due; hoping (if I may without vanity say it) they may be of as much use to Persons in your Lordship's place, as they are of little or none to me, which is no more then the fairest

Diamonds are to the *Journey-man Jeweller* that works them, or the poor *Labourer* that first dig'd them from the Earth. For with all humble submission to your Lordship, I conceive, That it doth not ill-become a *Peer of the Parliament*, or *Member of his Majestie's Council*, to consider how few starve of the many that beg: That the irreligious *Proposals* of some, to multiply People by *Polygamy*, is withall irrational, and fruitless: That the troublesome seclusions in the *Plague-time* is not a remedy to be purchased at vast inconveniences: That the greatest *Plagues* of the City are equally, and quickly repaired from the Country: That the wasting of *Males* by Wars, and Colonies do not prejudice the due proportion between them and *Females*: That the Opinions of *Plagues* accompanying the Entrance of *Kings* is false, and seditious: That *London*, the *Metropolis* of *England*, is perhaps a Head too big for the Body, and possibly too strong: That this Head grows three times as fast as the Body unto which it belongs, that is, It doubles its People in a third part of the time: That our *Parishes* are now grown madly disproportionable: That our *Temples* are not sutable to our *Religion*: That the *Trade*, and very *City of London* removes *Westward*: That the walled City is but a one fifth of the whole Pyle: That the old Streets are unfit for the present frequencie of *Coaches*: That the passage of *Ludgate* is a throat too straight for the Body: That the fighting men about *London*, are able to make three as great Armies as can be of use in this *Island*: That the number of Heads is such, as hath certainly much deceived some of our *Senatours* in their appointments of *Pole-money*, &c. Now, although your Lordship's most excellent Discourses have well informed me, That your Lordship is no stranger to all these *Positions*; yet because I knew not that your Lordship had ever deduced them from the *Bills of Mortality*; I hoped it might not be ungratefull to your Lordship, to see unto how much profit that one Talent might be improved, besides the many curiosities concerning the waxing, and waning of Diseases, the relation between *Healthfull*, and *fruitfull Seasons*, the difference between the City and Country *Air*, &c. All which, being new, to the best of my knowledge, and the whole Pamphlet, not two hours reading, I did make bold to trouble your Lordship with a perusal of it, and by this humble Dedication of it, let your Lordship and the world see the Wisdom of our City, in appointing, and keeping these Accompts, and with how much affection and success I am

My Lord,

Your Lordship's most obedient, and
most faithfull Servant,

Birchin-Lane,
   25 January 166½.

JOHN GRAUNT.

THE PREFACE

Having been born, and bred in the City of *London*, and having always observed, that most of them who constantly took in the weekly Bills of *Mortality*, made little other use of them, then to look at the foot, how the *Burials* increased, or decreased; And, among the *Casualties*, what had happened rare, and extraordinary in the week currant: so as they might take the same as a *Text* to talk upon, in the next Company; and withall, in the *Plague-time*, how the *Sickness* increased, or decreased, that so the *Rich* might judge of the necessity of their removal, and *Trades-men* might conjecture what doings they were like to have in their respective dealings:

Now, I thought that the Wisdom of our City had certainly designed the laudable practice of takeing, and distributing these Accompts, for other, and greater uses then those above-mentioned, or at least, that some other uses might be made of them: And thereupon I casting mine Eye upon so many of the General *Bills*, as next came to hand, I found encouragement from them, to look out all the *Bills* I could, and (to be short) to furnish my self with as much matter of that kind, even as the Hall of the *Parish-Clerks* could afford me; the which, when I had reduced into Tables (the Copies whereof are here inserted) so as to have a view of the whole together, in order to the more ready comparing of one *Year*, *Season, Parish*, or other *Division* of the City, with another, in respect of all the *Burials*, and *Christnings*, and of all the *Diseases*, and *Casualties* happening in each of them respectively; I did then begin, not onely to examine the Conceits, Opinions, and Conjectures, which upon view of a few scattered *Bills* I had taken up; but did also admit new ones, as I found reason, and occasion from my *Tables*.

Moreover, finding some *Truths*, and not commonly-believed Opinions, to arise from my Meditations upon these neglected *Papers*, I proceeded further, to consider what benefit the knowledge of the same would bring to the World; that I might not engage my self in idle, and useless Speculations, but like those Noble *Virtuosi of Gresham-Colledge* (who reduce their subtile Disquisitions upon Nature into downright Mechanical uses) present the World with some real fruit from those ayrie Blossoms.

How far I have succeeded in the Premisses, I now offer to the World's censure. Who, I hope, will not expect from me, not professing Letters, things demonstrated with the same certainty, wherewith Learned men determine in their *Scholes*; but will take it well, that I should offer at a new thing, and could forbear presuming to meddle where any of the Learned Pens have ever touched before, and that I have taken the pains, and been at the charge, of setting out those *Tables*, whereby all men may both correct my *Positions*, and raise others of their own: For herein I have, like a silly Scholeboy, coming to say my Lesson to the World (that Peevish, and Tetchie Master) brought a bundle of Rods wherewith to be whipt, for every mistake I have committed.

OF THE BILLS OF MORTALITY, THEIR BEGINNING, AND PROGRESS

The first of the continued weekly *Bills* of *Mortality* extant at the Parish-Clerks *Hall*, begins the 29. of *December*, 1603, being the first year of *James* his Reign; since when, a weekly Accompt hath been kept there of *Burials* and *Christnings*. It is true, There were *Bills* before, *viz.* for the years 1592, -93, -94. but so interrupted since, that I could not depend upon the sufficiencie of them, rather relying upon those Accompts which have been kept since, in order, as to all the uses I shall make of them.

I believe, that the rise of keeping these Accompts, was taken from the *Plague*: for the said *Bills* (for ought appears) first began in the said year 1592. being a time of great *Mortality*; And after some disuse, were resumed again in the year 1603, after the great *Plague* then happening likewise.

These *Bills* were Printed and published, not onely every week on *Thursdays*, but also a general Accompt of the whole Year was given in, upon the *Thursday* before *Christmas Day*: which said general Accompts have been presented in the several manners following, *viz.* from the Year 1603, to the Year 1624, *inclusive* . . .

We have hitherto described the several steps, whereby the *Bills* of *Mortality* are come up to their present state; we come next to shew how they are made, and composed, which is in this manner, *viz.* When any one dies, then, either by tolling, or ringing of a Bell, or by bespeaking of a Grave of the *Sexton*, the same is known to the *Searchers*, corresponding with the said *Sexton*.

The *Searchers* hereupon (who are antient Matrons, sworn to their Office) repair to the place, where the dead Corps lies, and by view of the same, and by other enquiries, they examine by what *Disease*, or *Casualty* the Corps died. Hereupon they make their Report to the *Parish-Clerk*, and he, every *Tuesday* night, carries in an Accompt of all the *Burials*, and *Christnings*, hapning that Week, to the *Clerk* of the *Hall*. On *Wednesday* the general Accompt is made up, and Printed, and on *Thursdays* published, and dispersed to the several Families, who will pay four shillings *per Annum* for them. . . .

24

The Diseases, and Casualties this year being 1632.

Abortive, and Stilborn ..	445
A Affrighted	1
Aged	628
Ague	43
Apoplex, and Meagrom	17
Bit with a mad dog.......	1
Bleeding	3
Bloody flux, scowring, and flux	348
Brused, Issues, sores, and ulcers, ...	28
Burnt, and Scalded........	5
Burst, and Rupture........	9
Cancer, and Wolf..........	10
Canker	1
Childbed	171
Chrisomes, and Infants.....	2268
Cold, and Cough..........	55
Colick, Stone, and Strangury	56
Consumption	1797
Convulsion	241
Cut of the Stone..........	5
Dead in the street, and starved	6
Dropsie, and Swelling......	267
Drowned	34
Executed, and prest to death	18
Falling Sickness..........	7
Fever	1108
Fistula	13
Flocks, and small Pox.....	531
French Pox...............	12
Gangrene	5
Gout	4
Grief	11
Jaundies	43
Jawfaln	8
Impostume	74
Kil'd by several accidents..	46
King's Evil..............	38
Lethargie	2
Livergrown	87
Lunatique	5
Made away themselves.....	15
Measles	80
Murthered	7
Over-laid, and starved at nurse	7
Palsie	25
Piles....................	1
Plague..................	8
Planet	13
Pleurisie, and Spleen......	36
Purples, and spotted Feaver	38
Quinsie	7
Rising of the Lights......	98
Sciatica	1
Scurvey, and Itch........	9
Suddenly	62
Surfet	86
Swine Pox	6
Teeth	470
Thrush, and Sore mouth...	40
Tympany	13
Tissick	34
Vomiting	1
Worms	27

Christened	Males....4994 / Females..4590 / In all....9584
Buried	Males....4932 / Females..4603 / In all....9535

Whereof, of the Plague.8

Increased in the Burials in the 122 Parishes, and at the Pest-house this year....................................... 993

Decreased of the Plague in the 122 Parishes, and at the Pest-house this year................................. 266 [10]

GENERAL OBSERVATIONS UPON THE CASUALTIES

In my Discourses upon these *Bills* I shall first speak of the *Casualties,* then give my Observations with reference to the *Places,* and *Parishes* comprehended in the *Bills;* and next of the *Years,* and *Seasons.*

1. There seems to be good reason, why the *Magistrate* should himself take notice of the numbers of *Burials,* and *Christnings,* viz. to see, whether the City increase or decrease in people; whether it increase proportionably with the rest of the Nation; whether it be grown big enough, or too big, &c. But why the same should be made known to the People, otherwise then to please them as with a curiosity, I see not.

2. Nor could I ever yet learn (from the many I have asked, and those not of the least *Sagacity*) to what purpose the distinction between *Males* and *Females* is inserted, or at all taken notice of; or why that of *Marriages* was not equally given in? Nor is it obvious to everybody, why the Accompt of *Casualties* (whereof we are now speaking) is made? The reason, which seems most obvious for this latter, is, That the state of health in the City may at all times appear.

3. Now it may be Objected, That the same depends most upon the Accompts of *Epidemical Diseases,* and upon the chief of them all, the *Plague;* wherefore the mention of the rest seems onely matter of curiosity.

4. But to this we answer; That the knowledg even of the numbers, which die of the *Plague,* is not sufficiently deduced from the meer Report of the *Searchers,* which onely the Bills afford; but from other Ratiocinations, and comparings of the *Plague* with some other *Casualties.*

5. For we shall make it probable, that in Years of *Plague* a quarter part more dies of that *Disease* then are set down; the same we shall also prove by the other *Casualties.* Wherefore, if it be necessary to impart to the World a good Accompt of some few *Casualties,* which since it cannot well be done without giving an Accompt of them all, then is our common practice of so doing very apt, and rational.

6. Now, to make these Corrections upon the perhaps, ignorant, and careless *Searchers* Reports, I considered first of what Authority they were in themselves, that is, whether any credit at all were to be given to their Distinguishments: and finding that many of the *Casualties* were but matter of sense, as whether a Childe were *Abortive,* or *Stilborn;* whether men were *Aged,* that is to say, above sixty years old, or thereabouts, when they died, without any curious determination, whether such *Aged* persons died purely of *Age,* as for that the *Innate heat* was quite extinct, or the *Radical moisture* quite dried up (for I have heard some Candid *Physicians* complain of the darkness, which themselves were in hereupon) I say, that

these Distinguishments being but matter of sense, I concluded the *Searchers* Report might be sufficient in the Case.

7. As for *Consumptions*, if the *Searchers* do but truly Report (as they may) whether the dead Corps were very lean, and worn away, it matters not to many of our purposes, whether the Disease were exactly the same, as *Physicians* define it in their Books. Moreover, In case a man of seventy five years old died of a *Cough* (of which had he been free, he might have possibly lived to ninety) I esteem it little errour (as to many of our purposes) if this Person be, in the Table of *Casualties*, reckoned among the *Aged*, and not placed under the Title of *Coughs*.

8. In the matter of *Infants* I would desire but to know clearly, what the *Searchers* mean by *Infants*, as whether Children that cannot speak, as the word *Infants* seems to signifie, or Children under two or three years old, although I should not be satisfied, whether the *Infant* died of *Winde*, or of *Teeth*, or of the *Convulsion*, &c. or were choak'd with *Phlegm*, or else of *Teeth*, *Convulsion*, and *Scowring*, apart or together, which, they say, do often cause one another: for, I say, it is somewhat, to know how many die usually before they can speak, or how many live past any assigned number of years.

9. I say, it is enough, if we know from the *Searchers* but the most predominant Symptomes; as that one died of the *Head-Ache*, who was sorely tormented with it, though the *Physicians* were of Opinion, that the Disease was in the *Stomach*. Again, if one died *suddenly*, the matter is not great, whether it be reported in the Bills, *Suddenly*, *Apoplexie*, or *Planet-strucken*, &c.

10. To conclude, In many of these cases the *Searchers* are able to report the Opinion of the *Physician*, who was with the Patient, as they receive the same from the Friends of the Defunct, and in very many cases, such as *Drowning*, *Scalding*, *Bleeding*, *Vomiting*, *making-away them selves*, *Lunatiques*, *Sores*, *Small-Pox*, &c. their own senses are sufficient, and the generality of the World, are able prettie well to distinguish the *Gowt*, *Stone*, *Dropsie*, *Falling-Sickness*, *Palsie*, *Agues*, *Plurisy*, *Rickets*, &c. one from another.

11. But now as for those Casualties, which are aptest to be confounded, and mistaken, I shall in the ensuing Discourse presume to touch upon them so far, as the Learning of these Bills hath enabled me.

12. Having premised these general Advertisements, our first Observation upon the *Casualties* shall be, that in twenty Years there dying of all diseases and *Casualties*, 229250. that 71124. dyed of the *Thrush*, *Convulsion*, *Rickets*, *Teeth*, and *Worms*; and as *Abortives*, *Chrysomes*, *Infants*, *Liver-grown*, and *Over-laid*; that is to say, that about ⅓. of the whole died

of those Diseases, which we guess did all light upon Children under four or five Years old.

13. There died also of the *Small-Pox, Swine-Pox,* and *Measles,* and of *Worms* without *Convulsions,* 12210. of which number we suppose likewise, that about ½. might be Children under six Years old. Now, if we consider that 16. of the said 229 thousand died of that extraordinary and grand *Casualty* the *Plague,* we shall finde that about thirty six *per centum* of all quick conceptions, died before six years old.

14. The second Observation is; That of the said 229250. dying of all Diseases, there died of acute Diseases (the *Plague* excepted) but about 50000. or ²⁄₉ parts. The which proportion doth give a measure of the state, and disposition of this *Climate,* and *Air,* as to health, these *acute,* and *Epidemical* Diseases happening suddenly, and vehemently, upon the like corruptions, and alterations in the *Air.*

15. The third Observation is, that of the said 229. thousand about 70. died of *Chronical* Diseases, which shews (as I conceive) the state, and disposition of the Country (including as well it's *Food,* as *Air*) in reference to health, or rather to *longœvity:* for as the proportion of *Acute* and *Epidemical* Diseases shews the aptness of the *Air* to suddain and vehement Impressions, so the *Chronical* Diseases shew the ordinary temper of the Place, so that upon the proportion of *Chronical* Diseases seems to hang the judgment of the fitness of the Country for *long Life.* For, I conceive, that in Countries subject to great *Epidemical* sweeps men may live very long, but where the proportion of the *Chronical* distempers is great, it is not likely to be so; because men being long sick, and always sickly, cannot live to any great age, as we see in several sorts of *Metalmen,* who although they are less subject to acute Diseases then others,

Table of notorious Diseases.		*Table of Casualties.*	
Apoplex	1306	Bleeding	069
Cut of the Stone	0038	Burnt, and Scalded	125
Falling Sickness	0074	Drowned	829
Dead in the Streets	0243	Excessive drinking	002
Gowt	0134	Frighted	022
Head-Ach	0051	Grief	279
Jaundice	0998	Hanged themselves	222
Lethargy	0067	Kil'd by several accidents	1021
Leprosy	0006	Murthered	0086
Lunatique	0158	Poysoned	014
Overlaid, and Starved	0529	Smothered	026
Palsy	0423	Shot	007
Rupture	0201	Starved	051
Stone and Strangury	0863	Vomiting	136
Sciatica	0005		
Sodainly	0454		

yet seldome live to be old, that is, not to reach unto those years, which *David* saies is the age of man.

16. The fourth Observation is; That of the said 229000. not 4000. died of outward Griefs, as of *Cancers, Fistulaes, Sores, Ulcers, broken and bruised Limbs, Impostumes, Itch, King's-evil, Leprosie, Scald-head, Swine-Pox, Wens,* &c. *viz.* not one in 60.

17. In the next place, whereas many persons live in great fear, and apprehension of some of the more formidable, and notorious diseases following; I shall onely set down how many died of each: that the respective numbers, being compared with the Total 229250, those persons may the better understand the hazard they are in.

OF PARTICULAR CASUALTIES

My first Observation is, That few are *starved*. This appears, for that of the 229250 which have died, we find not above fifty one to have been *starved*, excepting helpless *Infants* at Nurse, which being caused rather by carelessness, ignorance, and infirmity of the Milch-women, is not properly an effect, or sign of want of food in the Countrey, or of means to get it.

The Observation, which I shall add hereunto, is, That the vast numbers of *Beggars*, swarming up and down this City, do all live, and seem to be most of them healthy and strong; whereupon I make this Question, Whether, since they do all live by Begging, that is, without any kind of labour; it were not better for the State to keep them, even although they earned nothing; that so they might live regularly, and not in that Debauchery, as many Beggars do; and that they might be cured of their bodily Impotencies, or taught to work, &c. each according to his condition, and capacity; or by being employed in some work (not better undone) might be accustomed, and fitted for labour. . . .

My next Observation is; That but few are *Murthered*, viz. not above 86 of the 22950 [*sic*]. which have died of other diseases, and casualties; whereas in *Paris* few nights scape without their *Tragedie*.

The Reasons of this we conceive to be *Two:* One is the *Government*, and *Guard* of the City by *Citizens* themselves, and that alternately. No man settling into a Trade for that employment. And the other is, The natural, and customary abhorrence of that inhumane *Crime*, and all *Bloodshed* by most *Englishmen:* for of all that are *Executed* few are for *Murther*. Besides the great and frequent Revolutions, and Changes of *Government* since the year 1650, have been with little *bloodshed*; the *Usurpers* themselves having *Executed* few in comparison, upon the Accompt of disturbing their Innovations.

In brief, when any dead Body is found in *England,* no *Algebraist,* or *Uncipherer* of Letters, can use more subtile suppositions, and varietie of conjectures to finde out the Demonstration, or Cipher; then every common unconcerned Person doth to finde out the Murtherers, and that for ever, untill it be done.

The *Lunaticks* are also but few, *viz.* 158 in 229250. though I fear many more then are set down in our *Bills,* few being entred for such, but those who die at *Bedlam;* and there all seem to die of their *Lunacie,* who died *Lunaticks;* for there is much difference in computing the number of *Lunaticks,* that die (though of *Fevers,* and all other Diseases, unto which *Lunacie* is no *Supersedeas*) and those, that die by reason of their *Madness.*

So that, this *Casualty* being so uncertain, I shall not force my self to make any inference from the numbers, and proportions we finde in our Bills concerning it: onely I dare ensure any man at this present, well in his Wits, for one in the thousand, that he shall not die a *Lunatick* in *Bedlam,* within these seven years, because I finde not above one in about one thousand five hundred have done so.

The like use may be made of the Accompts of men, that made away themselves, who are another sort of Madmen, that think to ease them-selves of pain by leaping into *Hell;* or else are yet more Mad, so as to think there is no such place; or that men may go to rest by death, though they die in *self-murther,* the greatest Sin.

We shall say nothing of the numbers of those, that have been *Drowned, Killed by falls from Scaffolds,* or by *Carts running over them,* &c. because the same depends upon the casual Trade, and Employment of men, and upon matters, which are but circumstantial to the Seasons, and Re-gions we live in; and affords little of that Science, and Certainty we aim at.

We finde one *Casualty* in our Bills, of which though there be daily talk, there is little effect, much like our abhorrence of *Toads,* and *Snakes,* as most poisonous Creatures, whereas few men dare say upon their own knowledge, they ever found harm by either; and this *Casualty* is the *French-Pox,* gotten, for the most part, not so much by the intemperate use of *Venery* (which rather causeth the *Gowt*) as of many common Women.

I say, the Bills of *Mortality* would take off these Bars, which keep some men within bounds, as to these extravagancies: for in the afore-mentioned 229250 we finde not above 392 to haved died of the *Pox.* Now, forasmuch as it is not good to let the World be lulled into a security, and belief of Impunity by our Bills, which we intend shall not be onely as *Death's-heads* to put men in minde of their *Mortality,* but also as *Mercurial*

Statues to point out the most dangerous ways, that lead us into it, and misery. We shall therefore shew, that the *Pox* is not as the *Toads*, and *Snakes* afore-mentioned, but of a quite contrary nature, together with the reason, why it appears otherwise.

17. Forasmuch as by the ordinary discourse of the world it seems a great part of men have, at one time, or other, had some *species* of this disease, I wondering why so few died of it, especially because I could not take that to be so harmless, whereof so many complained very fiercely; upon inquiry I found that those who died of it out of the Hospitals (especially that of *King's-Land*, and the *Lock* in *Southwark*) were returned of *Ulcers*, and *Sores*. And in brief I found, that all mentioned to die of the *French-Pox* were retured by the *Clerks* of Saint *Giles's*, and Saint *Martin's in the Fields* onely; in which place I understood that most of the vilest, and most miserable houses of uncleanness were: from whence I concluded, that onely *hated* persons, and such, whose very *Noses* were eaten of, were reported by the *Searchers* to have died of this too frequent *Maladie*. . . .

OF THE DIFFERENCE BETWEEN THE NUMBERS
OF MALES, AND FEMALES

The next Observation is, That there be more *Males* then *Females*.

There have been *Buried* from the year 1628, to the year 1662, *exclusive*, 209436 *Males*, and but 190474 *Females:* but it will be objected, that in *London* it may indeed be so, though otherwise elsewhere; because *London* is the great Stage and Shop of business, wherein the *Masculine Sex* bears the greatest part. But we Answer, That there have been also *Christned* within the same time, 139782 *Males*, and but 130866 *Females*, and that the Country Accompts are consonant enough to those of *London* upon this matter.

What the Causes hereof are, we shall not trouble our selves to conjecture, as in other Cases, onely we shall desire, that Travellers would enquire whether it be the same in other Countries.

We should have given an Accompt, how in every Age these proportions change here, but that we have Bills of distinction but for 32 years, so that we shall pass from hence to some inferences from this Conclusion; as first,

I. That *Christian Religion*, prohibiting *Polygamy*, is more agreeable to the *Law of Nature*, that is, the *Law of God*, then *Mahumetism*, and others, that allow it; for one man his having many women, or wives by Law, signifies nothing, unless there were many women to one man in Nature also.

II. The obvious Objection hereunto is, That one *Horse, Bull,* or *Ram,* having each of them many *Females,* do promote increase. To which I Answer, That although perhaps there be naturally, even of these *species,* more *Males* then *Females,* yet *artificially,* that is, by making *Geldings, Oxen,* and *Weathers,* there are fewer. From whence it will follow, That when by experience it is found how many *Ews* (suppose twenty) one *Ram* will serve, we may know what proportion of *male-Lambs* to castrate, or geld, *viz.* nineteen, or thereabouts: for if you emasculate fewer, *viz.* but ten, you shall by promiscuous copulation of each of those ten with two *Females,* (in such as admit the *Male* after conception) hinder the increase so far, as the admittance of two *Males* will do it: but, if you castrate none at all, it is highly probable, that every of the twenty *Males* copulating with every of the twenty *Females,* there will be little, or no conception in any of them all.

III. And this I take to be the truest Reason, why *Foxes, Wolves,* and other *Vermin Animals* that are not gelt, increase not faster than *Sheep,* when as so many thousands of these are daily Butchered, and very few of the other die otherwise then of themselves.

We have hitherto said there are more *Males,* then *Females;* we say next, That the one exceed the other by about a thirteenth part; so that although more men die violent deaths then women, that is, more are *slain* in Wars, *killed* by mischance, *drowned* at *Sea,* and die by the *Hand of Justice.* Moreover, more men go to *Colonies,* and travel into foreign parts, then women. And lastly, more remain unmarried, then of women, as *Fellows* of *Colleges,* and *Apprentises,* above eighteen, &c. yet the said thirteenth part difference bringeth the business but to such a pass, that every woman may have an Husband, without the allowance of *Polygamy.*

Moreover, although a man be *Prolifique* fourty years, and a woman but five and twenty, which makes the *Males* to be as 560 to 325 *Females,* yet the causes above named, and the later marriage of the men, reduce all to an equality. . . .

It is a Blessing to Man-kind, that by this overplus of *Males* there is this natural Bar to *Polygamy*: for in such a state Women could not live in that parity, and equality of expence with their Husbands, as now, and here they do.

The reason whereof is, not, that the Husband cannot maintain as splendidly three, as one; for he might, having three Wives, live himself upon a quarter of his Income, that is in a parity with all three, as-well as, having but one, live in the same parity at half with her alone: but rather, because that to keep them all quiet with each other, and himself, he must keep them all in greater awe, and less splendor which power he having will probably use it to keep them all as low, as he pleases, and at no more cost then makes for his own pleasure; the poorest Subjects (such as this plurality of Wives must be) being most easily governed.

THE CONCLUSION

It may be now asked, to what purpose tends all this laborious buzzling, and groping? To know,

1. The number of the People?
2. How many *Males*, and *Females*?
3. How many Married, and single?
4. How many *Teeming* Women?
5. How many of every *Septenary*, or *Decad* of years in *age*?
6. How many *Fighting* Men?
7. How much *London* is, and by what steps it hath increased?
8. In what time the housing is replenished after a *Plague*?
9. What proportion die of each general and perticular *Casualties*?
10. What years are Fruitfull, and Mortal, and in what Spaces, and Intervals, they follow each other?
11. In what proportion Men neglect the Orders of the *Church*, and *Sects* have increased?
12. The disproportion of Parishes?
13. Why the Burials in *London* exceed the Christnings, when the contrary is visible in the Country?

To this I might answer in general by saying, that those, who cannot apprehend the reason of these Enquiries, are unfit to trouble themselves to ask them.

I might answer by asking; Why so many have spent their times, and estates about the Art of making Gold? which, if it were much known, would onely exalt Silver into the place, which Gold now possesseth; and if it were known but to some one Person, the same single *Adeptus* could not, nay, durst not enjoy it, but must be either a Prisoner to some Prince, and Slave to some Voluptuary, or else skulk obscurely up and down for his privacie, and concealment.

I might Answer; That there is much pleasure in deducing so many abstruse, and unexpected inferences out of these poor despised Bills of *Mortality*; and in building upon that ground, which hath lain waste these eighty years. And there is pleasure in doing something new, though never so little, without pestering the World with voluminous Transcriptions.

But, I Answer more seriously; by complaining, That whereas the Art of Governing, and the true *Politiques*, is how to preserve the Subject in *Peace*, and *Plenty*, that men study onely that part of it, which teacheth how to supplant, and over-reach one another, and how, not by fair out-running, but by tripping up each other's heels, to win the Prize.

Now, the Foundation, or Elements of this honest harmless *Policy* is to understand the Land, and the hands of the Territory to be governed, according to all their intrinsick, and accidental differences: as for example; It were good to know the *Geometrical* Content, Figure, and Scituation of all the Lands of a Kingdom, especially, according to its most natural, permanent, and conspicuous Bounds. It were good to know, how much Hay an Acre of every sort of Meadow will bear? how many Cattel the same weight of each sort of Hay will feed, and fatten? what quantity of Grain, and other Commodities the same Acre will bear in one, three, or seven years *communibus Annis?* unto what use each soil is most proper? All which particulars I call the intrinsick value: for there is also another value meerly accidental, or extrinsick, consisting of the Causes, why a parcel of Land, lying near a good Market, may be worth double to another parcel, though but of the same intrinsick goodness; which answers the Queries, why Lands in the North of *England* are worth but sixteen years purchase, and those of the West above eight and twenty. It is no less necessary to know how many People there be of each Sex, State, Age, Religion, Trade, Rank, or Degree, &c. by the knowledg whereof Trade, and Government may be made more certain, and Regular; for, if men knew the People as aforesaid, they might know the consumption they would make, so as Trade might not be hoped for where it is impossible. As for instance, I have heard much complaint, that Trade is not set up in some of the *South-western*, and *North-western* Parts of *Ireland*, there being so many excellent Harbours for that purpose, whereas in several of those Places I have also heard, that there are few other Inhabitants, but such as live *ex sponte creatis*, and are unfit Subjects of Trade, as neither employing others, nor working themselves.

Moreover, if all these things were clearly, and truly known (which I have but guessed at) it would appear, how small a part of the People work upon necessary Labours, and Callings, *viz.* how many Women, and Children do just nothing, onely learning to spend what others get? how many are meer Voluptuaries, and as it were meer Gamesters by Trade? how many live by puzling poor people with unintelligible Notions in Divinity, and Philosophie? how many by perswading credulous, delicate, and Litigious Persons, that their Bodies, or Estates are out of Tune, and in danger? how many by fighting as Souldiers? how many by Ministeries of Vice, and Sin? how many by Trades of meer Pleasure, or Ornaments? and how many in a way of lazie attendance, &c. upon others? And on the other side, how few are employed in raising, and working necessary food, and covering? and of the speculative men, how few do truly studie *Nature*, and *Things*? The more ingenious not advancing much further then to write, and speak wittily about these matters.

I conclude, That a clear knowledge of all these particulars, and many more, whereat I have shot but at rovers, is necessary in order to good, certain, and easie Government, and even to balance Parties, and factions both in *Church* and *State*. But whether the knowledge thereof be necessary to many, or fit for others, then the Sovereign, and his chief Ministers, I leave to consideration.

PHILOSOPHICAL
TRANSACTIONS:

Giving fome

ACCOUNT

OF THE

Prefent *Undertakings*, *Studies* and *Labours*

OF THE

INGENIOUS,

In many

Confiderable Parts of the WORLD.

VOL. XVII. For the Year 1693.

LONDON:

Printed for *S. Smith* and *B. Walford*, Printers to the *Royal Society*, at the *Prince*'s *Arms*, in St. *Paul*'s Church-yard. 1694.

Factual science may collect statistics and make charts. But its predictions are, as has been well said, but past history reversed. —JOHN DEWEY

2 First Life Insurance Tables

By EDMUND HALLEY

AN ESTIMATE OF THE DEGREES OF THE MORTALITY OF MANKIND, DRAWN FROM CURIOUS TABLES OF THE BIRTHS AND FUNERALS AT THE CITY OF BRESLAW; WITH AN ATTEMPT TO ASCERTAIN THE PRICE OF ANNUITIES UPON LIVES. BY MR. E. HALLEY, R.S.S.

THE Contemplation of the *Mortality* of *Mankind,* has besides the *Moral,* its *Physical* and *Political Uses,* both which have been some years since most judiciously considered by the curious Sir *William Petty,* in his *Natural* and *Political* Observations on the Bills of *Mortality* of *London,* owned by Captain *John Graunt.* And since in a like Treatise on the Bills of *Mortality* of *Dublin.* But the Deduction from those Bills of *Mortality* seemed even to their Authors to be defective: First, In that the *Number* of the People was wanting. Secondly, That the *Ages* of the People dying was not to be had. And Lastly, That both *London* and *Dublin* by reason of the great and casual Accession of *Strangers* who die therein, (as appeared in both, by the great Excess of the *Funerals* above the *Births*) rendred them incapable of being Standards for this purpose; which requires, if it were possible, that the People we treat of should not at all be changed, but die where they were born, without any Adventitious Increase from Abroad, or Decay by Migration elsewhere.

This *Defect* seems in a great measure to be satisfied by the late curious Tables of the Bills of *Mortality* at the City of *Breslaw,* lately communicated to this Honourable Society by Mr. *Justell,* wherein both the *Ages* and *Sexes* of all that die are monthly delivered, and compared with the number of the *Births,* for Five Years last past, *viz.* 1687, 88, 89, 90, 91, seeming to be done with all the Exactness and Sincerity possible.

This City of *Breslaw* is the Capital City of the Province of *Silesia*; or, as the *Germans* call it, *Schlesia,* and is scituated on the Western Bank of the River *Oder,* anciently called *Viadrus*; near the Confines of *Germany* and *Poland,* and very nigh the Latitude of *London.* It is very far from the Sea, and as much a *Mediterranean Place* as can be desired, whence the Confluence of Strangers is but small, and the Manufacture of Linnen employs chiefly the poor People of the place, as well as of the Country round about: whence comes that sort of Linnen we usually call your

Sclesie Linnen; which is the chief, if not the only Merchandize of the place. For these Reasons the People of this City seem most proper for a *Standard*; and the rather, for that the *Births* do, a small matter, exceed the *Funerals*. The only thing wanting is the Number of the whole People, which in some measure I have endeavoured to supply by comparison of the *Mortality* of the People of all Ages, which I shall from the said Bills trace out with all the Acuracy possible.

It appears that in the Five Years mentioned, *viz.* from 87 to 91 inclusive, there were *born* 6193 Persons, and *buried* 5869; that is, born *per Annum* 1238, and *buried* 1174; whence an *Encrease* of the People may be argued of 64 *per Annum*, or of about a 20th part, which may perhaps be ballanced by the Levies for the *Emperor*'s Service in his Wars. But this being contingent, and the Births certain, I will suppose the People of *Breslaw* to be encreased by 1238 *Births* annually. Of these it appears by the same Tables, that 348 do die *yearly* in the *first Year* of their *Age*, and that but 890 do arrive at a full *Years Age*; and likewise, that 198 do die in the *Five Years* between 1 and 6 compleat, taken at a *Medium*; so that but 692 of the Persons *born* do survive *Six* whole *Years*. From this *Age* the Infants being arrived at some degree of Firmness, grow less and less *Mortal*; and it appears that of the whole People of *Breslaw* there die *yearly*, as in the following Table, wherein the upper Line shews the *Age*, and the next under it the *Number* of Persons of that Age *dying yearly*.

```
 7 . 8 . 9 . .    14   . 18 . 21 . 27 . 28 . .  35 . 36 .
11 . 11 . 6 . 5½ . 2 . 3½ 5 6 4½ 6½ 9 . 8 . 7 . 7 . 8 .

    42 .    45      49 54 . 55 . 56 .    63    70 71 . 72
9½  8 . 9 . 7 . 7 . 10 11 . 9 . 9 . 10 . 12 9½ 14  9 . 11

    77      81      84 .    90   91 . 98 . 99 . 100.
9½  6 . 7 . 3 . 4 . 2 . 1 . 1 . 1 . 0 . ⅕ . ⅜
```

And where no *Figure* is placed over, it is to be understood of those that die between the Ages of the preceding and consequent *Column*.

From this Table it is evident, that from the Age of 9 to about 25 there does not die above 6 *per Annum* of each *Age*, which is much about one *per Cent.* of those that are of those *Ages:* And whereas in the 14, 15, 16, 17 *Years* there appear to die much fewer, as 2 and 3½, yet that seems rather to be attributed to Chance, as are the other Irregularities in the Series of Ages, which would rectifie themselves, were the number of Years much more considerable, as 20 instead of 5. And by our own Experience in *Christ-Church Hospital*, I am informed there die of the *Young Lads*, much about one *per Cent. per Annum*, they being of the foresaid *Ages*. From 25 to 50 there seem to die from 7 to 8 and 9 *per Annum* of each Age; and after that to 70, they growing more *crasie*, though the number be much diminished, yet the *Mortality encreases*, and there are found to die 10 or 11 of each Age *per Annum:* From thence the number of the *Living* being grown very small, they gradually decline till there be none left to *die*; as may be seen at one View in the Table.

From these Considerations I have formed the *adjoyned Table*, whose Uses are manifold, and give a more just *Idea* of the *State* and *Condition* of *Mankind*, than any thing yet extant that I know of. It exhibits the *Number* of *People* in the City of *Breslaw* of all Ages, from the *Birth* to extream *Old Age*, and thereby shews the Chances of *Mortality* at all *Ages*, and likewise how to make a certain Estimate of the value of *Annuities* for *Lives*, which hitherto has been only done by an imaginary *Valuation:* Also the *Chances* that there are that a *Person* of any *Age* proposed does live to any other *Age* given; with many more, as I shall hereafter shew. This *Table* does shew the *number* of *Persons* that are living in the *Age* current annexed thereto, as follows:

Age Curt.	Persons	Age Curt.	sons	Age Curt.	sons	Age Curt.	sons	Age Curt.	sons	Age Curt.	sons	Age	Persons
												7	5547
1	1000	8	680	15	628	22	586	29	539	36	481	14	4584
2	855	9	670	16	622	23	579	30	531	37	472	21	4270
3	798	10	661	17	616	24	573	31	523	38	463	28	3964
4	760	11	653	18	610	25	567	32	515	39	454	35	3604
5	732	12	646	19	604	26	560	33	507	40	445	42	3178
6	710	13	640	20	598	27	553	34	499	41	436	49	2709
7	692	14	634	21	592	28	546	35	490	42	427	56	2194
												63	1694
Age Curt.	sons	Age Curt.	sons	Age Curt.	sons	Age Curt.	sons	Age Curt.	sons	Age Curt.	sons	70	1204
												77	692
43	417	50	346	57	272	64	202	71	131	78	58	84	253
44	407	51	335	58	262	65	192	72	120	79	49	100	107
45	397	52	324	59	252	66	182	73	109	80	41		
46	387	53	313	60	242	67	172	74	98	81	34		34000
47	377	54	302	61	232	68	162	75	88	82	28		
48	367	55	292	62	222	69	152	76	78	83	23		Sum Total
49	357	56	282	63	212	70	142	77	68	84	20		

Thus it appears, that the whole People of *Breslaw* does consist of 34000 *Souls*, being the Sum *Total* of the Persons of all Ages in the *Table*: The first use hereof is to shew the Proportion of *Men* able to bear *Arms* in any *Multitude*, which are those between 18 and 56, rather than 16 and 60; the one being generally too weak to bear the *Fatigues* of *War* and the Weight of *Arms*, and the other too crasie and infirm from *Age*, notwithstanding particular Instances to the contrary. Under 18 from the *Table*, are found in this City 11997 Persons, and 3950 above 56, which together make 15947. So that the Residue to 34000 being 18053 are Persons between those *Ages*. At least one half thereof are Males, or 9027: So that the whole Force this City can raise of *Fencible Men*, as the *Scotch* call them, is about 9000, or ⁹⁄₃₄, or somewhat more than a quarter of the *Number* of *Souls*, which may perhaps pass for a Rule for all other places.

The *Second Use* of this Table is to shew the differing degrees of *Mortality*, or rather *Vitality* in all *Ages*; for if the number of Persons of any *Age* remaining after one year, be divided by the difference between that and the number of the Age proposed, it shews the *odds* that there is, that a Person of that Age does not die in a *Year*. As for Instance, a Person of 25 *Years* of *Age* has the odds of 560 to 7 or 80 to 1, that he does not *die* in a *Year*: Because that of 567, living of 25 years of Age, there do die no more than 7 in a *Year*, leaving 560 of 26 Years old.

So likewise for the *odds*, that any Person does not die before he attain any proposed *Age*: Take the *number* of the remaining Persons of the Age proposed, and divide it by the difference between it and the number of those of the *Age* of the Party proposed; and that shews the *odds* there is between the Chances of the Party's living or dying. As for Instance; What is the *odds* that a Man of 40 lives 7 Years: Take the number of Persons of 47 years, which in the Table is 377, and subtract it from the number of Persons of 40 years, which is 445, and the *difference* is 68: Which shews that the *Persons dying* in that 7 years are 68, and that it is 377 to 68 or 5½ to 1, that a Man of 40 does live 7 Years. And the like for any other *number* of *Years*.

Use III. But if it be enquired at what number of *Years*, it is an even Lay that a Person of any *Age* shall die, this Table readily performs it: For if the *number* of Persons *living* of the *Age* proposed be *halfed*, it will be found by the *Table* at what Year the said *number* is reduced to half by *Mortality*; and that is the Age, to which it is an even Wager, that a Person of the *Age* proposed shall arrive before he *die*. As for Instance; A Person of 30 Years of *Age* is proposed, the number of that Age is 531, the half thereof is 265, which number I find to be between 57 and 58 Years; so that a Man of 30 may reasonably expect to live between 27 and 28 Years.

Use IV. By what has been said, the *Price* of *Insurance* upon *Lives* ought to be regulated, and the difference is discovered between the *price* of ensuring the *Life* of a *Man* of 20 and 50, for Example: it being 100 to 1 that a Man of 20 dies not in a year, and but 38 to 1 for a Man of 50 Years of Age.

Use V. On this depends the Valuation of *Annuities* upon *Lives*; for it is plain that the *Purchaser* ought to pay for only such a part of the value of the *Annuity*, as he has Chances that he is living; and this ought to be computed yearly, and the Sum of all those yearly Values being added together, will amount to the value of the *Annuity* for the *Life* of the Person proposed. Now the present value of Money payable after a term of years, at any given rate of Interest, either may be had from Tables already computed; or almost as compendiously, by the Table of Logarithms: For the Arithmetical Complement of the Logarithm of Unity and its yearly Interest (that is, of 1, 06 for Six *per Cent.* being 9, 974694.) being multiplied by the number of years proposed, gives the present value of One Pound payable after the end of so many years. Then by the foregoing Proposition, it will be as the number of Persons living after that term of years, to the number dead; so are the Odds that any one Person is Alive or Dead. And by consequence, as the Sum of both or the number of Persons living of the Age first proposed, to the number remaining after so many years, (both given by the Table) so the present value of the yearly Sum payable after the term proposed, to the Sum which ought to be paid for the Chance the person has to enjoy such an Annuity after so many Years. And this being repeated for every year of the persons Life, the Sum of all the present Values of those Chances is the true Value of the Annuity. This will without doubt appear to be a most laborious Calculation, but it being one of the principal Uses of this Speculation, and having found some Compendia for the Work, I took the pains to compute the following Table, being the short Result of a not ordinary number of Arithmetical Operations; It shews the Value of Annuities for every Fifth Year of Age, to the Seventieth, as follows.

Age	Years Purchase	Age	Years Purchase	Age	Years Purchase
1	10,28	25	12,27	50	9,21
5	13,40	30	11,72	55	8,51
10	13,44	35	11,12	60	7,60
15	13,33	40	10,57	65	6,54
20	12,78	45	9,91	70	5,32

This shews the great Advantage of putting Money into the present *Fund* lately granted to their Majesties, giving 14 *per Cent. per Annum,* or at the rate of 7 years purchase for a Life; when young Lives, at the usual rate of Interest, are worth above 13 years Purchase. It shews likewise the Advantage of young Lives over those in Years; a Life of Ten Years being almost worth 13½ years purchase, whereas one of 36 is worth but 11.

Use VI. Two Lives are likewise valuable by the same Rule; for the number of Chances of each single Life, found in the Table, being multiplied together, become the Chances of the Two Lives. And after any certain Term of Years, the Product of the two remaining Sums is the Chances that both the Persons are living. The Product of the two Differences, being the numbers of the Dead of both Ages, are the Chances that both the Persons are dead. And the two Products of the remaining Sums of the one Age multiplied by those dead of the other, shew the

Chances that there are that each Party survives the other: Whence is derived the Rule to estimate the value of the Remainder of one Life after another. Now as the Product of the Two Numbers in the Table for the Two Ages proposed, is to the difference between that Product and the Product of the two numbers of Persons deceased in any space of time, so is the value of a Sum of Money to be paid after so much time, to the value thereof under the Contingency of Mortality. And as the aforesaid Product of the two Numbers answering to the Ages proposed, to the Product of the Deceased of one Age multiplied by those remaining alive of the other; So the Value of a Sum of Money to be paid after any time proposed, to the value of the Chances that the one Party has that he survives the other whose number of Deceased you made use of, in the second Term of the proportion. This perhaps may be better understood, by putting N for the number of the younger Age, and n for that of the Elder; Y, y the deceased of both Ages respectively, and R, r for the Remainders; and $R + Y = N$ and $r + y = n$. Then shall $N n$ be the whole number of Chances; $N n - Y y$ be the Chances that one of the two Persons is living, $Y y$ the Chances that they are both dead; $R y$ the Chances that the elder Person is dead and the younger living; and $r Y$ the Chances that the elder is living and the younger dead. Thus two Persons of 18 and 35 are proposed, and after 8 years these Chances are required. The Numbers for 18 and 35 are 610 and 490, and there are 50 of the First Age dead in 8 years, and 73 of the Elder Age. There are in all 610×490 or 298900 Chances; of these there are 50×73 or 3650 that they are both dead. And as 298900, to $298900 - 3650$, or 295250: So is the present value of a Sum of Money to be paid after 8 years, to the present value of a Sum to be paid if either of the two live. And as 560×73, so are the Chances that the Elder is dead, leaving the Younger; and as 417×50, so are the Chances that the Younger is dead, leaving the Elder. Wherefore as 610×490 to 560×73, so is the present value of a Sum to be paid at eight years end, to the Sum to be paid for the Chance of the Youngers Survivance; and as 610×490 to 417×50, so is the same present value to the Sum to be paid for the Chance of the Elders Survivance.

This possibly may be yet better explained by expounding these Products by Rectangular Parallelograms, as in *Figure 1*, wherein A B or C D represents the number of persons of the younger Age, and D E, B H those remaining alive after certain term of years; whence C E will answer the number of those dead in that time: So A C, B D may represent the number of the Elder Age; A F, B I the Survivors after the same term; and C F, D I, those of that Age that are dead at that time. Then shall the whole Parallelogram A B C D be N n, or the Product of the two Numbers of persons, representing such a number of Persons of the two Ages given; and by what was said before, after the Term proposed the Rectangle H D

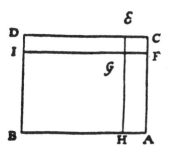

FIGURE 1

shall be as the number of Persons of the younger Age that survive, and the Rectangle *A E* as the number of those that die. So likewise the Rectangles *A I, F D* shall be as the Numbers, living and dead, of the other Age. Hence the Rectangle *H I* shall be as an equal number of both Ages surviving. The Rectangle *F E* being the Product of the deceased, or *Y y*, an equal number of both dead. The Rectangle *G D* or *R y*, a number living of the younger Age, and dead of the Elder: And the Rectangle *A G* or *r Y* a number living of the Elder Age, but dead of the younger. This being understood, it is obvious, that as the whole Rectangle *A D* or *N n* is to the *Gnomon F A B D E G* or *N n − Y y*, so is the whole number of Persons or Chances, to the number of Chances that one of the two Persons is living: And as *A D* or *N n* is to *F E* or *Y y*, so are all the Chances, to the Chances that both are dead; whereby may be computed the value of the Reversion after both Lives. And as *A D* to *G D* or *R y*, so the whole number of Chances, to the Chances that the younger is living and the other dead; whereby may be cast up what value ought to be paid for the Reversion of one Life after another, as in the case of providing for Clergy-mens Widows and others by such Reversions. And as *A D* to *A G* or *r Y*, so are all the Chances, to those that the Elder survives the younger. I have been the more particular, and perhaps tedious, in this matter, because it is the Key to the Case of Three Lives, which of it self would not have been so easie to comprehend.

VII. If Three Lives are proposed, to find the value of an Annuity during the continuance of any of those three Lives. The Rule is, *As the Product of the continual multiplication of the Three Numbers, in the Table, answering to the Ages proposed, is to the difference of that Product and of the Product of the Three Numbers of the deceased of those Ages, in any given term of Years; So is the present value of a Sum of Money to be paid certainly after so many Years, to the present value of the same Sum to be paid, provided one of those three Persons be living at the Expiration of that term.* Which proportion being yearly repeated, the Sum of all those present values will be the value of an Annuity granted for

three such Lives. But to explain this, together with all the Cases of Survivance in three Lives: Let N be the Number in the Table for the Younger
Age, n for the Second, and ν for the Elder Age; let Y be those dead of
the Younger Age in the term proposed, y those dead of the Second Age,
and υ those of the Elder Age; and let R be the Remainder of the younger
Age, r that of the middle Age, and $\int$ the Remainder of the Elder Age.
Then shall $R + Y$ be equal to N, $r + y$ to n, and $\int + \upsilon$ to ν, and the con-
tinual Product of the three Numbers $N\, n\, \nu$ shall be equal to the continual
Product of $R + Y \times r + Y \times \int + \upsilon,$[1] which being the whole number of
Chances for three Lives is compounded of the eight Products following.
(1) $R\, r\, \int$, which is the number of Chances that all three of the Persons
are living. (2) $r \int Y$, which is the number of Chances that the two Elder
Persons are living, and the younger dead. (3) $R\, \rho\, y$ the number of
Chances that the middle Age is dead, and the younger and Elder living.
(4) $R\, r\, \upsilon$ being the Chances that the two younger are living, and the
elder dead. (5) $\int Y\, y$ the Chances that the two younger are dead, and
the elder living. (6) $r\, Y\, \upsilon$ the Chances that the younger and elder are
dead, and the middle Age living. (7) $R\, y\, \upsilon$, which are the Chances that
the younger is living, and the two other dead. And Lastly and Eightly,
$Y\, y\, \upsilon$, which are the Chances that all three are dead. Which latter subtracted from the whole number of Chances $N\, n\, \nu$, leaves $N\, n\, \nu - Y\, y\, \upsilon$
the Sum of all the other Seven Products; in all of which one or more of
the three Persons are surviving:

To make this yet more evident, I have added *Figure 2*, wherein these
Eight several Products are at one view exhibited. Let the rectangled
Parallelepipedon $A\, B\, C\, D\, E\, F\, G\, H$ be constituted of the sides $A\, B$, $G\, H$,
&c. proportional to N the number of the younger Age; $A\, C$, $B\, D$, &c.
proportional to n; and $A\, G$, $C\, E$, &c. proportional to the number of the
Elder, or ν. And the whole Parallelepipedon shall be as the Product $N\, n\, \nu$,
or our whole number of Chances. Let $B\, P$ be as R, and $A\, P$ as Y: let $C\, L$
be as r, and $L\, n$ as y; and $G\, N$ as $\int$, and $N\, A$ as υ; and let the Plain
$P\, R\, e\, a$ be made parallel to the plain $A\, C\, G\, E$; the plain $N\, V\, b\, Y$ parallel
to $A\, B\, C\, D$; and the plain $L\, X\, T\, Q$ parallel to the plain $A\, B\, G\, H$. And
our first Product $R\, r\, \int$ shall be as the Solid $S\, T\, W\, I\, F\, Z\, e\, b$. The Second,
or $r \int Y$ will be as the Solid $E\, Y\, Z\, e\, Q\, S\, M\, l$. The Third, $R \int y$, as the
Solid $R\, H\, O\, V\, W\, I\, S\, T$. And the Fourth, $R\, r\, \upsilon$, as the Solid $Z\, a\, b\, D\, W$
$X\, I\, K$. Fifthly, $\int Y\, y$, as the Solid $G\, Q\, R\, S\, I\, M\, N\, O$. Sixthly, $r\, Y\, \upsilon$, as
$I\, K\, L\, M\, G\, Y\, Z\, A$. Seventhly, $R\, y\, \upsilon$, as the Solid $I\, K\, P\, O\, B\, X\, V\, W$. And
Lastly, $A\, I\, K\, L\, M\, N\, O\, P$ will be as the Product of the 3 numbers of per-
sons dead, or $Y\, y\, \upsilon$. I shall not apply this in all the cases thereof for
brevity sake; only to shew in one how all the rest may be performed, let

---

[1] EDITOR'S NOTE: The manuscript for the 9th line of this page contains a misprint.
It should read $R + Y \times r + y$, etc. Likewise, in line 13, $R\, \rho\, y$ should read $R \int y$.

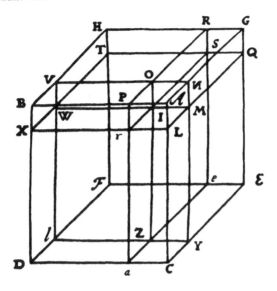

FIGURE 2

it be demanded what is the value of the Reversion of the younger Life after the two elder proposed. The proportion is as the whole number of Chances, or *N n v* to the Product *R y v*, so is the certain present value of the Sum payable after any term proposed, to the value due to such Chances as the younger person has to bury both the elder, by the term proposed; which therefore he is to pay for. Here it is to be noted, that

Years	Present value of 1 *l.*	Years	Present value of 1 *l.*	Years	Present value of 1 *l.*
1	0,9434	19	0,3305	37	0,1158
2	0,8900	20	0,3118	38	0,1092
3	0,8396	21	0,2941	39	0,1031
4	0,7921	22	0,2775	40	0,0972
5	0,7473	23	0,2618	45	0,0726
6	0,7050	24	0,2470	50	0,0543
7	0,6650	25	0,2330	55	0,0406
8	0,6274	26	0,2198	60	0,0303
9	0,5919	27	0,2074	65	0,0227
10	0,5584	28	0,1956	70	0,0169
11	0,5268	29	0,1845	75	0,0126
12	0,4970	30	0,1741	80	0,0094
13	0,4688	31	0,1643	85	0,0071
14	0,4423	32	0,1550	90	0,0053
15	0,4173	33	0,1462	95	0,0039
16	0,3936	34	0,1379	100	0,0029
17	0,3714	35	0,1301		
18	0,3503	36	0,1227		

the first term of all these Proportions is the same throughout, viz. *N n v.* The Second changing yearly according to the Decrease of *R, r,* ʃ, and Encrease of *Y, y, v.* And the third are successively ᴛhe present values of Money payable after one, two, three, &c. years, according to the rate of Interest agreed on. These numbers, which are in all cases of Annuities of necessary use, I have put into the following Table, they being the Decimal values of One Pound payable after the number of years in the Margent, at the rate of 6 *per Cent.* . . .

I. SOME FURTHER CONSIDERATIONS ON THE BRESLAW BILLS OF MORTALITY. BY THE SAME HAND, &C.

S I R,

What I gave you in my former Discourse on these Bills, was chiefly designed for the Computation of the Values of Annuities on Lives, wherein I believe I have performed what the short Period of my Observations would permit, in relation to exactness, but at the same time do earnestly desire, that their Learned Author Dr. *Newman* of *Breslaw* would please to continue them after the same manner for yet some years further, that so the casual Irregularities and apparent Discordance in the Table, [p. 1438] may by a certain number of Chances be rectified and ascertain'd.

Were this *Calculus* founded on the Experience of a very great number of Years, it would be very well worth the while to think of Methods for facilitating the Computation of the Value of two, three, or more Lives; which as proposed in my former, seems (as I am inform'd) a Work of too much Difficulty for the ordinary Arithmetician to undertake. I have sought, if it were possible, to find a Theorem that might be more concise than the Rules there laid down, but in vain; for all that can be done to expedite it, is by Tables of Logarithms ready computed, to exhibit the *Rationes* of *N* to *Y* in each single Life, for every third, fourth or fifth Year of Age, as occasion shall require; and these Logarithms being added to the Logarithms of the present Value of Money payable after so many Years, will give a Series of Numbers, the Sum of which will shew the Value of the Annuity sought. However for each Number of this Series two Logarithms for a single Life, three for two Lives, and four for three Lives, must necessarily be added together. If you think the matter, under the uncertainties I have mentioned, to deserve it, I shall shortly give you such a Table of Logarithms as I speak of, and an Example or two of the use thereof: But by Vulgar Arithmetick the labour of these Numbers were immense; and nothing will more recommend the useful Invention of Logarithms to all Lovers of Numbers, than the advantage of Dispatch in this and such like Computations.

Besides the uses mentioned in my former, it may perhaps not be an unacceptable thing to infer from the same Tables, how unjustly we repine at the shortness of our Lives, and think our selves wronged if we attain not Old Age; whereas it appears hereby, that the one half of those that are born are dead in Seventeen years time, 1238 being in that time reduced to 616. So that instead of murmuring at what we call an untimely Death, we ought with Patience and unconcern to submit to that Dissolution which is the necessary Condition of our perishable Materials, and of our nice and frail Structure and Composition: And to account it as a Blessing that we have survived, perhaps by many Years, that Period of Life, whereat the one half of the whole Race of Mankind does not arrive.

A second Observation I make upon the said Table, is that the Growth and Encrease of Mankind is not so much stinted by any thing in the Nature of the *Species*, as it is from the cautious difficulty most People make to adventure on the state of *Marriage*, from the prospect of the Trouble and Charge of providing for a Family. Nor are the poorer sort of People herein to be blamed, since their difficulty of subsisting is occasion'd by the unequal Distribution of Possessions, all being necessarily fed from the Earth, of which yet so few are Masters. So that besides themselves and Families, they are yet to work for those who own the Ground that feeds them: And of such does by very much the greater part of Mankind consist; otherwise it is plain, that there might well be four times as many Births as we now find. For by computation from the Table, I find that there are nearly 15000 Persons above 16 and under 45, of which at least 7000 are Women capable to bear Children. Of these notwithstanding there are but 1238 born yearly, which is but little more than a sixth part: So that about one in six of these Women do breed yearly; whereas were they all married, it would not appear strange or unlikely, that four of six should bring a Child every year. The Political Consequences hereof I shall not insist on, only the Strength and Glory of a King being in the multitude of his Subjects, I shall only hint, that above all things, Celibacy ought to be discouraged, as, by extraordinary Taxing and Military Service: And those who have numerous Families of Children to be countenanced and encouraged by such Laws as the *Jus trium Liberorum* among the *Romans*. But especially, by an effectual Care to provide for the Subsistence of the Poor, by finding them Employments, whereby they may earn their Bread, without being chargeable to the Publick.

The Law of
Large Numbers

I N Jacob Bernoulli's famous book, *Ars Conjectandi*, appears a theorem
of cardinal significance to the theory of probability. Usually called
Bernoulli's Theorem, it is also known as the Law of Large Numbers, a
name given to it by the French mathematician, Siméon Poisson (1781–
1840). This theorem was the first attempt to deduce statistical measures
from individual probabilities and Bernoulli claimed that it took him
twenty years to perfect it. The time was not ill spent considering the
central importance of the result, but mathematicians, scientists and philosophers have since then devoted many more than twenty years to examining and debating the exact meaning of the theorem and the proper range
of its application in statistics.

The theorem is quite simple to state. Indeed, on first seeing it you may
wonder what Bernoulli could have been stewing over for twenty years,
and why it should have stirred so much controversy later on. The fact is,
it is a nest of subtleties and traps; the harder one thinks about it, the
more one grows uneasy. Bernoulli, of course, had his hands full making
the machinery; he was entirely unaware of the logical and philosophical
snares he was setting.

Here, then, is a fairly simple statement of Bernoulli's Theorem: If the
probability of an event's occurrence on a single trial is p, and if a number
of trials are made, independently and under the same conditions, the
most probable *proportion* of the event's occurrences to the total number
of trials is also p; further, the probability that the proportion in question
will differ from p by less than a given amount, however small, increases
as the number of trials increases.[1] By "throwing mathematical discretion
to the winds," a well-known student of the subject arrives at this rough
everyday definition: "If the probability of an event is p, and if an infinity
of trials are made, the proportion of successes is sure to be p." [2] Nothing
bad will happen to you if you adopt this as a reference point in my discussion and in reading the selection from Bernoulli. Nevertheless, I should
like to make one more definition available. This one is a skillful com-
promise between a mathematical and a nonmathematical formulation:
" 'In a sufficiently large set of a things it is almost certain that the relative

---

[1] For two excellent definitions, from which mine has been synthesized, see John
Maynard Keynes, *A Treatise on Probability*, London, 1921, pp. 337–338; and Thorn-
ton C. Fry, *Probability and Its Engineering Uses*, New York, 1928, p. 100.
[2] Fry, *op. cit.*, p. 100.

frequency of β things will approximate to the probability of an α thing's being β within any degree of approximation which may be desired.' Here the phrase 'almost certain' is to be understood as a convenient way of saying that there is a probability as near as we like to 1." [3]

Each of these definitions raises several difficult questions. I cannot begin to state, let alone answer, them all adequately. But a few points should be set forth briefly to give the reader an inkling of the importance of Bernoulli's result and of the problems surrounding it.

1. The first thing to notice is that the theorem has to do with *proportions* or *frequencies* and not with the absolute number of times an event occurs. Increase the number of trials and it becomes increasingly likely (i.e., the probability approaches 1) that the *ratio* of successes to total trials will differ from p (the probability of the single event) by less than a fixed amount, *no matter how small*; but as the number of trials increases, it becomes increasingly unlikely (i.e., the probability approaches zero) that the *number* of successes will differ from p by less than a fixed amount, *no matter how large*.[4] This point is easily illustrated. In tossing a coin, one may assume that the probability of getting a head is ½. On this assumption the most probable number of heads in a series of trials is half the total number of tosses. It is not surprising, however, to find a deviation from this ratio: six heads in ten tosses, forty-eight in a hundred, and so on. The question is, what is the probability that in n successive throws the proportion of heads will differ from ½ by, for example, not more than ⅒? According to Bernoulli's theorem, the probability approaches 1 as the number n (of successive throws) increases. In other words the probability of a *relative deviation* of at least ⅒ "sinks rapidly." Six heads in ten exceeds our limit, but it is much less likely that the relative frequency of heads will differ from ½ by more than ⅒ in 100 throws, and even more unlikely in 1,000 throws. At the same time, it grows more probable as n increases that the absolute deviation of the number of heads from half the total number of throws will exceed any given number. For example, there is a much greater chance of a deviation of 1 in 100 tosses (i.e., 51 or 49 heads instead of 50), than of 1 in 10 tosses (i.e., 6 or 4 heads instead of 5); of 10 in 1,000 tosses than 10 in 100; and so on. That is the essence of the theorem.[5]

2. Bernoulli's theorem is the subject of gross misconceptions, some of

[3] William Kneale, *Probability and Induction*, Oxford, 1949, p. 139.
[4] See Fry, *op. cit.*, p. 101, for an instructive table comparing facts about number and proportion of successes.
[5] See Kneale, *op. cit.*, pp. 139–140; also Ernest Nagel, *Principles of the Theory of Probability*, in *International Encyclopaedia of Unified Science*, Vol. I, no. 6, p. 35. Each of these studies presents an admirable analysis not only of Bernoulli's theorem but of current problems of probability.

them due to the picturesque name coined by Poisson. It is often supposed, for example, that the theorem is "a mysterious law of nature which guarantees that in a sufficiently large number of trials a probability will be 'realized as a frequency.' " [6] This odd opinion derives apparently from the conviction that in the long run nature is bound to imitate man. The theorem is a part of the mathematical calculus of probability. The propositions of this calculus are not statements of fact or experience. They are formal arithmetical propositions, valid in their own domain, and neither capable of validating "facts" nor of being invalidated by them. [7] It is no more to be expected that the theorem can be proved by experiment than that the multiplication table or binomial theorem can be so proved.

Another fallacious inference from the theorem is the so-called "law of averages." This is an article of faith widely and fervently adhered to. It is the basis for the belief that when a player has had a bad run at cards, his luck is bound to turn; that after red has come up five times in a row at the roulette table, it is prudent to bet on black; that if a coin falls heads three or four times in succession, the chances are that tails will come up more frequently in the next three or four throws to "even things up." The only safe thing to say about these beliefs is that the player who acts on them consistently is more likely to be ruined than the player whose guide of action is erratic impulse. Bernoulli's Theorem is itself the sole ground for expecting a particular proportion of heads in the coin-tossing game, and it is an essential condition of the theorem that the trials be independent, i.e., without influence on each other. It is patently foolish, then, to invoke the theorem that sets out from the premise that the probability of a head at every toss is ½, to prove that the probability is less than ½ after a consecutive run of heads. Yet this is the muddleheaded idea underlying all gambling systems. "In a genuine game of chance there can be no system for improving one's chances of winning. That is part of what we mean by calling it a game of chance." [8]

3. While the theorem of Bernoulli (and certain generalizations of it by Poisson, Tchebycheff, Markoff and others) has proved exceptionally

---

[6] Kneale, op. cit., p. 139.

[7] "The calculus of probability has the same general function as a demonstrative geometry or a demonstrative arithmetic: given certain initial probabilities, the calculus of probability makes it possible to calculate the probabilities of certain properties which are related to the initial ones in various ways. Thus, arithmetic cannot tell us how many people live in either China or Japan; but if the population of China and the population of Japan are given, we can compute the combined populations of these countries. The calculus of probability functions in the same way." Nagel, op. cit., p. 27.

[8] The example is from Kneale, op. cit., p. 140. See also Julian Lowell Coolidge, "The Gambler's Ruin," Annals of Mathematics, Vol. X, Series 2, 1909; and the same author's An Introduction to Mathematical Probability, Oxford, 1925, pp. 52–59.

important in practical affairs—insurance, for example,[9] and in scientific research—e.g., the kinetic theory of gases, its limitations and weaknesses have repeatedly been stressed. Among the points raised most frequently are these: that the theorem applies only in special cases (e.g., where the events are truly independent) and under conditions which are the exception rather than the rule; "that a knowledge of what has occurred at some of the trials would not affect the probability of what may occur at any of the others"; [10] that the law of large numbers is useless as a tool of prediction concerning "sequences of observations" unless the concept of probability itself is defined in terms of relative frequency as against the formulation employed in classical probability theory.[11] None of these limitations, it should be observed, affect the *mathematical* validity of the theorem, nor do they depreciate either its contribution to the growth of statistical theory or its continuing indispensability as an instrument of inquiry. A sound grasp of the theorem is essential to an appreciation of the mathematics of probability and statistics—which, in turn, carries an increasingly responsible part in almost every branch of modern science, in industry, commerce, government and other activities. That is why I have devoted so much space to introducing the selection which follows.

[9] "As an illustration of the importance of the law of large numbers in practical affairs it will be sufficient to mention the business of insurance. Let us suppose that the probability that a man of a certain age and constitution will die within a year is 1/10. If such an individual considers insuring his life, this is the fraction which he should bear in mind and use in making his decision. But the insurance company which offers to cover the risk of his dying within the year considers another probability derived from this probability. If there are a great many people of the same characteristics insuring their lives with that company, there is a very high probability that the company will not have to pay claims on more than about one tenth of the policies. If, therefore, the company charges in each case a premium of rather more than one tenth of the amount of the policy, it is very likely that it will have enough over after all claims are paid to meet its administrative expenses and distribute a dividend to its shareholders. The greater the number of persons insuring with the company, the greater the probability that the company's finances will remain sound, provided always that its premiums are calculated in the way described. This is the all-important consideration which distinguishes the business of an insurance company from gambling." Kneale, *op. cit.*, p. 141.

[10] Keynes, *op. cit.*, p. 342; also, generally, pp. 341–345.

[11] A strong protagonist of this view argues his case persuasively in a most interesting book written for nonmathematicians: Richard von Mises, *Probability, Statistics and Truth*, 1939. See especially Lecture Four, "The Laws of Large Numbers," pp. 156–193.

Defendit numerus: There is safety in numbers. —AUTHOR UNKNOWN

The number is certainly the cause. The apparent disorder augments the grandeur. —EDMUND BURKE (*On the Sublime and the Beautiful*)

It doesn't depend on size, or a cow would catch a rabbit. —PENNSYLVANIA GERMAN PROVERB

3 The Law of Large Numbers[1]

By JACOB BERNOULLI

WE have now reached the point where it seems that, to make a correct conjecture about any event whatever, it is necessary only to calculate exactly the number of possible cases,[2] and then to determine how much more likely it is that one case will occur than another. But here at once our main difficulty arises, for this procedure is applicable to only a very few phenomena, indeed almost exclusively to those connected with games of chance. The original inventors of these games designed them so that all the players would have equal prospects of winning, fixing the number of cases that would result in gain or loss and letting them be known beforehand, and also arranging matters so that each case would be equally likely. But this is by no means the situation as regards the great majority of the other phenomena that are governed by the laws of nature or the will of man. In the game of dice, for instance, the number of possible cases [or throws] is known, since there are as many throws for each individual die as it has faces; moreover all these cases are equally likely when each face of the die has the same form and the weight of the die is uniformly distributed. (There is no reason why one face should come up more readily than any other, as would happen if the faces were of different shapes or part of the die were made of heavier material than the rest.) Similarly, the number of possible cases is known in drawing a white or a black ball from an urn, and one can assert that any ball is equally likely to be drawn: for it is known how many balls of each kind are in the jar, and there is no reason why this or that ball should be drawn more readily than any other. But what mortal, I ask, could ascertain the number of diseases, counting all possible cases, that afflict the human body in every one of its many parts and at every age, and say how much more likely

[1] Translated from "*Klassische Stücke der Mathematik*, ausgewählt von A. Speiser" (Zürich, 1925), pp. 90–95. The selection is from the German translation of the *Ars Conjectandi* by R. Haussner in Ostwald's *Klassiker der exakten Wissenschaften*, Leipzig, 1899, nr. 108.

[2] For "case," the correct translation of the German, one may read *result* or *outcome*. ED.

one disease is to be fatal than another—plague than dropsy, for instance, or dropsy than fever—and on that basis make a prediction about the relationship between life and death in future generations? Or who could enumerate the countless changes that the atmosphere undergoes every day, and from that predict today what the weather will be a month or even a year from now? Or again, who can pretend to have penetrated so deeply into the nature of the human mind or the wonderful structure of the body that in games which depend wholly or partly on the mental acuteness or the physical agility of the players he would venture to predict when this or that player would win or lose? These and similar forecasts depend on factors that are completely obscure, and which constantly deceive our senses by the endless complexity of their interrelationships, so that it would be quite pointless to attempt to proceed along this road.

There is, however, another way that will lead us to what we are looking for and enable us at least to ascertain *a posteriori* what we cannot determine *a priori*, that is, to ascertain it from the results observed in numerous similar instances. It must be assumed in this connection that, under similar conditions, the occurrence (or nonoccurrence) of an event in the future will follow the same pattern as was observed for like events in the past. For example, if we have observed that out of 300 persons of the same age and with the same constitution as a certain *Titius*, 200 died within ten years while the rest survived, we can with reasonable certainty conclude that there are twice as many chances that Titius also will have to pay his debt to nature within the ensuing decade as there are chances that he will live beyond that time. Similarly, if anyone has observed the weather over a period of years and has noted how often it was fair and how often rainy, or has repeatedly watched two players and seen how often one or the other was the winner, then on the basis of those observations alone he can determine in what ratio the same result will or will not occur in the future, assuming the same conditions as in the past.

This empirical process of determining the number of cases by observation is neither new nor unusual; in chapter 12 and following of *L'art de penser* [3] the author, a clever and talented man, describes a procedure that is similar, and in our daily lives we can all see the same principle at work. It is also obvious to everyone that it is not sufficient to take any single observation as a basis for prediction about some [future] event, but that a large number of observations are required. There have even been instances where a person with no education and without any previous instruction has by some natural instinct discovered—quite remarkably— that the larger the number of pertinent observations available, the smaller

[3] *La logique, ou L'art de penser*, by Antoine Arnauld and Pierre Nicole. 1662. (Makes use of Pascal, Fragment no. 14.) There are in fact *two* authors but Bernoulli makes it appear there is only one.

the risk of falling into error. But though we all recognize this to be the case from the very nature of the matter, the scientific proof of this principle is not at all simple, and it is therefore incumbent on me to present it here. To be sure I would feel that I were doing too little if I were to limit myself to proving this one point with which everyone is familiar. Instead there is something more that must be taken into consideration—something that has perhaps not yet occurred to anyone. *What is still to be investigated is whether by increasing the number of observations we thereby also keep increasing the probability that the recorded proportion of favorable to unfavorable instances will approach the true ratio, so that this probability will finally exceed any desired degree of certainty,* or whether the problem has, as it were, an asymptote. This would imply that there exists a particular degree of certainty that the true ratio has been found which can never be exceeded by any increase in the number of observations: thus, for example, we could never be more than one-half, two-thirds, or three-fourths certain that we had determined the true ratio of the cases. The following illustration will make clear what I mean: We have a jar containing 3000 small white pebbles and 2000 black ones, and we wish to determine empirically the ratio of white pebbles to the black—something we do not know—by drawing one pebble after another out of the jar, and recording how often a white pebble is drawn and often a black. (I remind you that an important requirement of this process is that you put back each pebble, after noting its color, before drawing the next one, so that the number of pebbles in the urn remains constant.) Now we ask, is it possible by indefinitely extending the trials to make it 10, 100, 1000, etc., times more probable (and ultimately "morally certain") that the ratio of the number of drawings of a white pebble to the number of drawings of a black pebble will take on the same value (3:2) as the actual ratio of white to black pebbles in the urn, than that the ratio of the drawings will take on a different value? If the answer is no, then I admit that we are likely to fail in the attempt to ascertain the number of instances of each case [i.e., the number of white and of black pebbles] by observation. But if it is true that we can finally attain moral certainty by this method [4] . . . then we can determine the number of instances *a posteriori* with almost as great accuracy as if they were known to us *a priori.* Axiom 9 [presented in an earlier chapter] shows that in our everyday lives, where moral certainty is regarded as absolute certainty, this consideration enables us to make a prediction about any event involving chance that will be no less scientific than the predictions made in games of chance. If, instead of the jar, for instance, we take the atmosphere or the human body, which conceal within themselves a multitude of the most varied processes or diseases, just as the jar conceals the pebbles, then for

[4] Bernoulli demonstrates that this is true in his next chapter.

these also we shall be able to determine by observation how much more frequently one event will occur than another.

Lest this matter be imperfectly understood, it should be noted that the ratio reflecting the actual relationship between the numbers of the cases—the ratio we are seeking to determine through observation—can never be obtained with absolute accuracy; for if this were possible, the ruling principle would be opposite to what I have asserted: that is, the more observations were made, the *smaller* the probability that we had found the correct ratio. The ratio we arrive at is only approximate: it must be defined by two limits, but these limits can be made to approach each other as closely as we wish. In the example of the jar and the pebbles, if we take two ratios, 301/200 and 299/200, 3001/2000 and 2999/2000, or any two similar ratios of which one is slightly less than 1½ and the other slightly more, it is evident that we can attain any desired degree of probability that the ratio found by our many repeated observations will lie between these limits of the ratio 1½, rather than outside them.

It is this problem that I decided to publish here, after having meditated on it for twenty years. . . .

. . . If all events from now through eternity were continually observed (whereby probability would ultimately become certainty), it would be found that everything in the world occurs for definite reasons and in definite conformity with law, and that hence we are constrained, even for things that may seem quite accidental, to assume a certain necessity and, as it were, fatefulness. For all I know that is what Plato had in mind when, in the doctrine of the universal cycle, he maintained that after the passage of countless centuries everything would return to its original state.

Statistics and the Lady
with a Fine Palate

STATISTICS has shot up like Jack's beanstalk in the present century. And as fast as the theory has developed, politics, economics, social affairs, business and science have taken it over for their special purposes. The anthologist must survey a literature so new and so vast that even the expert can scarcely comprehend it. It is impossible to know where to begin, let alone how to make a representative choice. I have tried to subdue the problem by a gross expedient. The selections which follow do no more than discuss a few fundamental concepts; the history of statistics, the great bulk of theoretical considerations and the applications of statistical method are scarcely mentioned. This procedure is justified on grounds other than expediency. Statistics, as Tippett observes, resembles arithmetic in its impact on science and human affairs.[1] But while arithmetic "is so woven into our thinking that we use it almost subconsciously," statistics is invariably regarded as a separate branch of study, and even its basic ideas are grasped by only a small proportion of educated persons. The material presented below may clarify a few essential principles for readers who, though they recognize the importance of the subject, would not dream of working through an entire book on statistics.

The first two selections are from two very good primers, L. H. C. Tippett's *Statistics*, and M. J. Moroney's *Facts from Figures*. Tippett, a leading British statistician, explains what is meant by sampling, theory of random errors, the nature of statistical laws. Two chapters are excerpted from Moroney's fat little volume in the Pelican series. They treat the concept of averages, scatter, mean and standard deviations. Both Tippett and Moroney—the latter is a British statistician, industrial consultant and lecturer in mathematics at Leicester College of Technology and Commerce—write for the nonmathematician and express themselves with commendable clarity.

The selection by R. A. Fisher is more advanced. It has to do with the design of experiments, a branch of scientific inquiry dealing with the nature of scientific inquiry itself. This study of method represents one of the most fruitful advances of scientific thought in the past two or three decades.

The theory of experimental design does not of course represent the

---

[1] L. H. C. Tippett, *Statistics*, Oxford University Press, 1944, p. 178.

first effort of scientists to examine and systematize their methods of observation and inference. It has long been known, for example, that scientific measurements, however careful and precise (such as those of astronomy), never yield the same results in successive determinations. These irregularities or "errors" are regarded as unavoidable accompaniments of experiment due to small, undetectable causes. Mathematicians have lumped together the various mysterious disturbances under the name of chance and have invented brilliant theories to control the mischief and confusion caused by chance. Laplace and Gauss, shortly after the opening of the nineteenth century, laid the foundation for the theory of errors of observation, a mathematical achievement of the first order. These matters are considered in the selections from Tippett and Moroney and I need dwell on them no further.

The design of experiments is also concerned with observational error, but it moves along a somewhat broader front than the classical mathematical attack on the problem. It may help to an appreciation of the difference in approach if I say that the theory of experimental design is less "complacent" than the theory of random errors. The latter thrives on errors, merely fixing limits within which experimental results are acceptable despite variations. The former worries the experiment itself to make certain that its structure is logical, that it is broad enough to serve as a foundation for inference, that the objects studied are fully and fairly exploited, that every recognizable and avoidable source of error, however small, has in fact been eliminated. Concretely, the designer must consider such factors as the variables of the system to be examined (e.g., do they form the simplest set compatible with the objective of telling the experimenter what he really wants to know?), the adequacy and representativeness of the sample, the sources of psychological bias in subjects, instruments or experimenter, the selection of suitable controls to serve as standards of comparison for the significant variables in the main experiment, the appropriate "level of significance" for the given test (i.e., what is the minimum probability the experimenter would require "before he would be willing to admit that his observations have demonstrated a positive result"?), the value of enlarging and repeating the experiment to increase its "sensitiveness." These and other points like them apply generally to all experiments, whether laid out in the physics, chemistry, biology or psychology laboratory, the hospital, the agricultural station, the schoolroom, or the lagoon at Bikini. One feels that the earlier experimental geniuses—Galileo, Faraday, Boyle, Galton, Pasteur, to name a few—avoided almost instinctively many of the pitfalls now carefully fenced off by the modern theories of experimental procedure. At the same time, one must recognize that they had less to fret about when they performed their experiments. The great increase in the complexity of experiments multiplies the opportunities for going astray as much as it enlarges the effectiveness and scope of scientific research.

The Fisher excerpt has to do with a specific example of testing design. Sir Ronald Fisher, one of the foremost statisticians of the century, is the pioneer of the theory of design of experiments. He is professor of genetics at Cambridge University, a Fellow of the Royal Society and a Foreign Associate of the U. S. National Academy of Sciences. His name is associated prominently with the development of elaborate experimental techniques in agriculture at the famous Rothamsted Experimental Station, with important advances in genetics and the mathematical theory of natural selection, with searching reforms and innovations of statistical method such as factorial design, the "confounding" procedure, the use of Latin Squares, the exploitation of small samples and substantial refinements of randomization. (For enlightenment as to these terms I refer the reader to the selection from Fisher and also to a recently published book by E. Bright Wilson: *Introduction to Scientific Research*, New York, 1952.) Fisher's writings include the standard textbook *Statistical Methods for Research Workers* [2] and *The Design of Experiments*, regarded as a classic work of statistics and scientific method. The second chapter of the latter book is entitled "The Principles of Experimentation, Illustrated by a Psycho-Physical Experiment." It concerns a lady who says that when a cup of tea is made with milk she is able to tell whether the tea or milk was first added to the cup. The surpassing nicety of taste displayed by this hypothetical lady provides Sir Ronald with the excuse for a most remarkable series of experiments. Fisher is not an easy writer; the presentation of this case, however, is a model of lucidity and requires no mathematics other than elementary arithmetic. It demands of the reader the ability to follow a closely reasoned argument, but it will repay the effort by giving a vivid understanding of the richness, complexity and subtlety of modern experimental method.

[2] Eleventh Revised Edition, 1951, Oliver and Boyd Ltd., Edinburgh.

It is then, but an exceeding little Way, and in but a very few Respects, that we can trace up the natural Course of things before us, to general Laws. And it is only from Analogy, that we conclude, the Whole of it to be capable of being reduced into them; only from our seeing, that Part is so. It is from our finding, that the Course of Nature, in some Respects and so far, goes on by general Laws, that we conclude this of the Whole.

—BISHOP BUTLER

4 Sampling and Standard Error

By L. C. TIPPETT

SAMPLING

THE practice of taking a small part of a large bulk to represent the whole is fairly generally understood and widely used. The housewife will 'sample' a piece of cheese at the shop before making a purchase; and a cotton spinner will buy a bale of cotton, having seen only a small sample of it. The sample is also a very important tool of the statistician.

There are two general reasons for working with samples instead of the bulk. (1) Some appraisals of the thing in question involve destructive tests, and there is no point in appraising it if the whole is destroyed in the process; the housewife cannot eat her cheese and have it. (2) It is very much more economical to investigate a sample than the whole bulk. In social and economic work, for example, it is usually prohibitively expensive to investigate the whole field of inquiry in any detail. Even the population census, which has behind it the financial and coercive resources of the state, is made only at infrequent intervals, and the questions asked are few and comparatively simple. If a sample inquiry is made, on the other hand, it is feasible to employ experienced field workers who can collect information that is comparatively detailed and elaborate, and can ensure that the records are reasonably accurate.

Unfortunately, however, the method of inquiry by sample is somewhat mistrusted, sometimes honestly and sometimes, I suspect, because a sample has in some instance given a result the sceptic does not like. Yet a sample *may* give reliable results. For example, an earthquake disaster in 1923 interrupted the tabulation of the results of the Japanese census of 1920 and interim figures were given based on a sample containing one family in every thousand. These results agreed well with those given later when the regular tabulations were completed. Nevertheless it must be agreed that samples do not represent the bulk exactly, and that they may sometimes be much in error.

In the present discussion [of sampling methods] I shall follow the usual practice of statisticians of referring to the bulk that is being sampled as the *population*. The population in this chapter is to be thought of specially as contrasting with the sample. I shall refer only to populations consisting of recognizably discrete individuals, e.g., men or electric lamps.

The ideal sample is the simple random one in which chance alone decides which of the individuals in the population are chosen. Suppose we wish to obtain a random sample of the people of England and Wales in order to make an estimate of their average height. To do this we may, in principle, take forty-odd million exactly similar cards, one for each person, and write each person's national registration number on the appropriate card. These cards may then be put in a large churn, thoroughly mixed, and (say) one thousand cards be drawn, somewhat in the way the names are drawn for the Irish sweepstake. The thousand people whose numbers are on the cards are a random sample, and we can measure their heights, find the average, and so obtain a figure which is an estimate of the average height for the population.

To investigate the error in the average so estimated we could, again in principle, subsequently measure the heights of all individuals in the population and so obtain the true average. An easier thing to do is to draw a number of samples, each of one thousand, and calculate the several averages. These will vary above and below the true, or population value, and the extent to which they vary gives some idea of the error with which any one sample estimates the true average.

To do such an experiment in fact requires far greater resources than I can command, but there are other experiments that are similar in principle and are easier to do. What we really want to know is how chance works in deciding the choice of the sample, and chance also operates in games of the table, with such things as cards, dice and roulette wheels. In these games, a population does not exist in the sense that the population of England and Wales does, but we may use the concept of a *hypothetical* population. Suppose, for example, we threw a perfectly balanced six-sided die millions of times. We should expect one-sixth of the throws to score aces, one-sixth to score twos, and so on, and the average score would be $\frac{1}{6}(1 + 2 + 3 + 4 + 5 + 6) = 3 \cdot 5$. These millions of throws are a population, and any thousand of them including the first thousand is a random sample. But the millions of throws need not, in fact, be made; they need only be imagined as a hypothetical population, of which any number of actual throws form a sample.

To illustrate the way in which random sampling errors arise I have made an experiment which I need not describe in exact detail. The experiment is equivalent to that described here, which is not quite so easy to perform but easier to imagine. The imagined apparatus consists of ten packs, each of ten cards, the cards in each pack being numbered respectively 1, 2, 3 . . . 10. The packs are shuffled separately, one card is drawn from each, and the ten numbers on the cards are added to give a score. For example, the numbers might be 2, 4, 2, 10, 2, 5, 9, 2, 9, 8 and the score would then be 53. Then the cards are put back in their packs, the packs are reshuffled, and again ten cards are drawn to give another score. This is repeated, so that a large number of scores results, which are individuals from a hypothetical population consisting of the very large number of scores that could conceivably be obtained. The lowest conceivable score is 10, resulting from ten aces; the highest is 100, resulting from ten tens; and the true average score is 55. Now let us consider the results of the experiment.

It would take too much space to give in full the results of a really extensive experiment, but enough are given in Table 1 to show the kind of thing that happens. The top part of the table gives the first thirty individual scores. Chance has not given a score as high as 100 or as low as 10, as it might have done, and presumably would have done had I continued long enough with the experiment. The first thirty scores vary between 36 and 72, the range being 36. Now, in order to see what happens when we take samples and find the averages, I took 30 samples, each of ten scores. Such samples are far too small for most statistical inquiries (although statisticians sometimes have to be content with small

TABLE 1

Individual Scores and Average Scores in Samples of Ten and Forty

Individual Scores									
52	46	72	53	36	55	42	56	61	53
56	65	48	54	62	65	48	65	61	60
58	42	58	46	63	61	68	53	54	43

Averages of Samples of Ten									
52·6	58·4	54·6	52·6	48·6	54·0	52·8	50·8	46·0	55·8
53·4	59·4	55·0	56·2	61·6	53·6	54·2	56·8	52·3	54·0
56·7	55·2	56·3	52·3	53·8	57·8	55·9	61·8	58·6	49·2

Averages of Samples of Forty									
54·6	51·6	53·6	56·6	54·3	55·1	57·3	54·4	56·0	55·4
55·3	54·1	55·8	55·4	56·0	53·2	55·1	54·3	54·8	54·2
54·3	57·2	53·2	56·0	54·5	51·5	53·7	56·0	54·8	55·4

samples) but they illustrate the errors of random sampling. The average scores are in the middle section of Table 1. The first average of 52·6 is obtained from the ten individual scores in the top row of the table. The thirty averages vary between 46·0 and 61·8, the range being 15·8, and no average differs from the population value of 55 by more than $9 \cdot 0$. In so far as these thirty samples show the variations we are likely to get in the averages of the millions of samples we could draw, we may say that the biggest error with which the average of any one sample of ten scores estimates the population average is $9 \cdot 0$. When I took larger samples, each of forty scores, I obtained results given in the lowest section of Table 1. They vary between $51 \cdot 5$ and $57 \cdot 3$ with a range of $5 \cdot 8$, and the biggest error with which any one sample of forty scores estimates the population average is $55 - 51 \cdot 5 = 3 \cdot 5$. Thus we see that the averages estimated from random samples vary among themselves and differ from the average for the population, but that the biggest error decreases as the size of the sample is increased from ten to forty; and you may take on trust that this tendency would have continued had I extended the experiment to deal with still larger samples. For example, by calculating the average of the thirty averages of samples of forty, we have the average of a single sample of 1,200 scores, which comes to $54 \cdot 8$—very close to the population value of 55.

These results are shown in the frequency distributions of Figure 1 where, instead of a frequency for each sub-range, there are dots, each dot representing an individual score or the average of a sample. Notice how the averages tend to be clustered more closely round the population value as the size of the sample is increased. A frequency distribution of sample averages for any given size of sample is called the *sampling distribution* of the average.

The errors of random sampling, which in an experiment like that just described show themselves as variations between sample means arise from the variation between the individuals in the original population. Other things being equal, such sampling errors are proportional to the amount of variation in the population. As an extreme example, it is easy to see that had there been no variation between the individual scores and they had all been 55, the means of all samples of all sizes would have been 55 and there would have been no sampling errors.

When the statistician thinks of the random error of the average of a sample he thinks of a whole collection of possible values of error, any one of which the given sample may have: of the sampling distribution of errors. The actual error of the given sample probably exceeds the smallest of these values; it may easily exceed the intermediate values; and it is unlikely to exceed the very largest values. There is a whole list of probabilities with which the various values of error are likely to be exceeded, and these can be calculated from a quantity called the *standard error*. The standard error is a measure of the variation in the sampling distribution analogous to the standard deviation [1] and for the statistician it sums

---

[1] [See selection by M. J. Moroney, p. 1487, ED.]

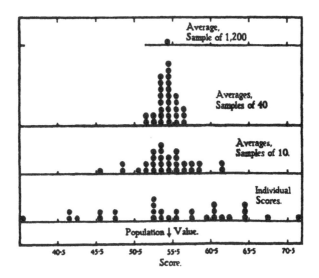

FIGURE 1—Frequency distributions giving results of sampling experiments.

up the whole distribution of errors. If the standard error of a sample is large, the errors to which that sample is liable are, as a whole, large; if the standard error is small, the likely errors are small. This quantity, carrying with it the idea of errors occurring with various probabilities, should replace the cruder 'biggest error' I introduced in describing the results of the experiment.

It is not usually necessary to do an actual experiment to measure sampling errors, as the mathematical theory of probability enables statisticians to deduce sampling distributions and standard errors theoretically. This method is better because it is less laborious and more exact, giving results as accurate as an experiment involving millions of samples. The results of the theoretical calculations are of the same kind as those given by the experiment, and in some instances they have been checked by very large-scale experiments.

I have considered only the sampling errors of the average, but the same principles apply to other statistical quantities such as ratios, and the measures of variation and correlation. The theoretical deduction of sampling distributions of the many statistical quantities in use is a very highly developed branch of mathematical statistics; and sometimes the problems have proved so difficult to solve that statisticians have had to fall back upon actual sampling experiments.

With the ability to calculate errors of sampling, statisticians can make allowances for them when making deductions from sample results. It is a standard procedure to examine the results of a sample to see how far they can be explained by random errors. This is called *testing the significance* of the results, and only such results as cannot reasonably be attributed to errors of random sampling are held to be *statistically significant*.

Before going on to the more practical problems of sampling, I will summarize the ground covered so far. When many samples of the same size are taken from a population of variable individuals, the sample averages show variation which may be described by a sampling distribution and measured by the standard error. A given sample of that size may have any one of the averages in the distribution, and the probability that its error will exceed any stated value can be calculated from the standard error. The standard error of the average is a measure of the errors to which a sample average is liable. For a sample of given size, this standard error increases as the variation between the individuals in the original population increases; for a given population, the standard error becomes smaller as the size of the sample is increased. (For the sake of the mathematically minded it may be stated that the standard error is inversely proportional to the square root of the number in the sample.) Consequently the random errors can be made as small as we please by making the sample large enough, and for a given population it is possible to calculate the size of sample necessary to reduce these errors to any desired value. Similar remarks apply to quantities other than the average.

The tendency for large samples from some population to have averages that vary little amongst themselves and differ but little from the population value is the reality behind the popular conception of the Law of Averages. This law does not operate, as some people think, so that an abnormally high individual score or run of scores is followed by an abnormally low score or run, correcting the average by compensation. In a random series, the scores following an abnormal score or run are quite unaffected by what has gone before; they tend to be nearer the general average than the abnormal scores are, i.e. to be more normal, so that when included in the average they reduce the effect of the abnormal scores. Averaging has more of a swamping than a compensating effect. Thus, if we may regard the days of weather as individuals from a population, the average weather for the population being the general type experienced at a given time of the year and place, the law of averages does not require that a very wet spell shall be followed by a very dry spell. For all I know, there may be a law to that effect, but if so, it is not the law of averages.

If the individuals in a statistical population are well mixed up, no known method of investigation can give more accurate results for a given cost than the method of purely random sampling just described, unless something is known about the individuals to enable some sort of selection to be made. Sometimes, however, a more complex form of random sample called the representative sample gives greater accuracy. Suppose, for example, that in a housing survey we wish to find the average number of rooms per family in some town. Some families at one end of the scale of wealth will live in one room each and at the other end there may be families that have say twelve rooms each; and this variation over a range of eleven rooms per family will give rise to a certain standard error in a simple random sample. Suppose further that we can divide the town into three districts—'poor,' 'middle-class,' and 'wealthy'—in each of which the total number of families is known, and that the range of variation of rooms per family is from one to seven in the poor, from four to ten in the middle class, and from six to twelve in the wealthy district. Then if we take a random sample from any one district, the district average is estimated with a smaller standard error than that just mentioned, resulting from a range of variation of six rooms per family (i.e. 7 minus 1, 10 minus 4, or 12 minus 6). Further, it can be proved that if a *representative* sample of the same size is taken, in which the proportion of families from each district is the same as in the whole town, the standard error of the average of that sample will be the same as the smaller error resulting from a range of variation of six rooms per family. This is because the proportion of families from each district is left to chance in the simple random sample; in the representative sample it is not, and that source of error is removed.

Random sampling is the basis of the representative sample, however, which is nothing more than a weighted combination of random subsamples.

Representative sampling is used in the Gallup polls of public opinion, where care is taken to see that the opinions of various classes of people are represented in appropriate proportions instead of leaving it to chance to determine what these proportions shall be.

If it is granted that the ideal random sample can be a reliable instrument of investigation, the questions remain: Can the ideal be attained? Are the actual samples that are used as reliable as random samples? As a random sample is increased in size, it gives a result that progressively comes closer to the population value, whereas samples taken in some of the ways that are used give results that progressively come closer to some value other than the population value, results that may for some kinds of sample be too high, or for others too low. A sample of this kind is said to be *biased*, and the difference between the value given by a very large sample and the corresponding population value is called an *error of bias*. A biased die, for example, is one for which the fraction of throws showing an ace, say, tends to a value other than one-sixth (the value for the hypothetical population), and the greater the number of throws, the clearer is it that the fraction of actual aces is not one-sixth. Errors of bias are added to the random errors, and since they follow no laws from which they can be calculated, they must be eliminated entirely or reduced so that they become unimportant. This may be difficult to do, and it is often necessary to use very elaborate sampling methods to avoid errors of bias.

It is nearly impossible for anyone to select individuals at random without some randomizing apparatus. If a teacher tries to select a few children from a class, he will tend to choose too many clever ones, or dull ones, or average ones; or if he tries to be random he may select too many clever and dull children and too few intermediate ones. In selecting a sample of houses 'at random,' the investigator will be very unlikely to select anything like the right proportions of large and small ones, shabby and smart ones, new and old ones, and so on. Bias almost inevitably will creep in. This is illustrated by the results of large experiments conducted on several thousands of school children in Lanarkshire in 1930 to measure the effect of feeding them with milk, on their growth during the period of the experiment—about six months. At each school the children were divided into two comparable groups; one group received the milk and the other did not, and the effect of the milk was to be measured by comparing the growth rates of the two groups. The results for a number of schools were combined. In an experiment of this kind, the accuracy depends very much on the two groups or samples of children being similar on the average before the feeding with milk begins, i.e. on one being unbiased with respect to the other. To secure this, the children were selected for the two groups either by ballot or by a system based on the alphabetical order of the names. Usually, these are both good ways of making unbiased random samples of the two groups, but the whole thing was spoilt by giving the teachers discretionary powers, where either method gave an undue proportion of well-fed or ill-nourished children, 'to substitute others to obtain a more level selection.' Presumably the substitution was not done on the basis of the actual weights of the children, but was left to the personal judgement of the teachers. The result was that at the start of the experiment, the children in the group that were later fed with milk were smaller than those in the other group, the average difference being an amount that represented three months of growth. It has been suggested that teachers tended, perhaps subconsciously, to allow their natural sympathies to cause them to put into the 'milk' group more of the children who looked as though they needed nourishment. This bias did not ruin the experiment, but unfortunately the interpretation of some of the results was left somewhat a matter of conjecture instead of relative certainty, and there was later a certain amount of controversy about some of the interpretations. The substitutions of the children could have been done without introducing bias had the actual weights been made the basis, and there would have been an improvement on the purely random sampling; but by unwittingly introducing the bias, it seems that the teachers actually made matters worse.

A sampling method that is very liable to give biased results, particularly when testing opinion on controversial matters, is that of accepting voluntary returns. An undue proportion of people with strong views one way or the other are likely to make the returns, and people with moderate views are not so likely to take the trouble to represent them. For this reason, the post-bags of newspapers and Members of Parliament do not give random samples of public opinion.

A spectacular example of a biased sample is provided by the attempt of the American magazine, the *Literary Digest*, to forecast the results of the Presidential election of 1936 by means of a 'straw vote.' Some ten millions of ballot post cards were sent to people whose names were in telephone directories and lists of motor-car owners, and several million cards were returned each recording a vote for one of the candidates. Of those votes, only $40 \cdot 9$ per cent were in favour of President Roosevelt, whereas a few weeks later in the actual election he actually polled $60 \cdot 7$ per cent of the votes. Those from among telephone users and motor-car owners who returned voting cards did not provide a random sample of American public opinion on this question.

Bias does not result only from obviously bad sampling methods; it may arise in more subtle ways when a perfectly satisfactory method is modified slightly, perhaps because practical conditions make this necessary. In some Ministry of Labour samples of the unemployed, a 1 per cent sample was made by marking every hundredth name in the register of claims, which was in alphabetical order. Bias was introduced by not confining the inquiry to the marked names; instead, the first claimant appearing at the Exchange whose name was marked or was among the five names on either side of the marked one, was interviewed to provide the necessary data. Claimants who are in receipt of benefit attend at the exchange several days in a week, whereas those whose claims are disallowed but who are maintaining registration only attend once a week. The effect of this and of the latitude allowed in the choice of persons for interview was that too many claimants in receipt of benefit were included in the sample. It was only when the existence of this bias was realized that some of the results that were apparently inconsistent with other known facts made sense. A similar kind of effect can arise in surveys of households if no one is at home when the investigator first calls at some house chosen to be one of the sample. Such houses are likely to contain small families with few or no young children, since in large households someone is almost certain to be at home to answer the door; and unless the houses with no one at home are re-visited, the sample will be biased in respect of size and character of household.

Although there is no general theory of errors of bias by which the amount of such errors can be calculated in any particular instance, as can be done for random errors, statisticians do not work entirely in the dark. Sometimes the sample gives, as part of its results, information that is also known accurately from a full census, and the sample is usually regarded as free from bias in all respects if in this one respect it agrees with the census. The soundness of the results of a sample inquiry may sometimes be checked by comparing them with data obtained in other ways, perhaps by other investigators. Where none of these checks are available, it may be necessary to rely on the statistician's general experience of sampling methods in deciding whether the sample in question is a good one. I have given enough examples to show that a good deal is known of the ways in which errors of bias arise, and what must be done to avoid them.

It is implicit in my definition of errors of bias that they cannot be 'drowned' by taking very large samples, in the way that random errors can; a fact that the experience of the *Literary Digest's* straw-vote on the American Presidential election of 1936 amply confirms. From this point of view, a good sample can be arrived at only by employing a good sampling method. I have already mentioned some methods incidentally, and it is only necessary here to give it as a warning that when a statistician advises adherence to an elaborate method with a closeness that may seem to the layman to be 'fussy,' that advice had better be followed; failure to do so has been known to lead to biased results.

Altogether, the method of inquiry by sample is difficult and full of pitfalls. But statisticians could not get on without it and experience of its use is both wide and deep, so that in competent hands the method is capable of giving results that are reasonably accurate. Moreover, the inevitable errors in the results can be estimated, and allowance can be made for them in arriving at conclusions.

TAKING ACCOUNT OF CHANCE

Chance operates in many fields besides that of random sampling, and many of its effects can be calculated by applying the same general methods as are used to calculate the errors of random sampling. Some of the further applications of those methods will be described in the present chapter.

The effects of chance can be calculated only because they follow certain laws, but these differ in kind from the *exact* laws of subjects like physics.

Events that follow exact laws can be described or predicted precisely; but we can only specify probabilities that chance events will occur, or specify limits within which chance variations will probable lie. Newton's laws of motion, for example, are exact because they describe exactly the relations between the motions of bodies and the forces acting upon them; the errors of random sampling follow chance laws because we cannot predict exactly what average a random sample will have; we can only state, as I have suggested on p. 1462, the probability that it will lie within certain limits.

I cannot embark upon a full discussion of what we mean by chance, but as a preliminary I shall indicate a few ideas associated with the word. Statisticians attribute to chance, phenomena (events or variations) that are not exactly determined, or do not follow patterns described by known exact laws, or are not the effects of known causes. That is to say, the domain of chance varies with our state of knowledge—or rather of ignorance. Such ignorance may be fundamental because the relevant exact laws or causes are unknowable; it may be non-essential or temporary, and exist because the exact laws do not happen to have been discovered; or the ignorance may be deliberately assumed because the known exact laws and causes are not of such a character that they can profitably be used in the particular inquiry in hand.

An example of ignorance that, according to present-day ideas, is fundamental, is in the Principle of Indeterminacy of modern physics; we do not and cannot know the precise motion of an electron. We do not know what determines the position of a shot on a target, but that ignorance is non-essential and in some degree temporary. The variations in the positions of the shots depend on a host of factors such as variations in the primary aim of the marksman, the steadiness of his hand, the weight, size, and shape of the bullets, the propelling charges, the force of the wind, and so on; but presumably these factors can be investigated and laws be discovered. Indeed, this has happened; and the history of gunnery shows the temporary character of the ignorance. Gunnery is much more of a science and more exact than it was in the days of the Battle of Waterloo, or even during the 1914–18 War; and as knowledge has increased, unpredictable variations in placing of shots have been reduced; but at each stage these variations are regarded as due to chance. Ignorance of causes is assumed by an insurance company in using its past experience of accident claims to establish future premiums for motor-car insurance. The company has considerable knowledge of the circumstances surrounding every accident on which a claim is made, but is unable to make more than limited use of that knowledge, and so treats accidents largely as chance events, except for a few special allowances such as 'no claims bonuses' or extra premiums charged to people with bad accident records.

Usually, events regarded as coming within the domain of chance are those governed by a complicated system of many causes, each of which produces only a small variation; and one frequent characteristic of such events is that small changes in the circumstances surrounding them make a big difference to the results.

Chance as I have described it operates in a very wide field, covering the whole of the unknown; but mathematical calculations can be made and chance laws be propounded only for comparatively simple systems covering a portion of this field. Nevertheless, such calculations have a wide range of usefulness, which the following examples will illustrate.

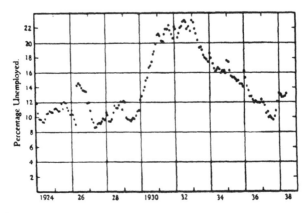

FIGURE 2—Percentage of insured workers unemployed, Great Britain 1924–38.

One use of chance calculations is for deciding which of the fluctuations in a time series are random and which are trends having some significance. As an example of a time series, consider the unemployment data represented in Figure 2. Readers will have no difficulty in recognizing the broad changes, viz. the minor waves in 1925 and towards the end of 1928, the large upward sweep in 1930, the improvement from the end of 1932 to the end of 1937, and the upward movement again in 1938. For the time being we shall omit 1926, the year of the General Strike, as being exceptional. These changes are reasonably attributed to fundamental causes that operate fairly slowly and may be represented by a smooth curve drawn through the actual points of the graph. There are a number of mathematically determined curves that have the property of changing in level in such a slow, regular way, and are of the nature of exact laws or descriptions.

Let us imagine such a curve to be drawn through the points of Figure 2. Then the actual points will be seen to deviate from this curve. They may in some degree follow a seasonal pattern (another exact law), but in Figure 2 that pattern is not very evident, and most of the deviations are irregular. Presumably many of them can be explained in terms of a minor strike in some industry, a political change in some country affecting another industry, an exceptionally hard winter, and so on; but we cannot bring such knowledge into a system, and so we assume ignorance. Hence, having tried all the known kinds of exact laws that are relevant, we consider the deviations to be due to a complex system of chance causes that operate we know not how; and we apply to them the same laws as describe the results of games of chance and sampling experiments. This is the argument for applying the theory of errors of random sampling to testing the statistical significance of fluctuations in time series. For example, there was a sudden and temporary rise in unemployment in the beginning of 1936; is it significant? It actually occurred, and therefore is real, but when we ask the above question we in effect ask: Can the rise be reasonably considered as a random fluctuation arising from that system of causes we have labelled *chance*, or has some unusual event happened? And so we apply the theory of random errors. If this theory had been applied to testing the statistical significance of the sudden rise in unemployment in 1926 it would have shown the operation of something unusual—we know that to be the General Strike.

This kind of application of the theory enables us, in retrospect, to decide whether any particular events with which we try to associate fluctuations have had important effects compared with the system of random fluctuations. When following changes week by week or month by month as they occur it is useful, too, to be able to decide whether the last increase or decrease is large enough to call for action, or whether it is random. At one period, for example, a local newspaper used to publish weekly figures of deaths due to road accidents in a certain town, and the number used to fluctuate about an average of four or five per week. Should we worry if between two particular weeks the number rises from three to six, or rejoice if it falls from five to two? No! Such changes are no greater than any that can be attributed to chance, and do not indicate a real change in conditions. Sometimes the chance coincidence of random fluctuations may give rise to several consecutive small increases or decreases, giving a spurious appearance of a trend. Sampling theory can show when such is the case.

To arrive at results of these kinds, it is necessary to analyse the time series so as to separate the random fluctuations from the secular movements; and additional complications occur if the system of random fluctuations changes. Some would say, for example, that in trade the random fluctuations during a slump and a boom are different. The whole analysis is only approximate, but it is based on ideas that are sufficiently close to reality to give useful results.

The theory of random errors was used for measuring the accuracy of astronomical measurements long before it was applied to statistical samples, and it is used somewhat in measuring the errors of experimental observations in general. When the astronomer measures, say, the position of a star, he finds that in spite of the precision of his apparatus, and the care with which he adjusts it and makes his observations, he does not get the same answer from successive determinations. He repudiates the idea that the position is varying and attributes the variations in his results to unavoidable errors of observation. The question arises: What is the true position? And if it cannot be measured exactly, how accurately can it be estimated? A similar situation arises in the other so-called exact sciences: e.g., in physics and chemistry. Several determinations have been made of the velocity of light, but they do not agree exactly; and a chemist would be very surprised if he got exactly the same result every time he measured an atomic weight.

This interpretation of experimental results as being due to an invariable quantity plus observational or experimental errors is purely a mental conception. The only reality is the set of observations, the characteristics of which can, if desired, be expressed by any statistical constants such as the average, or a measure of variation, or by a frequency distribution. For most experiments, however, it is useful and (within limits) valid to adopt the more common conception.

The errors do not follow any known exact laws, and so the laws of chance are sometimes used to describe them. In applying these laws, the results are regarded as a random sample from a hypothetical population of results, the average of this population being the true value. Then, the average of the sample is an estimate of the true value, and the error in that estimate can be calculated as for any statistical sample. Is this idea valid? On the face of things, it seems as reasonable to imagine the millions of results that would have been obtained had the experiment been repeated millions of times under the same conditions as to imagine the results of millions of tosses of a die. But it is not so certain that the variations between experimental results are entirely of the same kind as those we get when we toss dice.

On this question there are differences of opinion among experimentalists. Some refuse to admit any similarities between experimental and random errors. Others, faced with otherwise intractable results, use the theory of random errors as the only way out. Experimental errors are not, in general, random. There are 'personal' factors, and any one person shows a bias that changes from time to time. I prefer to regard a set of experimental results as a biased sample from a population, the extent of the bias varying from one kind of experiment and method of observation to another, from one experimenter to another, and, for any one experimenter, from time to time. If this view is accepted, experimental errors can be regarded as forming a chance system, but the system is not as simple as that assumed in calculating the errors of random sampling.

In general the bias cannot be estimated and the theory of random errors is therefore not enough. Sometimes, however, one can say that the bias is likely to be small compared with the random errors, and then the theory may give useful, if approximate, results. For example, if, say, five separate chemists were to determine the atomic weight of an element independently, in different times and places, and possibly by different methods, the results would vary because of the effects of random errors and bias. But the separate biases for the five chemists would differ and so would appear as the random errors between the results, the group as a whole would probably exhibit but little bias, and the theory of errors would provide a reasonably close measure of the precision with which the average of the five results estimates the true atomic weight. This might not be so, on the other hand, for the average of, say, twenty consecutive determinations made by one chemist in one laboratory.

Errors of bias are often relatively unimportant when the observed quantity is the difference between two similar quantities. In measuring the distance between two lines in a spectrum, for example, the main error is often due to the uncertainty of setting the cross-hairs of the measuring microscope on the centres of the lines. If there is a bias in doing this, it is likely to be similar for the two lines (provided they are not too dissimilar in width and appearance), and the difference in the two settings will probably be practically unbiased. The theory of errors gave a result that was at least qualitatively right, when applied to Lord Rayleigh's measurements of the density of nitrogen. He made a number of determinations on 'atmospheric' and 'chemical' nitrogen and found a difference in the two averages. Subsequent treatment by the theory of errors has shown that the difference is greater than can be attributed to random variations, and this result is in accordance with a real difference we now know to exist, owing to the presence of the rarer inert gases in 'atmospheric' nitrogen.

Where the bias is completely unknown, I doubt if it is possible to do more than hope that the true value lies somewhere between the highest and lowest of the actual values, and regard the average as an estimate of the true value, that is as good as, but no better than, any other single estimate that could be made from the data. It is, of course, the experimenter's job to reduce bias and random errors to a minimum.

To sum up, the theory of random errors may be usefully applied to some experimental observations, particularly of differences in values, but great caution must be observed on account of bias. Certainly such an application is no substitute for careful experimental control.

Much experimental work, particularly in biological subjects, is now done under conditions, many of which can be well controlled, and the observations can be made accurately; but the material is inherently variable and the results have to be treated statistically. The Lanarkshire experiment made to measure the effect of milk on the growth of children, already mentioned on p. 1466, is of this kind. The amount of milk fed can be controlled, children fed and not fed with milk can be kept in the same environment, and the changes in weight can be measured accurately; but it would not do to base conclusions on an experiment on, say, two children. Children vary, and it is necessary to observe a large number and take averages.

The problem of interpreting the results of such experiments is essentially statistical, and it has fallen to the lot of statisticians to study the general questions of arranging experiments with variable material, of drawing conclusions from the results, and of testing them. Under the leadership of Professor R. A. Fisher, who started this work at the Rothamsted Experimental Station (for agriculture), an elaborate technique for doing this has been developed and is very widely used. I propose to give some description of this subject.

There are three main principles to be observed in designing such an experiment; they are replication, randomization, and economy in arrangement.

The necessity for replication has already been stated. The problem first arose chiefly in agricultural field trials made to measure such things as the effects of various fertilizers on wheat yield. It was early seen that different plots treated in the same way gave different yields. Hence, it was not sufficient to have two plots, say, to treat one with a fertilizer, to grow the crops and measure the yields, and to regard the difference as measuring the effect of the fertilizer. The experiment had to be replicated by treating several plots in each way and measuring the difference between the average yields.

Even differences in such averages can be affected by variations between plots, as we can see from the results of the sampling experiment described in the last chapter; and it is desirable to estimate the accuracy of the observed difference. The only known way of doing this is by the theory of random errors. It was found, however, that variations in plot fertility were not random. There was usually a fertility pattern, e.g. a gradient in fertility across the field. In order that the theory of sampling could be applied, an element of randomization was introduced artificially by using some such device as a ballot to decide which plots should receive the various experiment treatments. This is a 'trick of the trade' for making fertility variations into a comparatively simple chance system. A statistician might apply this principle to the above-mentioned experiment of feeding milk to school children by tossing a coin once for each child, giving that child milk if the result is 'heads,' say, and no milk if the result is 'tails.'

The pattern in fertility differences between plots in a field was used to increase the accuracy of experimental comparisons. Adjacent plots tend to be more alike than those in different parts of the field, and by comparing the treatments on adjacent plots the random variations affecting the comparison were reduced, with an increase in accuracy. The other way of increasing accuracy is to increase the number of plots, and hence the expense of the experiments; the arrangement using adjacent plots is therefore more economical. In the same way, had it been possible in the Lanarkshire milk experiment to use identical twins, giving milk to one of each pair, far fewer children would have given the same accuracy as thousands chosen at random. This kind of arrangement can be made to satisfy the condition of randomness sufficiently for the application of the theory of random errors in an appropriate form.

The above are the elementary principles of the modern approach to the design of what I shall term *statistical-experimental* investigations. The whole subject has, however, become very complicated as several treatments of one kind have been included, and treatments of several kinds. Thus, experiments may be done with various quantities and combinations of several kinds of fertilizer on several varieties of wheat. Further complication arises when experiments are done on different farms and in different years, and it is necessary to consider to what extent results obtained on one group of farms in one year apply to other farms and other years.

In spite of the fact that sound methods are available, experimenters continue to work with variable material on non-statistical lines, and they get discordant results which they cannot fit into a system. Different workers sometimes get different results in the same subject, and controversies arise. When, in such circumstances, the experimenters turn to sound methods of statistical analysis, involving proper experimental arrangements, difficulties of these kinds tend to disappear. Then, experiments which were previously done on an inadequate scale are increased in size, often they are designed more economically than before, and the advancement of knowledge is made more orderly and certain.

Statistical methods are often regarded as applying only to very large numbers of observations, but that is no longer true. It would be far too costly to replicate some experiments hundreds and thousands of times, and statisticians have had to make do with small numbers. They have, however, developed the theory of errors to apply to small samples as well as to large ones.

There are many chance events that occur in life, to which the general theory of random errors may in some degree be applied.

For example, many telephone subscribers have access to one trunk line, and a multitude of causes determine how many will want to use it at any given instant, i.e. it is to some extent a question of chance whether more than one subscriber will want to use it at once and thus cause delay. In so far as this is true, the extent of delays of this kind can be calculated from the theory of probability which is the basis of the theory of errors. This is typical of a number of congestion problems that arise in telephony, in road and rail traffic, and so on; and although many of them are difficult mathematically, the theory is being applied.

Accidents have a large element of chance in their causation—the circumstances preceding a 'near shave' often differ by only a hairbreadth from those preceding a catastrophically fatal accident, and the theory of probability has been useful for studying accident problems in calculating the effects of chance and showing the importance of other factors. The following is an example.

Records were kept of the numbers of accidents that happened during the course of one year to 247 men workers engaged in moulding chocolate in a factory. Some of the men had no accident, some had one, some two, and so on, a few having as many as twenty-one accidents. The data are arranged in a frequency distribution in the first two columns of Table 2. Now we ask: Were all the variations between the men in the numbers of accidents they suffered due to chance, or were there differences between the men in their tendency to have accidents? Were the 42 men who had no accidents exceptionally skilful or just lucky; and were the 22 men who had ten accidents or more clumsy or unlucky? The average number of accidents per man is 3·94, and even if all the men were equally skilful in avoiding accidents, chance would give rise to some variation. It has been calculated from the extended theory of random sampling that this variation would result in the frequency distribution of the last column of figures of Table 2. This is very different from the actual distribution. We may say, roughly, that 5 of the 42 men with no accidents were lucky and the remaining 37 skilful; that one of the 22 men with ten or more accidents was unlucky and the remainder clumsy. Comparisons of this kind between actual and calculated chance distributions have led to investigations that have shown how people differ in 'accident proneness,' i.e. in their tendency under given circumstances to suffer accidents. The chance distribution given in Table 2 is calculated by assuming a very simple system of chance variations; more complicated systems taking into account variations in accident proneness have been used in the more advanced investigations on the subject.

TABLE 2

FREQUENCY DISTRIBUTION OF MEN WHO HAD
VARIOUS NUMBERS OF ACCIDENTS. COMPARISON
BETWEEN ACTUAL AND CHANCE DISTRIBUTIONS

(Data by E. M. Newbold. Report No. 34, *Industrial
Fatigue (Now Health) Research Board*)

Number of Accidents	Frequency of Men	
	Actual	*Chance*
0	42	5
1	44	21
2	30	40
3	30	50
4	25	48
5	11	37
6	12	23
7	15	13
8	8	6
9	8	3
10–15	19	1
15–21	3	–
Total	247	247

STATISTICAL LAWS

The central problem of statistics is dealing with groups variously described as collections, crowds, aggregates, masses, or populations, rather than with individual or discrete entities; with events that happen on the average or in the long run rather than with those that happen on particular occasions; with the general rather than with the particular. A fuller consideration of this aspect of statistics is the subject of the present chapter.

Again I shall use the language common in statistical writings and refer to *populations* of *individuals*. The population is regarded (in the discussion of sampling) as something from which samples are taken, but here as an aggregate of individuals, which will in most instances be represented by a sample, i.e. I shall not distinguish between the population and the sample.

The population has characteristics and properties of its own, which are essentially derived from and are an aggregate of those of the individuals, although the two sets of properties may be different in kind. In the population, the individuals merge and their individuality is dissolved, but from the dissolution rises a new entity like a phoenix from the flames. The population is at the same time less and more than the totality of the individuals.

This conception is not peculiar to statistics. Rousseau, for example, distinguishes in *The Social Contract* between the General Will and the wills of all the people:

'In fact, each individual, as a man, may have a particular will contrary or dissimilar to the general will which he has as a citizen. His particular interest may speak to him quite differently from the common interest.'

'There is often a great deal of difference between the will of all and the general will; the latter considers only the common interest, while the former takes private interest into account, and is no more than a sum of particular wills: but take away from these same wills the pluses and minuses that cancel one another, and the general will remains as the sum of the differences.'

The general idea is expressed in another way in the following passage from *Old Junk* by Mr. H. M. Tomlinson:

'His shop had its native smell. It was of coffee, spices, rock-oil, cheese, bundles of wood, biscuits and jute bags, and yet was none of these things, for their separate essences were so blended by old association that they made one indivisible smell, peculiar, but not unpleasant, when you were used to it.'

The loss of individuality results from the method of the statistician in confining his attention to only a few characteristics of the individuals and grouping them into classes. Consider a married couple, say Mr. and Mrs. Tom Jones. As a couple their individuality consists of a unique combination of a multitude of characteristics. Mr. Jones is tall and thin, is aged 52 years, has brown hair turning grey, and is a farmer. Mrs. Jones is called Mary and at 38 years is still handsome; she is blonde and is really a little too 'flighty' for a farmer's wife. The couple have been married for 16 years and have three children: two boys aged 14½ and 11 years, and a girl aged 2. In addition to these and similar attributes the couple have a number of moral and spiritual qualities that we may or may not be able to put down on paper. It is by all these, and a host of other qualities that their relatives and neighbours know Mr. and Mrs. Jones; the uniqueness of the combination of qualities is the individuality of the couple.

The statistician who is investigating, say, the ages of husbands and wives in England and Wales is interested only in the ages, and does not wish to describe even these accurately. So he puts our couple in that class for which the age of the husband is 45–55 years and that of the wife is 35–45 years. Mr. and Mrs. Jones are now merely one of a group of some 320,000 other couples, and are indistinguishable from the others in their group.

Statistical investigations are not always confined to one or two characters of the individuals, and elaborate methods have been developed for dealing with many attributes, e.g. the ages of married couples at marriage, income, number of children, fertility of the grandparents of the children, and so on, but however many attributes are included, they are very few compared with the number that make up the individuality of each couple.

A population of individuals is the most characteristic and simplest chance system the statistician has to deal with. We do not know, or do not take any account of, the causes of the differences between the individuals, and so we dismiss them as being due to chance, and fasten our attention on the population.

Statistics is essentially totalitarian because it is not concerned with individual values of even the few characters measured, but only with classes. However much we analyse the data to show the variation between the parts, we still deal with sub-groups and sub-averages; we never get back to the individuals. In studying the death rate of a country, for example, we may decompose the general average into sub-averages for the two sexes, for the separate age-groups, for different localities, industries, and social classes; but the death rate of an individual has no meaning. When we think of variation, we think of a mass of variable individuals rather than of one or two being very different from the remainder.

This part of statistical technique in selecting only a few characteristics for investigation, and in classifying the data, is not only necessary because of the limited power of the human brain to apprehend detail, but is a part of the general scientific method. It is an essential step in the development of general scientific laws. However much we know of Mr. and Mrs. Jones in particular, if we know nothing more we have no basis for drawing conclusions about married couples in general. It is only by paying attention to such features as individuals have in common with others that we can generalize. Individuals are important, as such, to themselves, to their neighbours and relations, and to professional consultants—the parson, the doctor, and the lawyer; they have no importance for the statistician, nor indeed for any scientist, except that they, with a host of other individuals, provide data.

Our first and, for most of us, our only reactions to our environment are individualistic. We *are* individuals, our experience is mostly with individuals, and even when considering a group we are conscious mostly of our personal relationship to it. The concept of the population as an entity does not come easily, and our ordinary education does little to correct this defect. The mental effort required to realize this concept is perhaps something like that necessary to appreciate a fugue with its contrapuntal pattern, as compared with the ease of following a tune with simple harmonies.

The characteristics of the population are described by frequencies and by the statistical constants and averages already described, but it is apparently so difficult to think of the reality behind these constants—the mass of individuals—that we personify the population and speak in such terms as 'the average man.' This is only possible because of a similarity between some of the measures of a population and those of an individual; the average height of a group of men is expressed in feet and inches, just as the height of one man is; but the similarity is only superficial.

We have already seen the inadequacy of the average as a description of variable material,[2] but the average individual sometimes is also a rather absurd figure. In 1938, for example, he was among those comparatively rare individuals who died at the age of 58 years. His age in England and Wales in 1921 was 29·9, and in 1938 it was 33·6 years; i.e., in 17 years the average man aged by only 3·7 years! The average family can have fractions of a person. Books on the upbringing of babies usually contain a curve showing the growth in weight of an average baby; but few actual curves are like that. The curve for a real baby may be above or below that for the average and it may have a different slope in various parts. It will also usually have 'kinks' due to teething troubles and minor illnesses, whereas the curve for the average baby is fairly smooth; this paragon among children has no troubles!

Variation is, of course, an important characteristic of populations that individuals cannot have. I have already been at pains to describe this,[3] and to point out how, for example, the deviations from any relationship shown by a contingency or correlation table are as characteristic of the data as the relationship itself. Indeed, without variation, a collection of individuals is scarcely a population in the statistical sense. A thousand exactly similar steel bearing balls (if such were possible) would be no more than one ball multiplied one thousand times. It is the quality of variation that makes it difficult at first to carry in mind a population in its complexity.

All the special properties of populations I have considered arise in aggregates of independent individuals, but there are additional characteristics due to interactions between individuals. The behaviour of men in the mass is often different from their behaviour as individuals. Some men affect (or 'infect') others and such phenomena as mass enthusiasms and panics arise. We speak of mass-psychology. Similarly the effect of an infectious disease on a community of people in close contact is different from its effect on a number of more or less isolated individuals. Statistical description can take account of interactions between individuals, but it is seldom necessary to do so.

---

[2] Discussed in an earlier chapter of Tippett's book. ED.
[3] Discussed in an earlier chapter of Tippett's book. ED.

Although the individuals in a population vary, the characteristics of the population itself are very stable. Sir Arthur Eddington has well said: 'Human life is proverbially uncertain; few things are more certain than the solvency of a life-insurance company.' This means that we do not know when any individual will die, but an insurance company can estimate the incidence of death in its population of policy-holders with great accuracy.

This contrast between individualistic variability and statistical stability, and the fact that the latter emerges from the former, this apparent paradox of order coming out of chaos, has from time to time given rise to metaphysical speculations. People in the eighteenth century, accustomed to considering the variations between individuals, seem to have been struck by the statistical regularities and saw evidences of a Divine order. Sir Arthur Eddington, on the other hand, presumably taking for granted the regularity of the laws of physics, is more struck by the compatibility with these laws of the unpredictable variation in the behaviour of individual electrons, and offers comfort to those who want to believe in free will and scientific law at the same time. The practical statistician may accept it as a fact requiring no special metaphysical explanation, that mass regularities can often be discerned where the individuals apparently follow no regular laws.

Galton writes of the regularity of form of the frequency distribution in the following terms:

'I know of scarcely anything so apt to impress the imagination as the wonderful form of cosmic order expressed by the "Law of Frequency of Error." The law would have been personified by the Greeks and deified, if they had known of it. It reigns with serenity and in complete self-effacement, amidst the wildest confusion. The huger the mob, and the greater the apparent anarchy, the more perfect is its sway. It is the supreme law of Unreason. Whenever a large sample of chaotic elements are taken in hand and marshalled in the order of their magnitude, an unsuspected and most beautiful form of regularity proves to have been latent all along.'

Let us re-examine the data from the sampling experiment described on pp. 1460–1462 and see if we can repeat Galton's experience and recapture something of his mood.

I have extended the experiment to obtain 4,000 scores altogether. The first thirty are given in the top part of Table 1 (p. 1461) in the order in which they occurred, and these together with the 3,970 other scores are the 'large sample of chaotic elements'—and chaotic they undoubtedly appear. I then proceeded to marshal the scores in the order of their magnitude by forming a frequency distribution, and stage by stage stopped to look at the result as the distribution began to grow. The results for 50, 200, 1,000, and 4,000 scores are in Figure 3. Since the scores are whole numbers, I have not grouped them into sub-ranges; the scales of the distributions in the vertical direction have been reduced as the numbers of scores have increased. At 50 scores, there is no sign of any regularity or form in the distribution, but at 200 scores, a vague suggestion of a form seems to be emerging; the scores show a slight tendency to pile up in

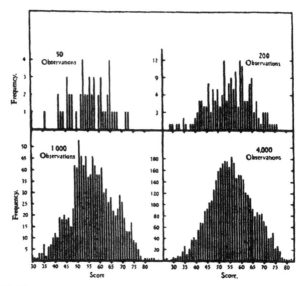

FIGURE 3—Frequency distribution of various numbers of observations from the same population.

the middle of the range. At 1,000 scores, the form is clearly apparent, although irregularities are still pronounced; but at 4,000 scores, the 'most beautiful form of regularity' is there, almost in perfection. It is not difficult to imagine the regularity that would be apparent were the sample so large as to be indistinguishable from the population.

The formulae and laws that describe populations and their behaviour as opposed to individuals are termed *statistical laws*. The various statistical constants (e.g. standard deviation, mean deviation, measures of association and correlation) are elementary statistical laws. Other laws of a higher order of complexity describe how populations change with time or place, or other circumstances. Laws of heredity, for example, are a way of describing how some characters in populations of plants or animals change from generation to generation.

Some statistical laws are discovered by simple observation of the population as a whole. For example, the change in the death rate for the country may be recorded from year to year, without any consideration being given to the changes in the chances of death from various causes, to which the individual is exposed. A public lighting authority could compare two batches of electric lamps by counting how many of each are burnt out after having been in use for, say, 500 hours. Or a colony of the banana fly may be kept in a bottle under standard conditions, and the growth in numbers observed. However, there is nothing necessarily statistical in the technique applied in such experiments, although investigations of this character are often classed as statistical in the widest sense of the word. The introduction of the concept of pieces of matter as populations of electrons or atoms does not necessarily turn an ordinary physical investigation of the macroscopic properties of matter into a statistical one.

Statistical methods and calculations are involved, however, when the laws for the population are deduced from those for individuals. The calculation of statistical constants is a case in point, and the estimation of some quality of a batch of electric lamps from calculations made on the full frequency distribution of lives is another. Estimates, made by demographers, of the size and age composition of the future population from a consideration of the characteristics of the present population and the various birth and death rates, are an important example of the statistical deduction of statistical laws. Such calculations may involve complicated mathematics.

It is implicit in all I have written that statistical laws have nothing to do with individuals. It is no exception to the statistical law that old men have old wives, on the average, if one old man of one's acquaintance has a young wife. A failure to recognize the distinction between the two types of laws sometimes leads to attempts to apply statistical laws to individuals, with paradoxical results.

We now return to the starting-point of this chapter—a consideration of individuals. They in the aggregate are the population, and from their characteristics we can calculate those of the population. We cannot perform the reverse process. Individuality is lost, as far as the statistician is concerned, for good and all. Does this mean we know absolutely nothing of the individual when we know the population? Not quite.

Consider a single electric lamp taken at random from the batch represented by the distribution of Table 3. Even if we do not know its life, we know that it will be an exceptional lamp if its life is greater than, say, 2,800 hours—it will be one of $\frac{4}{150}$ths of the batch. Indeed, it is more likely to be one of the $\frac{89}{150}$ths of the lamps with lives between, say, 1,000 and 2,000 hours.

We are used, in ordinary life, to dealing with data of this kind by

TABLE 3

LENGTH OF LIFE OF ELECTRIC LAMPS

(Data by E. S. Pearson, *Journal of the Royal
Statistical Society*, **96**, 1933, p. 21)

Life (hours)	Frequency of Lamps
0– 200	1
200– 400	3
400– 600	2
600– 800	10
800–1,000	19
1,000–1,200	21
1,200–1,400	23
1,400–1,600	18
1,600–1,800	17
1,800–2,000	10
2,000–2,200	8
2,200–2,400	5
2,400–2,600	5
2,600–2,800	4
2,800–3,000	2
3,000–3,200	1
3,200–3,400	1
Total	150

introducing the concept of probability. In the example quoted we would say that the probability of any one lamp having a life greater than 2,800 hours is $\frac{4}{150} = 0 \cdot 027$, and that the probability of the life being between 1,000 and 2,000 hours is $\frac{89}{150} = 0 \cdot 593$.

This is an application of what is commonly regarded as the statistician's definition of probability as a ratio of frequencies. Corresponding to any frequency distribution there can be calculated a whole series of probabilities of a random individual lying within various stated limits, and statistical probability is a device (a verbal trick!) for attaching to the random individual the characteristics of the whole distribution. In this way, a population is epitomized in an individual much more satisfactorily than in the concept of 'the average man.' But statistical probability does more than this. It corresponds closely to the more popular idea of probability as a measure of the strength of belief in a thing. Most people if asked what is the probability of a tossed penny falling heads uppermost would reflect that heads was as likely as tails and would reply: one-half. The statistician, if in a pedantic mood, would reply: in the hypothetical population of tosses, one-half of the total give heads, therefore the probability of a head is one-half. An alternative method of expression is to state that the chances of a head are even, or for the lamps, that they are 593 to 407 in favour of a life of between 1,000 and 2,000 hours.

Probability is, in ordinary life, also applied to events that do not occur as frequencies. We speak of the probability of, or the chances in favour of, a particular horse winning a race. Even in such instances, however, I think that people carry at the backs of their minds the idea of frequencies; they in effect imagine a lot of races, in a given proportion of which the particular horse wins. The idea is described in the following quotation from a lecture given by Karl Pearson in 1892:

'A friend is leaving us, say in Chancery Lane at 4 o'clock in the afternoon, and we tell him that he will find a Hansom cab at the Fleet Street corner. There is no hesitation in our assertion. We speak with knowledge, because an invariable experience has shown us Hansom cabs at 4 o'clock in Fleet Street. But given the like conditions within reach of a suburban cab-stand, and our statement becomes less definite. We hesitate to say absolutely that there will be a cab: "You are sure to find a cab," "I believe there will be a cab on the stand," "There is likely to be a cab on the stand," "There will possibly be a cab on the stand," "There might *perhaps* be a cab," "I don't expect there will be a cab," "It's very improbable," "You are sure not to find a cab," etc., etc. In each and every case we go through some rough kind of statistics, *once* we remember to have seen the stand without a cab; on occasions few and far between, "perhaps on an average once a month," "perhaps once a week," "every other day," "more often than not there has been no cab there." Certainty in the case of Fleet Street passes through every phase of belief to disbelief in the case of the suburban cab-stand. If once a month is the very maximum of times I have seen an empty cab-stand, my belief that my friend will find a cab there to-day is far stronger than if I have seen it vacant once a week. A measure of my belief in the occurrence of some event in the future is thus based upon my statistical experience of its occurrence or failure in the past.'

Thus probability in its most general use is a measure of our degree of confidence that a thing will happen. If the probability is $1 \cdot 0$, we know the thing will certainly happen, and if the probability is high, say $0 \cdot 9$, we feel that the event is likely to happen. A probability of $0 \cdot 5$ denotes that the event is as likely to happen as not, and one of zero means that it certainly will not. This interpretation, applied to statistical probabilities calculated from frequencies, is the only way of expressing what we know of the individual from our knowledge of the population.

Statistical laws, which describe the characters and behaviour of populations in one way or another, may be transformed into probabilities— i.e. from them the probabilities and frequencies in the population may be calculated. Thus, statistical laws are the chance laws referred to earlier.

It may have been noticed that probabilities have been calculated from the frequencies of a distribution, either known as for the lamps, or assumed as for the penny. In general, it is necessary to have some data on which to calculate probabilities. I am often asked what is the probability of some queer or interesting event, without being given any data. Statisticians do not evolve probabilities out of their inner consciousness, they merely calculate them.

*"Let us sit on this log at the roadside, says I, and forget the inhumanity
and the ribaldry of the poets. It is in the glorious columns of ascertained
facts and legalized measures that beauty is to be found. In this very log
that we sit upon, Mrs. Sampson, says I, is statistics more wonderful than
any poem. The ring shows it was sixty years old. At the depth of two thousand feet it would become coal in three thousand years. The deepest coal
mine in the world is at Killingworth, near Newcastle. A box four feet long,
three feet wide, and two feet eight inches deep will hold one ton of coal.
If an artery is cut, compress it above the wound. A man's leg contains
thirty bones. The Tower of London was burned in 1841."*

*"Go on, Mr. Pratt," says Mrs. Sampson. "Them ideas is so original and
soothing. I think statistics are just as lovely as they can be."*

—O. HENRY (*The Handbook of Hymen*)

*When Tennyson wrote The Vision of Sin, Babbage read it. After doing so,
it is said he wrote the following extraordinary letter to the poet:*

"In your otherwise beautiful poem, there is a verse which reads:

'*Every moment dies a man,*
Every moment one is born.'

*"It must be manifest that, were this true, the population of the world
would be at a standstill. In truth the rate of birth is slightly in excess of
that of death. I would suggest that in the next edition of your poem you
have it read:*

'*Every moment dies a man,*
Every moment 1¹⁄₁₆ is born.'

*"Strictly speaking this is not correct. The actual figure is a decimal so
long that I cannot get it in the line, but I believe 1¹⁄₁₆ will be sufficiently
accurate for poetry. I am etc."* —MATHEMATICAL GAZETTE

5 On the Average and Scatter

By M. J. MORONEY

ON THE AVERAGE

'The figure of $2 \cdot 2$ children per adult female was felt to be in some
respects absurd, and a Royal Commission suggested that the middle classes
be paid money to increase the average to a rounder and more convenient
number.' (*Punch*)

IN former times, when the hazards of sea voyages were much more
serious than they are today, when ships buffeted by storms threw a portion
of their cargo overboard, it was recognized that those whose goods were
sacrificed had a claim in equity to indemnification at the expense of those
whose goods were safely delivered. The value of the lost goods was paid
for by agreement between all those whose merchandise had been in the
same ship. This sea damage to cargo in transit was known as 'havaria'
and the word came naturally to be applied to the compensation money
which each individual was called upon to pay. From this Latin word

1487

derives our modern word *average*. Thus the idea of an average has its roots in primitive insurance. Quite naturally, with the growth of shipping, insurance was put on a firmer footing whereby the risk was shared, not simply by those whose goods were at risk on a particular voyage, but by large groups of traders. Eventually the carrying of such risks developed into a separate skilled and profit-making profession. This entailed the payment to the underwriter of a sum of money which bore a recognizable relation to the risk involved.

The idea of an average is common property. However scanty our knowledge of arithmetic, we are all at home with the idea of goal averages, batting and bowling averages, and the like. We realize that the purpose of an average is *to represent a group of individual values* in a simple and concise manner so that the mind can get a quick understanding of the general size of the individuals in the group, undistracted by fortuitous and irrelevant variations. It is of the utmost importance to appreciate this fact that the average is to act as a *representative*. It follows that it is the acme of nonsense to go through all the rigmarole of the arithmetic to calculate the average of a set of figures which do not in some real sense constitute a single family. Suppose a prosperous medical man earning £3,000 a year had a wife and two children none of whom were gainfully employed and that the doctor had in his household a maid to whom he paid £150 a year and that there was a jobbing gardener who received £40 a year. We can go through all the processes of calculating the average income for this little group. Six people between them earn £3,190 in the year. Dividing the total earnings by the number of people we may determine the average earnings of the group to be £531 13s. 4d. But this figure is no more than an impostor in the robes of an average. It represents not a single person in the group. It gives the reader a totally meaningless figure, because he cannot make one single reliable deduction from it. This is an extreme example, but mock averages are calculated with great abandon. Few people ask themselves: What conclusions will be drawn from this average that I am about to calculate? Will it create a false impression?

The idea of an average is so handy that it is not surprising that several kinds of average have been invented so that as wide a field as possible may be covered with the minimum of misrepresentation. We have a choice of averages; and we pick out the one which is appropriate both to our data and our purpose. We should not let ourselves fall into the error that because the idea of an average is easy to grasp there is no more to be said on the subject. Averages can be very misleading.

The simplest average is that which will be well known to every reader. This common or garden average is also called the *mean*, a word meaning 'centre.' (All averages are known to statisticians as 'measures of central tendency,' for they tell us the point about which the several different values cluster.) The *arithmetic mean* or average of a set of numbers is calculated by totalling the items in the set and dividing the total by the number of individuals in the set. No more need be said on this point, save that the items to be averaged must be of the same genus. We cannot, for example, average the wages of a polygamist with the number of his wives.

A second kind of average is the *harmonic mean*, which is the reciprocal [1] of the arithmetic mean of the reciprocals of the values we wish to average. The harmonic mean is the appropriate average to use when we are dealing with rates and prices. Consider the well-known academic example of the aeroplane which flies round a square whose side is 100 miles long, taking the first side at 100 m.p.h., the second side at 200 m.p.h., the third side at 300 m.p.h., and the fourth side at 400 m.p.h. What is the average speed of the plane in its flight around the square? If we average the speeds using the arithmetic average in the ordinary way, we get:

$$\text{Average speed} = \frac{100 + 200 + 300 + 400}{4} = 250 \text{ m.p.h.}$$

But this is not the correct result as may easily be seen as follows.

Time to travel along the first side   = 1 hour
Time to travel along the second side = 30 minutes
Time to travel along the third side   = 20 minutes
Time to travel along the fourth side   = 15 minutes
Hence total time to travel 400 miles = 2 hours 5 minutes
                                                                        = $25\frac{1}{12}$ hours

From this it appears that the average velocity is $40\%_1 \div 25\frac{1}{12} = 192$ m.p.h.

The ordinary arithmetic average, then, gives us the wrong result. A clue as to the reason for this will be found in the fact that the different speeds are not all maintained for the same time—only for the same distance. The correct average to employ in such a case is the harmonic mean.

In order to give the formula for this we shall here introduce a little more mathematical notation which will be of great benefit to us later in this book. In calculating averages we have to *add up* a string of items which make up the set whose average is required. The mathematician uses a shorthand sign to tell us when to add up. He calls adding up 'summing' and uses the Greek letter *S* which is written Σ and called 'sigma' to indicate when terms are to be added. (This is actually the capital sigma. Later we shall have a lot to say about the small letter sigma which is written σ.) Each of the numbers which have to be taken into account in our calcula-

---

[1] The reciprocal of a number is found by dividing that number into unity, e.g., the reciprocal of $4 = \frac{1}{4} = 0.25$.

tion is denoted by the letter x. If we wish to differentiate between the various quantities we can number them thus: x_1, x_2 x_3, x_4, etc., the labelling numbers being written as subscripts so that they will not be confused with actual numbers entering into the calculation. (This may sound as confusing to the novice as it will be boring to the learned. Let the learned turn over the pages till they find something interesting, while we explain this simple and useful shorthand to the novice.) Let us take as an example the calculation of the arithmetic average of the five numbers 5, 6, 8, 7, 6. We could, if there were any reason for keeping track of these, label them as follows:

$$x_1 = 5 \qquad x_2 = 6 \qquad x_3 = 8 \qquad x_4 = 7 \qquad x_5 = 6$$

Now the advantage of using algebraic notation (i.e., letters to stand for any numbers we care to substitute for them according to the problem in hand) is that we can write down in a very compact way the rules for performing the calculation which will give us the correct answer to the type of problem we are dealing with. In fact, a formula is nothing else than the *answer to every problem* of the type to which it applies. We solve the problem once and for all when we work out a formula. The formula *is* the answer. All we have to do is to substitute for the letters the actual quantities they stand for in the given problem. Suppose, now, we denote the number of quantities which are to be averaged in our problem by the letter n (in our case here, $n = 5$). To calculate the arithmetic average we have to add up all the five quantities thus: $5 + 6 + 8 + 7 + 6 = 32$. This adding part of the calculation would appear in algebraic form as $x_1 + x_2 + x_3 + x_4 + x_5$. The next step would be to divide the total by the number of items to be averaged, viz. 5, giving the result $6 \cdot 4$ for the average. In algebraic notation this would appear as

$$\text{Average} = \frac{x_1 + x_2 + x_3 + x_4 + x_5}{n}$$

This method of writing the formula would be very inconvenient if there were a large number of items to be averaged; moreover, there is no need to keep the individual items labelled, for in an average the identity of the individuals is deliberately thrown away as irrelevant. So we introduce the summation sign, Σ, and write our formula in the very compact form:

$$\text{Average} = \frac{\Sigma x}{n}$$

The formula thus tells us that to get the average we 'add up all the x values and divide their total by the number of items, n.'

In similar fashion, now, the *harmonic mean*, which we have said is the average to be used in averaging speeds and so on and which is defined as the reciprocal (the reciprocal of a number x is equal to $\frac{1}{x}$) of the arithmetic mean of the reciprocals of the values, x, which we wish to average, has the formula:

$$\text{Harmonic mean} = \frac{n}{\sum\left(\dfrac{1}{x}\right)}$$

To illustrate the use of this formula let us use it on our aeroplane problem. The four speeds, which were each maintained over the same distance, were 100, 200, 300, and 400 m.p.h. These are our x values. Since there are four of them the value of n in our formula is 4, and we get:

$$\text{Harmonic mean} = \frac{n}{\sum\left(\dfrac{1}{x}\right)} = \frac{4}{(\frac{1}{100} + \frac{1}{200} + \frac{1}{300} + \frac{1}{400})} = \frac{4}{(\frac{25}{1200})}$$

$$= \frac{4 \times 1200}{25} = 192 \text{ m.p.h.}$$

which we know to be the correct answer.

The reader should note carefully that the harmonic mean is here appropriate because the times were variable, with the distances constant. Had it been that times were constant and distances variable the ordinary arithmatic average would have been the correct one to use. The type of average which is appropriate always depends on the terms of the problem in hand. Formulae should never be applied indiscriminately.

Yet a third type of average is the *geometric mean*. This is the appropriate average to use when we wish to average quantities which are drawn from a situation in which they follow what W. W. Sawyer in *Mathematician's Delight* calls the 'gangster law of growth,' i.e., a geometric progression or the exponential law. Many quantities follow this type of law. For example, the population of a city, given a stable birth-rate and death-rate with no migration, will increase at a rate proportional to the number of people in the city. Suppose that in the year 1940 a certain city had a population of 250,000 and that in the year 1950 its population were 490,000. If we wished to estimate the population in the year 1945 (estimating populations at various times between successive censuses is an important matter in public health statistics) then we might, as a rough approximation, take the average of the populations at the two known dates, thus:

$$\text{Population at } 1945 = \frac{250,000 + 490,000}{2} = 370,000$$

This would only be a sensible method if we were able to assume that the population increased by the same number every year. This is not likely, however, for, as the city grows in size, so the number of citizens is likely to grow at an ever increasing rate (see Figure 1). A better estimate is likely to be obtained, in normal circumstances, by calculating the *geometric mean* of the population at the two known dates. To calculate the geometric mean, we multiply together all the quantities which it is desired to average. Then, if there are *n* such quantities, we find the *n*th root of the product. Denoting our *n* quantities by x_1, x_2, x_3, ... x_n, we may write the formula for the geometric mean as follows:

$$\text{Geometric mean} = \sqrt[n]{x_1 \times x_2 \times x_3 \times \ldots x_n}$$

Applying this to the problem given above where we wish to estimate the population of a city in 1945, given that in 1940 the population was 250,000 and in 1950 was 490,000, we have $n = 2$ items to average, and we find:

$$\text{Geometric mean} = \sqrt[2]{250,000 \times 490,000} = 350,000$$

as our estimate for the population at 1945. This result, it will be noted, is appreciably lower than we obtained using the arithmetic average (370,000). If the reader considers Figure 1 he will see that it is the more likely estimate.

Collecting together, at this point, our three different averages, we have:
Arithmetic Mean (usually denoted as $\bar{x}$ and called *x*-bar)

$$\bar{x} = \frac{\Sigma x}{n}$$

Harmonic Mean (usually denoted by H)

$$H = \frac{n}{\Sigma\left(\dfrac{1}{x}\right)}$$

Geometric Mean (usually denoted by G)

$$G = \sqrt[n]{x_1 \times x_2 \times x_3 \times \ldots x_n}$$

Each of these measures of central tendency has its own special applications. All of them are obtained by simple arithmetical processes which take into account the magnitude of every individual item.

We emphasized the important idea of any average or measure of central tendency as the *representative* of a homogeneous group in which the

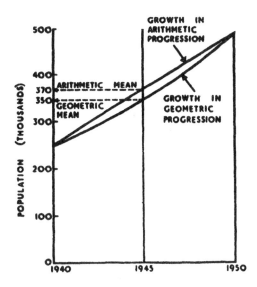

FIGURE 1—Comparison of Interpolation by Arithmetic Mean and Geometric Mean. The population of a city often grows according to the exponential law. This would certainly be true with stable birth-rate and death-rate and in absence of migration. Under these conditions, the geometric average would be more appropriate than the arithmetic average to interpolate the population at a given date between two dates at which the population was known.

members are recognizably similar. Now many distributions, while being undoubtedly homogeneous in the sense that there is continuity between the various members of the group, nevertheless are such that very great differences exist between the largest and smallest members, and, moreover, exhibit a marked lack of symmetry, the family tending to cluster much nearer to one extreme than the other. Figure 2 is a typical example. It shows the way in which annual income is distributed. There is certainly continuity, but small incomes are the norm. The reader will appreciate at once that to calculate averages for distributions of this type using the arithmetic mean would be very misleading. The relatively few people with extremely high incomes would pull up the average appreciably, so that it could not be taken as truly representative of the population in general. Figure 3, which shows the relative frequency of different sizes of family, presents the same difficulty. Some families are *very* well off for children and the calculation of an arithmetic average might well be misleading— particularly if our purpose is purely descriptive.

It is evident that what we need in such cases is a measure of central tendency which is unaffected by the relatively few extreme values in the 'tail' of the distribution. Two ideas suggest themselves. The first is that if

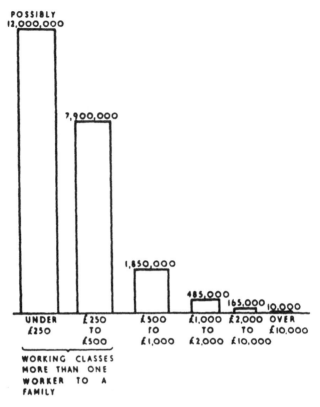

FIGURE 2—Numbers of people in different income ranges forming a positively skew distribution.

we were to take all our families and set them down in a long column starting with the smallest and working up to the largest, we could then use the size of that family which came halfway down the column as our measure of central tendency. This measure is called the *median* (meaning 'middle item'). Half of all families would have a size not less than that of the median family and half not more than that of the median family. Notice that in this way we do not take account at all of the actual numbers of children except for ranking purposes. It is evident that the number of children in the largest family could be increased to 50,000 without in any way disturbing our measure of central tendency, which would still be the middle item.

A second method of getting a measure of central tendency which is not upset by extreme values in the distribution is to use the *most commonly occurring value*. This is the fashionable value, the value à la mode, so to say. It is called the *mode* or *modal value*. For example, in Figure 3 the modal value for the size of family is seen to be two children. This is really

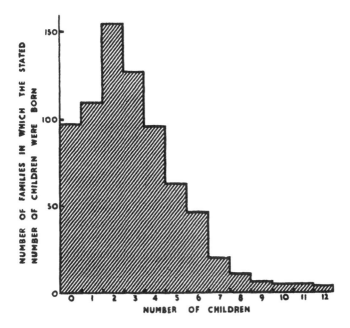

FIGURE 3—Distribution of number of children per family is also positively skewed.

a typical value and seems real to us compared with the arithmetic average which in this case works out to 2·96. It is difficult to imagine 2·96 children. Notice that the arithmetic mean is markedly affected by the relatively few very large families. Which is correct? Neither and both. Both averages serve a purpose. The mode would form a very poor basis for any further calculations of an arithmetical nature, for it has deliberately excluded arithmetical precision in the interests of presenting a typical result. The arithmetic average, on the other hand, excellent as it is for numerical purposes, has sacrificed its desire to be typical in favour of numerical accuracy. In such a case it is often desirable to quote *both* measures of central tendency. Better still, go further and present a histogram of the distribution as in Figure 3.

A problem which not infrequently arises is to make an estimate of the median value of a distribution when we do not have the actual values of each individual item given, but only the numbers of items in specified ranges.

We shall now say a few words about frequency distributions. If we have a large group of items each of which has connected with it some numerical value indicative of its magnitude, which varies as between one member of the group and another (as, for example, when we consider the heights of men or the amount of income tax paid by them), and if we draw up a table or graph showing the relative frequency with which members of the group have the various possible values of the variable quantity (e.g., proportion of men at each different height, or proportions

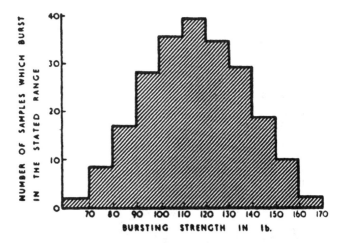

FIGURE 4—Distribution for bursting strength of samples of vinyl coated nylon exhibiting symmetry.

of the population, falling into various income tax groups), then we have what is called a *frequency distribution* for the variable quantity in question. This is usually called simply the *distribution*. Thus we have distributions for height, weight, chest size, income, living rooms per person, and so on. Similarly we have distributions for the number of deaths according to age for different diseases, number of local government areas with specified birthrates and deathrates and so on. The quantity which varies

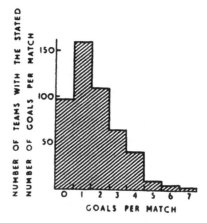

FIGURE 5—The number of goals scored per team per match gives a positively skewed distribution of a discontinuous variable.

(height, birthrate, income, and so on) is called the *variate*. Some variates
are *continuous*, i.e., they can assume *any* value at all within a certain
range. Income, height, birth-rate, and similar variates are continuous.

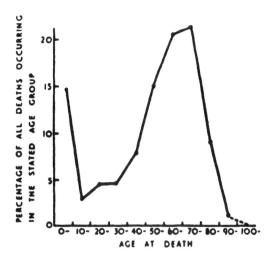

FIGURE 6—Example of a Bimodal (double peaked) Distribution. The peak in the first years of
life reflects the serious loss of potential life due to the infantile mortality rate. (From
the "Registrar General's Report, Years 1930–32," quoted by M. G. Kendall in "Advanced Statistics.")

Other variates are said to be *discontinuous*, because they can only assume
isolated values. For example, the number of children in a family can only
be a whole number, fractions being impossible. Families grow in distinct
jumps. An addition to the family is an event. Goals scored in football
matches, articles lost in buses, the number of petals on a flower—all such
variable quantities are discontinuous.

When we collect together information for the purposes of statistical
analysis it is rare that we have information about all the individuals in a
group. Census data are perhaps the nearest to perfection in this sense;
but even in this case the information is already getting out of date as it is
collected. We may say that the census count in a certain country taken
on a certain day came to 43,574,205, but it would be nothing short of
silly to keep quoting the last little figure 5 for the next ten years—or even
the next ten minutes. Such accuracy would be spurious. In general it is
not possible to investigate the whole of a population. We have to be content with a *sample*. We take a sample with the idea of making inferences
from it about the population from which it was drawn, believing, for example, that the average of a good sample is closely related to the average
of the whole population. The word *population* is used in statistics to refer
not simply to groups of people, but, by a natural extension, to groups of

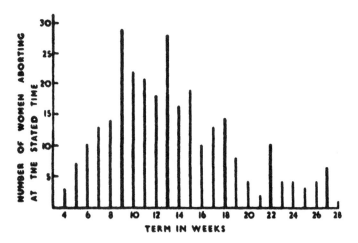

FIGURE 7—Abortion in women. Data given by T. V. Pearce (1930) and quoted by M. G. Kendall, "Advanced Statistics." The reader may care to speculate about possible periodicity in these data. Is there reasonable suggestion of a cycle whose duration is roughly one month? What other conclusion can you draw?

measurements associated with any collection of inanimate objects. By drawing a sufficiently large sample of measurements, we may arrive at a frequency distribution for any population. Figures 4–8 give examples of various types of distribution.

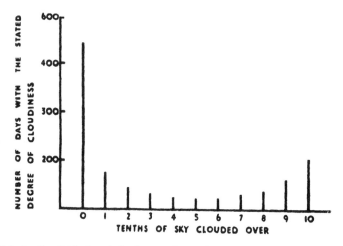

FIGURE 8—Peculiar distribution of cloudiness at Greenwich. Based on data given by Gertrude Pearse (1928) for month of July 1890–1904 (excluding 1901) and quoted by M. G. Kendall, "Advanced Statistics," Vol. I. Note tendency for sky to be either very clear or very cloudy.

Some distributions, as will be seen from the diagrams, are *symmetrical* about their central value. Other distributions have marked asymmetry and are said to be *skew*. Skew distributions are divided into two types. If the 'tail' of the distribution reaches out into the larger values of the variate, the distribution is said to show *positive skewness*; if the tail extends towards the smaller values of the variate, the distribution is called *negatively skew*. In the next chapter we shall take up the question of the concentration of the members of the distribution about their central value, for it is clearly a matter of the greatest importance to be able to measure the degree to which the various members of a population may differ from each other.

Figure 9 illustrates an interesting relationship which is found to hold approximately between the median, mode, and mean of moderately skew distributions. Figures 10 and 11 illustrate geometrical interpretations of the three measures of central tendency.

We shall close this chapter with an elementary account of Index Numbers, which are really nothing more than a special kind of average. The

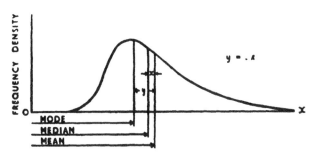

FIGURE 9—Mean, Median and Mode in moderately skew cases. For moderately skew distributions we have the simple approximate relation: Mean — Mode = 3 (Mean — Median). For a perfectly symmetrical distribution they all coincide.

best known index number is the Cost of Living Index, which, as readers will know, is a rough measure of the average price of the basic necessities of life. In many industries, the Cost of Living Index is a strong chain which keeps a man's reward tied strictly to his necessity rather than to his ambition. But index numbers are a widespread disease of modern life, or, we might better say, a symptom of the modern disease of constantly trying to keep a close check on everything. We have index numbers for exports, for imports, for wage changes and for consumption. We have others for wholesale and retail prices. The Board of Trade has an index. The Ministry of Labour has an index. *The Economist* has another. It is scarcely possible to be respectable nowadays unless one owns at least one index number. It is a corporate way of 'keeping up with the Joneses'—the private individual having been forced by taxation to give up this inspiring aim long ago.

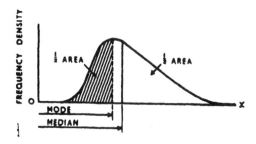

FIGURE 10—Geometrical interpretation of Mode and Median. The vertical line at the median value divides the area under the frequency curve into halves (area is proportional to frequency). The vertical line at the modal value passes through the peak of the curve, i.e., it is the value at which the frequency density is a maximum.

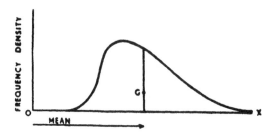

FIGURE 11—Geometrical interpretation of the Mean The vertical line at the mean will pass through the centre of gravity of a sheet of uniform thickness and density cut to the shape of the distribution. The mean is the abscissa of the centre of gravity "G."

It is really questionable—though bordering on heresy to put the question—whether we would be any the worse off if the whole bag of tricks were scrapped. So many of these index numbers are so ancient and so out of date, so out of touch with reality, so completely devoid of practical value when they have been computed, that their regular calculation must be regarded as a widespread compulsion neurosis. Only lunatics and public servants with no other choice go on doing silly things and liking it. Yet, since we become more and more the servants of our servants, and since they persist in tying us down to this lugubrious system whereby the housewife, the business man, and the most excellent groups of the citizenry have all their difficulties compressed into the brevity of an index number, we reluctantly ask the reader to bear with us while we explain, briefly, this academic tomfoolery of telling us in cryptic form what we knew already from hard experience: namely, that the cost of living has risen in the last couple of months, sufficiently for us to be able to submit a humble claim for higher wages to offset part of our increased burden.

Consider the question of the changes which take place in retail prices. As every housewife knows, the price we are asked to pay bears only the faintest resemblance in many cases to the worth of the article. She knows, too, that for many commodities it is more accurate to speak of prices rather than price. Tomatoes in one shop may be 6d. per pound; the same tomatoes in another shop may be 10d. or 1s. Some people are well enough off to be able to shop by price. They like lots of service and servility and are willing to pay for it. Yet, even if these sections of the community are excluded, there still remains a fair variation between one district and another for the same article, things like fish and fruit being notorious in this respect. In addition to this variation in the price of the articles, we have to recognize that different families have different spending patterns. If cheese were made as dear as gold it would not matter one iota to the family that hates cheese like poison. Conscientious vegetarians would probably regard it as an excellent thing if the price of meat rose to prohibitive levels. Total abstainers positively loathe the idea of beer and spirits being cheap. Non-smokers love to see the Chancellor raise the money by piling the tax on 'non-essentials' like tobacco. It is evident that we shall get nowhere if all this individuality is to run riot. It is far too inconvenient for the statistician.

We get over the difficulty by shutting our eyes to it. All we have to do is to invent a 'standard family.' [2] We might, for example, choose the standard urban working-class family. We then do a sample survey, to find out what quantities of the various articles we are considering they consume in a week under normal conditions, and draw up a table as follows:

EXPENDITURE OF THE STANDARD WORKING-CLASS FAMILY
(1949)

	Quantity	Price	Expenditure	Weight
Bread and Flour	39 lb.	4d./lb.	156d.	31·2
Meat	7 lb.	24d./lb.	168d.	33·6
Potatoes	35 lb.	2d./lb.	70d.	14·0
Tea	1 lb.	36d./lb.	36d.	7·2
Sugar	2 lb.	5d./lb.	10d.	2·0
Butter	1 lb.	18d./lb.	18d.	3·6
Margarine	1 lb.	12d./lb.	12d.	2·4
Eggs	1 doz.	30d./doz.	30d.	6·0
		Total	500d.	100·0

[2] Composed of one underpaid male, one overworked female, and 2·2 underfed children.

Now, it is a relatively simple matter to keep track of the changes in prices as time goes on. It would be very much more troublesome to keep a check on whether the spending pattern, as indicated by the amounts of the various items bought by the standard family, was tending to change. One line of approach would be to assume that our standard family will not change its demands from year to year. Suppose for the year 1950 the prices were as in the following table.

EXPENDITURE OF THE STANDARD WORKING-CLASS FAMILY
(1950)

	Quantity	Price	Expenditure	Weight
Bread and Flour	39 lb.	5d./lb.	194d.	30·1
Meat	7 lb.	30d./lb.	210d.	32·6
Potatoes	35 lb.	3d./lb.	105d.	16·3
Tea	1 lb.	36d./lb.	36d.	5·6
Sugar	2 lb.	6d./lb.	12d.	1·9
Butter	1 lb.	27d./lb.	27d.	4·2
Margarine	1 lb.	15d./lb.	15d.	2·3
Eggs	1 doz.	45d./doz.	45d.	7·0
		Total	644d.	100·0

The reader should ignore, for the moment, the last column, headed 'Weight,' in each table. The obvious thing, at once, is that to buy the same quantities of the same articles, and therefore to get the same 'satisfaction,' as the economists have it, cost the standard family 644d. in 1950 as against 500d. in 1949, i.e., the cost in 1950 as compared with 1949 was $^{644}/_{500} \times 100 = 128 \cdot 8\%$. We could then say that the index of retail prices, as represented by this group of items, stood at 129 in 1950 (1949 = 100).

We could get a similar indication of the rise in retail prices as follows. Consider, first, the amount of money our standard family spent on the various items in our 'base year, 1949.' These can be reduced to percentages of the total expenditure (on the group of items considered in the index). For instance, out of a total expenditure of 500d., bread and flour claimed 156d. or 31·2%. Similarly, meat took 33·6% of the total expenditure, potatoes 14·0%, and so on. These figures are entered in the column headed 'Weight' since they tell us the relative importance of the different items in the household budget. Meat is a very heavy item, sugar a relatively small one. These weights give us a pattern of expenditure as it actually appeared to the standard housewife in the base year. They take account of both quantity and price. The first thing that is obvious from this pattern of weights is that, while a 50% increase in the cost of sugar is not a matter of great importance to the housewife, even a 10% increase in the price of meat would be a serious extra burden to carry in the standard family where income is usually closely matched to expenditure. We must remember that our standard family is a standardized family. Its wants are not supposed to change. It is supposed to be devoid of ambition. It only gets a rise in salary when such a rise is absolutely necessary.

Now while it is true (in the absence of subsidies and purchase tax or price fixing by combines) that all commodities tend to rise in price together, nevertheless, superimposed on this general tendency, there will be a certain irregularity. Comparing the price of bread and flour in our two years we find that the 'price relative,' as it is called, of this item is ¾ × 100 = 125% in 1950 as compared with the base year, 1949.

The following table shows the 'prices relative' for the several items, together with the weights corresponding to the base year. The weights have been quoted to the first decimal place, further places being condemned as coming under the heading 'delusions of accuracy.'

	Price relative	Base year weight	Price-rel. × weight
Bread and Flour	125	31·2	3,900
Meat	125	33·6	4,200
Potatoes	150	14·0	2,100
Tea	100	7·2	720
Sugar	120	2·0	240
Butter	150	3·6	540
Margarine	125	2·4	300
Eggs	150	6·0	900
Total		100·0	12,900

If, now, we divide the total of the 'prices relative × weight' by the total of the weights, we get the average prices of the commodities in 1950, as compared with the base year, 1949, equals 129·00, which we certainly quote no more accurately than 129. This would now be our index of retail prices. For every hundred pennies spent in 1949 we need to spend 129 in 1950 to get the same amount of 'satisfaction.' Evidently, every succeeding year—or month, for that matter—can be compared with our base year.

The economists, of course, have great fun—and show remarkable skill —in inventing more refined index numbers. Sometimes they use geometric averages instead of arithmetic averages (the advantage here being that the geometric average is less upset by extreme oscillations in individual items), sometimes they use the harmonic average. But these are all refinements of the basic idea of the index number which we have indicated in this chapter. Most business men seem to thrive without understanding this simple matter. Perhaps they half realize that it doesn't mean a lot, except in regard to wage negotiations between themselves and Trade Unions— and in such cases experts on both sides of the fence do all the statistics required. The employer and employee don't much mind how much of this arithmetic goes on, so long as the final agreement is reasonably fair to both sides.

The snags in this index number game will be apparent to the reader. First of all, if he will inspect the pattern of weights in the tables for 1949 and 1950, he will see that they are not identical. Over a reasonable period of years the pattern can change appreciably. Then, again, if we try to measure the cost of living of our standard family by including heating, lighting, rent, beer, cigarettes, football pools, and the rest, we soon get into deep water. For example, if we find that in the base year the standard family spends one-tenth of its income on football pools, are we to argue that since this is a heavy item of expenditure it shall be supported somehow in the cost of living calculations? Until very recently the cost of living index in this country took account of the cost of paraffin and candles for lighting purposes, and assumed that no working-class family had heard of electricity. Then there is the difficulty that the standard family tends to become a standardized family in so far as its wages are tied to an index which is slow to recognize the right of its standard family to be anything but standard in its requirements from year to year. The reader should consider carefully the full implication of 'subsidies on essentials' (included in cost of living index) and 'purchase tax on non-essentials' (not included in the index or only modestly represented). The pernicious nature of tying wages to cost of living indexes while this jiggery-pokery is official policy will be apparent. The whole scheme is positively Machiavellian in its acceptance of deception as a necessity in politics. And does it really work so well, after all? The truth is that it is too inefficient even to keep the worker standardized. As new items are available from manufacturers, the public has to be given the power to purchase them, whether they are included in the cost of living index or not. Shall we ask the economists: What good do your indexes do—really?

SCATTER

'The words figure and fictitious both derive from the same Latin root, fingere. Beware!'—M. J. M.

We have discussed various ways of measuring the central tendency of distributions and have seen that such measures are characteristic of the distribution of any quantity, so that different populations are distinguished from each other by different values of these measures. For example, the average value for the height of women differs from the average height for men. Numerical characteristics of populations are called *parameters*. Having dealt with parameters of central tendency, we now turn to the no less

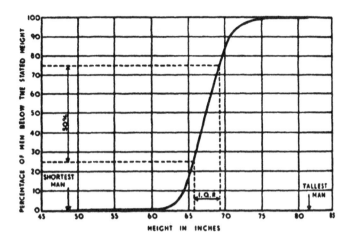

FIGURE 12—Ogive for heights for young men (I.Q.R. = interquartile range). Based on W. T. Martin, "Physique of the Young Male," by permission of H.M.S.O.)

important matter of parameters of dispersion. According to Memorandum No. 20 issued by the Medical Research Council (W. J. Martin: *The Physique of Young Males*) the height of young males, aged between 20 and 21 years, has an average value of 5 feet 7½ inches. This is information. But we should like to know more,[3] for it is evident that not all the young men were exactly of this height. The adjoining ogive (Figure 12) shows the percentages of men less than stated heights in a total of 91,163 who were measured. Figure 13 shows the data displayed in histogram form. It is evident that very considerable variability exists, so that, whilst the great majority of men differ relatively little from the average height, very noticeable departures from it are not at all infrequent. How are we to get a measure of the variability about the mean value?

The easiest way is to state the height of the tallest man seen and the shortest, thus. Tallest: 6 feet 9 inches. Average: 5 feet 7½ inches. Shortest: 4 feet 0 inches. Alternatively, we might state the *range*, i.e., the difference between the tallest and the shortest, viz. 6 feet 9 inches *minus* 4 feet 0 inches = 2 feet 9 inches. This is not a very good way. A moment's thought will make it clear that we might very easily not have met these two extreme heights. It might well have been that we should have found the shortest man to be 4 feet 4 inches and the tallest 6 feet 6 inches. This would give us a range of 6 feet 6 inches *minus* 4 feet 4 inches = 2 feet 2 inches—a result which is appreciably different from the previous one. Again, it might have happened that among those examined in this group for military service were the giant and the dwarf from some circus.

―――――――

[3] The author does not disappoint us in this desire.

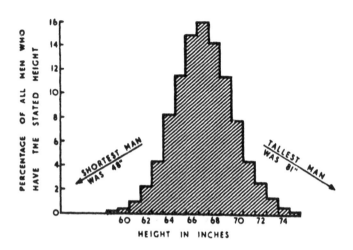

FIGURE 13—Histogram corresponding to the ogive of Figure 12.

Supposing the giant to be 9 feet 7 inches and the dwarf 3 feet 2 inches, we should have obtained for our range the value 6 feet 5 inches. It is obviously undesirable to have a measure which will depend entirely on the value of any freaks that may occur. It is impossible for a measure based on freaks to speak as the *representative* of the ordinary population. The range, then, although it is used in certain circumstances, is not ideal as a measure of dispersion.[4] It would be better to have a parameter less likely to be upset by extreme values.

We may tackle this problem by devising a measure for dispersion along the same line that we took for the median when we were discussing measures of central tendency. The median was the value above which 50% of the population fell and below which the other 50% fell. Suppose, now, we divide the population, after it has been set out in order of size, into *four equal groups*. The value above which only 25% of the population falls we call the *upper quartile*, and the value below which only 25% of the population falls we call the *lower quartile*. Evidently, 50% of the population falls between the upper and lower quartile values. The reader may care to check for himself that the upper and lower quartiles, for the table of heights we are using as an example, are roughly 5 feet 9 inches and 5 feet 6 inches respectively. Thus, we may see at once that roughly 50% of the population differ in height by amounts not exceeding three inches, despite the fact that the tallest man observed was no less than 2 feet 9 inches taller than the shortest man. This, of course, is a consequence of the way in which the large majority of heights cluster closely to the average. This is a very common effect. Intelligence Quotients be-

---

[4] The range is very efficient when the samples contain very few items.

have in the same sort of way. Most people are little removed from average intelligence, but geniuses and morons tend to occur in splendid isolation. (We may recall here that the modal ('fashionable') value tends to coincide with the arithmetic mean when the distribution is fairly symmetrical.) Thus the *interquartile range*, i.e., the difference between the upper and lower quartile values, makes a good measure of dispersion. It is immune from the disturbances occasioned by the incidence of extreme values. It is easy to calculate. It has a simple and meaningful significance in that it tells us the range of variability which is sufficient to contain 50% of the population. The interquartile range is frequently used in economic and commercial statistics for another reason. Often, data are collected in such a way that there are indeterminate ranges at one or both

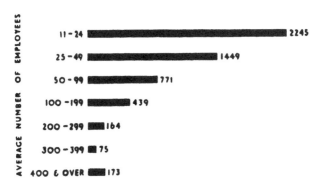

FIGURE 14—Showing numbers of firms with the stated number of employees in the food, drink, and tobacco trades of Great Britain. (Based on Census of Production 1930, quoted by M. G. Kendall, "Advanced Statistics," Vol. I.)

ends of the table. An example is shown in Figure 14. The largest group is labelled '400 and over.' This is vague, and it would obviously be impossible to do a precise calculation for any measure depending on arithmetical processes involving the actual values in the unbounded upper class. (We shall show in the next chapter how the limited vagueness in the other *bounded* classes is dealt with.) The median and the interquartile range provide us with measures of central tendency and scatter respectively in such cases.

Median and quartiles are simply special cases of a quite general scheme for dividing up a distribution by *quantiles*. Thus, we may arrange our distribution in order of size and split it up into *ten* groups containing equal numbers of the items. The values of the variable at which the divisions occur are known then as the first, second, third, and so on, *deciles*. This idea is used by educational psychologists to divide pupils into 'top 10%, second 10%, third 10%,' and so on, with regard to inherent intelligence in so far as that characteristic may be measured by tests.

Yet another measure of dispersion, which depends on all the measurements, is the *mean deviation*. In order to calculate this parameter, we first of all find the arithmetic mean of the quantities in the distribution. We then find the difference between each of the items and this average, calling all the differences positive. We then add up all the differences thus obtained and find the average difference by dividing by the number of differences. Thus the mean deviation is the average difference of the several items from their arithmetic mean. In mathematical form we have

$$\text{Mean Deviation} = \frac{\Sigma|x - \bar{x}|}{n}$$

where as before the symbol $\bar{x}$ stands for the arithmetic mean of the various values of x. The sign $|x - \bar{x}|$ indicates that we are to find the difference between x and the average of the x values, ignoring sign. The sign Σ means 'add up all the terms like.'

Example. Find the arithmetic mean and mean deviation for the set of numbers: 11, 8, 6, 7, 8.

Here we have $n = 5$ items to be averaged. As previously shown, the average of the items is

$$\bar{x} = \frac{\Sigma x}{n} = \frac{11 + 8 + 6 + 7 + 8}{5} = \frac{40}{5} = 8$$

In order to get the mean difference, we calculate the several differences of the items from their average value of 8 and sum them, thus:

$$|11 - 8| + |8 - 8| + |6 - 8| + |7 - 8| + |8 - 8|$$
$$= \quad 3 \quad + \quad 0 \quad + \quad 2 \quad + \quad 1 \quad + \quad 0 \quad = 6$$

We then calculate the mean deviation by dividing this total of the deviations by $n = 5$, and so find the mean deviation as $\%_6 = 1 \cdot 2$.

The mean deviation is frequently met with in economic statistics.

The measures so far suggested are often used in elementary work on account of their being easy to calculate and easy to understand. They are, however, of no use in more advanced work because they are extremely difficult to deal with in sampling theory, on which so much of advanced work depends. The most important measure of dispersion is the *standard deviation*, which is a little more difficult to calculate and whose significance is less obvious at first sight. Calculation and interpretation, however, soon become easy with a little practice, and then the standard deviation is the most illuminating of all the parameters of dispersion. The standard deviation will be familiar to electrical engineers and mathematicians as the *root-mean-square deviation*.[5] The general reader will do well to remember this phrase as it will help him to remember exactly how the

[5] It is strictly analogous to radius of gyration in the theory of moments of inertia.

standard deviation is calculated. We shall detail the steps for the calculation of the standard deviation of a set of values thus:

Step 1. Calculate the arithmetic average of the set of values.

Step 2. Calculate the differences of the several values from their arithmetic average.

Step 3. Calculate the squares of these differences (the square of a number is found by multiplying it by itself. Thus the square of 4 is written 4^2 and has the value $4 \times 4 = 16$).

Step 4. Calculate the sum of the squares of the differences to get a quantity known as the *sample sum of squares.*

Step 5. Divide this 'sample sum of squares' by the number of items, n, in the set of values. This gives a quantity known as the *sample variance.*

Step 6. Take the square root of the variance and so obtain the standard deviation. (The square root of any number, x, is a number such that when it is multiplied by itself it gives the number x. Thus, if the square root of x is equal to a number y then we shall have $y^2 = y \times y = x$.)

This sounds much more complicated than it really is. Let us work out an example, step by step.

Example. Find the standard deviation of the set of values 11, 8, 6, 7, 8.

Step 1. We calculated the arithmetic average previously as $\bar{x} = 8$.

Step 2. The differences of the items from this average (sign may be ignored) are: 3, 0, 2, 1, 0.

Step 3. The squares of these differences are:
$3 \times 3 = 9 \qquad 0 \times 0 = 0 \qquad 2 \times 2 = 4 \qquad 1 \times 1 = 1 \qquad 0 \times 0 = 0$

Step 4. The sample sum of squares is: $9 + 0 + 4 + 1 + 0 = 14$.

Step 5. Dividing the sample sum of squares by the number of items, $n = 5$, we get the sample variance as $s^2 = {}^{14}/_5 = 2 \cdot 8$ (s^2 is the accepted symbol for sample variance).

Step 6. The standard deviation is found as the square root of the sample variance thus: $s = \sqrt{2 \cdot 8} = 1 \cdot 673$.

The formula for the standard deviation is:

$$s = \sqrt{\frac{\Sigma(x - \bar{x})^2}{n}}$$

We have seen how to calculate the standard deviation. What use is it to us in interpretation? Actually it is very easy to visualize. If we are given *any* distribution which is reasonably symmetrical about its average and which is *unimodal* (i.e., has one single hump in the centre, as in the histogram shown in Figure 13) then we find that we make very little error in assuming that two-thirds of the distribution lies less than one standard deviation away from the mean, that 95% of the distribution lies less than two standard deviations away from the mean, and that less than

1% of the distribution lies more than three standard deviations away from the mean. This is a rough rule, of course, but it is one which is found to work very well in practice. Let us suppose, for example, that we were told no more than that the distribution of intelligence, as measured by Intelligence Quotients (a person's I.Q. is defined as $\dfrac{\text{Mental Age}}{\text{Chronological Age}} \times 100$)

has an average value $\bar{x} = 100$, with standard deviation $s = 13$. Then we might easily picture the distribution as something like the rough sketch shown in Figure 15.

The reader may care to compare the rough picture thus formed from a simple knowledge of the two measures $\bar{x}$ and s with the histogram shown in Figure 16 which is based on results obtained by L. M. Terman and

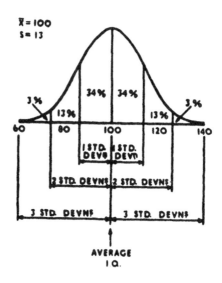

FIGURE 15—Knowing only that we have a fairly symmetrical, unimodal distribution whose mean value is I.Q. 100 units and whose standard deviation is I.Q. 13 units, we can at once picture in our minds that the distribution looks something as shown.

quoted by J. F. Kenney from his book *The Measurement of Intelligence*. This is typical of the use of measures of central tendency and dispersion in helping us to carry the broad picture of a whole distribution (provided it be reasonably symmetrical and unimodal) in the two values $\bar{x}$ and s. Such measures properly may be said to *represent* the distribution for which they were calculated.

The measures of dispersion which we have so far dealt with are all expressed in terms of the units in which the variable quantity is measured. It sometimes happens that we wish to ask ourselves whether one distribution is relatively more variable than another. Let us suppose, for example, that for the heights of men in the British Isles we find a mean value 57 inches with standard deviation 2·5 inches, and that for Spaniards the mean height is 54 inches with standard deviation 2·4 inches. It is evident that British men are taller than Spaniards and also slightly more variable in height. How are we to compare the *relative* variability bearing in mind

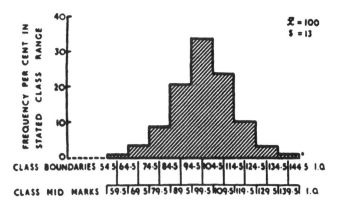

FIGURE 16—Distribution of Intelligence Quotient (compare with Figure 15). Distribution of I.Q. with $\bar{x} = 100$, $s = 13$. Based on data by L. M. Terman and quoted by J. F. Kenney, "Mathematics of Statistics," Vol. I).

that the Spaniards are shorter in height than the British? Karl Pearson's *coefficient of variation* is the most commonly used measure in practice for such a case.

It is defined as: $v = \dfrac{100s}{\bar{x}}$

If we calculate the coefficient of variation for our two cases, we get:

$$\text{British} \qquad v = \frac{100 \times 2\cdot 5}{67} = 37\cdot 3\%$$

$$\text{Spaniards} \quad v = \frac{100 \times 2\cdot 4}{64} = 37\cdot 5\%$$

We conclude that, though the British are more variable in an absolute sense, the variability of the Spaniards, expressed as a percentage of the mean height, is just slightly greater.

> For her own breakfast she'll project a scheme
> Nor take her tea without a stratagem.   —EDWARD YOUNG (1683–1765)

> "Come little girl, you seem
> To want my cup of tea
> And will you take a little cream?
> Now tell the truth to me"
> She had a rustic woodland grin
> Her cheek was soft as silk,
> And she replied, "Sir, please put in
> A little drop of milk."   —BARRY PAIN (The Poets at Tea)

6   Mathematics of a Lady Tasting Tea

By SIR RONALD A. FISHER

STATEMENT OF EXPERIMENT

A LADY declares that by tasting a cup of tea made with milk she can discriminate whether the milk or the tea infusion was first added to the cup. We will consider the problem of designing an experiment by means of which this assertion can be tested. For this purpose let us first lay down a simple form of experiment with a view to studying its limitations and its characteristics, both those which appear to be essential to the experimental method, when well developed, and those which are not essential but auxiliary.

Our experiment consists in mixing eight cups of tea, four in one way and four in the other, and presenting them to the subject for judgment in a random order. The subject has been told in advance of what the test will consist, namely that she will be asked to taste eight cups, that these shall be four of each kind, and that they shall be presented to her in a random order, that is in an order not determined arbitrarily by human choice, but by the actual manipulation of the physical apparatus used in games of chance, cards, dice, roulettes, etc., or, more expeditiously, from a published collection of random sampling numbers purporting to give the actual results of such manipulation. Her task is to divide the 8 cups into two sets of 4, agreeing, if possible, with the treatments received.

INTERPRETATION AND ITS REASONED BASIS

In considering the appropriateness of any proposed experimental design, it is always needful to forecast all possible results of the experiment, and to have decided without ambiguity what interpretation shall be placed upon each one of them. Further, we must know by what argument this interpretation is to be sustained. In the present instance we may argue as follows. There are 70 ways of choosing a group of 4 objects out of 8. This may be demonstrated by an argument familiar to students of "permutations and combinations," namely, that if we were to choose the 4 objects in succession we should have successively 8, 7, 6, 5 objects to choose from, and could make our succession of choices in $8 \times 7 \times 6 \times 5$, or 1680 ways. But in doing this we have not only chosen every possible set of 4, but every possible set in every possible order; and since 4 objects can be arranged in order in $4 \times 3 \times 2 \times 1$, or 24 ways, we may find the number of possible choices by dividing 1680 by 24. The result, 70, is essential to our interpretation of the experiment. At best the subject can judge rightly with every cup and, knowing that 4 are of each kind, this amounts to choosing, out of the 70 sets of 4 which might be chosen, that particular one which is correct. A subject without any faculty of discrimination would in fact divide the 8 cups correctly into two sets of 4 in one trial out of 70, or, more properly, with a frequency which would approach 1 in 70 more and more nearly the more often the test were repeated. Evidently this frequency, with which unfailing success would be achieved by a person lacking altogether the faculty under test, is calculable from the number of cups used. The odds could be made much higher by enlarging the experiment, while, if the experiment were much smaller even the greatest possible success would give odds so low that the result might, with considerable probability, be ascribed to chance.

THE TEST OF SIGNIFICANCE

It is open to the experimenter to be more or less exacting in respect of the smallness of the probability he would require before he would be willing to admit that his observations have demonstrated a positive result. It is obvious that an experiment would be useless of which no possible result would satisfy him. Thus, if he wishes to ignore results having probabilities as high as 1 in 20—the probabilities being of course reckoned from the hypothesis that the phenomenon to be demonstrated is in fact absent—then it would be useless for him to experiment with only 3 cups of tea of each kind. For 3 objects can be chosen out of 6 in only 20 ways, and therefore complete success in the test would be achieved without sensory discrimination, *i.e.*, by "pure chance," in an average of 5 trials out of 100. It is usual and convenient for experimenters to take 5 per cent. as a standard level of significance, in the sense that they are prepared to ignore all results which fail to reach this standard, and, by this means, to eliminate from further discussion the greater part of the fluctuations which chance causes have introduced into their experimental results. No such selection can eliminate the whole of the possible effects of chance coincidence, and if we accept this convenient convention, and agree that an event which would occur by chance only once in 70 trials is decidedly "significant," in the statistical sense, we thereby admit that no isolated experiment, however significant in itself, can suffice for the experimental demonstration of any natural phenomenon; for the "one chance in a million" will undoubtedly occur, with no less and no more than its appropriate frequency, however surprised we may be that it should occur to *us*. In order to assert that a natural phenomenon is experimentally demonstrable we need, not an isolated record, but a reliable method of procedure. In relation to the test of significance, we may say that a phenomenon is experimentally demonstrable when we know how to conduct an experiment which will rarely fail to give us a statistically significant result.

Returning to the possible results of the psycho-physical experiment, having decided that if every cup were rightly classified a significant positive result would be recorded, or, in other words, that we should admit that the lady had made good her claim, what should be our conclusion if, for each kind of cup, her judgments are 3 right and 1 wrong? We may take it, in the present discussion, that any error in one set of judgments will be compensated by an error in the other, since it is known to the subject that there are 4 cups of each kind. In enumerating the number of ways of choosing 4 things out of 8, such that 3 are right and 1 wrong, we may note that the 3 right may be chosen, out of the 4 available, in 4 ways and, independently of this choice, that the 1 wrong may be chosen, out of the 4 available, also in 4 ways. So that in all we could make a selection of the kind supposed in 16 different ways. A similar argument shows that, in each kind of judgment, 2 may be right and 2 wrong in 36 ways, 1 right and 3 wrong in 16 ways and none right and 4 wrong in 1 way only. It should be noted that the frequencies of these five possible results of the experiment make up together, as it is obvious they should, the 70 cases out of 70.

It is obvious, too, that 3 successes to 1 failure, although showing a bias, or deviation, in the right direction, could not be judged as statistically significant evidence of a real sensory discrimination. For its frequency of chance occurrence is 16 in 70, or more than 20 per cent. Moreover, it is not the best possible result, and in judging of its significance we must take account not only of its own frequency, but also of the frequency for any better result. In the present instance "3 right and 1 wrong" occurs 16 times, and "4 right" occurs once in 70 trials, making 17 cases out of 70 as good as or better than that observed. The reason for including cases better than that observed becomes obvious on considering what our conclusions would have been had the case of 3 right and 1 wrong only 1 chance, and the case of 4 right 16 chances of occurrence out of 70. The rare case of 3 right and 1 wrong could not be judged significant merely because it was rare, seeing that a higher degree of success would frequently have been scored by mere chance.

THE NULL HYPOTHESIS

Our examination of the possible results of the experiment has therefore led us to a statistical test of significance, by which these results are divided into two classes with opposed interpretations. Tests of significance are of many different kinds, which need not be considered here. Here we are only concerned with the fact that the easy calculation in permutations which we encountered, and which gave us our test of significance, stands for something present in every possible experimental arrangement; or, at least, for something required in its interpretation. The two classes of results which are distinguished by our test of significance are, on the one hand, those which show a significant discrepancy from a certain hypothesis; namely, in this case, the hypothesis that the judgments given are in no way influenced by the order in which the ingredients have been added; and on the other hand, results which show no significant discrepancy from this hypothesis. This hypothesis, which may or may not be impugned by the result of an experiment, is again characteristic of all experimentation. Much confusion would often be avoided if it were explicitly formulated when the experiment is designed. In relation to any experiment we may speak of this hypothesis as the "null hypothesis," and it should be noted that the null hypothesis is never proved or established, but is possibly disproved, in the course of experimentation. Every experiment may be said to exist only in order to give the facts a chance of disproving the null hypothesis.

It might be argued that if an experiment can disprove the hypothesis that the subject possesses no sensory discrimination between two different sorts of object, it must therefore be able to prove the opposite hypothesis, that she can make some such discrimination. But this last hypothesis, however reasonable or true it may be, is ineligible, as a null hypothesis to be tested by experiment, because it is inexact. If it were asserted that the subject would never be wrong in her judgments we should again have an exact hypothesis, and it is easy to see that this hypothesis could be disproved by a single failure, but could never be proved by any finite amount of experimentation. It is evident that the null hypothesis must be exact, that is free from vagueness and ambiguity, because it must supply the basis of the "problem of distribution," of which the test of significance is the solution. A null hypothesis may, indeed, contain arbitrary elements, and in more complicated cases often does so: as, for example, if it should assert that the death-rates of two groups of animals are equal, without specifying what these death-rates usually are. In such cases it is evidently the equality rather than any particular values of the death-rates that the experiment is designed to test, and possibly to disprove.

In cases involving statistical "estimation" these ideas may be extended to the simultaneous consideration of a series of hypothetical possibilities. The notion of an error of the so-called "second kind," due to accepting the null hypothesis "when it is false" may then be given a meaning in reference to the quantity to be estimated. It has no meaning with respect to simple tests of significance, in which the only available expectations are those which flow from the null hypothesis being true.

RANDOMISATION; THE PHYSICAL BASIS OF THE VALIDITY OF THE TEST

We have spoken of the experiment as testing a certain null hypothesis, namely, in this case, that the subject possesses no sensory discrimination whatever of the kind claimed; we have, too, assigned as appropriate to this hypothesis a certain frequency distribution of occurrences, based on the equal frequency of the 70 possible ways of assigning 8 objects to two classes of 4 each; in other words, the frequency distribution appropriate to a classification by pure chance. We have now to examine the physical conditions of the experimental technique needed to justify the assumption that, if discrimination of the kind under test is absent, the result of the experiment will be wholly governed by the laws of chance. It is easy to see that it might well be otherwise. If all those cups made with the milk first had sugar added, while those made with the tea first had none, a very obvious difference in flavour would have been introduced which might well ensure that all those made with sugar should be classed alike. These groups might either be classified all right or all wrong, but in such a case the frequency of the critical event in which all cups are classified correctly would not be 1 in 70, but 35 in 70 trials, and the test of significance would be wholly vitiated. Errors equivalent in principle to this are very frequently incorporated in otherwise well-designed experiments.

It is no sufficient remedy to insist that "all the cups must be exactly alike" in every respect except that to be tested. For this is a totally impossible requirement in our example, and equally in all other forms of experimentation. In practice it is probable that the cups will differ perceptibly in the thickness or smoothness of their material, that the quantities of milk added to the different cups will not be exactly equal, that the strength of the infusion of tea may change between pouring the first and the last cup, and that the temperature also at which the tea is tasted will change during the course of the experiment. These are only examples of the differences probably present; it would be impossible to present an exhaustive list of such possible differences appropriate to any one kind of experiment, because the uncontrolled causes which may influence the result are always strictly innumerable. When any such cause is named, it is usually perceived that, by increased labour and expense, it could be largely eliminated. Too frequently it is assumed that such refinements constitute improvements to the experiment. Our view, which will be much more fully exemplified in later sections, is that it is an essential characteristic of experimentation that it is carried out with limited resources, and an essential part of the subject of experimental design to ascertain how these should be best applied; or, in particular, to which causes of disturbance care should be given, and which *ought* to be deliberately ignored. To ascertain, too, for those which are not to be ignored, to what *extent* it is worth while to take the trouble to diminish their magnitude. For our present purpose, however, it is only necessary to recognise that, whatever degree of care and experimental skill is expended in equalising the conditions, other than the one under test, which are liable to affect the result, this equalisation must always be to a greater or less extent incomplete, and in many important practical cases will certainly be grossly defective. We are concerned, therefore, that this inequality, whether it be great or small, shall not impugn the exactitude of the frequency distribution, on the basis of which the result of the experiment is to be appraised.

THE EFFECTIVENESS OF RANDOMISATION

The element in the experimental procedure which contains the essential safeguard is that the two modifications of the test beverage are to be prepared "in random order." This, in fact, is the only point in the experimental procedure in which the laws of chance, which are to be in exclusive control of our frequency distribution, have been explicitly introduced. The phrase "random order" itself, however, must be regarded as an incomplete instruction, standing as a kind of shorthand symbol for the full procedure of randomisation, by which the validity of the test of significance may be guaranteed against corruption by the causes of disturbance which have not been eliminated. To demonstrate that, with satisfactory randomisation, its validity is, indeed, wholly unimpaired, let us imagine all causes of disturbance—the strength of the infusion, the quantity of milk, the temperature at which it is tasted, etc.—to be predetermined for each cup; then since these, on the null hypothesis, are the only causes influencing classification, we may say that the probabilities of each of the 70 possible choices or classifications which the subject can make are also predetermined. If, now, after the disturbing causes are fixed, we assign, strictly at random, 4 out of the 8 cups to each of our experimental treatments, then every set of 4, whatever its probability of being so classified, will certainly have a probability of exactly 1 in 70 of *being* the 4, for example, to which the milk is added first. However important the causes of disturbance may be, even if they were to make it certain that one particular set of 4 should receive this classification, the probability that the 4 so classified and the 4 which ought to have been so classified should be the same, must be rigorously in accordance with our test of significance.

It is apparent, therefore, that the random choice of the objects to be treated in different ways would be a complete guarantee of the validity of the test of significance, if these treatments were the last in time of the stages in the physical history of the objects which might affect their experimental reaction. The circumstance that the experimental treatments cannot always be applied last, and may come relatively early in their history, causes no practical inconvenience; for subsequent causes of differentiation, if under the experimenter's control, as, for example, the choice of different pipettes to be used with different flasks, can either be predetermined before the treatments have been randomised, or, if this has not been done, can be randomised on their own account; and other causes of differentiation will be either (*a*) consequences of differences already randomised, or (*b*) natural consequences of the difference in treatment to be tested, of which on the null hypothesis there will be none, by definition, or (*c*) effects supervening by chance independently from the treatments applied. Apart, therefore, from the avoidable error of the experimenter himself introducing with his test treatments, or subsequently, other differences in treatment, the effects of which the experiment is not intended to study, it may be said that the simple precaution of randomisation will suffice to guarantee the validity of the test of significance, by which the result of the experiment is to be judged.

THE SENSITIVENESS OF AN EXPERIMENT. EFFECTS OF ENLARGEMENT AND REPETITION

A probable objection, which the subject might well make to the experiment so far described, is that only if every cup is classified correctly will she be judged successful. A single mistake will reduce her performance below the level of significance. Her claim, however, might be, not that she could draw the distinction with invariable certainty, but that, though sometimes mistaken, she would be right more often than not; and that the experiment should be enlarged sufficiently, or repeated sufficiently often, for her to be able to demonstrate the predominance of correct classifications in spite of occasional errors.

An extension of the calculation upon which the test of significance was based shows that an experiment with 12 cups, six of each kind, gives, on the null hypothesis, 1 chance in 924 for complete success, and 36 chances for 5 of each kind classified right and 1 wrong. As 37 is less than a twentieth of 924, such a test could be counted as significant, although a pair of cups have been wrongly classified; and it is easy to verify that, using larger numbers still, a significant result could be obtained with a still higher proportion of errors. By increasing the size of the experiment, we can render it more sensitive, meaning by this that it will allow of the detection of a lower degree of sensory discrimination, or, in other words, of a quantitatively smaller departure from the null hypothesis. Since in every case the experiment is capable of disproving, but never of proving this hypothesis, we may say that the value of the experiment is increased whenever it permits the null hypothesis to be more readily disproved.

The same result could be achieved by repeating the experiment, as originally designed, upon a number of different occasions, counting as a success all those occasions on which 8 cups are correctly classified. The chance of success on each occasion being 1 in 70, a simple application of the theory of probability shows that 2 or more successes in 10 trials would occur, by chance, with a frequency below the standard chosen for testing significance; so that the sensory discrimination would be demonstrated, although, in 8 attempts out of 10, the subject made one or more mistakes. This procedure may be regarded as merely a second way of enlarging the experiment and, thereby, increasing its sensitiveness, since in our final calculation we take account of the aggregate of the entire series of results, whether successful or unsuccessful. It would clearly be illegitimate, and would rob our calculation of its basis, if the unsuccessful results were not all brought into the account.

QUALITATIVE METHODS OF INCREASING SENSITIVENESS

Instead of enlarging the experiment we may attempt to increase its sensitiveness by qualitative improvements; and these are, generally speaking, of two kinds: (*a*) the reorganisation of its structure, and (*b*) refinements of technique. To illustrate a change of structure we might consider that, instead of fixing in advance that 4 cups should be of each kind, determining by a random process how the subdivision should be effected, we might have allowed the treatment of each cup to be determined independently by chance, as by the toss of a coin, so that each treatment has an equal chance of being chosen. The chance of classifying correctly 8 cups randomised in this way, without the aid of sensory discrimination, is 1 in 2^8, or 1 in 256 chances, and there are only 8 chances of classifying 7 right and 1 wrong; consequently the sensitiveness of the experiment has been increased, while still using only 8 cups, and it is possible to score a significant success, even if one is classified wrongly. In many types of experiment, therefore, the suggested change in structure would be evidently advantageous. For the special requirements of a psycho-physical experiment, however, we should probably prefer to forego this advantage, since it would occasionally occur that all the cups would be treated alike, and this, besides bewildering the subject by an unexpected occurrence, would deny her the real advantage of judging by comparison.

Another possible alteration to the structure of the experiment, which would, however, decrease its sensitiveness, would be to present determined, but unequal, numbers of the two treatments. Thus we might arrange that 5 cups should be of the one kind and 3 of the other, choosing them properly by chance, and informing the subject how many of each to expect. But since the number of ways of choosing 3 things out of 8 is only 56, there is now, on the null hypothesis, a probability of a completely correct classification of 1 in 56. It appears in fact that we cannot by these means do better than by presenting the two treatments in equal numbers, and the choice of this equality is now seen to be justified by its giving to the experiment its maximal sensitiveness.

With respect to the refinements of technique, we have seen above that these contribute nothing to the validity of the experiment, and of the test of significance by which we determine its result. They may, however, be important, and even essential, in permitting the phenomenon under test to manifest itself. Though the test of significance remains valid, it may be that without special precautions even a definite sensory discrimination would have little chance of scoring a significant success. If some cups were made with India and some with China tea, even though the treatments were properly randomised, the subject might not be able to discriminate the relatively small difference in flavour under investigation, when it was confused with the greater differences between leaves of different origin. Obviously, a similar difficulty could be introduced by using in some cups raw milk and in others boiled, or even condensed milk, or by adding sugar in unequal quantities. The subject has a right to claim, and it is in the interests of the sensitiveness of the experiment, that gross differences of these kinds should be excluded, and that the cups should, not as far as *possible*, but as far as is practically convenient, be made alike in all respects except that under test.

How far such experimental refinements should be carried is entirely a matter of judgment, based on experience. The validity of the experiment is not affected by them. Their sole purpose is to increase its sensitiveness, and this object can usually be achieved in many other ways, and particularly by increasing the size of the experiment. If, therefore, it is decided that the sensitiveness of the experiment should be increased, the experimenter has the choice between different methods of obtaining equivalent results; and will be wise to choose whichever method is easiest to him, irrespective of the fact that previous experimenters may have tried, and recommended as very important, or even essential, various ingenious and troublesome precautions.

The Scientific Aptitude of Mr. George Bernard Shaw

BERNARD SHAW was not at his best as a scientific thinker. Science interested him but he was inclined to be erratic. Though science offered a fertile field for the exercise of his talents as a controversialist, a foe of pretense and a joker, he was often unable to distinguish between the stuffed robe and the honest scientist, between theories that merited serious attention and theories that were pure humbug. Moreover he himself espoused the most incredible nonsense. He fought vivisection and vaccination; he had a low opinion of medical knowledge and an even lower opinion of its practitioners; he had his own astonishing theories of biology, physiology, bacteriology and hygiene, and nothing would persuade him that the sun was burning itself out (since he expected to live longer than Methuselah he felt he had a personal stake in the catastrophe); he dismissed laboratory experiments generally as mere "put-up jobs," performances rigged for the purpose of proving preconceived theories regardless of the weight of evidence.

But for all his prejudices and eccentric notions, Shaw did not close his mind to the important works of science. He followed the advances of research in fields as varied as Pavlov's work on dogs and the Michelson-Morley interferometer experiments on ether drift. He "liked visiting laboratories and peeping at bacteria through the microscope." [1] He was curious about how things work: automobiles, radios, machine tools, motorcycles, phonographs. He was an enthusiastic photographer and camera tinkerer. Every efficient labor-saving device won his admiration but "for old-fashioned factory machinery his contempt was boundless: he said a louse could have invented it all if it had been keen enough on profits." [2]

Shaw and Jonathan Swift were much alike in their attitudes to science. Both men lived in periods of great scientific advance; both respected science; neither had any special aptitude for it. Both approached the subject as social reformers and satirists; both despised pretentiousness; neither had much use for science as a purely speculative activity. Swift aimed his wit at mathematics, which in its advanced forms seemed to him completely trivial; Shaw waged war on biological practices which he thought

---

[1] Hesketh Pearson, *G.B.S., A Full Length Portrait*, New York, 1942, p. 270. I have drawn on this biography for many of the details of this sketch; also on Bernard Shaw, *Sixteen Self-Sketches*, New York, 1949.

[2] Pearson, *op. cit.*, p. 270.

cruel and stupid. That he exaggerated is understandable. He enjoyed exaggeration and he regarded it as an essential tool of reform. "If you do not say a thing in an irritating way, you may just as well not say it at all, since nobody will trouble themselves about anything that does not trouble them." (The grammar *is* bizarre, even for G.B.S.)

There was, however, one branch of science which Shaw neither tilted at nor enlarged with theories of his own. The subject he spared was mathematics. He did not minimize its importance and he admitted, dropping at least this once the pose of omniscience, that he knew very little about it. He blamed his ignorance on the wretched instruction he received at the Wesleyan Connexional School. "Not a word was said to us about the meaning or utility of mathematics: we were simply asked to explain how an equilateral triangle could be constructed by the intersection of two circles, and to do sums in a, b, and x instead of in pence and shillings, leaving me so ignorant that I concluded that a and b must mean eggs and cheese and x nothing, with the result that I rejected algebra as nonsense, and never changed that opinion until in my advanced twenties Graham Wallas and Karl Pearson convinced me that instead of being taught mathematics I had been made a fool of."

The influence of these distinguished men was highly beneficial. To be sure Shaw never became unduly proficient as a calculator: "I never used a logarithm in my life, and could not undertake to extract the square root of four without misgivings." But he learned to appreciate the importance of at least one division of higher mathematics, the theory of probability and statistics. The following selection presents a Shavian version of the development and practical application of the calculus of chance. It is a delightful account and very sensible. No one else, so far as I know, has treated the history of mathematics in this way. I suspect that if there were more Shaws teaching the subject it would become popular. But the mathematical probability of this compound circumstance is admittedly small.

Let the king prohibit gambling and betting in his kingdom, for these are vices that destroy the kingdoms of princes. —THE CODE OF MANU (c. 100)

In play there are two pleasures for your choosing—
The one is winning, and the other losing.   —BYRON

7   The Vice of Gambling and the Virtue of Insurance

By GEORGE BERNARD SHAW

INSURANCE, though founded on facts that are inexplicable, and risks that are calculable only by professional mathematicians called actuaries, is nevertheless more· congenial as a study than the simpler subjects of banking and capital. This is because for every competent politician in our country there must be at least a hundred thousand gamblers who make bets every week with turf bookmakers. The bookmaker's business is to bet against any horse entered for a race with anybody who thinks it will win and wants to bet that it will. As only one horse can win, and all the rest must lose, this business would be enormously lucrative if all the bets were for even money. But the competition among bookmakers leads them to attract customers by offering "odds," temptingly "long," against horses unlikely to win: whilst giving no odds at all on the most likely horse, called the favorite. The well-known cry, puzzling to novices, of "two to one bar one" means that the bookmaker will bet at odds of two to one against any horse in the race except the favorite. Mostly, however, he will bet at odds of ten to one or more against an "outsider." In that case, if, as sometimes happens, the outsider wins, the bookmaker may lose on his bet against it all that he gained on his bets against the favorites. On the scale between the possible extremes of gain and loss he may come out anywhere according to the number of horses in the race, the number of bets made on each of them, and the accuracy of his judgment in guessing the odds he may safely offer. Usually he gains when an outsider wins, because mostly there is more money laid on favorites and fancies than on outsiders; but the contrary is possible; for there may be several outsiders as well as several favorites; and, as outsiders win quite often, to tempt customers by offering too long odds against them is gambling; and a bookmaker must never gamble, though he lives by gambling. There are practically always enough variable factors in the game to tax the bookmaker's financial ability to the utmost. He must budget so as to come out at worst still solvent. A bookmaker who gambles will ruin himself as certainly as a licensed victualler (publican) who drinks, or a picture dealer who cannot bear to part with a good picture.

The question at once arises, how is it possible to budget for solvency in dealing with matters of chance? The answer is that when dealt with in sufficient numbers matters of chance become matters of certainty, which is one of the reasons why a million persons organized as a State can do things that cannot be dared by private individuals. The discovery of this fact nevertheless was made in the course of ordinary private business.

In ancient days, when travelling was dangerous, and people before starting on a journey overseas solemnly made their wills and said their prayers as if they were going to die, trade with foreign countries was a risky business, especially when the merchant, instead of staying at home and consigning his goods to a foreign firm, had to accompany them to their destination and sell them there. To do this he had to make a bargain with a ship owner or a ship captain.

Now ship captains, who live on the sea, are not subject to the terrors it inspires in the landsman. To them the sea is safer than the land; for shipwrecks are less frequent than diseases and disasters on shore. And ship captains make money by carrying passengers as well as cargo. Imagine then a business talk between a merchant greedy for foreign trade but desperately afraid of being shipwrecked or eaten by savages, and a skipper greedy for cargo and passengers. The captain assures the merchant that his goods will be perfectly safe, and himself equally so if he accompanies them. But the merchant, with his head full of the adventures of Jonah, St. Paul, Odysseus, and Robinson Crusoe, dares not venture. Their conversation will be like this:

CAPTAIN. Come! I will bet you umpteen pounds that if you sail with me you will be alive and well this day year.

MERCHANT. But if I take the bet I shall be betting you that sum that I shall die within the year.

CAPTAIN. Why not if you lose the bet, as you certainly will?

MERCHANT. But if I am drowned you will be drowned too; and then what becomes of our bet?

CAPTAIN. True. But I will find you a landsman who will make the bet with your wife and family.

MERCHANT. That alters the case of course; but what about my cargo?

CAPTAIN. Pooh! The bet can be on the cargo as well. Or two bets: one on your life, the other on the cargo. Both will be safe, I assure you. Nothing will happen; and you will see all the wonders that are to be seen abroad.

MERCHANT. But if I and my goods get through safely I shall have to pay you the value of my life and of the goods into the bargain. If I am not drowned I shall be ruined.

CAPTAIN. That also is very true. But there is not so much for me in it as you think. If you are drowned I shall be drowned first; for I must be the last man to leave the sinking ship. Still, let me persuade you to venture. I will make the bet ten to one. Will that tempt you?

MERCHANT. Oh, in that case—

The captain has discovered insurance just as the goldsmiths discovered banking.

It is a lucrative business; and, if the insurer's judgment and information are sound, a safe one. But it is not so simple as bookmaking on the turf, because in a race, as all the horses but one must lose and the bookmaker gain, in a shipwreck all the passengers may win and the insurer be ruined. Apparently he must therefore own, not one ship only, but several, so that, as many more ships come safely to port than sink, he will win on half a dozen ships and lose on one only. But in fact the marine insurer need no more own ships than the bookmaker need own horses. He can insure the cargoes and lives in a thousand ships owned by other people without his having ever owned or even seen as much as a canoe. The more ships he insures the safer are his profits; for half a dozen ships may perish in the same typhoon or be swallowed by the same tidal wave; but out of a thousand ships most by far will survive. When the risks are increased by war the odds on the bets can be lowered.

When foreign trade develops to a point at which marine insurers can employ more capital than individual gamesters can supply, corporations like the British Lloyds are formed to supply the demand. These corporations soon perceive that there are many more risks in the world than the risk of shipwreck. Men who never travel nor send a parcel across the seas, may lose life or limb by accident, or have their houses burnt or robbed. Insurance companies spring up in all directions; and the business extends and develops until there is not a risk that cannot be insured. Lloyds will bet not only against shipwreck but against almost any risk that is not specifically covered by the joint stock companies, provided it is an insurable risk: that is, a safe one.

This provision is a contradiction in terms; for how can a safe transaction involve a risk or a risk be run safely?

The answer takes us into a region of mystery in which the facts are unreasonable by any method of ratiocination yet discovered. The stock example is the simplest form of gambling, which is tossing a coin and betting on which side of it will be uppermost when it falls and comes to rest. Heads or tails they call it in England, head or harp in Ireland. Every time the coin is tossed, each side has an equal chance with the other of winning. If head wins it is just as likely to win the next time and the next and so on to the thousandth; so that on reasonable grounds a thousand heads in succession are possible, or a thousand tails; for the fact that head wins at any toss does not raise the faintest reasonable probability that tails will win next time. Yet the facts defy this reasoning. Anyone who possesses a halfpenny and cares to toss it a hundred times may find the same side turning up several times in succession; but the total result will be fifty-fifty or as near thereto as does not matter. I happen to have in my pocket ten pennies; and I have just spilt them on the floor ten times. Result: forty nine heads and fifty one tails, though five-five occurred only twice in the ten throws, and heads won three times in succession to begin with. Thus though as between any two tosses the result is completely uncertain, in ten throws it may be six-four or seven-three often enough to make betting a gamble; but in a hundred the result will certainly be close enough to fifty-fifty to leave two gamblers, one crying heads and the other tails every time, exactly or very nearly where they were when they started, no richer and no poorer, unless the stakes are so high that only players out of their senses would hazard them.

An insurance company, sanely directed, and making scores of thousands of bets, is not gambling at all; it knows with sufficient accuracy at what age its clients will die, how many of their houses will be burnt every year, how often their houses will be broken into by burglars, to what extent their money will be embezzled by their cashiers, how much compensation they will have to pay to persons injured in their employment, how many accidents will occur to their motor cars and themselves, how much they will suffer from illness or unemployment, and what births and deaths will cost them: in short, what will happen to every thousand or ten thousand or a million people even when the company cannot tell what will happen to any individual among them.

In my boyhood I was equipped for an idle life by being taught to play whist, because there were rich people who, having nothing better to do, escaped from the curse of boredom (then called ennui) by playing whist every day. Later on they played bezique instead. Now they play bridge. Every gentleman's club has its card room. Card games are games of chance; for though the players may seem to exercise some skill and judgment in choosing which card to play, practice soon establishes rules by which the stupidest player can learn how to choose correctly: that is, not to choose at all but to obey the rules. Accordingly people who play every day for sixpences or shillings find at the end of the year that they have neither gained nor lost sums of any importance to them, and have killed time pleasantly instead of being bored to death. They have not really been gambling any more than the insurance companies.

At last it is discovered that insurers not only need not own ships or horses or houses or any of the things they insure, but that they need not exist. Their places can be taken by machines. On the turf the bookmaker, flamboyantly dressed and brazenly eloquent, is superseded by the Totalizator (Tote for short) in which the gamblers deposit the sums they are prepared to stake on the horses they fancy. After the race all the money staked on the winner is divided among its backers. The machine keeps the rest. On board pleasure ships young ladies with more money than they know what to do with drop shillings into gambling machines so constructed that they occasionally return the shilling ten or twentyfold. These are the latest successors of the roulette table, the "little horses," the dice casters, and all other contraptions which sell chances of getting money for nothing. Like the Tote and the sweepstake, they do not gamble: they risk absolutely nothing, though their customers have no certainty except that in the lump they must lose, every gain to Jack and Jill being a loss to Tom and Susan.

How does all this concern the statesman? In this way. Gambling, or the attempt to get money without earning it, is a vice which is economically (that is, fundamentally) ruinous. In extreme cases it is a madness which persons of the highest intelligence are unable to resist: they will stake all they possess though they know that the chances are against them. When they have beggared themselves in half an hour or half a minute, they sit wondering at the folly of the people who are doing the same thing, and at their own folly in having done it themselves.

Now a State, being able to make a million bets whilst an individual citizen can afford only one, can tempt him or her to gamble without itself running the slightest risk of losing financially; for, as aforesaid, what will happen in a million cases is certain, though no one can foresee what will happen in any one case. Consequently governments, being continually in pressing need of money through the magnitude of their expenses and the popular dislike of taxation, are strongly tempted to replenish the Treasury by tempting their citizens to gamble with them.

No crime against society could be more wickedly mischievous. No public duty is more imperative than the duty of creating a strong public conscience against it, making it a point of bare civic honesty not to spend without earning nor consume without producing, and a point of high civic honor to earn more than you spend, to produce more than you consume, and thus leave the world better off than you found it. No other real title to gentility is conceivable nowadays.

Unfortunately our system of making land and capital private property not only makes it impossible for either the State or the Church to inculcate these fundamental precepts but actually drives them to preach just the opposite. The system may urge the energetic employer to work hard and develop his business to the utmost; but his final object is to become a member of the landed gentry or the plutocracy, living on the labor of others and enabling his children to do the same without ever having worked at all. The reward of success in life is to become a parasite and found a race of parasites. Parasitism is the linchpin of the Capitalist apple-cart: the main Incentive without which, we are taught, human society would fall to pieces. The boldest of our archbishops, the most democratic of our finance ministers, dares not thunder forth that parasitism, for peers and punters alike, is a virus that will rot the most powerful civilization, and that the contrary doctrine is diabolical. Our most eminent churchmen do not preach very plainly and urgently against making selfishness the motive power of civilization; but they have not yet ventured to follow Ruskin and Proudhon in insisting definitely that a citizen who is neither producing goods nor performing services is in effect either a beggar or a thief. The utmost point yet reached in England is the ruling out of State lotteries and the outlawing of the Irish sweepstakes.

But here again the matter is not simple enough to be disposed of by counsels of Socialist perfection in the abstract. There are periods in every long lifetime during which one must consume without producing. Every baby is a shamelessly voracious parasite. And to turn the baby into a highly trained producer or public servant, and make its adult life worth living, its parasitism must be prolonged well into its teens. Then again old people cannot produce. Certain tribes who lay an excessive stress on Manchester School economics get over this difficulty easily by killing their aged parents or turning them out to starve. This is not necessary in modern civilization. It is quite possible to organize society in such a manner as to enable every ablebodied and ableminded person to produce enough not only to pay their way but to repay the cost of twenty years education and training, making it a first-rate investment for the community, besides providing for the longest interval between disablement by old age and natural death. To arrange this is one of the first duties of the modern statesman.

Now childhood and old age are certainties. What about accidents and illnesses, which for the individual citizen are not certainties but chances? Well, we have seen that what are chances for the individual are certainties for the State. The individual citizen can share its certainty only by gambling with it. To insure myself against accident or illness I must make a bet with the State that these mishaps will befall me; and the State must accept the bet, the odds being fixed by the State actuaries mathematically. I shall at once be asked Why with the State? Why not with a private insurance company? Clearly because the State can do what no private company can do. It can compel every citizen to insure, however improvident or confident in his good luck he may happen to be, and thus, by making a greater number of bets, combine the greatest profit with the greatest certainty, and put the profit into the public treasury for the general good. It can effect an immense saving of labor by substituting a single organization for dozens of competing ones. Finally it can insure at cost price, and, by including the price in the general rate of taxation, pay for all accidents and illnesses directly and simply without the enormous clerical labor of collecting specific contributions or having to deal in any way with the mass of citizens who lose their bets by having no accidents nor illnesses at any given moment.

The oddity of the situation is that the State, to make insurance certain and abolish gambling, has to compel everyone to gamble, becoming a Supertote and stakeholder for the entire population.

As ship insurance led to life insurance, life insurance to fire insurance and so on to insurance against employer's liability, death duties, and unemployment, the list of insurable risks will be added to, and insurance policies will become more comprehensive from decade to decade, until no risks that can worry a reasonably reckless citizen are left uncovered. And when the business of insurance is taken on by the State and lumped into the general taxation account, every citizen will be born with an unwritten policy of insurance against all the common risks, and be spared the painful virtues of providence, prudence, and self-denial that are now so oppressive and demoralizing, thus greatly lightening the burden of middle-class morality. The citizens will be protected whether they like it or not, just as their children are now educated and their houses now guarded by the police whether they like it or not, even when they have no children to be educated nor houses to be guarded. The gain in freedom from petty cares will be immense. Our minds will no longer be crammed and our time wasted by uncertainty as to whether there will be any dinners for the family next week or any money left to pay for our funerals when we die.

There is nothing impossible or even unreasonably difficult in all this. Yet as I write, a modest and well thought-out plan of national insurance by Sir William Beveridge, whose eminence as an authority on political science nobody questions, is being fiercely opposed, not only by the private insurance companies which it would supersede, but by people whom it would benefit; and its advocates mostly do not understand it and do not know how to defend it. If the schooling of our legislators had included a grounding in the principles of insurance the Beveridge scheme would pass into law and be set in operation within a month. As it is, if some mutilated remains of it survive after years of ignorant squabbling we shall be lucky, unless, indeed, some war panic drives it through Parliament without discussion or amendment in a few hours. However that may be, it is clear that nobody who does not understand insurance and comprehend in some degree its enormous possibilities is qualified to meddle in national business. And nobody can get that far without at least an acquaintance with the mathematics of probability, not to the extent of making its calculations and filling examination papers with typical equations, but enough to know when they can be trusted, and when they are cooked. For when their imaginary numbers correspond to exact quantities of hard coins unalterably stamped with heads and tails, they are safe within certain limits; for here we have solid certainty and two simple possibilities that can be made practical certainties by an hour's trial (say one constant and one variable that does not really vary); but when the calculation is one of no constant and several very capricious variables, guesswork, personal bias, and pecuniary interests, come in so strongly that those who began by ignorantly imagining that statistics cannot lie end by imagining, equally ignorantly, that they never do anything else.

PART IX

The Supreme Art of Abstraction: Group Theory

Certain Important Abstractions

THE Theory of Groups is a branch of mathematics in which one does something to something and then compares the result with the result obtained from doing the same thing to something else, or something else to the same thing. This is a broad definition but it is not trivial. The theory is a supreme example of the art of mathematical abstraction. It is concerned only with the fine filigree of underlying relationships; it is the most powerful instrument yet invented for illuminating structure.

The term *group* was first used in a technical sense by the French mathematician Évariste Galois in 1830. He wrote his brilliant paper on the subject at the age of twenty, the night before he was killed in a stupid duel.[1] The concept was strongly developed in the nineteenth century by leading mathematicians, among them Augustin-Louis Cauchy (1789–1857),[2] Sir Arthur Cayley (see p. 341), Camille Jordan (1838–1922), and two eminent Norwegians, Ludwig Sylow (1832–1918) and Marius Sophus Lie (1842–1899). In a little more than a century it has effected a remarkable unification of mathematics, revealing connections between parts of algebra and geometry that were long considered distinct and unrelated. "Wherever groups disclosed themselves, or could be introduced, simplicity crystallized out of comparative chaos." [3] Group theory has also helped physicists penetrate to the basic structure of the phenomenal world, to catch glimpses of innermost pattern and relationship. This is as deep, it should be observed, as science is likely to get. Even if we do not accept the idea that the ultimate essence of things *is* pattern, we may conclude with Bertrand Russell that any other essence is an individuality "which always eludes words and baffles description, but which, for that very reason, is irrelevant to science."

Let us return briefly to our somewhat dreamy definition and make it more concrete and explicit. The best plan, perhaps, is to give a specific example of a group and then to erase most of its details until nothing is left but essentials. This is the famous Carrollian method of defining a grin as what remains after the Cheshire cat, the vehicle of the grin, has vanished.

The class or set of all the positive and negative integers, including zero,

---

[1] The story of his tragic life is dramatically told by E. T. Bell in *Men of Mathematics*, New York, 1937.

[2] "To Cauchy has been given the credit of being the founder of the theory of groups of finite order, even though fundamental results had been previously reached by J. L. Lagrange, Pietro Abbati, P. Ruffini, N. H. Abel, and Galois." Florian Cajori, *A History of Mathematics*, New York, 1919, p. 352.

[3] E. T. Bell, *Mathematics, Queen and Servant of Science*, New York, 1951, p. 164.

in conjunction with the ordinary arithmetic operation of addition, constitutes a familiar group. Its defining properties are these: (1) The sum of any two integers of the set is an integer of the set; (2) in adding three (or more) integers, any set of them may be added first without varying the result; you will recognize this as the associative rule of arithmetic (e.g., $(3 + 7) + 9 = 3 + (7 + 9)$); (3) the set contains an "identity" or "unit" element (namely zero) such that the sum of this element and any other element in the set is again the latter element (e.g., $4 + 0 = 4$, $0 + 8 = 8$, etc.); (4) every integer in the set has an inverse or reciprocal, such that the sum of the two is the identity element (e.g., $2 + (-2) = 0$, $-77 + 77 = 0$, etc.). These are the attributes of our particular group. Now for some erasures. (1) The elements of the set may be arithmetic objects (e.g., numbers), geometric objects (e.g., points), physical entities (e.g., atoms), or they may be undefined; [4] (2) their number may be finite or infinite; (3) the operation or rule of combining the elements may be an arithmetic process (e.g., addition, multiplication), a geometric process (e.g., rotation, translation), or it may be undefined. Two further conditions are essential: (4) the combining rule must be associative; (5) every element of the set must have an inverse. Besides these five conditions, a set may be Abelian or non-Abelian according as the combining rule is commutative or noncommutative (i.e., for addition, either $2 + 3 = 3 + 2$ or $2 + 3 \neq 3 + 2$, and, for multiplication, either $2 \cdot 3 = 3 \cdot 2$ or $2 \cdot 3 \neq 3 \cdot 2$). These are the bare bones of the group concept. It is hard to believe how much unification of bewildering details has been achieved by the theory; "what a wealth, what a grandeur of thought may spring from what slight beginnings." [5]

A few words should be added concerning two other fundamental mathematical terms which arise frequently in group theory. The first is *transformation*, which embodies the idea of change or motion. An algebraic expression is transformed by changing it to another having different form, by substituting for the variables their values in terms of another set of variables; a geometric figure is transformed by changing its co-ordinates, by mapping one space on another, by moving the figure pursuant to a procedural rule, e.g., projection, rotation, translation. [6] More generally,

---

[4] It should be emphasized that while both the elements and operations of a group may theoretically be undefined, if the group is to be useful in science they must in some way correspond to elements and operations of observable experience. Otherwise manipulating the group amounts to nothing more than a game, and a pretty vague and arid game at that, suitable only for the most withdrawn lunatics.

[5] The British geometer H. F. Baker, as quoted by Florian Cajori in *A History of Mathematics*, New York, 1919, p. 283.

[6] This definition of transformation suffices for our purposes, but the concept is much more comprehensive than I have indicated. Any problem, process or operation, as Keyser says (see next note), having to do with ordinary functions is a problem, process or operation having to do with relations or transformations. A pairing or coupling is a transformation; so is a relation, a function, a mathematical calculation, a deductive inference.

any object of thought may be transformed by associating it with or converting it into another object of thought.

The second important term is *invariance*. The invariant properties of an algebraic expression, geometric figure, class, or other object of thought are those which remain the same under transformations. Suppose the elements of the class of positive integers ($1, 2, 3 \cdots$) are transformed by the rule of doubling each element

$$
\begin{array}{ccccccc}
1 & 2 & 3 & 4 & \cdot & \cdot & \cdot \\
\downarrow & \downarrow & \downarrow & \downarrow & & & \\
2 & 4 & 6 & 8 & \cdot & \cdot & \cdot
\end{array}
$$

The transforms constitute the class of even integers: $2, 4, 6, \cdots$. Since the integers are transformed into integers, the property of being an integer is evidently preserved; which is to say it is an invariant under the doubling transformation. However, the *value* of each integer of the original class is not preserved; it is doubled; thus value is not an invariant under this transformation. Another example of an invariant in our case are the ratios of the elements of each class: if the elements of the first class are called x's and those of the transformed class y's then under the rule $y = 2x$,

$$
\frac{y_1}{y_2} = \frac{2x_1}{2x_2} = \frac{x_1}{x_2}.\text{[7]}
$$

A geometric invariant is similar. Take a rigid object such as a glass paper-weight and move it by sliding from one end of the table to the other. This is a transformation. The paper-weight retains its shape and dimensions; the retained properties are therefore invariants. The mathematician describes these facts by saying that the metric properties of rigid bodies are invariant under the transformation of motion. Since the paper-weight's position and distance from an object such as the mirror on the wall or the Pole Star are changed by the transformation, they are not invariant. If the object moved were a blob of mercury it is unlikely that it would retain its shape or dimensions, but its mass would probably be invariant and certainly its atomic structure. (For a further discussion of this point see pp. 581–598, selection on topology.)

Group theory has to do with the invariants of groups of transformations. One studies the properties of an object, the features of a problem unaffected by changes of condition. The more drastic the changes, the fewer the invariants. What better way to get at the fundamentals of structure than by successive transformations to strip away the secondary properties. It is a method analogous to that used by the archaeologist who clears away hills to get at cities, digs into houses to uncover ornaments, utensils and potsherds, tunnels into tombs to find sarcophagi, the winding

[7] The example is from Cassius J. Keyser, *Mathematical Philosophy*, New York, 1922, pp. 183–185.

sheets they hold and the mummies within. Thus he reconstructs the features of an unseen society; and so the mathematician and scientist create a theoretical counterpart of the unseen structure of the phenomenal world. Whitehead has characterized these efforts in a famous observation: "To see what is general in what is particular and what is permanent in what is transitory is the aim of scientific thought."

*   *   *   *   *

I have selected two essays to illustrate group theory. The first is a chapter from Cassius J. Keyser's lectures on mathematical philosophy.[8] Keyser, a prominent American mathematician, was born in Rawson, Ohio, in 1862. He was educated in Ohio schools and at Missouri University; for a time he studied law but then turned to mathematics and earned his graduate degrees at Columbia. After five years as superintendent of schools in Ohio and Montana (1885–90), he became professor of mathematics at the New York State Normal School, and in 1897 joined the staff of Columbia. He was appointed Adrain professor in 1904, serving in this post until 1927 when he was made emeritus. He died in 1947, aged eighty-five. Keyser had broad interests in mathematics, as a geometer, historian and philosopher. He was much admired as a teacher for the care he took to his lectures, their breadth, clarity and honesty. He is a little old-fashioned in his style and a trifle long-winded. Keyser was not the man to drop a point until he had squeezed it dry both as to its scientific content and cultural bearings; the reader may also come to feel a little squeezed. But he had an unfailingly interesting and reflective mind, and I have nowhere found a better survey of the group concept than in the selection below. The second essay, by Sir Arthur Eddington, is one of the Messenger Lectures given at Cornell University, appearing in a book titled *New Pathways of Science*. The discussion of groups exhibits the usual dazzling Eddington virtuosity; it is one of his best pieces of popularization in one of his best books.

[8] *Mathematical Philosophy, A Study of Fate and Freedom*, New York, 1922.

Mazes intricate.
Eccentric, interwov'd, yet regular
Then most, when most irregular they seem. —MILTON

1   The Group Concept

By CASSIUS J. KEYSER

I INVITE your attention during the present hour to the notion of group. Even if I were a specialist in group theory,—which I am not,—I could not in one hour give you anything like an extensive knowledge of it, nor facility in its technique, nor a sense of its intricacy and proportions as known to its devotees, the priests of the temple. But the hour should suffice to start you on the way to acquiring at least a minimum of what a respectable philosopher should know of this fundamental subject; and such a minimum will include: a clear conception of what the term "group" means; ability to illustrate it copiously by means of easily under-stood examples to be found in all the cardinal fields of interest—number, space, time, motion, relation, play, work, the world of sense-data and the world of ideas; a glimpse of its intimate connections with the ideas of transformation and invariance; an inkling of it both as subject-matter and as an instrument for the delimitation and discrimination of doctrines; and discernment of the concept as vaguely prefigured in philosophic specula-tion from remote antiquity down to the present time.

I believe that the best way to secure a firm hold of the notion of group is to seize upon it first in the abstract and then, by comparing it with concrete examples, gradually to win the sense of holding in your grasp a living thing. In presenting the notion of group in the abstract, it is con-venient to use the term *system.* This term has many meanings in mathe-matics and so at the outset we must clearly understand the sense in which the term is to be employed here. The sense is this: as employed in the definition of group, the term system means some definite class of things together with some definite rule, or way, in accordance with which any member of the class can be combined with any member of it (either with itself or any other member). For a simple example of such a system we may take for the class the class of ordinary whole numbers and for the rule of combination the familiar rule of addition. You should note that there are three and only three respects in which two systems can differ: by having different classes, by having different rules of combination, and by differing in both of these ways.

The definition of the term "group" is as follows.

Let S denote a system consisting of a class C (whose members we will denote by a, b, c and so on) and of a rule of combination (which rule we will denote by the symbol o, so that by writing, for example, aob, we shall mean the result of combining b with a). The system S is called a group if and only if it satisfies the following four conditions:

(*a*) If a and b are members of C, then aob is a member of C; that is, $aob = c$, where c is some member of C.

(*b*) If a, b, c are members of C, then $(aob)oc = ao(boc)$; that is, combining c with the result of combining b with a yields the same as combining with a the result of combining c with b; that is, the rule of combination is associative.

(*c*) The class C contains a member i (called the identical member or element) such that if a be a member of C, then $aoi = ioa = a$; that is, C has a member such that, if it be combined with any given member, or that member with it, the result is the given member.

(*d*) If a be a member of C, then there is a member a' (called the reciprocal of a) such that $aoa' = a'oa = i$; that is, each member of C is matched by a member such that combining the two gives the identical member.

Other definitions of the term "group" have been proposed and are sometimes used. The definitions are not all of them equivalent but they all agree that to be a group a system must satisfy condition (*a*).

Systems satisfying condition (*a*) are many of them on that account so important that in the older literature of the subject they are called groups, or closed systems, and are now said to have "the group property," even if they do not satisfy conditions (*b*), (*c*) and (*d*). The propriety of the term "closed system" is evident in the fact that a system satisfying (*a*) is such that the result of combining any two of its members is itself a member—a thing *in* the system, not out of it.

Various Simple Examples of Groups and of Systems that Are Not Groups.—You observe that by the foregoing definition of group every group is a system; groups, as we shall see, are infinitely numerous; yet it is true that relatively few systems are groups or have even the group property—so few relatively that, if you select a system at random, it is highly probable you will thus hit upon one that is neither a group nor has the group property.

Take, for example, the system S_1 whose class C is the class of integers from 1 to 10 inclusive and whose rule of combination is that of ordinary multiplication $\times$; $3 \times 4 = 12$; 12 is not a member of C, and so S_1 is not closed—it has not the group property.

Let S_2 have for its C the class of all the ordinary integers, 1, 2, 3, . . . *ad infinitum*, and let o be $\times$ as before; as the product of any two integers is an integer, (a) is satisfied—S_2 is closed, has the group property; (b), too, is evidently satisfied, and so is (c), the identity element being 1 for, if n be any integer, $n \times 1 = 1 \times n = n$; but ($d$) is not satisfied—none of the integers (except 1) composing C has a reciprocal in C—there is, for example, no integer n such that $2 \times n = n \times 2 = 1$; and so S_2, though it has the group property, is not a group.

Let S_3 be the system consisting of the class C of all the positive and negative integers including zero and of addition as the rule of combination; you readily see that S_3 is a group, zero being the identical element, and each element having its own negative for reciprocal.

A group is said to be finite or infinite according as its C is a finite or an infinite class and it is said to be Abelian or non-Abelian according as its rule of combination is or is not commutative—according, that is, as we have or do not have $aob = boa$, where a and b are arbitrary members of C. You observe that the group S_3 is both infinite and Abelian.

For an example of a group that is finite and Abelian it is sufficient to take the system S_4 whose C is composed of the four numbers, 1, -1, i, $-i$, where i is $\sqrt{-1}$, and whose rule of combination is multiplication; you notice that the identical element is 1, that 1 and -1 are each its own reciprocal and that i and $-i$ are each the other's reciprocal.

Let S_5 have the same C as S_3 and suppose o to be subtraction instead of addition; show that S_5 has the group property but is not a group. Show the like for S_6 in which C is the same as before and o denotes multiplication. Show that S_7 where C is the same as before and o means the rule of division, has not even the group property.

Consider S_8 where C is the class of all the rational numbers (that is, all the integers and all the fractions whose terms are integers, it being understood that zero can not be a denominator) and where o denotes $+$; you will readily find that S_8 is a group, infinite and Abelian. Examine the systems obtained by keeping the same C and letting o denote subtraction, then multiplication, then division. Devise a group system where o means division.

If S and S' be two groups having the same rule of combination and if the class C of S be a proper part of the class C' of S' (*i.e.*, if the members of C are members of C' but some members of C' are not in C), then S is said to be a sub-group of S'. Observe that S_3 is a sub-group of S_8.

Show that S_9 is a group if its C is the class of all real numbers and its o is $+$; note that S_8 is a sub-group of S_9 and hence that S_3 is a sub-group of a sub-group of a group. Is S_9 itself a sub-group? If so, of what group or groups? Examine the systems derived from S_9 by altering the rule of combination.

The most difficult thing that teaching has to do is to give a worthy sense of the meaning and scope of a great idea. A great idea is always generic and abstract but it has its living significance in the particular and concrete—in a countless multitude of differing instances or examples of it; each of these sheds only a feeble light upon the idea, leaving the infinite range of its significance in the dark; whence the necessity of examining and comparing a large number of widely differing examples in the hope that many little lights may constitute by union something like a worthy illustration; but to present these numerous examples requires an amount of time and a degree of patience that are seldom at one's disposal, and so it is necessary to be content with a selected few. And now here is the difficulty: if the examples selected be complex and difficult, they repel; if they be simple and easy, they are not impressive; in either case, the significance of the general concept in question remains ungrasped and unappreciated. I am going, however, to take the risk—to the foregoing illustrations of the group concept I am going to add a few further ones,—some of them very simple, some of them more complex,—trusting that the former may not seem to you too trivial nor the latter too hard.

Every one has seen the pretty phenomenon of a grey squirrel rapidly rotating a cylindrical wire cage enclosing it. It may rotate the cage in either of two opposite ways, senses or directions. Let us think of rotation in only one of the ways, and let us call any rotation, whether it be much or little, a *turn*. Each turn carries a point of the cage along a circle-arc of some length, short or long. Denote by R the special turn (through $360°$) that brings each point of the cage back to its starting place. Let S_{10} be the system whose C is the class of all possible turns and whose o is addition of turns so that aob shall be the whole turn got by following turn a by turn b. You see at once that S has the group property for the sum of any two turns is a turn; it is equally evident that the associative law— condition (b)—is satisfied. Note that R is equivalent to no turn,—equivalent to rest,—equivalent to a zero turn, if you please; note that, if a be a turn greater than R and less than $2R$, then a is equivalent to a's excess over R; that, if a be greater than $2R$ and less than $3R$, then a is equivalent to a's excess over $2R$; and so on; thus any turn greater than R and not equal to a multiple of R is equivalent to a turn less than R; let us regard any turn that is thus greater than R as identical with its equivalent less than R; we have, then, to consider no turns except R and those less than R—of which there are infinitely many; you see immediately that, if a be any turn, $aoR = Roa = a$, which means that condition (c) is satisfied with R for identical element. Next notice that for any turn a there is a turn a' such that $aoa' = a'oa = R$. Hence S_{10} is, as you see, a group. Show it to be Abelian. You will find it instructive to examine the system derived from S_{10} by letting C be the class of all turns (forward or backward).

Perhaps, you will consider the system suggested by the familiar spectacle of a ladybug or a measuring-worm moving round the rim or edge of a circular tub; or the system suggested by motions along the thread of an endless screw; or that suggested by the turns of the earth upon its axis; or that suggested by the motions of a planet in its orbit.

Do such examples give the meaning of the group concept? Each one gives it somewhat as a water-drop gives the meaning of ocean, or a burning match the meaning of the sun, or a pebble the meaning of the Rocky Mountains. Are they, therefore, to be despised? Far from it. Taken singly, they tell you little; but taken together, if you allow your imagination to play upon them, noting their differences, their similitudes, and the variety of fields they represent, they tell you much. Let us pursue them further, having a look in other fields.

Consider the field of the data of sense,—a field of universal interest,—and fundamental. We are here in the domain of sights and sounds and motions among other things. Are there any groups to be found here? Who, except the blind-born, are not lovers of color? Do the colors constitute a group? I mean sensations of color,—color sensations,—including all shades thereof and white and black. Denote by S_{11} the system whose C is the class of all such sensations and whose rule of combination is, let us say, the *mixing* of such sensations. But what are we to understand by the mixing of two color sensations? Suppose we have two small boxes of powder,—say of finely pulverized chalk,—a box of, say, red powder and a box of blue; one of the powders gives us the sensation red, the other the sensation blue; let us thoroughly mix the powders; the mixture gives us a color sensation; we agree to say that we have mixed the sensations and that the new sensation is the result of mixing the old ones. As the combination of any two color sensations is a color sensation, S_{11} has, you see, the group property. Is it a group? Evidently condition (b) is satisfied. Are conditions (c) and (d) also satisfied?

Let us pass from colors to figures or shapes,—to figures or shapes, I mean, of physical or material objects,—rocks, chairs, trees, animals and the like,—as known to sense-perception. No doubt what we ordinarily call perception of an object's figure or shape is genetically complex, a result of experience contributed to by two or more senses, as sight, touch, motion; let us not, however, try to analyze it thus; let us take it at its face value—let us regard it as being, what it appears to be before analytic reflection upon it, a sense-given datum; and let us confine ourselves to the sense of sight. Here is a dog; its ears have shape; so, too, its eyes, its nose and the other features of its head; these shapes combine to make the shape or figure of the head; each other one of its visible organs has a shape of its own; these shapes all of them combine to make that thing which we call the shape or figure of the dog. Yonder is a table; it has a shape, and this is due to some sort of combination of the shapes of its parts—legs, top, and so on; upon it are several objects—a picture frame, a candlestick, some vases; each has a shape; the table and the other things together constitute one object—disclosed as such to a single glance of the eye; this object has a figure or shape due to the combined presence of the other shapes. In speaking of the dog and the table, I have been using the word "combination" in a very general sense. Can it, in this connection, be made precise enough for our use? Is it possible to find or frame a rule by which, any two visible shapes being given, these can be combined? If so, is the result of the combination a visible shape? If so, the system consisting of the rule and the total class of shapes has the group property. Does the system satisfy the remaining three conditions for a group?

And what of sounds—sensations of sound? Are sounds combinable? Is the result always a sound or is it sometimes silence? If we agree to regard silence as a species of sound,—as the zero of sound,—has the system of sounds the property of a group? There is the question of thresholds: sound is a vibrational phenomenon; if the rate of vibration be too slow or too great,—say, 100,000 per second,—no sound is heard. If you disregard the thresholds, has the sound system the group property? Is it a group? If so, what is the identical element? And what would you say is the reciprocal of a given sound or tone?

Consider other vibrational phenomena—as those of light or electricity. Can you so conceive them as to get group systems? Sharpen your questions and then carry them to physicists. You need have no hesitance—the service is apt to be mutual.

The Infinite Abelian Group of Angel Flights.—We are accustomed to think of ourselves as being in a boundless universe of space filled with what we call points any two of which are joined by what we call a straight line. Imagine one of those curious creatures which are to-day for most of us hardly more than figures of speech but which for many hundreds of years were very real and very lovely or very terrible things for millions of men, women and children and were studied and discoursed about seriously by men of genius: I mean angels. Angels can fly. Let us confine their flights to straight lines but impose no other restrictions. I am going to ask you to understand a flight as having nothing but length, direction and sense; if it is parallel to a given straight line, it has that line's direction; if it goes from A towards B, it has that sense; if from B toward A, the opposite sense; a flight from A to B and one from C to D are the same if they agree in length, direction and sense. Consider a flight a from point A to point B followed by a flight b from B to C; you readily see that a and b are two adjacent sides of a parallelogram, one of whose diagonals is the direct flight d from A to C; d is called the resultant or flight-sum of a and b because d tells us how far the angel has finally got from the starting place; and so we write $aob = d$. If flight b' goes from P to Q but agrees with b in length, direction and sense, we write $aob' = d$

as before for, as already said, b and b' are one and the same. Now let S_{12} denote the system whose C is the class of all possible angel flights including rest, or zero flight, and whose rule of combination is flight summation as above explained; you see at once that S_{12} is a closed system, has the group property, for the combination of any two flights is a flight; if a, b and c be three flights, we may suppose them to go respectively from A to B, from B to C, and from C to D; consider $(aob)oc$; $aob = d$, the flight from A to C; $doc = e$, the flight from A to D; now consider $ao(boc)$; $boc = d'$, flight from B to D; $aod' = e'$, flight from A to D; so $e = e'$ and $(aob)oc = ao(boc)$; hence summation of flights is associative —condition (b) is satisfied. Condition (c) is satisfied with zero (0) flight for i; for, if a be any flight, it is plain that $ao0 = 0oa = a$. And condition (d) is satisfied for it is evident that $aoa' = a'oa = 0$ where a' and a agree in length and direction but are opposite in sense. Hence the system of angel flights is a group. And it is easy to see that it is both infinite and Abelian.

What I have here called an angel flight is known in mathematics and in physics as a *vector*; a vector has no position—it has its essential and complete being in having a length, a direction and a sense. And so, you see, the system composed of the vectors of space and of vector addition as a rule of combination is an infinite Abelian group.

Connection of Groups with Transformations and Invariants.—Let us have another look at our angel flights, or vectors. I am going to ask you to view them in another light. Let V be any given vector—that is, a vector of given length, sense and direction; where does it begin and where does it end? A moment's reflection will show you that every point in the universe of space is the beginning of a vector identical with V and the end of a vector identical with V. Though these vectors are but one, it is convenient to speak of them as many equal vectors—having the same length, direction and sense. Let the point P be the beginning of a V and let the point Q be its end. Let us now associate every such P with its $Q(P \rightarrow Q)$; we have thus transformed our space of points into itself in such wise that the *end* of each V is the *transform* of its *beginning*; call the transformation T; let us follow it with a transformation T' converting the beginnings of all vectors equal to a given vector V' into their corresponding ends. What is the result? Notice that T converted P into Q and that T' then converted Q into Q', the end of the V' beginning at Q; now there is a vector beginning at P and running direct to Q'; and so there is a transformation T'' converting P into Q'; it is this T'' that we shall mean by ToT'. Without further talk, you see that our group of angel flights, or vectors, now appears as an infinite Abelian group of *transformations* (of our space of points into itself). Such transformations do not involve motion in fact; it is customary, however, for mathematicians to call them motions, or *translations*, of space; T, for example, being thought of and spoken of as a translation of the whole of space (as a rigid body) in the direction and sense of V and through a distance equal to V's length. In accordance with this stimulating, albeit purely figurative, way of speaking, the group in question is the group of the translations of our space.

We are now in a good position to glimpse the very intimate connection between the idea of group and the idea of invariance. Suppose we are given a group of transformations; one of the big questions to be asked regarding it is this: what things remain unaltered,—remain invariant,— under each and all the transformations of the group? In other words, what property or properties are common to the objects transformed and their transforms? Well, we have now before us a certain group of space transformations—the group of translations. Denote it by G. Each translation in G converts (transforms, carries, moves) any point into a point, and so converts any configuration F of points,—any geometric figure,—into some configuration F'. What remains unchanged? What are the invariants? It is obvious that one of the invariants,—a very important one,—is distance; that is, if P and Q be any two points and if their transforms under some translation be respectively P' and Q', then the distance between P' and Q' is the same as that between P and Q; another is order among points—if Q is between P and R, Q' is between P' and R'; you see at once that angles, areas, volumes, shapes are all of them unchanged: in a word, congruence is invariant—if a translation convert a figure F into a figure F', F and F' are congruent. Of course congruence is invariant under all the translations having a given direction. Do these constitute a group? Obviously they do, and this group G' is a sub-group of G. Congruence is invariant under G'; it is also invariant under G; G' is included in G; it is natural, then, to ask whether G itself may not be included in a still larger group having congruence for an invariant. The question suggests the inverse of the one with which we set out. The former question was: given a group, what are its invariants? The inverse question is: given an invariant, what are its groups and especially its largest group? This question is as big as the other one. Consider the example in hand. The given invariant is congruence. Is G,—the group of translations,—the largest group of space transformations leaving congruence unchanged? Evidently not; for think of those space transformations that consist in rotations of space (as a rigid body) about a fixed line (as axis); if such a rotation converts a figure F into F', the two figures are congruent. It is clear that the same is true if a transformation be a *twist*—that is, a simultaneous rotation about, and translation along, a fixed line. All such rotations and twists together with the translations constitute a group called the group of *displacements* of space; it includes all transformations leaving congruence invariant. This group, as a little reflection will show you, has many sub-groups, infinitely many;

the set of displacements leaving a specified point invariant is such a sub-group; the set leaving two given points unchanged is another. How is the latter related to the sub-group leaving only one of the two points invariant? Is there a displacement leaving three non-collinear points invariant? Do the displacements leaving a line unchanged constitute a group? Such questions are but samples of many that you will find it profitable to ask and to try to answer.

For the sake of emphasis, permit me to repeat the two big questions: (1) Given a group of transformations, what things are unchanged by them? (2) Given something—an object or property or relation, no matter what—that is to remain invariant, what are the groups of transformations, and especially the largest group, that leave the thing unaltered? You may wish to say: I grant that the questions are interesting, and I do not deny that they are big—big in the sense of giving rise to innumerable problems and big in the sense that many of the problems are difficult; but I do not see that they are big with importance. Why should I bother with them? In reply I shall not undertake to say why you *should* bother with them; it is sufficient to remind you that as human beings you cannot help it and you do not desire to do so. In the preceding lecture, we saw that the sovereign impulse of Man is to find the answer to the question: what abides? We saw that Thought,—taken in the widest sense to embrace art, philosophy, religion, science, taken in *their* widest sense,—is the quest of invariance in a fluctuant world. We saw that the craving and search for things eternal is the central binding thread of human history. We saw that the passion for abiding reality is itself the supreme invariant in the life of reason. And we saw that the bearings of the mathematical theory of invariance upon the universal enterprise of Thought are the bearings of a prototype and guide. It is evident that the same is true of the mathematical theory of groups. Our human question is: what abides? As students of thought and the history of thought, we have learned at length that the question can not be answered fully at once but only step by step in an endless progression. And now what are the steps? You can scarcely fail to see, if you reflect a little, that each of them,—whether taken by art or by science or by philosophy,—consists virtually in ascertaining either the invariants under some group of transformations or else the groups of transformations that leave some thing or things unchanged.

Groups as Instruments for Defining, Delimiting, Discriminating and Classifying Doctrines.—The foregoing question (2) has another aspect, which I believe to be of profound interest to all students except those, if there be such, who are insensate to things philosophical. I mean that, if and whenever you ascertain the group of all the transformations that leave invariant some specified object or objects of thought, you thereby define perfectly some actual (or potential) branch of science—some actual (or potential) doctrine. I will endeavor to make this fact evident by a few simple examples, and I will choose them from the general field of geometry, though, as you will perceive, such examples might be taken from other fields.

For a first example, consider the above-mentioned group D of the displacements of our space. I say that this group defines a geometry of space, which may be called the geometry of displacements. It defines it by defining, or delimiting, its subject-matter. What is its subject-matter? What does the geometry study? The two questions are not equivalent. It studies all the figures in space but it does not study all their properties. Its subject-matter consists of those properties which it does study. What are these? They are those and only those properties (of figures) that remain invariant under all displacements but under no other transformations of space. The geometry of displacements might be called congruence geometry. It includes the greater part of the ordinary geometry of high school but not all of it, for the latter deals, for example, with similarity of figures; similarity is indeed invariant under displacements, but it is also invariant under other transformations—the so-called similitude transformations, to be mentioned presently.

For a second example, consider the following. I may wish to confine my study of spatial figures to their *shape*. The doctrine thus arising may be called the geometry of shape, or shape geometry. If I tell you that I am studying shape geometry and you ask me what I mean by the geometry of shape, there are two ways in which I may answer your question. One of the ways requires me to define the term shape—shape of a geometric figure; the other way,—the group way,—does not. Let us examine them a little. I have never seen a mathematical definition of shape, but it may, I believe, be precisely defined as follows. We must distinguish the three things: sameness of shape; shape of a given figure; and shape of a figure. I will define the first; then the second in terms of the first; and, finally, the third in terms of the second. Two figures F and F' will be said to have the same shape if and only if it is possible to set up a one-to-one correspondence between the points of F and those of F', such that, AB and CD being any distances between points of F, and $A'B'$ and $C'D'$ being the distances between the corresponding points of F', $AB/CD = A'B'/C'D'$. Two figures having the same shape will be said to be similar, and conversely. Having defined sameness of shape, or similarity, of figures, I will define the term "shape of a given figure" as follows: if F be a given, or specific, figure, the shape of F is the class σ of all figures similar to F; it is evident that, if F and F' are not similar, the class σ and the class σ'—the shape of F'—have no figures in common; it is evident, moreover, that there are as many σ's as there are figure shapes. And now what do we mean by the general term shape, or—what is tantamount—shape of a figure? What the answer must be is pretty obvious: shape is the class Σ of all the σ's. Note that Σ is a class of classes and that any σ is a class of (similar) figures. Having defined the general term shape, I have, you see, virtually answered your question: what is the geometry of shape?

Let us now see how the question may be answered by means of the group concept. Two congruent figures are clearly similar, and so similarity is invariant under the group of displacements. But you readily see that there are many other transformations under which similarity is invariant. Let O be a point; consider the bundle of straight lines,—all the lines O,—having O for its vertex; every point of space is on some line of the bundle; let k be any real number (except zero); let P be any point and let P' be such a point on the line OP that the segment $OP' = k \times$ segment OP; you see that each point P is transformed into a point P'; the transformation is called *homothetic*; its effect, if k be positive and exceed 1, is a uniform expansion of space from O outward in all directions; if k be positive and less than 1, the effect is contraction toward O; if k be negative, the effect is such an expansion or contraction, followed or preceded by reflection in O as in a mirror; distances are clearly not preserved, but distance ratios are; that is, if A, B, C be any three points and if their respective transforms under some homothetic transformation be A', B', C', it is evident that $AB/BC = A'B'/B'C'$; accordingly, if F' be the transform of a figure F, the figures are similar,— they have the same shape but not the same size,—they are not congruent: similarity is, then, invariant under all homothetic transformations, and hence under combinations of them with one another and with displacements; the displacements and the homothetic transformations together with all such combinations constitute a group called the group of *similitude transformations*; it contains all and only such space transformations as leave similarity unchanged. Here, then, is our group definition of shape geometry: namely, *the geometry of shape is the study of that property of figures which is common to every figure and its transforms under each and all transformations of the similitude group.* Observe that this definition, unlike the former one, employs neither the notion of shape in general, nor that of the shape of a given figure; it employs only the notion of similarity—sameness of shape.

We ought, I think, to consider one more example of how a group of transformations serves to determine the nature and limits of a doctrine and hereby to discriminate the doctrine from all others. I will again take a geometric example, but for the sake of simplicity I will choose it from the geometry of the plane (instead of space). Before presenting it, let me adduce a yet simpler example of the same kind taken from the geometry of points in a straight line. In a previous lecture I explained what is meant by a projective line—an ordinary line endowed with an "ideal" point, or point at infinity, where the line meets all lines parallel to it. Let L

be a projective line. In the preceding lecture, we gained some acquaintance with the transformations of the form

$$(1)\ x' = \frac{ax + b}{cx + d}$$

where the coefficients are any real numbers such that $ad - bc \neq 0$; we saw that there are ∞^3 such transformations and that each of them converts the points of L into the points of L in such a way that the anharmonic ratio of any four points is equal to the anharmonic ratio of their transforms. Distances are not preserved; neither are the ordinary ratios of distances preserved; hence neither congruence nor similarity is invariant; no relation among points—that is, no property of figures (for here a figure is simply a set of points on L)—is invariant except anharmonic ratio and properties expressible in terms of the latter; no other transformations leave these properties invariant. By a little finger work you can prove in a formal way that these transformations constitute a group. I will merely indicate the procedure, leaving it to you to carry it out if you desire to do so. The transformations differ only in their coefficients. Let (a_1, b_1, c_1, d_1), (a_2, b_2, c_2, d_2), (a_3, b_3, c_3, d_3) be any three of the transformations; consider the first and second; the rule o of combination is to be: operate with the first and then on the result with the second. The first converts point x into point x':

$$(2)\ x' = \frac{a_1 x + b_1}{c_1 x + d_1};$$

the second converts x' into point x'':

$$(3)\ x'' = \frac{a_2 x' + b_2}{c_2 x' + d_2};$$

in (3) replace x' by its value given by (2), simplify and then notice that you have a transformation of form (1) converting x directly into x''. This shows that the set of transformations have the group property. To show that they obey the associative law, it is sufficient to perform the operations

$$(4)\ (a_1, b_1, c_1, d_1) \circ [(a_2, b_2, c_2, d_2) \circ (a_3, b_3, c_3, d_3)],$$
$$(5)\ [(a_1, b_1, c_1, d_1) \circ (a_2, b_2, c_2, d_2)] \circ (a_3, b_3, c_3, d_3),$$

and then to observe that the results are the same. The identical element i is $(a, 0, 0, a)$—that is, the transformation, $x' = x$. The inverse of any transformation (a, b, c, d) is $(-d, b, c, -a)$ for you can readily show that combination of these gives $(a, 0, 0, a)$.

The fact to be specially noted is that this group of so-called homographic transformations defines a certain kind of geometry in the line L—

namely, its projective geometry. In a line there are various geometries; among these the projective geometry is characterized by its subject-matter, and its subject-matter consists of such properties of point sets, or figures, as remain invariant under its homographic group.

And now I come to the example alluded to a moment ago—the one to be taken from geometries in (or of) a plane. The foregoing homographic group—in a line, a one-dimensional space—has an analogue in a projective plane, another in ordinary 3-dimensional projective space, another in a projective space of four dimensions, and so on *ad infinitum*. What is the analogous group for a plane? In a chosen plane take a pair of axes and consider the pair of equations

$$(1) \begin{cases} x' = \dfrac{Ax + By + C}{Gx + Hy + I}, \\ y' = \dfrac{Dx + Ey + F}{Gx + Hy + I}, \end{cases}$$

where the coefficients are any real numbers such that

$$(1') \quad \begin{vmatrix} A & B & C \\ D & E & F \\ G & H & I \end{vmatrix} \neq 0;$$

i.e., such that

$$(1') \quad AEI - CEG + CDH - BDI + BFG - AFH \neq 0.$$

The coefficients furnish eight independent ratios,—called "parameters,"— and so we have ∞^8 equation pairs of form (1); choose any one of them and notice that it is a transformation converting the number pair (x, y) into a number pair (x', y'), and so converting the point (x, y) into a point (x', y'); owing to the inequality (1'), any point (x, y) is transformed into a definite point (x', y'). In any line $ax' + by' + c = 0$, replace x' and y' by their values given by (1), and simplify; the resulting equation is of first degree in x and y and hence represents a line; hence, you see, points of a straight line are converted into points of a straight line—the relation, collinearity, is preserved; so is copunctality—a pencil of lines has a pencil for its transform; you can readily show that order is not preserved, nor distances nor ordinary distance-ratios, nor angles; hence, if the figure F' be the transform of F, the two figures are, in general, neither congruent nor similar; we say, however, that F and F' are projective because, as can be proved, the anharmonic ratio of any 4 points (or lines) of either is equal to that of the corresponding (transform) points (or lines) of the other. By the method indicated for the homographic transformations of a line, you can prove that the ∞^8 transformations of form (1) constitute a group.

Just as a *point* of the plane has two coordinates (x, y), so a *line* depends on two coordinates; there are various ways to see that such is the case; an easy way is this: the line, $ax + by + c = 0$, depends solely upon the ratios $(a : b : c)$ of the coefficients; these three ratios are not independent—two of them determine the third one; you thus see that the line depends upon only two independent variables—it has, like the point, two coordinates; let us denote them by (u, v). Now consider the transformations

$$(2) \begin{cases} u' = \dfrac{Ju + Kv + \cdot L}{Pu + Qv + R}, \\ v' = \dfrac{Mu + Nv + O}{Pu + Qv + R}, \end{cases}$$

where the coefficients are subject to a relation like (1'). We saw that a transformation (1) converts points into points directly and lines into lines indirectly; just so, a transformation (2) converts lines into lines directly and points into points indirectly; hence the group of line-to-line transformations (2) is essentially the same as the foregoing group of point-to-point transformations (1). This latter group is called the group of *collineations* of (or in) the plane.

I am going now to ask you to notice an ensemble of transformations (of the plane) that are neither point-to-point nor line-to-line transformations but are at once point-to-line and line-to-point transformations. These are represented by the pair of formulas

$$(3) \begin{cases} u = \dfrac{ax + by + c}{gx + hy + i}, \\ v = \dfrac{dx + ey + f}{gx + hy + i}, \end{cases}$$

where the coefficients are again subject to a relation like (1'). Any such transformation converts a point (x, y) into a line (u, v); now operate on the points of this line by the same transformation or another one of form (3); the points are converted into lines constituting a pencil having a vertex, say (x', y'); thus the combination converts point (x, y) into point (x', y')—it is a point-to-point transformation and hence belongs to the group of collineations; you thus see that the set of transformations (3) is not a group; but this set and the collineations together constitute a group including the collineations as a subgroup. This large group is called the *Group of Projective Transformations of the Plane*. Why? Because every transformation in it and no other transformation leaves all anharmonic ratios unchanged.

What is the projective geometry of the plane? The group now in hand enables us to answer the question perfectly. The answer is: Projective plane geometry is that geometry which studies such and only such properties of plane figures as remain invariant under the group of projective transformations.

In reading the essays of the late Henri Poincaré you have met the statement: "Euclidean space is simply a group." The foregoing examples should enable you to understand its meaning. And they should lead you to surmise —what is true—that answers like the foregoing ones are available for similar questions regarding all geometries of a space of any number of dimensions and—what is more—regarding mathematical doctrines in general. Whatsoever things are invariant under all and only the transformations of some group constitute the peculiar subject-matter of some (actual or potential) branch of knowledge. And you see that every such group-defined science views its subject matter under the aspect of eternity.

A Question for Psychologists.—Before closing this lecture, I desire to speak briefly of two additional matters connected with the notion of group: one of the matters is psychological: the other is historical. Being students of philosophy, you are obliged to have at least a good secondary interest in psychology. I wish to propose for your future consideration a psychological question—one which psychologists (I believe) have not considered and which, though it has haunted me a good deal from time to time in recent years, I am not yet prepared to answer confidently. The question is: Is mind a group? Let us restrict the question and ask: Is mind a closed system—that is, has it the group property? Some of the difficulties are immediately obvious. In order that the question shall have definite meaning, it is necessary to think of mind as a system composed of a class of things and a rule, or law, of combination by which each of the things can be combined with itself and with each of the other things. We may make a beginning by saying that the required class is the class of mental phenomena. But what does the class include? What phenomena are members of it? Some phenomena,—feeling, for example, seeing, hearing, tasting, thinking, believing, doubting, craving, hoping, expecting,—are undoubtedly mental; others seem not to be—as what I see, for example, what I taste, what I believe, and the like. Here are difficulties. You will find a fresh and suggestive treatment of them in Bertrand Russell's *The Ultimate Constituents of Matter*, found in the author's *Mysticism and Logic* and especially in his *Analysis of Mind*. Let us suppose that, despite the difficulties in the way, you have decided what you are going to call mental phenomena. You have then to consider the question of their combination. We do habitually speak of combining mental things: hoping, for example, is, in some sense, a union, or combination, of desiring and expecting; the feeling called patriotism is evidently a combination of a pretty large variety of feelings; in the realm of ideas,—which you will probably desire to include among mental phenomena,—we have seen that, for example, the idea named "vector" is a union, or combination, of the ideas of direction, sense and length; and so on—examples abound. But does combination as a process or operation have the same meaning in all such cases? It seems not. What, then, is your rule o to be? Possibly the difficulty could be surmounted as follows: if you discovered that some mental phenomena are combinable by a rule o_1, others by a rule o_2, still others by a rule o_3, and so on, thus getting a finite number of particular rules, you could then take for a more general rule o the disjunction, or so-called logical sum, of the particular ones; that is, you could say that rule o is to be: o_1 or o_2 or o_3 or . . . or o_n; so that two phenomena would be combinable by o if they were combinable by one or more of the rules $o_1, o_2, . . ., o_n$. If you thus found a rule by which every two of the phenomena you had decided to call *mental* admitted of combination, then your final question would be: is the result of every such combination a mental phenomenon? That is not quite the question; for under the rule two phenomena might be combinable in two or more ways, and some of the results might (conceivably) be mental and the others not; so your question would be: can every two mental phenomena be combined under your rule so as to yield a mental phenomenon? If so, then mind, as you had defined it, would have the group property under some rule of combination. If you found mind to have the group property under some rule or rules but not under others, you would be at once confronted with a further problem, which I will not tarry to state.

We have been speaking of mind—of mind in general. Similar questions, —perhaps easier if not more fruitful questions,—can be put respecting particular minds—your mind, mine, John Smith's. Has every individual mind the group property? Has no such mind the property? Have some of them the property and others not?

It seems very probable that the answer to the first of the questions must be negative. There are at all events some minds having (presenting, containing) mental phenomena that are definitely combinable in a way to yield mental phenomena that nevertheless do not belong to those minds. What is meant is this: a given mind may possess certain ideas which are combinable so as to form another idea; it may happen that the mind in question is incapable of grasping the new idea. Such minds have no doubt come under the observation of every experienced teacher. I myself have seen many such cases and remember one of them very vividly: that of a young woman who had made a brilliant record in undergraduate collegiate mathematics including the elements of analytical geometry and calculus; and who, encouraged by this initial success, aspired to the mathematical doctorate and entered seriously upon higher studies essential thereto; it was necessary for her to grasp more and more complicated concepts formed by combining ideas she already possessed; after no long time she reached the limit of her ability in this matter,—a fact first noticed by her instructors and then by herself,—and being a woman of good sense, she abandoned the pursuit of higher mathematics. I may add that subsequently she gained the doctorate in history. It may be that some minds are not thus limited. It may be that a genius of the so-called universal type,—an Aristotle, for example, or a Leibniz or a Leonardo da Vinci,—is one whose mind has the group property. May I leave the questions for your consideration in the days to come?

The Group Concept Dimly Adumbrated in Early Philosophic Speculation.—The mathematical theory of groups is immense and manifold; in the main it is a work of the last sixty years; even the germ of it seems not to antedate Ruffini and Lagrange. Why so modern? Why did not the con cept of a closed system,—of a system having the group property,—come to birth many centuries earlier? The elemental constituents of the concept,—the idea of a class of things, the idea of anything being or not being a member of a class, the idea of a rule or law of combination,—all these were as familiar thousands of years ago as they are now. The question is one of a host of similar questions whose answers, if ever they be found, will constitute what in a previous lecture I called the yet unwritten history of the development of intellectual curiosity. Who will write that history? And when?

The fact that the precise formation of the mathematical concept of group is of so recent date is all the more curious because an idea closely resembling that of group has haunted the minds of a long line of thinkers and is found stalking like a ghost in the mist of philosophic speculation from remote antiquity down even to Herbert Spencer. I refer to those worldwide, age-long, philosophic speculations which, because of their peculiar views of the universe, may be fitly called the *Philosophy of the Cosmic Cycle or Cosmic Year.* This philosophy, despite the spell of a certain beauty inherent in it, has lost its vogue. To-day we are accustomed to thinking of the universe as undergoing a beginningless and endless evolution in course of which no aspect or event ever was or ever will be exactly repeated. In sharpest contrast with that conception, the philosophy of the cosmic cycle regards all the changes of which the universe is capable as constituting an immense indeed but finite and closed system of transformations, which follow each other in definite succession, like the spokes of a gigantic revolving wheel, until all possible changes have occurred in the lapse of a long but finite period of time—called a cosmic cycle or cosmic year—whereupon everything is repeated precisely, and so on and on without end. This philosophy, I have said, has lost its vogue; but, if the philosophy be true, it will regain it, for, if true, it belongs to the cosmic cycle and hence will recur. The history of the philosophy of the cosmic year is exceedingly interesting and it would, I believe, be an excellent subject for a doctor's dissertation. The literature is wide-ranging in kind, in place and in time. Let me cite a little of it as showing how closely its central idea resembles the mathematical concept of a cyclic group.

In his *Philosophie der Griechen* (Vol. III, 2nd edition) Zeller, speaking of the speculations of the Stoics, says:

Out of the original substance the separate things are developed according to an inner law. For inasmuch as the first principle, according to its definition, is the creative and formative power, the whole universe must grow out of it with the same necessity as the animal or the plant from the seed. The original fire, according to the Stoics and Heraclitus, first changes to "air" or vapor, then to water; out of this a portion is precipitated as earth, another remains water, a third evaporates as atmospheric air, which again kindles the fire, and out of the changing mixture of these four elements there is formed,—from the earth as center,—the world. . . . Through this separation of the elements there arises the contrast of the active and the passive principle: the soul of the world and its body. . . . But as this contrast came in time, so it is destined to cease; the original substance gradually consumes the matter, which is segregated out of itself as its body, till at the end of this world-period a universal world conflagration brings everything back to the primeval condition. . . . But when everything has thus returned to the original unity, and the great world-year has run out, the formation of a new world begins again, which is so exactly like the former one that in it all things, persons and phenomena, return exactly as before; and in this wise the history of the world and the deity . . . moves in an endless cycle through the same stages.

A similar view of cosmic history is present in the speculations of Empedocles, for whom a cycle consists of four great periods: Predominant Love—a state of complete aggregation; decreasing Love and increasing Hate; predominant Strife—complete separation of the elements; decreasing Strife and increasing Love. At the end of this fourth period, the cycle is complete and is then repeated—the history of the universe being a continuous and endlessly repeated vaudeville performance of a single play.

Something like the foregoing seems to be implicit in the following statement by Aristotle in the Metaphysics:

Every art and every kind of philosophy have probably been invented many times up to the limits of what is possible and been again destroyed.

And in Ecclesiastes (III, 15):

That which hath been is now; and that which is to be hath already been.

Even Herbert Spencer at the close of his *First Principles* speaks as follows:

Thus we are led to the conclusion that the entire process of things, as displayed in the aggregate of the visible universe, is analogous to the entire process of things as displayed in the smallest aggregates. Motion as well as matter being fixed in quantity, it would seem that the change in the distribution of matter which motion effects, coming to a limit in whatever direction it is carried, the indestructible motion necessitates a reverse redistribution. Apparently the universally coexistent forces of attraction and repulsion, which necessitate rhythm in all minor changes throughout the universe, also necessitates rhythm in the totality of changes—alternate eras of evolution and dissolution. And thus there is suggested the conception of a past during which there have been successive evolutions analogous to that which is now going on; and a future during which successive other evolutions may go on—ever the same in principle but never the same in concrete result.

Spencer was, you know, but poorly informed in the history of thought and he was probably not aware that the main idea in the lines just now quoted was ancient two thousand years before he was born. You should note that the Spencerian universe of transformations narrowly escapes being a closed system—escapes by the last six words of the foregoing quotation. The cosmic cycles do indeed follow each other in an endless sequence—"ever the same in principle but never the same in concrete result." The repetitions are such "in principle" only, not in result—there is always something new.

One of the very greatest works of man is the *De Rerum Natura* of Lucretius—immortal exposition of the thought of Epicurus, "who surpassed in intellect the race of man and quenched the light of all as the ethereal sun arisen quenches the stars." Neither a student of philosophy nor a student of natural science can afford to neglect the reading of that book. For, although it contains many,—very, very many,—errors of detail,—some of them astonishing to a modern reader,—yet there are at least four great respects in which it is unsurpassed among the works that have come down from what we humans, in our ignorance of man's real antiquity, have been wont to call the ancient world: it is unsurpassed, I mean, in scientific spirit; in the union of that spirit with literary excellence; in the magnificence of its enterprise; and in its anticipation of concepts among the most fruitful of modern science. For such as can not read it in the original there are, happily, two excellent English translations of it—one by H. A. J. Munro and a later one by Cyril Bailey. Of this work I hope to speak further in a subsequent lecture of this course. My purpose in citing it here is to signalize it as being perhaps the weightiest of all contributions to what I have called the philosophy of the cosmic year. The Lucretian universe though not a finite system, is indeed a closed system, of transformations: any event, whether great or small, that has occurred in course of the beginningless past has occurred infinitely many times and will recur infinitely many times in course of an unending future; and nothing can occur that has not occurred,—there never has been and there never will be aught that is new,—every occurrence is a recurrence. Let me say parenthetically, in passing, that such a concept of the universe is damnably depressing but not more so than the regnant mechanistic hypothesis of modern natural science. In relation to this hypothesis you should by no means fail to read and digest Professor W. B. Smith's great address: "Push or Pull?" (*Monist*, Vol. XXIII, 1913). See also Smith's "Are Motions Emotions?" (*Tulane Graduates' Magazine*, Jan., 1914). And you should read J. S. Haldane's *Life, Mechanism and Personality*.

Should you desire to pursue the matter further either with a view to noting speculative adumbrations of the group concept or, as I hope, with the larger purpose of writing a historical monograph on the philosophy of the Cosmic Cycle, the following references may be of some service as a clue.

"The Dream of Scipio" in Cicero's *Republic* (Hardingham's translation).

Michael Foster's *Physiology*.

Lyell's *Principles of Geology*.

The fourth *Eclogue* of Virgil (verses 31–36).

Rückert's poem *Chidher*.

Moleschott's *Kreislauf des Lebens*.

Clifford's "The First and Last Catastrophe" in his *Lectures and Essays*.

Inge's *The Idea of Progress* (being the Romanes Lecture, 1920).

Les mathématiciens n'étudient pas des objets, mais des relations entre les objets; il leur est donc indifférent de remplacer ces objets par d'autres, pourvu que les relations ne changent pas. La matière ne leur importe pas, la forme seule les intéresse. —HENRI POINCARÉ

2   The Theory of Groups

By SIR ARTHUR STANLEY EDDINGTON

There has been a great deal of speculation in traditional philosophy which might have been avoided if the importance of structure, and the difficulty of getting behind it, had been realised. For example, it is often said that space and time are subjective, but they have objective counterparts; or that phenomena are subjective, but are caused by things in themselves, which must have differences *inter se* corresponding with the differences in the phenomena to which they give rise. Where such hypotheses are made, it is generally supposed that we can know very little about the objective counterparts. In actual fact, however, if the hypotheses as stated were correct, the objective counterparts would form a world having the same structure as the phenomenal world. . . . In short, every proposition having a communicable significance must be true of both worlds or of neither: the only difference must lie in just that essence of individuality which always eludes words and baffles description, but which, for that very reason, is irrelevant to science.

BERTRAND RUSSELL, *Introduction to Mathematical Philosophy*, p. 61.

I

LET us suppose that a thousand years hence archaeologists are digging over the sites of the forgotten civilisation of Great Britain. They have come across the following literary fragment, which somehow escaped destruction when the abolition of libraries was decreed—

'Twas brillig, and the slithy toves
    Did gyre and gimble in the wabe,
All mimsy were the borogoves
    And the mome raths outgrabe.

This is acclaimed as an important addition to the scanty remains of an interesting historical period. But even the experts are not sure what it means. It has been ascertained that the author was an Oxford mathematician; but that does not seem wholly to account for its obscurity. It is certainly descriptive of some kind of activity; but what the actors are, and what kind of actions they are performing, remain an inscrutable mystery. It would therefore seem a plausible suggestion that Mr. Dodgson was expounding a theory of the physical universe.

Support for this explanation might be found in a further fragment of the same poem—

One, two! One, two! and through and through
The vorpal blade went snicker-snack!

"One, two! One, two!" Out of the unknown activities of unknown agents mathematical numbers emerge. The processes of the external world cannot be described in terms of familiar images; whether we describe them by words or by symbols their intrinsic nature remains unknown. But they are the vehicle of a scheme of relationship which can be described by numbers, and so give rise to those numerical measures (pointer-readings) which are the data from which all knowledge of the external universe is inferred.

Our account of the external world (when purged of the inventions of the story teller in consciousness) must necessarily be a "Jabberwocky" of unknowable actors executing unknowable actions. How in these conditions can we arrive at any knowledge at all? We must seek a knowledge which is neither of actors nor of actions, but of which the actors and actions are a vehicle. The knowledge we can acquire is knowledge of a structure or pattern contained in the actions. I think that the artist may partly understand what I mean. (Perhaps that is the explanation of the Jabberwockies that we see hung on the walls of Art exhibitions.) In mathematics we describe such knowledge as knowledge of group structure.

It does not trouble the mathematician that he has to deal with unknown things. At the outset in algebra he handles unknown quantities x and y. His quantities are unknown, but he subjects them to known operations—addition, multiplication, etc. Recalling Bertrand Russell's famous definition, the mathematician never knows what he is talking about, nor whether what he is saying is true; but, we are tempted to add, at least he does know what he is doing. The last limitation would almost seem to disqualify him for treating a universe which is the theatre of unknowable actions and operations. We need a super-mathematics in which the operations are as unknown as the quantities they operate on, and a super-mathematician who does not know what he is doing when he performs these operations. Such a super-mathematics is the Theory of Groups.

The Theory of Groups is usually associated with the strictest logical treatment. I doubt whether anyone hitherto has committed the sacrilege of wrenching it away from a setting of pure mathematical rigour. But it is now becoming urgently necessary that it should be tempered to the understanding of a physicist, for the general conceptions and results are beginning to play a big part in the progress of quantum theory. Various mathematical tools have been tried for digging down to the basis of physics, and at present this tool seems more powerful than any other. So with rough argument and make-shift illustration I am going to profane the temple of rigour.

My aim, however, must be very limited. At the one end we have the phenomena of observation which are somehow conveyed to man's consciousness via the nerves in his body; at the other end we have the basal entities of physics—electrons, protons, waves, etc.—which are believed to be the root of these phenomena. In between we have theoretical physics, now almost wholly mathematical. In so far as physical theory is complete it claims to show that the properties assigned to, and thereby virtually defining, the basal entities are such as to lead inevitably to the laws which we see obeyed in the phenomena accessible to our senses. If further the properties are no more than will suffice for this purpose and are stated in the most non-committal form possible, we may take the converse point of view and say that theoretical physics has analysed the universe of observable phenomena into these basal entities. The working out of this connection is the province of the mathematician, and it is not our business to discuss it here. What I shall try to show is how mathematics first gets a grip on the basal entities whose nature and activities are essentially unknowable. We are to consider where the material for the mathematician comes from, and not to any serious extent how he manipulates the material.

This limitation may unfortunately give to the subject an appearance of triviality. We express mathematically ideas which, so far as we develop them, might just as well have been expressed non-mathematically. But that is the only way to begin. We want to see where the mathematics jumps off. As soon as the mathematics gets into its stride, it leaves the non-technical author and reader panting behind. I shall not be altogether apologetic if the reader begins to pant a little towards the end of the chapter. It is my task to show how a means of progress which begins with trivialities can work up momentum sufficient for it to become the engine of the expert. So in the last glimpse we shall have of it, we see it fast disappearing into the wilds.

II

In describing the behaviour of an atom reference is often made to the jump of an electron from one orbit to another. We have pictured the atom as consisting of a heavy central nucleus together with a number of light and nimble electrons circulating round it like the planets round the sun. In the solar system any change of the orbit of a planet takes place gradually, but in the atom the electron can only change its orbit by a jump. Such jumps from one orbit to an entirely new orbit occur when an atom absorbs or emits a quantum of radiation.

You must not take this picture too literally. The orbits can scarcely refer to an actual motion in space, for it is generally admitted that the ordinary conception of space breaks down in the interior of an atom; nor is there any desire nowadays to stress the suddenness or discontinuity conveyed by the word "jump." It is found also that the electron cannot be localised in the way implied by the picture. In short, the physicist draws up an elaborate plan of the atom and then proceeds critically to erase each detail in turn. What is left is the atom of modern physics!

I want to explain that if the erasure is carefully carried out, our conception of the atom need not become entirely blank. There is not enough left to form a picture; but something is left for the mathematician to work on. In explaining how this happens, I shall take some liberties by way of simplification; but if I can show you the process in a system having some distant resemblance to an actual atom, we may leave it to the mathematician to adapt the method to the more complex conditions of Nature.

For definiteness, let us suppose that there are nine main roads in the atom—nine possible orbits for the electron. Then on any occasion there are nine courses open to the electron; it may jump to any of the other eight orbits, or it may stay where it is. That reminds us of another well-known jumper—the knight in chess. He has eight possible squares to move to, or he may stay where he is. Instead of picturing the atom as containing a particle and nine roads or orbits, why should we not picture it as containing a knight and a chess-board? "You surely do not mean that literally!" Of course not; but neither does the physicist mean the particle and the orbits to be taken literally. If the picture is going to be rubbed out, is it so very important that it should be drawn one way rather than another?

It turns out that my suggestion would not do at all. However metaphorical our usual picture may be, it contains an essential truth about the behaviour of the atom which would not be preserved in the knight-chessboard picture. We have to formulate this characteristic in an abstract or mathematical way, so that when we rub out the false picture we may still have that characteristic—the something which made the orbit picture not so utterly wrong as the knight picture—to hand over to the mathematician. The distinction is this. If the electron makes two orbit jumps in succession it arrives at a state which it could have reached by a single jump; but if a knight makes two moves it arrives at a square which it could not have reached by a single move.

Now let us try to describe this difference in a regular symbolic way. We must find invent a notation for describing the different orbit jumps. The simplest way is to number the orbits from 1 to 9, and to imagine the numbers placed consecutively round a circle so that after 9 we come to 1 again. Then the jump from orbit 2 to orbit 5 will be described as moving on 3 places, and from orbit 7 to orbit 2 as moving on 4 places. We shall call the jump or operation of moving on one place P_1, of moving on two places P_2, and so on. We shall then have nine different operators P, including the stay-as-you-were or identical operator P_0.

We shall use the symbol A to denote the atom in some initial state, which we need not specify. Suppose that it undergoes the jump P_2. Then we shall call the atom in the new state P_2A; that is to say, the atom in the new state is the result of performing the operation P_2 on the system described as A. If the atom makes another jump P_4, the atom in the resulting state will be described as P_4P_2A, since that denotes the result of the operation P_4 on the system described as P_2A. If we do not want to mention the particular jumps, but to describe an atom which has made two jumps from the original state A, we shall call it correspondingly P_bP_aA; a and b stand for two of the numbers 0, 1, 2, . . . 8, but we do not disclose which.

We have seen that two orbit jumps in succession give a state which could have been reached by a single jump. If the state had been reached by a single jump we should have called the atom in that state P_cA, where c is one of the numbers 0, 1, 2, . . . 8. Thus we obtain a characteristic property of orbit jumps, viz. they are such that

$$P_bP_aA = P_cA.$$

Since it does not matter what was the initial state of the atom, and we do not pretend to know more about the atom than that it is the theatre of the operations P, we will divide the equation through by A, leaving

$$P_bP_a = P_c.$$

This division by A may be regarded as the mathematical equivalent of the rubbing out of the picture.

To treat the knight's moves similarly we may first distinguish them as directed approximately towards the points of the compass N.N.E., E.N.E., E.S.E., and so on, and denote them in this order by the operators Q_1, Q_2, . . . Q_8. Q_0 will denote stay-as-you-were. Then since two knight's moves are never equivalent to one knight's move, our result will be [1]

$$Q_bQ_a \neq Q_c \text{ (unless } a, b \text{ or } c = 0).$$

We have to exclude $c = 0$, because two moves might bring the knight back where it was originally.

Let us spend a few moments contemplating this first result of our activities as super-mathematicians. The P's represent activities of an unknown kind occurring in an entity (called an atom) of unknown nature. It is true that we started with a definite picture of the atom with electrons jumping from orbit to orbit and showed that the equation $P_aP_b = P_c$ was true of it. But now we have erased the picture; A has disappeared from

[1] The sign $\neq$ means "is not equal to."

the formula. Without the picture, the operations P which we preserve are of entirely unknown nature. An ordinary mathematician would want to be doing something definite—to multiply, take square roots, differentiate, and so on. He wants a picture with numbers in it so that he can say for example that the electron has jumped to an orbit of double or n times the former radius. But we super-mathematicians have no idea what we are doing to the atom when we put the symbol P before A. We do not know whether we are extending it, or rotating it, or beautifying it. Nevertheless we have been able to express some truth or hypothesis about the activities of the atom by our equation $P_b P_a = P_c$. That our equation is not merely a truism is shown by the fact that when we start with a knight moving on a chess-board and make similar erasures we obtain just the opposite result $Q_b Q_a \neq Q_c$.

It happens that the property expressed by $P_b P_a = P_c$ is the one which has given the name to the Theory of Groups. A set of operators such that the product of any two of them always gives an operator belonging to the set is called a *Group*. Knight's moves do not form a Group. I am not going to lead you into the ramifications of the mathematical analysis of groups and subgroups. It is sufficient to say that what physics ultimately finds in the atom, or indeed in any other entity studied by physical methods, is the *structure of a set of operations*. We can describe a structure without specifying the materials used; thus the operations that compose the structure can remain unknown. Individually each operation might be anything; it is the way they interlock that concerns us. The equation $P_b P_a = P_c$ is an example of a very simple kind of interlocking.

The mode of interlocking of the operations, not their nature, is responsible for those manifestations of the external universe which ultimately reach our senses. According to our present outlook this is the basal principle in the philosophy of science.

I must not mislead you into thinking that physics can derive no more than this one equation out of the atom, or indeed that this is one of the most important equations. But whatever is derived in the actual (highly difficult) study of the atom is knowledge of the same type, i.e., knowledge of the structure of a set of unknown operators.

III

A very useful kind of operator is the *selective operator*. In my school-days a foolish riddle was current—"How do you catch lions in the desert?" Answer: "In the desert you have lots of sand and a few lions; so you take a sieve and sieve out the sand, and the lions remain." I recall it because it describes one of the most usual methods used in quantum theory for obtaining anything that we wish to study.

Let Z denote the zoo, and S_l the operation of sieving out or selecting lions; then $S_l Z = L$, where L denotes lions—or, as we might more formally say, L denotes a pure ensemble having the leonine characteristic. These pure selective operators have a rather curious mathematical property, viz.

$$S_l^2 = S_l \qquad \text{(A)}.$$

For S_l^2 (an abbreviation for $S_l S_l$) indicates that having selected all the lions, you repeat the operation, selecting all the lions from what you have obtained. Putting through the sieve a second time makes no difference; and in fact, repeating it n times, you have $S_l^n = S_l$. The property expressed by equation (A) is called *idempotency*.

Now let S_t be the operation of selecting tigers. We have

$$S_t S_l = 0 \qquad \text{(B)}.$$

For if you have first selected all the lions, and go on to select from these all the tigers, you obtain nothing.

Now suppose that the different kinds of animals in the zoo are numbered in a catalogue from 1 to n and we introduce a selective operator for each; then

$$S_1 + S_2 + S_3 + S_4 + \ldots + S_n = I \qquad \text{(C)},$$

where I is the stay-as-you-were operator. For if you sieve out each constituent in turn and add together the results, you get the mixture you started with.

A set of operators which satisfies (A), (B) and (C) is called a *spectral set*, because it analyses any aggregation into pure constituents in the same way that light is analysed by a prism or grating into the different pure colours which form the spectrum. The three equations respectively secure that the operators of a spectral set are idempotent, non-overlapping and exhaustive.

Let us compare the foregoing method of obtaining lions from the zoo with the method by which "heavy water" is obtained from ordinary water. In the decomposition of water into oxygen and hydrogen by electrolysis, the heavy water for some reason decomposes rather more slowly than the ordinary water. Consequently if we submit a large quantity of water to electrolysis, so that the greater part disappears into gas, the residue contains a comparatively high proportion of heavy water. This process of "fractionating" is a selective operation, but it is not pure selection such as we have been considering. If taking the residue we again perform the operation of electrolysis we shall still further concentrate the heavy water. A fractionating operator F is not idempotent ($F^2 \neq F$), and this distinguishes it from a pure selective operator S.

The idea of analysing things into pure constituents and of distinguishing mixtures from pure ensembles evidently plays an important part in physical conceptions of reality. But it is not very easy to define just what we mean by it. We think of a pure ensemble as consisting of a number of individuals all exactly alike. But the lions at the zoo are not exactly alike; they are only alike from a certain point of view. Are the molecules of heavy water all exactly alike? We cannot speak of their intrinsic nature, because of that we know nothing. It is their relations to, or interactions with, other objects which define their physical properties; and in an interrelated universe no two things can be exactly alike in all their relations. We can only say then that the molecules of heavy water are alike in some common characteristic. But that is not sufficient to secure that they form a pure ensemble; the molecules which form any kind of mixture are alike in one common characteristic, viz. that they are molecules.

If we have a difficulty in defining purity of things for which we have more or less concrete pictures, we find still more difficulty with regard to the more recondite quantities of physics. Nevertheless it is clear that the idea of distinguishing pure constituents from mixtures contains a germ of important truth. It is the duty of the mathematician to save that germ out of the dissolving picture; and he does this by directing attention not to the nature of what we get by the operation but to the nature of the selective operation itself. He shows that those observational effects which reach our perceptions, generally attributed to the fact that we are dealing with an assembly of like individuals, are deducible more directly from the fact that the assembly is obtainable by a kind of operation which, once performed, can be repeated any number of times without making any difference. He thus substitutes a perfectly definite mathematical property of the operator, viz. $S_l^2 = S_l$, for a very vaguely defined property of the result of the operation, viz. a certain kind of likeness of the individuals which together form L. He thus frees his results from various unwarranted hypotheses that may have been introduced in trying to form a picture of this property of L.

In the early days of atomic theory, the atom was defined as an indivisible particle of matter. Nowadays dividing the atom seems to be the main occupation of physicists. The definition contained an essential truth; only it was wrongly expressed. What was really meant was a property typically manifested by indivisible particles but not necessarily confined to indivisible particles. That is the way with all models and pictures and familiar descriptions; they show the property that we are interested in, but they connect it with irrelevant properties which may be erroneous and for which at any rate we have no warrant. You will see that the mathematical method here discussed is much more economical of hypothesis. It says no more about the system than that which it is actually going to embody in the formulae which yield the comparison of theoretical physics with observation. And, in so far as it can surmount the difficulties of investigation, its assertions about the physical universe are the exact systematised equivalent of the observational results on which they are based. I think it may be said that hypotheses in the older sense are banished from those parts of physical science to which the group method has been extended. The modern physicist makes mistakes, but he does not make hypotheses.

One effect of introducing selective operators is that it removes the distinction between operators and operands. In considering the "jump" operators P, we had to introduce an operand A, for them to work on. We must furnish some description of A, and A is then whatever answers to that description. Let S_a be the operation of selecting whatever answers to the description A, and let U be the universe. Then evidently $A = S_a U$; and instead of $P_b P_a A$ we can write $P_b P_a S_a U$. Thus special operands, as distinct from operators, are not required. We have a large variety of operators, some of them selective, and just one operand—the same in every formula—namely the universe.

This mathematical way of describing everything with which we deal emphasizes, perhaps inadvertently, an important physical truth. Usually when we wish to consider a problem about a hydrogen atom, we take a blank sheet of paper and mark in first the proton and then the electron. That is all there is in the problem unless or until we draw something else that we suppose to be present. The atom thus presents itself as a work of creation—a creation which can be stopped at any stage. When we have created our hydrogen atom, we may or may not go on to create a universe for it to be part of. But the real hydrogen atoms that we experiment on are something selected from an always present universe, often selected or segregated experimentally, and in any case selected in our thoughts. And we are learning to recognise that a hydrogen atom would not be what it is, were it not the result of a selective operation performed on that maze of interrelatedness which we call the universe.

In Einstein's theory of relativity the observer is a man who sets out in quest of truth armed with a measuring-rod. In quantum theory he sets out armed with a sieve.

IV

I am now going to introduce a set of operations with which we can accomplish something rather more ambitious. They are performed on a set of four things which I will represent by the letters A, B, C, D. We begin with eight operations; after naming (symbolically) and describing each operation, I give the result of applying it to ABCD:

S_α. Interchange the first and second, also the third and fourth. BADC.

S_β. Interchange the first and third, also the second and fourth. CDAB.

S_γ. Interchange the first and fourth, also the second and third. DCBA.

S_δ. Stay as you were. ABCD.

D_α. Turn the third and fourth upside down. ABƆᗡ

D_β. Turn the second and fourth upside down. AᗺCᗡ.

D_γ. Turn the second and third upside down. AᗺƆD.

D_δ. Stay as you were. ABCD.

We also use an operator denoted by the sign − which means "turn them all upside down."

We can apply two or more of these operations in succession. For example, $S_\alpha S_\beta$ means that, having applied the operation S_β which gives CDAB, we perform on the result the further operation S_α which interchanges the first and second and also the third and fourth. The result is DCBA. This is the same as the result of the single operation S_γ; consequently

$$S_\alpha S_\beta = S_\gamma.$$

Sometimes, but not always, it makes a difference which of the two operations is performed first. For example,

Taking the result of the operation D_γ, viz. AᗺƆD, and performing on it the operation S_α, we obtain ᗺAƆᗡ.

But taking the result of the operation S_α, viz. BADC, and performing on it the operation D_γ, we obtain BⱯꓷC.

Thus the double operation $S_\alpha D_\gamma$ is not the same as $D_\gamma S_\alpha$. There is, however, a simple relation. ᗺAƆᗡ is obtained by inverting each letter in BⱯꓷC, that is to say, by applying the operation which we denote by the sign −. Thus

$$S_\alpha D_\gamma = -D_\gamma S_\alpha.$$

Operators related in this way are said to *anticommute*. On examination we find that S_α, S_β commute, and so do D_α, D_β; so also do S_α and D_α. It is only a combination of an S and a D with different suffixes α, β, γ (but not δ) which exhibits anticommutation.

We can make up sixteen different operators of the form $S_a D_b$, where a and b stand for any of the four suffixes α, β, γ, δ. It is these combined operators which chiefly interest us. I will call them E-operators and denote them by $E_1, E_2, E_3, \ldots E_{16}$. They form a Group, which (as we have seen) means that the result of applying two operations of the Group in succession can equally be obtained by applying a single operation of the Group.

I should, however, mention that the operation − is here regarded as thrown in gratuitously.[2] We may not by a single operation E_c be able to get the letters into the same arrangement as that given by E_bE_a; but if not, we can get the same arrangement with all the letters inverted. This property of the E-operators is accordingly expressed by

$$E_aE_b = \pm E_c.$$

We now pick out *five* of the E-operators. Our selection at first sight seems a strange one, because it has no apparent connection with their constitution out of S- and D-operators. It is as follows—

$$E_1 = S_aD_\delta, \text{ which gives BADC.}$$
$$E_2 = S_\delta D_\beta \qquad " \qquad \text{AᗺCᗡ.}$$
$$E_3 = S_\gamma D_\gamma \qquad " \qquad \text{DᗡᗺA.}$$
$$E_4 = S_a D_\gamma \qquad " \qquad \text{ᗺADᗡ.}$$
$$E_5 = S_\gamma D_\beta \qquad " \qquad \text{ᗡᗺᗺA.}$$

These five are selected because they all anticommute with one another; that is to say, $E_1E_2 = -E_2E_1$, and so on for all the ten pairs. You can verify this by operating with the four letters, though, of course, there are mathematical dodges for verifying it more quickly. We call a set like this a *pentad*.

There are six different ways of choosing our pentad, obtained by ringing the changes on the suffixes a, β, γ. But it is not possible to find more than five E-operators each of which anticommutes with *all* the others. That is why we have to stop at pentads.

Another important property must be noticed. You will see at once that $E_1{}^2 = 1$; for E_1 is the same as S_a, and a repetition of the interchange expressed by S_a restores the original arrangement. But consider $E_5{}^2$. In the operation E_5 we turn the second and fourth letters upside down, and then reverse the order of the letters. The letters left right way up are thereby brought into the second and fourth places, so that in repeating the operation they become turned upside down. Hence the letters come back to their original order, but are all upside down. This is equivalent to the operation −. So that we have $E_5{}^2 = -1$. In this way we find that

$$E_1{}^2 = E_2{}^2 = E_3{}^2 = 1, \; E_4{}^2 = E_5{}^2 = -1.$$

A pentad always consists of three operators whose square is 1, and two operators whose square is −1.

With regard to the symbols 1 and −1, I should explain that 1 here

² To obtain a Group according to the strict definition we should have to take 32 operators, viz. the above 16, and the 16 obtained by prefixing −.

stands for the stay-as-you-were operator. Since that is the effect of the number 1 when it is used as an operator (a multiplier) in arithmetic, the notation is appropriate. (We have also denoted the stay-as-you-were operator by S_δ and D_δ, so that we have $S_\delta = D_\delta = 1$.) Since the operator 1 makes no difference, the operators "$-$" and "-1" are the same; so we sometimes put in a 1, when $-$ by itself would look lonely. Repetition of the operation $-$ restores the original state of things; consequently $(-)^2$ or $(-1)^2$ is equal to 1. Although the symbol, as we have here defined it, has no connection with "minus," it has in this respect the same property as $-$ and -1 in algebra.

I have told you that the proper super-mathematician never knows what he is doing. We, who have been working on a lower plane, know what we have been doing. We have been busy rearranging four letters. But there is a super-mathematician within us who knows nothing about this aspect of what we have been studying. When we announce that we have found a group of sixteen operations, certain pairs of which commute and the remaining pairs anticommute, some of which are square roots of 1 and the others square roots of -1, he begins to sit up and take notice. For he can grasp this kind of structure of a group of operations, not referring to the nature of the operations but to the way they interlock. He is interested in the arrangement of the operators to form six pentads. That is his ideal of knowledge of a set of operations—knowledge of its distinctive kind of structure. A great many other properties of E-operators have been found, which I have not space to examine in detail. There are pairs of triads, such that members of the same triad all anticommute but each commutes with the three members of the opposite triad. There are anti-triads composed of three mutually commuting operators, which become anti-tetrads if we include the stay-as-you-were operator.

All this knowledge of structure can be expressed without specifying the nature of the operations. And it is through recognition of a structure of this kind that we can have knowledge of an external world which from an ordinary standpoint is essentially unknowable.

Some years ago I worked out the structure of this group of operators in connection with Dirac's theory of the electron. I afterwards learned that a great deal of what I had written was to be found in a treatise on Kummer's quartic surface. There happens to be a model of Kummer's quartic surface in my lecture-room, at which I had sometimes glanced with curiosity, wondering what it was all about. The last thing that entered my head was that I had written (somewhat belatedly) a paper on its structure. Perhaps the author of the treatise would have been equally surprised to learn that he was dealing with the behaviour of an electron. But then, you see, we super-mathematicians never do know what we are doing.

V

As the result of a game with four letters we have been able to describe a scheme of structure, which can be detached from the game and given other applications. When thus detached, we find this same structure occurring in the world of physics. One small part of the scheme shows itself in a quite elementary way, as we shall presently see; another part of it was brought to light by Dirac in his theory of the electron; by further search the whole structure is found, each part having its appropriate share in physical phenomena.

When we seek a new application for our symbolic operators E, we cannot foresee what kind of operations they will represent; they have been identified in the game, but they have to be identified afresh in the physical world. Even when we have identified them in the familiar story of consciousness, their ultimate nature remains unknown; for the nature of the activity of the external world is beyond our apprehension. Thus armed with our detached scheme of structure we approach the physical world with an open mind as to how its operations will manifest themselves in our experience.

I shall have to refer to an elementary mathematical result. Consider the square of $(2E_1 + 3E_2)$, that is to say the operation which is equivalent to twice performing the operation $(2E_1 + 3E_2)$. We have not previously mixed numbers with our operation; but no difficulty arises if we understand that in an expression of this kind 2 stands for the operation of *multiplying* by 2, 3 for the operation of multiplying by 3, as in ordinary algebra. We have

$$(2E_1 + 3E_2)(2E_1 + 3E_2) = 4E_1{}^2 + 6E_1E_2 + 6E_2E_1 + 9E_2{}^2.$$

We have had to attend to a point which does not arise in ordinary algebra. In algebra we should have lumped together the two middle terms and have written $12E_1E_2$ instead of $6E_1E_2 + 6E_2E_1$. But we have seen (p. 1568) that the operation E_2 followed by the operation E_1 is not the same as the operation E_1 followed by the operation E_2; in fact we deliberately chose these operators so that $E_2E_1 = -E_1E_2$. Consequently the two middle terms cancel one another and we are left with

$$(2E_1 + 3E_2)^2 = 4E_1{}^2 + 9E_2{}^2.$$

But we have also seen that $E_1{}^2 = 1$, $E_2{}^2 = 1$. Thus

$$(2E_1 + 3E_2)^2 = 4 + 9 = 13.$$

In other words $(2E_1 + 3E_2)$ is the square root of 13, or rather *a* square root of 13.

Suppose that you move to a position 2 yards to the right and 3 yards forward. By the theorem of Pythagoras your resultant displacement is

$\sqrt{(2^2 + 3^2)}$ or $\sqrt{13}$ yards. It suggests itself that when the super-mathematician (not knowing what kind of operations he is referring to) says that $(2E_1 + 3E_2)$ is a square root of 13, he may mean the same thing as the geometer who says that a displacement 2 yards to the right and 3 yards forward is square-root-of-13 yards. Actually the geometer does not know what kind of operations he is referring to either; he only knows the familiar story teller's description of them. He can render himself independent of the imaginations of the familiar story teller by becoming a super-mathematician. He will then say:

What the familiar story teller calls displacement to the right is an operation whose intrinsic nature is unknown to me and I will denote it by E_1; what he calls displacement forward is another unknown operation which I will denote by E_2. The kind of knowledge of the properties of displacement which I have acquired by experience is contained in such statements as "a displacement 2 yards to the right and 3 yards forward is square-root-of-13 yards." In my notation this becomes "$2E_1 + 3E_2$ is a square-root of 13." Super-mathematics enables me to boil down these statements to the single conclusion that displacement to the right and displacement forward are two operations of the set whose group structure has been investigated in Section IV.[3]

Similarly we can represent a displacement of 2 units to the right, 3 units forward and 4 units upward by $(2E_1 + 3E_2 + 4E_3)$. Working out the square of this expression in the same way, the result is found to be 29, which agrees with the geometrical calculation that the resultant displacement is $\sqrt{(2^2 + 3^2 + 4^2)} = \sqrt{29}$. The secret is that the super-mathematician expresses by the *anticommutation* of his operators the property which the geometer conceives as *perpendicularity* of displacements. That is why on p. 1568 we singled out a pentad of anticommuting operators, foreseeing that they would have an immediate application in describing the property of perpendicular directions without using the traditional picture of space. They express the property of perpendicularity without the picture of perpendicularity.

Thus far we have touched only the fringe of the structure of our set of sixteen E-operators. Only by entering deeply into the theory of electrons could I show the whole structure coming into evidence. But I will take you one small step farther. Suppose that you want to move 2 yards to the right, 3 yards forward, 4 yards upward, and 5 yards perpendicular to all three—in a fourth dimension. By this time you will no doubt have learned the trick, and will write down readily $(2E_1 + 3E_2 + 4E_3 + 5E_4)$ as the operator which symbolises this displacement. But there is a break-down.

---

[3] He will, of course, require more than a knowledge relating to two of the operators to infer the group structure of the whole set. The immediate inference at this stage is such portion of the group structure as is revealed by the equations $E_1{}^2 = E_2{}^2 = 1$, $E_2E_1 = -E_1E_2$.

The trouble is that we have exhausted the members of the pentad whose square is 1, and have to fall back on E_4 whose square is -1 (p. 1568). Consequently

$$(2E_1 + 3E_2 + 4E_3 + 5E_4)^2 = 2^2 + 3^2 + 4^2 - 5^2 = 4.$$

Thus our displacement is a square root of 4, whereas Pythagoras's theorem would require that it should be the square root of $2^2 + 3^2 + 4^2 + 5^2$, or 54. Thus there is a limitation to our representation of perpendicular directions by E-operators; it is only saved from failure in practice because in the actual world we have no occasion to consider a fourth perpendicular direction. How lucky!

It is not luck. The structure which we are here discussing is claimed to be the structure of the actual world and the key to its manifestations in our experience. The structure does not provide for a fourth dimension of space, so that there cannot be a fourth dimension in a world built in that way. Our experience confirms this as true of the actual universe.

If we wish to introduce a fourth direction of displacement we shall have to put up with a minus sign instead of a plus sign, so that it will be a displacement of a somewhat different character. It was found by Minkowski in 1908 that "later" could be regarded in this way as a fourth direction of displacement, differing only from ordinary space displacements in the fact that its square combines with a minus instead of a plus sign. Thus 2 yards to the right, 3 yards forward, 4 yards upward and 5 "yards" later [4] is represented by the operator $(2E_1 + 3E_2 + 4E_3 + 5E_4)$. We have calculated above that it is a square root of 4, so that it amounts to a displacement of 2 yards. When, as here, we consider displacement in time as well as in space, the resultant amount is called the *interval*. The value of the interval in the above problem according to Minkowski's formula is 2 yards, so that our results agree. Minkowski introduced the minus instead of the plus sign in the fourth term, regarding the change as expressing the mathematical distinction between time and space; we introduce it because we cannot help it—it is forced on us by the group structure that we are studying. Minkowski's interval afterwards became the starting point of the general theory of relativity.

Thus the distinction between space and time is already foretold in the structure of the set of E-operators. Space can have only three dimensions, because no more than three operators fulfil the necessary relationship of perpendicular displacement. A fourth displacement can be added, but it has a character essentially different from a space displacement. Calling it a time displacement, the properties of its associated operator secure that the relation of a time displacement to a space displacement shall be precisely that postulated in the theory of relativity.

---

[4] A "yard" of time is to be interpreted as the time taken by light to travel a yard.

I do not suggest that the distinction between the fourth dimension and the other three is something that we might have predicted entirely by a priori reasoning. We had no reason to expect a priori that a scheme of structure which we found in a game with letters would have any importance in the physical universe. The agreement is only impressive if we have independent reason to believe that the world-structure is based on this particular group of operators. We must recall therefore that the E-operators were first found to be necessary to physics in Dirac's wave equation of an electron. Dirac's great achievement in introducing this structure was that he thereby made manifest a recondite property of the electron, observationally important, which is commonly known as its "spin." That is a problem which seems as far removed as possible from the origin of the distinction of space and time. We may say that although the distinction of space and time cannot be predicted for a universe of unknown nature, it can be predicted for a universe whose elementary particles are of the character described in modern wave mechanics.

PART X
Mathematics of Infinity

See skulking Truth to her old cavern fled,
Mountains of Casuistry heap'd o'er her head!
Philosophy, that lean'd on Heav'n before,
Shrinks to her second cause, and is no more.
Physic of Metaphysic begs defence,
And Metaphysic calls for aid on Sense!
See Mystery to Mathematics fly!   —ALEXANDER POPE

1   Mathematics and the Metaphysicians

By BERTRAND RUSSELL*

THE nineteenth century, which prided itself upon the invention of steam and evolution, might have derived a more legitimate title to fame from the discovery of pure mathematics. This science, like most others, was baptised long before it was born; and thus we find writers before the nineteenth century alluding to what they called pure mathematics. But if they had been asked what this subject was, they would only have been able to say that it consisted of Arithmetic, Algebra, Geometry, and so on. As to what these studies had in common, and as to what distinguished them from applied mathematics, our ancestors were completely in the dark.

Pure mathematics was discovered by Boole, in a work which he called the *Laws of Thought* (1854). This work abounds in asseverations that it is not mathematical, the fact being that Boole was too modest to ,suppose his book the first ever written on mathematics. He was also mistaken in supposing that he was dealing with the laws of thought: the question how people actually think was quite irrelevant to him, and if his book had really contained the laws of thought, it was curious that no one should ever have thought in such a way before. His book was in fact concerned with formal logic, and this is the same thing as mathematics.

Pure mathematics consists entirely of assertions to the effect that, if such and such a proposition is true of *anything*, then such and such another proposition is true of that thing. It is essential not to discuss whether the first proposition is really true, and not to mention what the anything is, of which it is supposed to be true. Both these points would belong to applied mathematics. We start, in pure mathematics, from certain rules of inference, by which we can infer that *if* one proposition is true, then so is some other proposition. These rules of inference consti-

* For a biographical note about Bertrand Russell, see p. 377.

tute the major part of the principles of formal logic. We then take any hypothesis that seems amusing, and deduce its consequences. *If* our hypothesis is about *anything*, and not about some one or more particular things, then our deductions constitute mathematics. Thus mathematics may be defined as the subject in which we never know what we are talking about, nor whether what we are saying is true. People who have been puzzled by the beginnings of mathematics will, I hope, find comfort in this definition, and will probably agree that it is accurate.

As one of the chief triumphs of modern mathematics consists in having discovered what mathematics really is, a few more words on this subject may not be amiss. It is common to start any branch of mathematics—for instance, Geometry—with a certain number of primitive ideas, supposed incapable of definition, and a certain number of primitive propositions or axioms, supposed incapable of proof. Now the fact is that, though there are indefinables and indemonstrables in every branch of applied mathematics, there are none in pure mathematics except such as belong to general logic. Logic, broadly speaking, is distinguished by the fact that its propositions can be put into a form in which they apply to anything whatever. All pure mathematics—Arithmetic, Analysis, and Geometry—is built up by combinations of the primitive ideas of logic, and its propositions are deduced from the general axioms of logic, such as the syllogism and the other rules of inference. And this is no longer a dream or an aspiration. On the contrary, over the greater and more difficult part of the domain of mathematics, it has been already accomplished; in the few remaining cases, there is no special difficulty, and it is now being rapidly achieved. Philosophers have disputed for ages whether such deduction was possible; mathematicians have sat down and made the deduction. For the philosophers there is now nothing left but graceful acknowledgments.

The subject of formal logic, which has thus at last shown itself to be identical with mathematics, was, as every one knows, invented by Aristotle, and formed the chief study (other than theology) of the Middle Ages. But Aristotle never got beyond the syllogism, which is a very small part of the subject, and the schoolmen never got beyond Aristotle. If any proof were required of our superiority to the mediæval doctors, it might be found in this. Throughout the Middle Ages, almost all the best intellects devoted themselves to formal logic, whereas in the nineteenth century only an infinitesimal proportion of the world's thought went into this subject. Nevertheless, in each decade since 1850 more has been done to advance the subject than in the whole period from Aristotle to Leibniz. People have discovered how to make reasoning symbolic, as it is in Algebra, so that deductions are effected by mathematical rules. They have discovered many rules besides the syllogism, and a new branch of logic, called the Logic of Relatives,[1] has been invented to deal with topics that wholly surpassed the powers of the old logic, though they form the chief contents of mathematics.

It is not easy for the lay mind to realise the importance of symbolism in discussing the foundations of mathematics, and the explanation may perhaps seem strangely paradoxical. The fact is that symbolism is useful because it makes things difficult. (This is not true of the advanced parts of mathematics, but only of the beginnings.) What we wish to know is, what can be deduced from what. Now, in the beginnings, everything is self-evident; and it is very hard to see whether one self-evident proposition follows from another or not. Obviousness is always the enemy to correctness. Hence we invent some new and difficult symbolism, in which nothing seems obvious. Then we set up certain rules for operating on the symbols, and the whole thing becomes mechanical. In this way we find out what must be taken as premiss and what can be demonstrated or defined. For instance, the whole of Arithmetic and Algebra has been shown to require three indefinable notions and five indemonstrable propositions. But without a symbolism it would have been very hard to find this out. It is so obvious that two and two are four, that we can hardly make ourselves sufficiently sceptical to doubt whether it can be proved. And the same holds in other cases where self-evident things are to be proved.

But the proof of self-evident propositions may seem, to the uninitiated, a somewhat frivolous occupation. To this we might reply that it is often by no means self-evident that one obvious proposition follows from another obvious proposition; so that we are really discovering new truths when we prove what is evident by a method which is not evident. But a more interesting retort is, that since people have tried to prove obvious propositions, they have found that many of them are false. Self-evidence is often a mere will-o'-the-wisp, which is sure to lead us astray if we take it as our guide. For instance, nothing is plainer than that a whole always has more terms than a part, or that a number is increased by adding one to it. But these propositions are now known to be usually false. Most numbers are infinite, and if a number is infinite you may add ones to it as long as you like without disturbing it in the least. One of the merits of a proof is that it instils a certain doubt as to the result proved; and when what is obvious can be proved in some cases, but not in others, it becomes possible to suppose that in these other cases it is false.

The great master of the art of formal reasoning, among the men of our own day, is an Italian, Professor Peano, of the University of Turin.[2] He

---

[1] This subject is due in the main to Mr. C. S. Peirce.
[2] I ought to have added Frege, but his writings were unknown to me when this article was written. [Note added in 1917.]

has reduced the greater part of mathematics (and he or his followers will, in time, have reduced the whole) to strict symbolic form, in which there are no words at all. In the ordinary mathematical books, there are no doubt fewer words than most readers would wish. Still, little phrases occur, such as *therefore, let us assume, consider,* or *hence it follows.* All these, however, are a concession, and are swept away by Professor Peano. For instance, if we wish to learn the whole of Arithmetic, Algebra, the Calculus, and indeed all that is usually called pure mathematics (except Geometry), we must start with a dictionary of three words. One symbol stands for *zero,* another for *number,* and a third for *next after.* What these ideas mean, it is necessary to know if you wish to become an arithmetician. But after symbols have been invented for these three ideas, not another word is required in the whole development. All future symbols are symbolically explained by means of these three. Even these three can be explained by means of the notions of *relation* and *class*; but this requires the Logic of Relations, which Professor Peano has never taken up. It must be admitted that what a mathematician has to know to begin with is not much. There are at most a dozen notions out of which all the notions in all pure mathematics (including Geometry) are compounded. Professor Peano, who is assisted by a very able school of young Italian disciples, has shown how this may be done; and although the method which he has invented is capable of being carried a good deal further than he has carried it, the honour of the pioneer must belong to him.

Two hundred years ago, Leibniz foresaw the science which Peano has perfected, and endeavoured to create it. He was prevented from succeeding by respect for the authority of Aristotle, whom he could not believe guilty of definite, formal fallacies; but the subject which he desired to create now exists, in spite of the patronising contempt with which his schemes have been treated by all superior persons. From this "Universal Characteristic," as he called it, he hoped for a solution of all problems, and an end to all disputes. "If controversies were to arise," he says, "there would be no more need of disputation between two philosophers than between two accountants. For it would suffice to take their pens in their hands, to sit down to their desks, and to say to each other (with a friend as witness, if they liked), 'Let us calculate.' " This optimism has now appeared to be somewhat excessive; there still are problems whose solution is doubtful, and disputes which calculation cannot decide. But over an enormous field of what was formerly controversial, Leibniz's dream has become sober fact. In the whole philosophy of mathematics, which used to be at least as full of doubt as any other part of philosophy, order and certainty have replaced the confusion and hesitation which formerly reigned. Philosophers, of course, have not yet discovered this fact, and continue to write on such subjects in the old way. But mathematicians, at last in Italy, have now the power of treating the principles of mathematics in an exact and masterly manner, by means of which the certainty of mathematics extends also to mathematical philosophy. Hence many of the topics which used to be placed among the great mysteries—for example, the natures of infinity, of continuity, of space, time and motion—are now no longer in any degree open to doubt or discussion. Those who wish to know the nature of these things need only read the works of such men as Peano or Georg Cantor; they will there find exact and indubitable expositions of all these quondam mysteries.

In this capricious world, nothing is more capricious than posthumous fame. One of the most notable examples of posterity's lack of judgment is the Eleatic Zeno. This man, who may be regarded as the founder of the philosophy of infinity, appears in Plato's Parmenides in the privileged position of instructor to Socrates. He invented four arguments, all immeasurably subtle and profound, to prove that motion is impossible, that Achilles can never overtake the tortoise, and that an arrow in flight is really at rest. After being refuted by Aristotle, and by every subsequent philosopher from that day to our own, these arguments were reinstated, and made the basis of a mathematical renaissance, by a German professor, who probably never dreamed of any connection between himself and Zeno. Weierstrass,[3] by strictly banishing from mathematics the use of infinitesimals, has at last shown that we live in an unchanging world, and that the arrow in its flight is truly at rest. Zeno's only error lay in inferring (if he did infer) that, because there is no such thing as a state of change, therefore the world is in the same state at any one time as at any other. This is a consequence which by no means follows; and in this respect, the German mathematician is more constructive than the ingenious Greek. Weierstrass has been able, by embodying his views in mathematics, where familiarity with truth eliminates the vulgar prejudices of common sense, to invest Zeno's paradoxes with the respectable air of platitudes; and if the result is less delightful to the lover of reason than Zeno's bold defiance, it is at any rate more calculated to appease the mass of academic mankind.

Zeno was concerned, as a matter of fact, with three problems, each presented by motion, but each more abstract than motion, and capable of a purely arithmetical treatment. These are the problems of the infinitesimal, the infinite, and continuity. To state clearly the difficulties involved, was to accomplish perhaps the hardest part of the philosopher's task. This was done by Zeno. From him to our own day, the finest intellects of each generation in turn attacked the problems, but achieved, broadly speaking, nothing. In our own time, however, three men—Weier-

[3] Professor of Mathematics in the University of Berlin. He died in 1897.

strass, Dedekind, and Cantor—have not merely advanced the three problems, but have completely solved them. The solutions, for those acquainted with mathematics, are so clear as to leave no longer the slightest doubt or difficulty. This achievement is probably the greatest of which our age has to boast; and I know of no age (except perhaps the golden age of Greece) which has a more convincing proof to offer of the transcendent genius of its great men. Of the three problems, that of the infinitesimal was solved by Weierstrass; the solution of the other two was begun by Dedekind, and definitively accomplished by Cantor.

The infinitesimal played formerly a great part in mathematics. It was introduced by the Greeks, who regarded a circle as differing infinitesimally from a polygon with a very large number of very small equal sides. It gradually grew in importance, until, when Leibniz invented the Infinitesimal Calculus, it seemed to become the fundamental notion of all higher mathematics. Carlyle tells, in his *Frederick the Great*, how Leibniz used to discourse to Queen Sophia Charlotte of Prussia concerning the infinitely little, and how she would reply that on that subject she needed no instruction—the behaviour of courtiers had made her thoroughly familiar with it. But philosophers and mathematicians—who for the most part had less acquaintance with courts—continued to discuss this topic, though without making any advance. The Calculus required continuity, and continuity was supposed to require the infinitely little; but nobody could discover what the infinitely little might be. It was plainly not quite zero, because a sufficiently large number of infinitesimals, added together, were seen to make up a finite whole. But nobody could point out any fraction which was not zero, and yet not finite. Thus there was a deadlock. But at last Weierstrass discovered that the infinitesimal was not needed at all, and that everything could be accomplished without it. Thus there was no longer any need to suppose that there was such a thing. Nowadays, therefore, mathematicians are more dignified than Leibniz: instead of talking about the infinitely small, they talk about the infinitely great—a subject which, however appropriate to monarchs, seems, unfortunately, to interest them even less than the infinitely little interested the monarchs to whom Leibniz discoursed.

The banishment of the infinitesimal has all sorts of odd consequences, to which one has to become gradually accustomed. For example, there is no such thing as the next moment. The interval between one moment and the next would have to be infinitesimal, since, if we take two moments with a finite interval between them, there are always other moments in the interval. Thus if there are to be no infinitesimals, no two moments are quite consecutive, but there are always other moments between any two. Hence there must be an infinite number of moments between any two; because if there were a finite number one would be nearest the first of the two moments, and therefore next to it. This might be thought to be a difficulty; but, as a matter of fact, it is here that the philosophy of the infinite comes in, and makes all straight.

The same sort of thing happens in space. If any piece of matter be cut in two, and then each part be halved, and so on, the bits will become smaller and smaller, and can theoretically be made as small as we please. However small they may be, they can still be cut up and made smaller still. But they will always have *some* finite size, however small they may be. We never reach the infinitesimal in this way, and no finite number of divisions will bring us to points. Nevertheless there *are* points, only these are not to be reached by successive divisions. Here again, the philosophy of the infinite shows us how this is possible, and why points are not infinitesimal lengths.

As regards motion and change, we get similarly curious results. People used to think that when a thing changes, it must be in a state of change, and that when a thing moves, it is in a state of motion. This is now known to be a mistake. When a body moves, all that can be said is that it is in one place at one time and in another at another. We must not say that it will be in a neighbouring place at the next instant, since there is no next instant. Philosophers often tell us that when a body is in motion, it changes its position within the instant. To this view Zeno long ago made the fatal retort that every body always is where it is; but a retort so simple and brief was not of the kind to which philosophers are accustomed to give weight, and they have continued down to our own day to repeat the same phrases which roused the Eleatic's destructive ardour. It was only recently that it became possible to explain motion in detail in accordance with Zeno's platitude, and in opposition to the philosopher's paradox. We may now at last indulge the comfortable belief that a body in motion is just as truly where it is as a body at rest. Motion consists merely in the fact that bodies are sometimes in one place and sometimes in another, and that they are at intermediate places at intermediate times. Only those who have waded through the quagmire of philosophic speculation on this subject can realise what a liberation from antique prejudices is involved in this simple and straightforward commonplace.

The philosophy of the infinitesimal, as we have just seen, is mainly negative. People used to believe in it, and now they have found out their mistake. The philosophy of the infinite, on the other hand, is wholly positive. It was formerly supposed that infinite numbers, and the mathematical infinite generally, were self-contradictory. But as it was obvious that there were infinities—for example, the number of numbers—the contradictions of infinity seemed unavoidable, and philosophy seemed to have wandered into a "cul-de-sac." This difficulty led to Kant's antinomies, and hence, more or less indirectly, to much of Hegel's dialectic method.

Almost all current philosophy is upset by the fact (of which very few philosophers are as yet aware) that all the ancient and respectable contradictions in the notion of the infinite have been once for all disposed of. The method by which this has been done is most interesting and instructive. In the first place, though people had talked glibly about infinity ever since the beginnings of Greek thought, nobody had ever thought of asking, What is infinity? If any philosopher had been asked for a definition of infinity, he might have produced some unintelligible rigmarole, but he would certainly not have been able to give a definition that had any meaning at all. Twenty years ago, roughly speaking, Dedekind and Cantor asked this question, and, what is more remarkable, they answered it. They found, that is to say, a perfectly precise definition of an infinite number or an infinite collection of things. This was the first and perhaps the greatest step. It then remained to examine the supposed contradictions in this notion. Here Cantor proceeded in the only proper way. He took pairs of contradictory propositions, in which both sides of the contradiction would be usually regarded as demonstrable, and he strictly examined the supposed proofs. He found that all proofs adverse to infinity involved a certain principle, at first sight obviously true, but destructive, in its consequences, of almost all mathematics. The proofs favourable to infinity, on the other hand, involved no principle that had evil consequences. It thus appeared that common sense had allowed itself to be taken in by a specious maxim, and that, when once this maxim was rejected, all went well.

The maxim in question is, that if one collection is part of another, the one which is a part has fewer terms than the one of which it is a part. This maxim is true of finite numbers. For example, Englishmen are only some among Europeans, and there are fewer Englishmen than Europeans. But when we come to infinite numbers, this is no longer true. This breakdown of the maxim gives us the precise definition of infinity. A collection of terms is infinite when it contains as parts other collections which have just as many terms as it has. If you can take away some of the terms of a collection, without diminishing the number of terms, then there are an infinite number of terms in the collection. For example, there are just as many even numbers as there are numbers altogether, since every number can be doubled. This may be seen by putting odd and even numbers together in one row, and even numbers alone in a row below:—

$$1, 2, 3, 4, 5, \textit{ad infinitum.}$$
$$2, 4, 6, 8, 10, \textit{ad infinitum.}$$

There are obviously just as many numbers in the row below as in the row above, because there is one below for each one above. This property, which was formerly thought to be a contradiction, is now transformed into a harmless definition of infinity, and shows, in the above case, that the number of finite numbers is infinite.

But the uninitiated may wonder how it is possible to deal with a number which cannot be counted. It is impossible to count up *all* the numbers, one by one, because, however many we may count, there are always more to follow. The fact is that counting is a very vulgar and elementary way of finding out how many terms there are in a collection. And in any case, counting gives us what mathematicians call the *ordinal* number of our terms; that is to say, it arranges our terms in an order or series, and its result tells us what type of series results from this arrangement. In other words, it is impossible to count things without counting some first and others afterwards, so that counting always has to do with order. Now when there are only a finite number of terms, we can count them in any order we like; but when there are an infinite number, what corresponds to counting will give us quite different results according to the way in which we carry out the operation. Thus the ordinal number, which results from what, in a general sense may be called counting, depends not only upon how many terms we have, but also (where the number of terms is infinite) upon the way in which the terms are arranged.

The fundamental infinite numbers are not ordinal, but are what is called *cardinal*. They are not obtained by putting our terms in order and counting them, but by a different method, which tells us, to begin with, whether two collections have the same number of terms, or, if not, which is the greater.[4] It does not tell us, in the way in which counting does, *what* number of terms a collection has; but if we define a number as the number of terms in such and such a collection, then this method enables us to discover whether some other collection that may be mentioned has more or fewer terms. An illustration will show how this is done. If there existed some country in which, for one reason or another, it was impossible to take a census, but in which it was known that every man had a wife and every woman a husband, then (provided polygamy was not a national institution) we should know, without counting, that there were exactly as many men as there were women in that country, neither more nor less. This method can be applied generally. If there is some relation which, like marriage, connects the things in one collection each with one of the things in another collection, and vice versa, then the two collections have the same number of terms. This was the way in which we found that there are as many even numbers as there are numbers. Every number can be doubled, and every even number can be halved, and each process gives just one number corresponding to the one that is doubled or halved. And

---

[4] [Note added in 1917.] Although some infinite numbers are greater than some others, it cannot be proved that of any two infinite numbers one must be the greater.

in this way we can find any number of collections each of which has just as many terms as there are finite numbers. If every term of a collection can be hooked on to a number, and all the finite numbers are used once, and only once, in the process, then our collection must have just as many terms as there are finite numbers. This is the general method by which the numbers of infinite collections are defined.

But it must not be supposed that all infinite numbers are equal. On the contrary, there are infinitely more infinite numbers than finite ones. There are more ways of arranging the finite numbers in different types of series than there are finite numbers. There are probably more points in space and more moments in time than there are finite numbers. There are exactly as many fractions as whole numbers, although there are an infinite number of fractions between any two whole numbers. But there are more irrational numbers than there are whole numbers or fractions. There are probably exactly as many points in space as there are irrational numbers, and exactly as many points on a line a millionth of an inch long as in the whole of infinite space. There is a greatest of all infinite numbers, which is the number of things altogether, of every sort and kind. It is obvious that there cannot be a greater number than this, because, if everything has been taken, there is nothing left to add. Cantor has a proof that there is no greatest number, and if this proof were valid, the contradictions of infinity would reappear in a sublimated form. But in this one point, the master has been guilty of a very subtle fallacy, which I hope to explain in some future work.[5]

We can now understand why Zeno believed that Achilles cannot overtake the tortoise and why as a matter of fact he can overtake it. We shall see that all the people who disagreed with Zeno had no right to do so, because they all accepted premises from which his conclusion followed. The argument is this: Let Achilles and the tortoise start along a road at the same time, the tortoise (as is only fair) being allowed a handicap. Let Achilles go twice as fast as the tortoise, or ten times or a hundred times as fast. Then he will never reach the tortoise. For at every moment the tortoise is somewhere and Achilles is somewhere; and neither is ever twice in the same place while the race is going on. Thus the tortoise goes to just as many places as Achilles does, because each is in one place at one moment, and in another at any other moment. But if Achilles were to catch up with the tortoise, the places where the tortoise would have been would be only part of the places where Achilles would have been. Here, we must suppose, Zeno appealed to the maxim that the whole has more terms than the part.[6] Thus if Achilles were to overtake the tortoise, he would have been in more places than the tortoise; but we saw that he must, in any period, be in exactly as many places as the tortoise. Hence we infer that he can never catch the tortoise. This argument is strictly correct, if we allow the axiom that the whole has more terms than the part. As the conclusion is absurd, the axiom must be rejected, and then all goes well. But there is no good word to be said for the philosophers of the past two thousand years and more, who have all allowed the axiom and denied the conclusion.

The retention of this axiom leads to absolute contradictions, while its rejection leads only to oddities. Some of these oddities, it must be confessed, are very odd. One of them, which I call the paradox of Tristram Shandy, is the converse of the Achilles, and shows that the tortoise, if you give him time, will go just as far as Achilles. Tristram Shandy, as we know, employed two years in chronicling the first two days of his life, and lamented that, at this rate, material would accumulate faster than he could deal with it, so that, as years went by, he would be farther and farther from the end of his history. Now I maintain that, if he had lived for ever, and had not wearied of his task, then, even if his life had continued as eventfully as it began, no part of his biography would have remained unwritten. For consider: the hundredth day will be described in the hundredth year, the thousandth in the thousandth year, and so on. Whatever day we may choose as so far on that he cannot hope to reach it, that day will be described in the corresponding year. Thus any day that may be mentioned will be written up sooner or later, and therefore no part of the biography will remain permanently unwritten. This paradoxical but perfectly true proposition depends upon the fact that the number of days in all time is no greater than the number of years.

Thus on the subject of infinity it is impossible to avoid conclusions which at first sight appear paradoxical, and this is the reason why so many philosophers have supposed that there were inherent contradictions in the infinite. But a little practice enables one to grasp the true principles of Cantor's doctrine, and to acquire new and better instincts as to the true and the false. The oddities then become no odder than the people at the antipodes, who used to be thought impossible because they would find it so inconvenient to stand on their heads.

The solution of the problems concerning infinity has enabled Cantor to solve also the problems of continuity. Of this, as of infinity, he has given a perfectly precise definition, and has shown that there are no contradic-

---

[6] This must not be regarded as a historically correct account of what Zeno actually had in mind. It is a new argument for his conclusion, not the argument which influenced him. On this point, see e.g., C. D. Broad, "Note on Achilles and the Tortoise," *Mind*, N.S., Vol. XXII, pp. 318–19. Much valuable work on the interpretation of Zeno has been done since this article was written. [Note added in 1917.]

tions in the notion so defined. But this subject is so technical that it is impossible to give any account of it here.

The notion of continuity depends upon that of *order*, since continuity is merely a particular type of order. Mathematics has, in modern times, brought order into greater and greater prominence. In former days, it was supposed (and philosophers are still apt to suppose) that quantity was the fundamental notion of mathematics. But nowadays, quantity is banished altogether, except from one little corner of Geometry, while order more and more reigns supreme. The investigation of different kinds of series and their relations is now a very large part of mathematics, and it has been found that this investigation can be conducted without any reference to quantity, and, for the most part, without any reference to number. All types of series are capable of formal definition, and their properties can be deduced from the principles of symbolic logic by means of the Algebra of Relatives. The notion of a limit, which is fundamental in the greater part of higher mathematics, used to be defined by means of quantity, as a term to which the terms of some series approximate as nearly as we please. But nowadays the limit is defined quite differently, and the series which it limits may not approximate to it at all. This improvement also is due to Cantor, and it is one which has revolutionised mathematics. Only order is now relevant to limits. Thus, for instance, the smallest of the infinite integers is the limit of the finite integers, though all finite integers are at an infinite distance from it. The study of different types of series is a general subject of which the study of ordinal numbers (mentioned above) is a special and very interesting branch. But the unavoidable technicalities of this subject render it impossible to explain to any but professed mathematicians.

Geometry, like Arithmetic, has been subsumed, in recent times, under the general study of order. It was formerly supposed that Geometry was the study of the nature of the space in which we live, and accordingly it was urged, by those who held that what exists can only be known empirically, that Geometry should really be regarded as belonging to applied mathematics. But it has gradually appeared, by the increase of non-Euclidean systems, that Geometry throws no more light upon the nature of space than Arithmetic throws upon the population of the United States. Geometry is a whole collection of deductive sciences based on a corresponding collection of sets of axioms. One set of axioms is Euclid's; other equally good sets of axioms lead to other results. Whether Euclid's axioms are true, is a question as to which the pure mathematician is indifferent; and, what is more, it is a question which it is theoretically impossible to answer with certainty in the affirmative. It might possibly be shown, by very careful measurements, that Euclid's axioms are false; but no measurements could ever assure us (owing to the errors of observation) that they are exactly true. Thus the geometer leaves to the man of science to decide, as best he may, what axioms are most nearly true in the actual world. The geometer takes any set of axioms that seem interesting, and deduces their consequences. What defines Geometry, in this sense, is that the axioms must give rise to a series of more than one dimension. And it is thus that Geometry becomes a department in the study of order.

In Geometry, as in other parts of mathematics, Peano and his disciples have done work of the very greatest merit as regards principles. Formerly, it was held by philosophers and mathematicians alike that the proofs in Geometry depended on the figure; nowadays, this is known to be false. In the best books there are no figures at all. The reasoning proceeds by the strict rules of formal logic from a set of axioms laid down to begin with. If a figure is used, all sorts of things seem obviously to follow, which no formal reasoning can prove from the explicit axioms, and which, as a matter of fact, are only accepted because they are obvious. By banishing the figure, it becomes possible to discover *all* the axioms that are needed; and in this way all sorts of possibilities, which would have otherwise remained undetected, are brought to light.

One great advance, from the point of view of correctness, has been made by introducing points as they are required, and not starting, as was formerly done, by assuming the whole of space. This method is due partly to Peano, partly to another Italian named Fano. To those unaccustomed to it, it has an air of somewhat wilful pedantry. In this way, we begin with the following axioms: (1) There is a class of entities called *points*. (2) There is at least one point. (3) If *a* be a point, there is at least one other point besides *a*. Then we bring in the straight line joining two points, and begin again with (4), namely, on the straight line joining *a* and *b*, there is at least one other point besides *a* and *b*. (5) There is at least one point not on the line *ab*. And so we go on, till we have the means of obtaining as many points as we require. But the word *space*, as Peano humorously remarks, is one for which Geometry has no use at all.

The rigid methods employed by modern geometers have deposed Euclid from his pinnacle of correctness. It was thought, until recent times, that, as Sir Henry Savile remarked in 1621, there were only two blemishes in Euclid, the theory of parallels and the theory of proportion. It is now known that these are almost the only points in which Euclid is free from blemish. Countless errors are involved in his first eight propositions. That is to say, not only is it doubtful whether his axioms are true, which is a comparatively trivial matter, but it is certain that his propositions do not follow from the axioms which he enunciates. A vastly greater number of axioms, which Euclid unconsciously employs, are required for the proof of his propositions. Even in the first proposition of all, where he constructs an equilateral triangle on a given base, he uses two circles which are assumed to intersect. But no explicit axiom assures us that they do so, and in some kinds of spaces they do not always intersect. It is quite doubtful whether our space belongs to one of these kinds or not. Thus Euclid fails entirely to prove his point in the very first proposition. As he is certainly not an easy author, and is terribly longwinded, he has no longer any but an historical interest. Under these circumstances, it is nothing less than a scandal that he should still be taught to boys in England.[7] A book should have either intelligibility or correctness; to combine the two is impossible, but to lack both is to be unworthy of such a place as Euclid has occupied in education.

The most remarkable result of modern methods in mathematics is the importance of symbolic logic and of rigid formalism. Mathematicians, under the influence of Weierstrass, have shown in modern times a care for accuracy, and an aversion to slipshod reasoning, such as had not been known among them previously since the time of the Greeks. The great inventions of the seventeenth century—Analytical Geometry and the Infinitesimal Calculus—were so fruitful in new results that mathematicians had neither time nor inclination to examine their foundations. Philosophers, who should have taken up the task, had too little mathematical ability to invent the new branches of mathematics which have now been found necessary for any adequate discussion. Thus mathematicians were only awakened from their "dogmatic slumbers" when Weierstrass and his followers showed that many of their most cherished propositions are in general false. Macaulay, contrasting the certainty of mathematics with the uncertainty of philosophy, asks who ever heard of a reaction against Taylor's theorem? If he had lived now, he himself might have heard of such a reaction, for this is precisely one of the theorems which modern investigations have overthrown. Such rude shocks to mathematical faith have produced that love of formalism which appears, to those who are ignorant of its motive, to be mere outrageous pedantry.

The proof that all pure mathematics, including Geometry, is nothing but formal logic, is a fatal blow to the Kantian philosophy. Kant, rightly perceiving that Euclid's propositions could not be deduced from Euclid's axioms without the help of the figures, invented a theory of knowledge to account for this fact; and it accounted so successfully that, when the fact is shown to be a mere defect in Euclid, and not a result of the nature of geometrical reasoning, Kant's theory also has to be abandoned. The whole doctrine of *a priori* intuitions, by which Kant explained the possibility of pure mathematics, is wholly inapplicable to mathematics in its present form. The Aristotelian doctrines of the schoolmen come nearer in spirit to

---

[7] Since the above was written, he has ceased to be used as a textbook. But I fear many of the books now used are so bad that the change is no great improvement. [Note added in 1917.]

the doctrines which modern mathematics inspire; but the schoolmen were hampered by the fact that their formal logic was very defective, and that the philosophical logic based upon the syllogism showed a corresponding narrowness. What is now required is to give the greatest possible development to mathematical logic, to allow to the full the importance of relations, and then to found upon this secure basis a new philosophical logic, which may hope to borrow some of the exactitude and certainty of its mathematical foundation. If this can be successfully accomplished, there is every reason to hope that the near future will be as great an epoch in pure philosophy as the immediate past has been in the principles of mathematics. Great triumphs inspire great hopes; and pure thought may achieve, within our generation, such results as will place our time, in this respect, on a level with the greatest age of Greece.[8]

[8] The greatest age of Greece was brought to an end by the Peloponnesian War. [Note added in 1917.]

COMMENTARY ON

HANS HAHN

SINCE ancient times philosophers, theologians and mathematicians have occupied themselves with the subject of infinity. Zeno of Elea invented a group of famous paradoxes whose difficulties are connected with the concept; in their time such leading thinkers as Aristotle, Descartes, Leibniz and Gauss grappled with the infinity problem without making any notable contributions to its clarification. The subject is admittedly complex and undeniably important. A firm grasp of the problems of infinity is essential to an understanding of the revolution in ideas that paved the way for the triumphant advance of modern mathematics, with important consequences to physics, cosmology and related sciences.

The selection following is a slightly condensed version of a lecture on infinity by a noted Austrian mathematician, Hans Hahn, delivered some years ago before a general audience in Vienna. This is the first translation of the lecture into English. Hahn was born in Vienna in 1879 and after receiving his doctorate from the University of Vienna taught there and also at the universities of Innsbruck (where he took over Otto Stolz's post) and Czernowitz. During this period (1902–1915) he published a number of papers on the calculus of variations, function theory and the theory of sets, and worked in seminars with leading scientists, among them Ludwig Boltzmann, David Hilbert, Felix Klein and Hermann Minkowski. In 1915 Hahn entered the Austrian Army and soon thereafter was severely wounded in action and decorated for bravery. For a time he was professor at Bonn but in the spring of 1921 he returned to the University of Vienna to accept a chair in mathematics, which he held until he died of cancer in 1934 at the age of fifty-five. Hahn was a gifted and prolific investigator and a brilliant teacher. His main interest, pursued with great energy especially during the second part of his career, was in philosophy and the foundations of mathematics. In philosophical outlook he stood at many points close to Bertrand Russell and Ludwig Wittgenstein. He believed that experience and observation provide the only sound basis for knowledge about the physical universe while rational thought itself consists of nothing but "tautological transformations." Mathematics and logic are transformations of this kind and thus differ from a science such as physics whose propositions are rooted in the circumstances of the outside world. Mathematical propositions, he said, are like the propositions of logic in that neither is concerned with the things we wish to discuss

1591

but rather with the way we wish to discuss them.[1] One recognizes in these opinions Hahn's inclination to the philosophy of logical positivism and he was in fact a leading member of the celebrated Vienna Circle, a group of positivistic philosophers and scientists whose founders included Moritz Schlick, Karl Menger, Kurt Goedel, Otto Neurath and Rudolf Carnap. The Circle annually presented popular lectures on science, and this survey by Hahn of the concept of infinity is one of the best of the series.

Hahn began his lecture with a historical résumé (here omitted) and then launched his discussion with a description of the work of the founder of the modern mathematical theory of infinity, Georg Cantor.

[1] Hans Hahn, *"Die Bedeutung der Wissenschaftlichen Weltauffassung, Insbesondere für Mathematik und Physik,"* Erkenntnis 1, 1930, p. 97. Hahn's mathematical work and philosophical opinions are described in an obituary by K. Mayrhofer, Monatshefte für Mathematik und Physik, vol. 41 (1934), pp. 221–238.

In two words: IM-POSSIBLE.—SAMUEL GOLDWYN (*Quoted in Alva Johnson's*
The Great Goldwyn)

I have found you an argument; but I am not obliged to find you an under-
standing. —SAMUEL JOHNSON

The infinite! No other question has ever moved so profoundly the spirit
of man. —DAVID HILBERT (1921)

2 Infinity

By HANS HAHN

IT was Georg Cantor, who in the years 1871–84 created a completely
new and very special mathematical discipline, the theory of sets, in which
was founded, for the first time in a thousand years of argument back
and forth, a theory of infinity with all the incisiveness of modern mathematics.

Like so many other new creations this one began with a very simple
idea. Cantor asked himself: "What do we mean when we say of two finite
sets that they consist of equally many things, that they have the same
number, that they are equivalent?" Obviously nothing more than this,
that between the members of the first set and those of the second a correspondence can be effected by which each member of the first set matches
exactly a member of the second set, and likewise each member of the
second set matches one of the first. A correspondence of this kind is called
"reciprocally unique," or simply "one-to-one." The set of the fingers of
the right hand is equivalent to the set of fingers of the left hand, since
between the fingers of the right hand and those of the left hand a one-to-
one pairing is possible. Such a correspondence is obtained, for instance,
when we place the thumb on the thumb, the index finger on the index
finger, and so on. But the set of both ears and the set of the fingers of one
hand are not equivalent, since in this instance a one-to-one correspondence
is obviously impossible; for if we attempt to place the fingers of one hand
in correspondence with our ears, no matter how we contrive there will
necessarily be some fingers left over to which no ears correspond. Now
the number (or cardinal number) of a set is obviously a characteristic
that it has in common with all equivalent sets, and by which it distinguishes itself from every set not equivalent to itself. The number 5, for
instance, is the characteristic which all sets equivalent to the set of the
fingers of one hand have in common, and which distinguishes them from
all other sets.

Thus we have the following definitions. Two sets are called equivalent if between their respective members a one-to-one correspondence is possible; and the characteristic that one set has in common with all equivalent sets, and by which it distinguishes itself from all other sets not equivalent to itself is called the (cardinal) number of that set. And now we make the fundamental assertion that in these definitions the finiteness of the sets considered is in no sense involved; the definitions can be applied as readily to infinite sets as to finite sets. The concepts "equivalent" and "cardinal number" are thereby transferred to sets of infinitely many objects. The cardinal numbers of finite sets, i.e., the numbers 1, 2, 3, . . . , are called natural numbers; the cardinal numbers of infinite sets Cantor calls "transfinite cardinal numbers" or "transfinite powers."

But are there really any infinite sets? We can convince ourselves of this at once by a very simple example. There are obviously infinitely many different natural numbers; hence the set of all the natural numbers contains infinitely many members; it is an infinite set. Now then, those sets that are equivalent to the set of all natural numbers, whose members can be paired in one-to-one correspondence with the natural numbers, are called *denumerably infinite sets*. The meaning of this designation can be explained as follows. A set with the cardinal number "five" is a set whose members can be put in one-to-one correspondence with the first five natural numbers, that is, a set that can be numbered by the integers 1, 2, 3, 4, 5, or counted off with the aid of the first five natural numbers. A denumerably infinite set is a set whose members can be put in one-to-one correspondence with the totality of natural numbers. According to our definitions all denumerably infinite sets have the same cardinal number; this cardinal number must now be given a name, just as the cardinal number of the set of the fingers on one hand was earlier given the name 5. Cantor gave this cardinal number the name "aleph-null," written $\aleph_0$. (Why he gave it this rather bizarre name will become clear later.) The number $\aleph_0$ is thus the first example of a transfinite cardinal number. Just as the statement "a set has the cardinal number 5" means that its members can be put in one-to-one correspondence with the fingers of the right hand, or—what amounts to the same thing—with the integers 1, 2, 3, 4, 5, so the statement "a set has the cardinal number $\aleph_0$" means that its elements can be put in one-to-one correspondence with the totality of natural numbers.

If we look about us for examples of denumerably infinite sets we arrive immediately at some highly surprising results. The set of all natural numbers is itself denumerably infinite; this is self-evident, for it was from this set that we defined the concept "denumerably infinite." But the set of all even numbers is also denumerably infinite, and has the same cardinal number $\aleph_0$ as the set of all natural numbers, though we would be inclined to think that there are far fewer even numbers than natural numbers. To prove this proposition we have only to look at the correspondence diagrammed in Figure 1, that is, to put each natural number opposite its double. It may clearly be seen that there is a one-to-one correspondence between all natural and all even numbers, and thereby our point is established. In exactly the same way it can be shown that the set of all odd numbers is denumerably infinite. Even more surprising is the fact that the

1	2	3	4	5	6	. . .
↕	↕	↕	↕	↕	↕	
2	4	6	8	10	12	. . .

FIGURE 1

FIGURE 2

1	2	3	4	5	6	7	8	9
↕	↕	↕	↕	↕	↕	↕	↕	↕
(1,1)	(2,1)	(1,2)	(1,3)	(2,2)	(3,1)	(4,1)	(3,2)	(2,3)
10	11	12	13	14	15	. . .		
(1,4)	(1,5)	(2,4)	(3,3)	(4,2)	(5,1)	. . .		

FIGURE 3

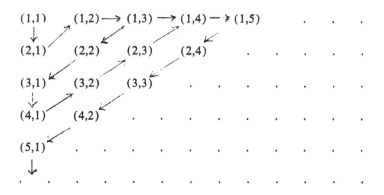

set of *all pairs of natural numbers* is denumerably infinite. In order to understand this we have merely to arrange the set of all pairs of natural numbers "diagonally" as indicated in Figure 2, whereupon we at once obtain the one-to-one correspondence shown in Figure 3 between all natural numbers and all pairs of natural numbers. From this follows the conclusion, which Cantor discovered while still a student, that the set of all rational fractions (i.e., the quotients of two whole numbers, like ½, ⅔, etc.) is also denumerably infinite, or equivalent to the set of all natural numbers, though again one might suppose that there are many, many more fractions than there are natural numbers. What is more, Cantor was able to prove that the set of all so-called algebraic numbers, that is, the set of all numbers that satisfy an algebraic equation of the form

$$a_0 x^n + a_1 x^{n-1} + \ldots + a_{n-1} x + a_n = 0$$

with integral coefficients $a_0, a_1, \ldots a_n$, is denumerably infinite.

At this point, the reader may ask whether, in the last analysis, *all* infinite sets are not denumerably infinite—that is, equivalent? If this were so, we should be sadly disappointed; for then, alongside the finite sets there would simply be infinite ones which would all be equivalent, and there would be nothing more to say about the matter. But in the year 1874 Cantor succeeded in proving that there are also infinite sets that are not denumerable; that is to say, there are other infinite numbers, transfinite cardinal numbers differing from aleph-null. Specifically, Cantor proved that the set of all so-called real numbers (i.e., composed of all whole numbers, plus all fractions, plus all irrational numbers) is nondenumerably infinite. The proof is so simple that I can give his reasoning here. It is sufficient to show that the set of all real numbers between 0 and 1 is not denumerably infinite, for then the set of *all* real numbers obviously cannot be denumerably infinite. In the proof we make use of the familiar fact that every real number between 0 and 1 can be expressed as an infinite decimal fraction. The statement that is to be proved, that the set of all real numbers between 0 and 1 is not denumerably infinite, can also be phrased as follows: "a *denumerably* infinite set of real numbers between 0 and 1," or thus: "Given a denumerably infinite set of real numbers between 0 and 1, there will always be a real number between 0 and 1 that does not belong to the given set." To prove this let us imagine a denumerably infinite set of real numbers between 0 and 1 as given; then, since this set is denumerably infinite, the real numbers that occur in it can be put in one-to-one correspondence with the natural numbers. Let us write down as a continuing decimal fraction the real number that in this arrangement corresponds to the natural number 1, and under it the real number that corresponds to the natural number 2, under it again the real number that corresponds to the natural number 3, and so on. We would get something like this:

$$0.20745 \ . \ . \ .$$

$$0.16238 \ . \ . \ .$$

$$0.97126 \ . \ . \ .$$

. . . . . .

Now one can, in fact, at once write down a real number between 0 and 1 which does *not* occur in the given denumerably infinite set of real numbers between 0 and 1. Take as its first digit one differing from the first digit of the decimal in the first row, say 3; as its second digit one differing from the second digit of the decimal in the second row, say 2; as its third digit one differing from the third digit of the decimal in the third row, say 5; and so on. It is clear that by this procedure we obtain a real number between 0 and 1 which differs from all the given infinitely many real numbers of our set, and this is precisely what we sought to prove was possible.[1]

It has thus been shown that the set of natural numbers and the set of real numbers are not equivalent; that these two sets have different cardinal numbers. The cardinal number of the set of real numbers Cantor called the "power of the continuum"; we shall designate it by c. Earlier it was noted that the set of all algebraic numbers is denumerably infinite, and we just now saw that the set of all real numbers is not denumerably infinite, hence there must be real numbers that are not algebraic. These are the so-called transcendental numbers, whose existence is demonstrated in the simplest way conceivable by Cantor's brilliant train of reasoning.

It is well known that the real numbers can be put in one-to-one correspondence with the points of a straight line; hence c is also the cardinal number of the set of all points of a straight line. Surprisingly Cantor was also able to prove that a one-to-one pairing is possible between the set of

---

[1] A few words further to clarify this point may be permitted. The essence of Cantor's proof is that no comprehensive counting procedure can be devised for the entire set of real numbers, nor even for one of its proper subsets, such as all the real numbers lying between 0 and 1. By various ingenious methods certain infinite sets such as all rational fractions or all algebraic numbers can be paired off with the natural numbers; every attempt, however, to construct a formula for counting the all-inclusive set of real numbers is invariably frustrated. No matter what counting scheme is adopted it can be shown that some of the real numbers in the set so considered remain uncounted, which is to say that the scheme fails. It follows that an infinite set for which no counting method can be devised is noncountable, in other words nondenumerably infinite. ED.

all points of a plane and the set of all points of a straight line. These two sets are thus equivalent, that is to say, c is also the cardinal number of the set of all points of a plane, though here too we should have thought that a plane would contain a great many more points than a straight line. In fact, as Cantor has shown, c is the cardinal number of *all* points of three-dimensional space, or even of a space of any number of dimensions.

We have discovered two different transfinite cardinal numbers, $\aleph_0$ and c, the power of the denumerably infinite sets and the power of the continuum. Are there yet others? Yes, there certainly are infinitely many different transfinite cardinal numbers; for given any set M, a set with a higher cardinal number can at once be indicated, since the set of all possible subsets of M has a higher cardinal number than the set M itself. Take, for example, a set of three elements, such as the set of the three figures 1, 2, 3. Its partial sets are the following: 1; 2; 3; 1, 2; 2, 3; 1, 3; thus the number of the partial sets is more than three. Cantor has shown that this is generally true,[2] even for infinite sets. For example, the set of all possible point-sets of a straight line has a higher cardinal number than the set of all points of the straight line, that is to say, its cardinal number is greater than c.

What is now desired is a general view of all possible transfinite cardinal numbers. As regards the cardinal numbers of finite sets, the natural numbers, the following simple situation prevails: Among such sets there is one that is the smallest, namely 1; and if a finite set M with the cardinal number m is given, a set with the next-larger cardinal number can be formed by adding one more object to the set M. What is the rule in this respect with regard to infinite sets? It can be shown without difficulty that among the transfinite cardinal numbers, as well as among the finite ones, there is one that is the smallest, namely $\aleph_0$, the power of denumerably infinite sets (though we must not think that this is self-evident, for among all positive fractions, for instance, there is none that is the smallest). It is, however, not so easy as it was in the case of finite sets to form the next-larger to a transfinite cardinal number; for whenever we add one more member to an infinite set we do not get a set of greater cardinality, only one of equal cardinality. But Cantor also solved this difficulty, by showing that there is a next-larger to every transfinite cardinal number (which again is by no means self-evident: there is no next-larger to a fraction, for instance) and by showing how it is obtained. We cannot go into his proof here, since this would take us too far into the realm of pure mathematics. It is enough for us to recognize the fact that there is a smallest transfinite cardinal number, namely $\aleph_0$; after this there is a next-larger,

---

[2] For sets, M, consisting of one or two members, the rule holds only if one adds to the proper partial sets the empty set (null-set) possessing no members and the set M itself.

which is called $\aleph_1$; after this there is again a next-larger, which is $\aleph_2$; and so on. But this still does not exhaust the class of transfinite cardinals; for if it be assumed that we have formed the cardinal numbers $\aleph_0$, $\aleph_1$, $\aleph_2$, $\cdots \aleph_{10}$, $\cdots \aleph_{100}$, $\cdots \aleph_{1000}$, $\cdots$ that is, all alephs ($\aleph_n$) whose index n is a natural number, then there is again a first transfinite cardinal number larger than any of these—Cantor called it $\aleph_w$—and a next-larger successor $\aleph_{w+1}$, and so on and on.

The successive alephs formed in this manner represent all possible transfinite cardinal numbers, and hence the power c of the continuum must occur among them. The question is which aleph is the power of the continuum? This is the famous problem of the continuum. We already know that it cannot be $\aleph_0$, since the set of all real numbers is non-denumerably infinite, that is to say, not equivalent to the set of natural numbers. Cantor took $\aleph_1$ to be the power of the continuum. This question, however, remains open, and for the present we see no trace of a path to its solution.

I must here call attention to a detail of logic that was first noticed by Ernst Zermelo some time after the formulation of the theory of sets by Cantor. In the proof that $\aleph_0$ is the smallest transfinite cardinal number, as well as in the proof that every transfinite cardinal number occurs in the aleph series, one makes an assumption that is also used in other mathematical proofs without its being explicitly recognized. This assumption, which, as it seems, cannot be derived from the other principles of logic, is known as the postulate, or the axiom of choice, and may be stated as follows: Given a set of M sets, no two of which have a common member, then there exists a set that has exactly one member in common with each of the sets M. The postulate raises no difficulty when dealing with a finite collection of sets, for one can select a member from each of these sets in a finite number of operations. When one member has been selected from each of the given sets, our task is done for we then have formed a new set that has exactly one member in common with each of the given sets. Neither is there any difficulty if infinitely many sets are given and at the same time a rule is provided that distinguishes one member of each of these sets. The set composed of the objects denoted by the rule is the one we seek, since it also has exactly one member in common with each of the given sets. But if infinitely many sets are given and no rule is formulated that distinguishes one member in each of them, we cannot proceed as we did in the case of a finite number of sets. For if we started to select arbitrarily one member from each of the given sets, we could of course never complete the task and thus never could obtain a set having one member in common with each of the given sets.[3] Hence the assertion

---

[3] The defining rule, in other words, makes it possible to form an infinite set in a finite operation; in the absence of such a rule the selection requires an infinite number of steps and is therefore impossible. ED.

that such a set exists in every case represents a special logical postulate. The famous English logician and philosopher, Bertrand Russell, has made this clear in an ingenious and amusing illustration. In civilized countries shoes are so designed that the right and left shoe of each pair can readily be distinguished; but the distinction between right and left cannot be made for pairs of stockings. Let us imagine an infinitely rich man (a millionaire or a billionaire will not do, he must be an "infinitillionaire") who is infinitely eccentric and owns a collection of infinitely many pairs of shoes and infinitely many pairs of stockings. This fortunate man, it may be observed, has at his disposal a set of shoes that contains exactly one shoe from each pair; for instance, the infinite set composed of the right shoes, one from each pair. But how is he to obtain a set that contains exactly one stocking from each pair of stockings? This is, of course, only a facetious illustration, yet the principle itself represents a serious and important logical discovery, which bears on many problems. For the moment, however, let us leave it; we shall return to it.

On the basis of this rather sketchy description of the structure of the theory of sets the answer to the question "Is there an infinity?" appears to be an unqualified "Yes." There are not only, as Leibniz had already asserted, infinite sets, but there are even what Leibniz had denied, infinite numbers, and it can also be shown that it is quite possible to operate with them, in a manner similar to, if not identical with that used for finite natural numbers.

But now we must look with a critical eye at what has been accomplished. When existence is asserted of some entity in mathematics by the words "there is" or "there are," and non-existence by the words "there is/are not," evidently some thing entirely different is meant than when these expressions are used in everyday life, or in geography, say, or natural history. Let us make this clear by a few examples. Since about the time of Plato it has been known that there are regular bodies with four sides (tetrahedrons), with six sides (hexahedrons or cubes), with eight sides (octahedrons), with twelve sides (dodecahedrons), and with twenty sides (icosahedrons). There are no regular bodies other than these five. Now it is obvious, that the expressions "there are" and "there are no" have a different meaning as used in the two preceding sentences about mathematics than in a geographical statement, such as: there are mountains over 25,000 feet high, but there are no mountains 35,000 feet high. For even though mathematics teaches that there are cubes and icosahedrons, yet in the sense that there are mountains over 25,000 feet high, that is, in the sense of physical existence, there are no cubes and no icosahedrons. The most beautiful rock-salt crystal is not an exact mathematical cube, and a model of an icosahedron, however well constructed, is not an icosahedron in the mathematical sense. While it is fairly clear what is meant by the expressions "there is" or "there are" as used in the sciences dealing with the physical world, it is not at all clear what mathematics means by such existence statements. On this point indeed there is no agreement whatever among scholars, whether they be mathematicians or philosophers. So many different interpretations are represented among them that one might almost say: "quot capita, tot sententiae" ("as many meanings as individuals"). However, if we stick to essentials we can perhaps distinguish three basically different points of view on this subject. These I shall describe briefly.

The first can be designated the realistic or the Platonic position. It ascribes to the objects of mathematics a real existence in the world of ideas; the physical world we may note is merely an imperfect image of the world of ideas. Thus, there are no perfect cubes in the physical world, only in the world of ideas; through our senses we can comprehend only the physical world, but it is in thought that we comprehend the world of ideas. A mathematical concept derives its existence from the real object corresponding to it in the world of ideas; a mathematical statement is true if it correctly represents the real relationship of the corresponding objects in the world of ideas. The second view which we can call the intuitionistic or the Kantian position, is to the effect that we possess pure intuition; mathematics is construction by pure intuition, and a mathematical concept "exists" if it is constructable by pure intuition. The philosophical formulation of this idea might be popularly expressed somewhat as follows: "If there is indeed no perfect cube in the physical world I can at least imagine a perfect cube." The third view is best called the logistic position. If one sought to trace its historical roots one could perhaps connect it with the nominalist school of scholastic philosophy; thus it may be called the nominalist position. According to this view mathematics is a purely logical discipline and, like logic, is carried on entirely within the confines of language; it has nothing whatever to do with reality, or with pure intuition; on the contrary it deals exclusively with the use of signs or symbols. These signs or symbols can be used as we like, in conformity with rules that we have set. The only restriction on our freedom is that we may under no circumstances contradict the self-established rules. The final criterion of mathematical existence thus becomes freedom from contradiction; that is, mathematical existence can be ascribed to every concept whose use does not enmesh us in contradiction.

Let us now examine a little more critically these three points of view. The realistic position is, in my opinion, untenable since it ascribes to the mind abilities that it does not possess; our thinking consists of tautological transformation, it is incapable of comprehending a reality. Plato assumed a mystical recollection (*anamnesis*) by the soul of a state in which it beheld the ideas face to face, as it were. In any event, this first position is entirely metaphysical and seems wholly unsuitable as the foundation of mathematics. Nonetheless, it is still the source, though often unsuspected, of a great deal of confusion in research on the foundations of mathematics.

With regard to the second, the intuitionistic position, I have attempted on other occasions to explain that there is no such thing as pure intuition. To be sure, Kant assigned it a very broad role, but in light of the development of mathematics since his day this view cannot possibly be maintained. Hence the recent supporters of the second position have, in fact, become more modest in their claims. But as to what this allegedly pure intuition can and cannot do, what is consistent with it and what is not—about these matters there is no agreement whatever among the supporters of the intuitionistic position. This is shown very clearly by their answers to the question that concerns us here, "Are there infinite sets and infinite numbers?" Some intuitionists would say that arbitrarily large numbers can perhaps be constructed by pure intuition, but not the set of *all* natural numbers. This group, in other words, would flatly deny the existence of infinite sets. Others of this school hold that while intuition suffices for the construction of the set of all natural numbers, non-denumerably infinite sets are beyond intuition's reach; which is to say they deny the existence of the set of all real numbers. Still others ascribe constructability, and thereby existence, to certain non-denumerable sets. The intuitionist doctrine is thus seen to rest on very uncertain ground; in glaring contrast to this uncertainty is the gruffness with which the supporters of this position declare meaningless everything that in their opinion is not constructable by pure intuition.

Having rejected the two first positions we must then turn to the third, the logistic interpretation. But before discussing the question "Are there infinite sets and infinite numbers?" from the logistic point of view, it may be useful to point out the difference between the three positions with respect to the axiom of choice, mentioned earlier. A representative of the realistic stand would say: "Whether we are to accept or reject the axiom of choice in logic depends on how reality is constituted; if it is constituted as the axiom of choice asserts, then we must accept it, but if reality is not so constituted we shall have to reject the axiom. Unfortunately we do not know which is the case, and because of the inadequacy of our means of perception we shall—again unfortunately—never know." An intuitionist would perhaps say: "We must consider whether a set of the kind required by the axiom of choice (that is to say, a set having exactly one member in common with each set of a given system of sets) can be constructed by pure intuition. For this purpose one would have to select from each set of the given system one member; if the system consisted of infinitely many sets the task would involve infinitely many separate operations;

since these cannot possibly be carried out, the axiom of choice is to be rejected." A consistent representative of the logistic position would say: "If the axiom of choice is in truth independent of the other principles of logic, that is to say, if the statement of the axiom of choice as well as the contrary statement is consistent with the other principles of logic, then we are free to accept it or replace it with its contradictory. That is, we can as well operate with a mathematics in which the axiom of choice is taken as a basic principle—a 'Zermelo mathematics'—as with a mathematics in which a contrary axiom is taken as the basis—a 'non-Zermelo mathematics.' The entire question has nothing to do with the nature of reality, as the realists think, or with pure intuition, as the intuitionists think. The question is rather in what sense we decide to use the word 'set'; it is a matter of determining the syntax of the word 'set.' "

Let us return to the problem we are mainly concerned with, and consider what answer an adherent of the logistic school would give to the question "Are there infinite sets and infinite numbers?" He would perhaps reply: "Yes, infinite sets and infinite numbers can be said to exist, provided it is possible to operate with them without contradiction." What is the situation, then, with regard to this freedom from contradiction?

Various philosophers have, in fact, repeatedly raised the objection against Cantor's theory that it would lead to contradictions. The objection might be phrased as follows: "According to Cantor the set of all natural numbers is equivalent to the set of even numbers; this however contradicts the axiom that the whole can never be equal to one of its parts." To refute the objection we must examine the meaning of this alleged axiom. Certainly it cannot mean that reality itself is constituted as the axiom asserts; for that would be a reversion to the metaphysical realistic position. Its meaning must rather be in a syntatic determination of how we are to use the words "whole," "part," "equal." We must establish, in other words, that this determination does not correspond completely with the usage of everyday speech or with the language of science. Yet note that even in ordinary linguistic usage it is meaningful to say that the whole may be equal to a part—as, for example, with respect to color; why, then, should it be forbidden to say that the whole is equal to one of its parts with respect to quantity? The truth is, however, this axiom is used neither in the construction of logic nor of mathematics. Hence no contradiction confronts us here, but merely the fact that as to certain aspects of behavior infinite sets differ from finite sets. A finite set cannot be equivalent to one of its parts, but an infinite set can be. It can be shown that every infinite set has parts to which it is equivalent, and one can indeed make use of this very fact—as Dedekind did—in defining the concept "infinite set."

It has been suggested that many other contradictions lurk in the concepts of infinite sets and infinite numbers. These objections, as in the case above, usually consist of showing that certain properties that necessarily belong to finite sets and finite numbers do not belong to infinite sets and infinite numbers. For example, every number must be either even or odd: but Cantor's transfinite cardinal numbers are neither even nor odd; or, every number becomes larger when 1 is added to it: but this is not true of Cantor's transfinites. That transfinite numbers have properties differing from those of finite numbers affords no contradiction; it is not even to be wondered at; it must be so. For if transfinites were in every respect indistinguishable from finite numbers the need for a separate transfinite category would vanish; transfinite numbers could be classed simply as oversized finite numbers. It is the same when we discover a new species of animal—it must differ in some way from the known ones, otherwise it would not be a new species. After disposing of these arguments a single meaningful question remains: "Since they differ so greatly from ordinary numbers is it perhaps not inappropriate to designate transfinites as numbers?" Like many so-called philosophical problems, we are free to consider this one as turning on a simple issue of terminology; though it was for the very purpose of avoiding such purely terminological controversies that Cantor gave his transfinite cardinals the neutral and relatively non-committal name of "powers." The controversy, however, intrigues us in much the same way as the once celebrated dispute over whether "one" is a number, or numbers only begin with "two." Let us content ourselves with the indisputable statement that the term "transfinite numbers" has shown itself wholly suited to its purpose.

Cantor had no difficulty in dealing with objections of this sort; they could not endanger his edifice. But the refutation of a few inadequate proofs of supposed contradictions clearly does not demonstrate that none exist; and serious contradictions have, in fact, been discovered in Cantor's structure. Contradictions have appeared in certain set formations that are sweepingly inclusive, such as the set of all objects, the set of all sets, and the set of all infinite cardinal numbers. Note, however, that the concept of infinity was not the source of these contradictions; instead they arose from certain deficiencies of classical logic. Thus it became evident that what was needed was a reform in logic. This reform consisted mainly in a more careful use of the word "all," as taught in Russell's theory of logical types.[4] Thereupon the contradictions that had appeared in the theory of sets were successfully explained and eliminated. There is no longer any known contradiction in the present formulation of the theory of sets.

But from the fact that no contradiction is known, it does not necessarily

---

[4] See A. N. Whitehead and B. Russell, *Principia Mathematica*; Cambridge, 1925, Vol. I, second edition, p. 37 *et seq.*

follow that none exist, any more than the fact that in 1900 no okapi was known proved that none existed. We face the question, then, "Can any proof be given of freedom from contradiction?" On the basis of present knowledge it may be said that an absolute proof of freedom from contradiction is probably unattainable; every such proof is relative; we can do no more than to relate the freedom from contradiction of one system to that of another. But is not this concession fatal to the logistic position, according to which mathematical existence depends entirely on freedom from contradiction? I think not. For here, as in every sphere of thought, the demand for absolute certainty of knowledge is an exaggerated demand: in no field is such certainty attainable. Even the evidence adduced by many philosophers—the evidence of immediate inner perception exhibited in a statement such as "I now see something white" affords no example of certain knowledge. For even as I formulate and utter the statement "I see something white" it describes a past event and I can never know whether in the period of time that has elapsed, however short, my memory has not deceived me.

There is, then, no absolute proof of freedom from contradiction for the theory of sets and thus no absolute proof of the mathematical existence of infinite sets and infinite numbers. But neither is there any such proof for the arithmetic of finite numbers, nor for the simplest parts of logic. It is a fact, however, that no contradiction is known in the theory of sets, and not a trace of evidence can be found that such a contradiction may turn up. Hence we can ascribe mathematical existence to infinite sets and to Cantor's transfinite numbers with approximately the same certainty as we ascribe existence to finite numbers.

So far we have dealt only with the question whether there are infinite sets and infinite numbers; but no less important, it would appear, is the question whether there are infinite extensions. This is usually phrased in the form: "Is space infinite?" Let us begin by treating this question also from a purely mathematical standpoint.

We must recognize at the outset that mathematics deals with very diverse kinds of space. Here, however, we are interested only in the so-called Riemann spaces, and in particular, in the three-dimensional Riemann spaces. Their exact definition does not concern us; it is sufficient to make the point that such a Riemann space is a set of elements, or points, in which certain subsets, called lines, are the objects of attention. By a process of calculation there can be assigned to every such line a positive number, called the length of the line, and among these lines there are certain ones of which every sufficiently small segment AB is shorter than every other line joining the points A,B. These lines are called the geodesics, or the straight lines of the space in question. Now it may be that in any particular Riemann space there are straight lines of arbitrarily great length; in that case we shall say that this space is of infinite extension. On the other hand, it may also be that in this particular Riemann space the length of all straight lines remains less than a fixed number; then we say that the space is of finite extension. Until the end of the 18th century only a single mathematical space was known and hence it was simply called "space." This is the space whose geometry is taught in school and which we call Euclidean space, after the Greek mathematician Euclid who was the first to develop the geometry of this space systematically. And from our definition above, this Euclidean space is of infinite extension.

There are, however, also three-dimensional Riemann spaces of finite extension; the best known of these are the so-called spherical spaces (and the closely related elliptical ones), which are three-dimensional analogues of a spherical surface. The surface of a sphere can be conceived as two-dimensional Riemann space, whose geodesics, or "straight" lines, are arcs of great circles. (A great circle is a circle cut on the surface of a sphere by a plane passing through the center of the sphere, as for instance, the equator and the meridians of longitude on the earth.) If r is the radius of the sphere, then the full circumference of a great circle is $2\pi r$; that is to say, no great circle can be longer than $2\pi r$. Hence the sphere considered as two-dimensional Riemann space is a space of finite extension. With regard to three-dimensional spherical space the situation is fully analogous; this also is a space of finite extension. Nevertheless, it has no boundaries; it is unbounded as the surface of a sphere is unbounded; one can keep walking along one of its straight lines without ever being stopped by a boundary of the space. After a finite time one simply comes back to the starting point, exactly as if one had kept moving farther and farther along a great circle of a spherical surface. In other words, we can make a circular tour of spherical space, just as easily as we can make a circular tour of the earth.

Thus we see that in a mathematical sense there are spaces of infinite extension (e.g., Euclidean space) and spaces of finite extension (e.g., spherical and elliptical spaces). Yet, this is not at all what most persons have in mind when they ask "Is space infinite?" They are asking, rather, "Is the space in which our experience and in which physical events take place of finite or of infinite extension?"

So long as no mathematical space other than Euclidean space was known, everyone naturally believed that the space of the physical world was Euclidean space, infinitely extended. Kant, who explicitly formulated this view, held that the arrangement of our observations in Euclidean space was an intuitional necessity; the basic postulates of Euclidean geometry are synthetic, *a priori* judgments.

But when it was discovered that in a purely mathematical sense spaces other than Euclidean also "existed," (that is, led to no logical contradictions) it became possible to question the view that the space of the physical world must be Euclidean space. And the idea developed that it was a question of experience, that is, a question that must be decided by experiment, whether the space of the physical world was Euclidean or not. Gauss actually made such experiments. But after the work of Henri Poincaré, the great mathematician of the end of the 19th century, we know that the question expressed in this way has no meaning. To a considerable extent we have a free choice of the kind of mathematical space in which we arrange our observations. The question does not acquire meaning until it is decided how this arrangement is to be carried out. For the important thing about Riemann space is the manner in which each of its lines is assigned a length, that is, how lengths are measured in it. If we decide that measurements of length in the space of physical events shall be made in the way they have been made from earliest times, that is, by the application of "rigid" measuring rods, then there is meaning in the question whether the space of physical events, considered as a Riemann space, is Euclidean or non-Euclidean. And the same holds for the question whether it is of finite or of infinite extension.

The answer that many perhaps are prompted to give, "Of course, by this method of measurement physical space becomes a mathematical space of infinite extension," would be somewhat too hasty. As background for a brief discussion of this problem we must first give a short and very simple statement of certain mathematical facts. Euclidean space is characterized by the fact that the sum of the three angles of a triangle in such space is 180 degrees. In spherical space the sum of the angles of every triangle is greater than 180 degrees, and the excess over 180 degrees is greater the larger the triangle in relation to the sphere. In the two-dimensional analogue of spherical space, the surface of a sphere, this point is presented to us very clearly. On the surface of a sphere, as already mentioned, the counterpart of the straight-line triangle of spherical space is a triangle whose sides are arcs of great circles, and it is a well-known proposition of elementary geometry that the sum of the angles of a spherical triangle is greater than 180 degrees, and that the excess over 180 degrees is greater the larger the surface area of the triangle. If a further comparison be made of spherical triangles of equal area on spheres of different sizes, it may be seen at once that the excess of the sum of the angles over 180 degrees is greater the smaller the diameter of the sphere, which is to say, the greater the curvature of the sphere. This gave rise to the adoption of the following terminology (and here it is simply a matter of terminology, behind which nothing whatever secret is hidden): A mathematical space is called "curved" if there are triangles in it the sum of whose angles deviates from 180 degrees. It is "positively curved" if the sum of the angles of every triangle in it (as in elliptical and spherical spaces) is greater than 180 degrees, and "negatively curved" if the sum is less than 180 degrees—as is the case in the "hyperbolic" spaces discovered by Bolyai and Lobachevsky.

From the mathematical formulations of Einstein's General Theory of Relativity it now follows that, if the previously mentioned method of measurement is used as a basis, space in the vicinity of gravitating masses must be curved in a "gravitational field." The only gravitational field immediately accessible, that of the earth, is much too weak for us to be able to test this assertion directly. It has been possible, however, to prove it indirectly by the deflection of light rays—as determined during total eclipses—in the much stronger gravitational field of the sun. So far as our present experience goes, we can say that if, by using the measuring methods mentioned above, we turn the space of physical events into a mathematical Riemann space, this mathematical space will be curved, and its curvature will, in fact, vary from place to place, being greater in the vicinity of gravitating masses and smaller far from them.

To return to the question that concerns us: Can we now say whether this space will be of finite or infinite extension? What has been said so far is not sufficient to give the answer; it is still necessary to make certain rather plausible assumptions. One such assumption is that matter is more or less evenly distributed throughout the entire space of the universe: that is to say, in the universe as a whole there is a spatially uniform density of mass. The observations of astronomers to date can, at least with the help of a little good will, be brought into harmony with this assumption. Of course it can be true only when taken in the sense of a rough average, in somewhat the same sense as it can be said that a piece of ice has on the whole the same density throughout. Just as the mass of the ice is concentrated in a great many very small particles, separated by intervening spaces that are enormous in relation to the size of these particles, so the stars in world-space are separated by intervening spaces that are enormous in relation to the size of the stars. Let us make another quite plausible assumption, namely, that the universe, taken by and large, is stationary in the sense that this average constant density of mass remains unchanged. We consider a piece of ice stationary, even though we know that the particles that constitute it are in active motion; we may likewise deem the universe to be stationary, even though we know the stars to be in active motion. With these assumptions, then, it follows from the principles of the General Theory of Relativity that the mathematical space in which we are to interpret physical events must on the whole have the same curvature throughout. Such a space, however, like the surface of a sphere in two dimensions, is necessarily of finite extension. In other words, if we use as a basis the usual method of measuring length and wish to arrange physical events in a mathematical space, and if we make the two plausible assumptions mentioned above, the conclusion follows that this space must be of finite extension.

I said that the first of our assumptions, that of the equal density of mass throughout space, conforms somewhat with observations. Is this also true of the second assumption, as to the constant density of mass with respect to time? Until recently this opinion was tenable. Now, however, certain astronomical observations seem to indicate—again speaking in broad terms—that all heavenly bodies (fixed stars and nebulae) are moving away from us with a velocity that increases the greater their distance from us, the velocity of those that are the farthest away from us, but still within reach of study, being quite fantastic. But if this is so, the average density of mass of the universe cannot possibly be constant in time; instead it must continually become smaller. Then if the remaining features of our picture of the universe are maintained, it would mean that we must assume that the mathematical space in which we interpret physical events is variable in time. At every instant it would be a space with (on the average) a constant positive curvature, that is to say, of finite extension, but the curvature would be continually decreasing while the extension would be continually increasing. This interpretation of physical events occurring in an expanding space turns out to be entirely workable and in accord with the General Theory of Relativity.

But is this the only theory consistent with our experience to date? I said before that the assumption that the space of the universe was on the whole of uniform density could fairly well be brought into harmony with astronomical observations. At the same time these observations do not contradict the entirely different assumption that we and our system of fixed stars are situated in a region of space where there is a strong con- centration of mass, while at increasing distances from this region the distribution of mass keeps getting sparser. This would lead us—still using the ordinary method of measuring length—to conceive of the physical world as situated in a space that has a certain curvature in the neighbor- hood of our fixed star system, a curvature, however, that grows smaller and smaller further away. Such a space can of course be of infinite exten- sion. Similarly the phenomenon that the stars are in general receding from us, with greater velocity the farther away they are, can be quite simply explained as follows: Assume that at some time many masses with completely different velocities were concentrated in a relatively small region of space, let us say in a sphere K. In the course of time these masses will then, each with its own particular velocity, move out of this region of the space. After a sufficient time has elapsed, those that have the greatest velocities will have moved farthest away from the sphere K, those with lesser velocities will be nearer to K, and those with the lowest velocities will still be very close to K or even within K. Then an observer within K, or at least not too far removed from K, will see the very picture of the stellar world that we have described above. The masses will on the whole be moving away from him, and those farthest away will be moving with the greatest velocities. We would thus have an interpretation of the physical world in an entirely different kind of mathematical space—that is to say, in an infinitely extended space.[5]

In summary we might very well say that the question: "Is the space of our physical world of infinite or of finite extension?" has no meaning as it stands. It does not become meaningful until we decide how we are to go about getting the observed events of the physical world into a mathematical space, that is, what assumptions must be made and what logical requirements must be satisfied. And this in turn leads to the question: "Is a finite or an infinite mathematical space better adapted for the arrangement and interpretation of physical events?" At the present stage of our knowledge we cannot give any reasonably well-founded answer to this question. It appears that mathematical spaces of finite and of infinite extension are almost equally well suited for the interpretation of the observational data thus far accumulated.

Perhaps at this point confirmed "finitists" will say: "If this is so, we prefer the scheme based on a space of finite extension, since any theory incorporating the concept of infinity is wholly unacceptable to us." They are free to take this view if they wish, but they must not imagine thereby to have altogether rid themselves of infinity. For even the finitely extended Riemann spaces contain infinitely many points, and the mathematical treatment of time is such that each time-interval, however small, contains infinitely many time-points.

Must this necessarily be so? Are we in truth compelled to lay the scene of our experience in a mathematical space or in a mathematical time that consists of infinitely many points? I say no. In principle one might very well conceive of a physics in which there were only a finite number of space points and a finite number of time points—in the language of the theory of relativity, a finite number of "world points." In my opinion neither logic nor intuition nor experience can ever prove the impossibility of such a truly finite system of physics. It may be that the various theories of the atomic structure of matter, or today's quantum physics, are the first foreshadowings of a future finite physics. If it ever comes, then we shall have returned after a prodigious circular journey to one of the starting points of western thought, that is, to the Pythagorean doctrine that everything in the world is governed by the natural numbers. If the famous theorem of the right-angle triangle rightly bears the name of Pythagoras then it was Pythagoras himself who shook the foundations of his doctrine that everything was governed by the natural numbers. For from the theorem of the right triangle there follows the existence of line segments that are incommensurable, that is, whose relationship with one another cannot be expressed by the natural numbers. And since no distinction was made between mathematical existence and physical existence, a finite physics appeared impossible. But if we are clear on the point that mathematical existence and physical existence mean basically different things; that physical existence can never follow from mathematical existence; that physical existence can in the last analysis only be proved by observation; and that the mathematical difference between rational and irrational forever transcends any possibility of observation—then we shall scarcely be able to deny the possibility in principle of a finite physics. Be that as it may, whether the future produces a finite physics or not, there will remain unimpaired the possibility and the grand beauty of a logic and a mathematics of the infinite.

PART XI

Mathematical Truth and the Structure of Mathematics

The Foundations of Mathematics

THE seven selections which follow deal with the nature of mathematical propositions, the structure of mathematical systems, mathematical inference and the concept of mathematical truth. There is no more interesting and important branch of the literature of mathematics than the study of foundations, to which these topics belong.

If one were to attempt to explain to someone who had never heard of mathematics what the subject is about it would be easy to give examples of mathematical statements but very difficult to give a general definition. The difficulty arises not only from the abstract character of the subject but also from the generality and lack of content of its propositions. It is hard to know what you are talking about in mathematics, yet no one questions the validity of what you say. There is no other realm of discourse half so queer.

Let us take an illustration. If I say "1 + 1 = 2," I am making a statement which everyone will agree is "true." (I cloak the word in quotes to concede that opinions differ as to the meaning of truth and that mathematicians and philosophers are no less troubled on this score, and no closer to a final conclusion, than was Pilate.) Why does this statement win universal acceptance? Obviously its power of conviction does not derive from my reputation as an expert. My authority on this point is neither smaller than Isaac Newton's nor greater than that of a performing crow. The statement is not to be proved true by appeals to experience. John Stuart Mill and certain other philosophers have suggested that mathematical propositions do not differ from other scientific laws, but are merely exceptionally "well-founded empirical generalizations." I regard this opinion as false and you will see it discredited in the selections below. The fact that one apple added to one apple invariably gives two apples helps in the teaching of arithmetic but has no bearing on the truth of the proposition about the sum of 1 + 1. Is the persuasiveness of the proposition adequately explained by pointing out that it is self-evident? This view does not lack supporters, but it is a purely subjective criterion and we cannot seriously argue that mathematics owes its persuasiveness to fervor of conviction or to everyman's common sense. Moreover, there are many mathematical propositions which, though as unassailable as our simple proposition, not only affront any sober man's common sense but even strain the uncommon sense of the specialist. To say that a proposition is believed because it is self-evident is only to baptize the difficulty—

as Poincaré remarked—not to solve it. Why is the statement self-evident? We are circling in the forest.

Mathematical statements are compelling, but their force is of a special kind; they are true, but their truth is uniquely defined. Mathematical reasoning is rigorous and deductive and mathematical propositions are simply the consequences of applying this reasoning to certain primitive axioms. Yet this ingrown, self-contained, iron-disciplined method is unlimited in its creativeness, unbounded in its freedom. Neither arithmetic, algebra and analysis on the one hand, nor geometry on the other hand, are empirical sciences. Mathematics cannot be validated by physical facts, nor its authority impugned or subverted by them. Yet there is a vital connection between the propositions of mathematics and the facts of the physical world. Even the symbols of pure mathematics correspond to some aspect of reality. Abstractions are after all made by men, not by other abstractions. Mathematical ideas are born of experience and are in many cases borne out by experience. It is no accident that a 25-cent piece covers the cost exactly of five five-cent candy bars, and that $5 \times 5 = 25$; nor that a straight-edge ruler is an accurate measure of length and that a straight line is a geodesic of a surface of zero curvature. Counting and measuring in the everyday world invariably parallel mathematical propositions but it is essential to distinguish between mathematical propositions and the results of counting and measuring.

Professor Carl Hempel's two essays are concerned with the nature of mathematical truth and the relation between geometry and empirical science. They treat a subtle and complex problem with uncommon clarity. Hempel, born in Germany in 1905, and a graduate of the University of Berlin, has since 1948 been professor of philosophy at Yale; earlier he taught at the College of the City of New York and Queens College. His special interest is in logic, philosophy of science, probability theory and the foundations of mathematics. He has published many papers on these topics.

From Raymond Wilder's excellent book, *Introduction to the Foundations of Mathematics*, I have taken two chapters which examine mathematical reasoning and the axiomatic method. They are first-rate pieces of exposition. Mr. Wilder, Research Professor of Mathematics at the University of Michigan, has also taught at Brown, the University of Texas and Ohio State, and has done research in the foundations of mathematics and topology at the Institute for Advanced Study and at the California Institute of Technology. (It is a commentary on the curious times in which we live that Wilder's most unmilitary study was financed by a grant from the Office of Naval Research.)

The essay by Ernest Nagel and myself on Goedel's Proof may be regarded as a companion to the Wilder selection. It is an attempt to explain to the nonspecialist what is generally regarded as the most brilliant, most difficult, and most stunning sequence of reasoning in modern logic. Goedel set out to show that the axiomatic method which has served mathematics so long and so well has limitations; in particular, that it is impossible within the framework of even a relatively simple mathematical system—ordinary whole-number arithmetic, for example—to demonstrate the internal consistency (non-contradictoriness) of the system without using principles of inference whose own consistency is as much open to question as that of the principles of the system being tested. In this endeavor he was successful; thus we reach a dead end so far as one of the major branches of mathematical research is concerned. Formal deduction has as its crowning achievement proved its own incapacity to make certain formal deductions. In a sense, therefore, formal deduction may be said to have refuted itself. If a mathematical system is to be proved flawless, other methods than the axiomatic will have to be devised for the task. But as the essay points out, Goedel's proof is occasion for neither despair nor mystery-mongering; on the contrary, it justifies "a renewed appreciation of the powers of creative reason."

For the more advanced reader I have included a selection from a classic book by Oswald Veblen and John Wesley Young, *Projective Geometry*. Veblen, one of the country's leading mathematicians, was born in Iowa in 1880, studied at Harvard and the University of Chicago, taught at Chicago and Princeton, and in 1932 became professor at the Institute for Advanced Study. He has made notable contributions to differential and projective geometry, foundations of geometry and topology. John Wesley Young (1879–1932), a native of Columbus, Ohio, was educated in German and American schools, receiving his doctorate at Cornell University. He was on the staff of Northwestern University, Princeton, and the universities of Illinois and Chicago, and for many years served as professor of mathematics at Dartmouth. His two best-known works are the volume on projective geometry with Veblen, and his *Lectures on Fundamental Concepts of Algebra and Geometry* (1911), an exemplary introduction to the subject, accessible to the general reader as well as to advanced students.

The discussion in the Veblen-Young treatise of a simple example of a mathematical science offers an entry into the structure of mathematical thinking and the deductive method. Professor Douglas Gasking of the University of Melbourne presents a quite charming philosophical analysis of the relation of mathematical propositions to the everyday world of counting and measuring. He finds an ingenious way of demonstrating that mathematics cannot be validated by experience by showing that the particular system we use is not dictated by physical events but rather selected for convenience. We could, for example, count and measure entirely differently than we usually do and use a queer multiplication table—a system, say, in which $2 \times 4 = 6$, $2 \times 8 = 10$, $4 \times 4 = 9$, $4 \times 6 = 12$; yet every activity in which these methods and this table were employed could be executed accurately. Our books would balance, our floors would be level, our bridges would stand, our bombs would explode as prettily as ever. The conventional system is of course simpler than many others and, having grown up with the human race, fits its psychological needs. But it is not the only possible system. And as we delve into strange things the need arises for strange tools. It might happen "that we found our physical laws getting very complicated indeed, and might discover that, by changing our mathematical system, we could effect a very great simplification in our physics. In such a case we might decide to use a different mathematical system." This, as Gasking reminds us, is exactly what we have done in certain important, disagreeably complex branches of contemporary physics. What is worth remembering is that mathematics has its pragmatic aspects: we should choose the mathematics that helps us understand the world; and we are free to choose it because physical events do not impose a system upon us.

The final selection is by the late Richard E. von Mises, a distinguished contemporary mathematician and a foremost exponent of the philosophy of positivism. From his *Positivism: A Study in Human Understanding* I have excerpted three sections dealing with axiomatics, logistics and the foundations of mathematics.[1] The material overlaps to some extent other selections in the group and elsewhere in the book, but I do not think you will find this objectionable. It is desirable to hear from a positivist on these matters, especially from one so moderate and knowledgeable as Von Mises. His book considers in turn the application of positivistic theory to the problems of language and communication, to mathematics and logic, to the physical sciences, philosophy and metaphysics, social studies, literature, law, ethics, poetry, religion and the fine arts. In each case his position is open and reasonable; in each case he makes a contribution to understanding. The selections given here sweep away a good deal of muddle and confusion in mathematical thinking. Von Mises merits particularly close attention when he speaks of the axiomatic formulation of a discipline and the practical purposes the axioms should serve; the relation between the so-called tautologies of mathematics (e.g., $2 + 2 = 4$) and the common experience of mathematical truths (e.g., 2 cows + 2 cows = 4 cows); the inadequacies of the three principal interpretations of the foundations of mathematics. For him, "there is no difference in principle between the disciplines of arithmetic, geometry, mechanics,

---

[1] The comment here follows on one or two points a brief review by me of Von Mises' book in *Scientific American*, February 1952, p. 79.

thermodynamics, optics, electricity, etc. . . . The foundations and basic assumptions of arithmetic are debatable in the same sense as those of any part of physics, i.e., on the one hand, as to the internal questions of tautological structure, and on the other hand, as to the relations with the world of experience." In these sections, as elsewhere in the book, the pointedness and polish of the arguments make Von Mises' brand of positivism almost irresistible. It is an outlook at once more logical and less positive (as another writer has characterized the "more chastened" form of this philosophy) than the logical positivism of the founders. No one who reads the book, says Ernest Nagel, "will credit the frequent but tiresome charge that positivism marks a failure of nerve or a decadence in thought." [2]

Dr. von Mises was born in Austria in 1883 and received his graduate degree from the University of Vienna in 1908. He taught applied mathematics at the Brno German Technical University and at the universities of Strassburg and Dresden, and was director (1920–33) of the Berlin Institute of Applied Mathematics. During the First World War he served with the Austro-Hungarian Air Force and in 1918 built the first 600 h.p. military plane. After holding a chair in mathematics at Istanbul from 1933 to 1939, he became a lecturer at Harvard and in 1943 was appointed Gordon McKay Professor of Aerodynamics and Applied Mathematics. In 1946 he became an American citizen. His research extended into many fields including the theory of flight, mechanics, theory of fluids, elasticity, probability and statistics; his writings comprise more than 100 scientific papers, two books, *Theory of Flight* and *Probability, Statistics and Truth*, essays on the philosophy of science, and a bibliography of Rainer Maria Rilke. Von Mises died after an operation in 1953.

[2] In a review of the book appearing in the *New Republic,* November 26, 1951, p. 20.

Mathematics is the most exact science, and its conclusions are capable of absolute proof. But this is so only because mathematics does not attempt to draw absolute conclusions. All mathematical truths are relative, conditional. —CHARLES PROTEUS STEINMETZ (1923)
(Quoted by E. T. Bell, Men of Mathematics)

1 On the Nature of Mathematical Truth

By CARL G. HEMPEL

1. THE PROBLEM

IT is a basic principle of scientific inquiry that no proposition and no theory is to be accepted without adequate grounds. In empirical science, which includes both the natural and the social sciences, the grounds for the acceptance of a theory consist in the agreement of predictions based on the theory with empirical evidence obtained either by experiment or by systematic observation. But what are the grounds which sanction the acceptance of mathematics? That is the question I propose to discuss in the present paper. For reasons which will become clear subsequently, I shall use the term "mathematics" here to refer to arithmetic, algebra, and analysis—to the exclusion, in particular, of geometry.

2. ARE THE PROPOSITIONS OF MATHEMATICS SELF-EVIDENT TRUTHS?

One of the several answers which have been given to our problem asserts that the truths of mathematics, in contradistinction to the hypotheses of empirical science, require neither factual evidence nor any other justification because they are "self-evident." This view, however, which ultimately relegates decisions as to mathematical truth to a feeling of self-evidence, encounters various difficulties. First of all, many mathematical theorems are so hard to establish that even to the specialist in the particular field they appear as anything but self-evident. Secondly, it is well known that some of the most interesting results of mathematics—especially in such fields as abstract set theory and topology—run counter to deeply ingrained intuitions and the customary kind of feeling of self-evidence. Thirdly, the existence of mathematical conjectures such as those of Goldbach and of Fermat, which are quite elementary in content and yet undecided up to this day, certainly shows that not all mathematical truths can be self-evident. And finally, even if self-evidence were attributed only to the basic postulates of mathematics, from which all other mathematical propositions can be deduced, it would be pertinent to remark that judgments as to what may be considered as self-evident are subjective; they may vary from person to person and certainly cannot constitute an adequate basis for decisions as to the objective validity of mathematical propositions.

3. IS MATHEMATICS THE MOST GENERAL EMPIRICAL SCIENCE?

According to another view, advocated especially by John Stuart Mill, mathematics is itself an empirical science which differs from the other branches such as astronomy, physics, chemistry, etc., mainly in two respects: its subject matter is more general than that of any other field of scientific research, and its propositions have been tested and confirmed to a greater extent than those of even the most firmly established sections of astronomy or physics. Indeed, according to this view, the degree to which the laws of mathematics have been borne out by the past experiences of mankind is so overwhelming that—unjustifiably—we have come to think of mathematical theorems as qualitatively different from the well-confirmed hypotheses or theories of other branches of science: we consider them as certain, while other theories are thought of as at best "very probable" or very highly confirmed.

But this view, too, is open to serious objections. From a hypothesis which is empirical in character—such as, for example, Newton's law of gravitation—it is possible to derive predictions to the effect that under certain specified conditions certain specified observable phenomena will occur. The actual occurrence of these phenomena constitutes confirming evidence, their non-occurrence disconfirming evidence for the hypothesis. It follows in particular that an empirical hypothesis is theoretically disconfirmable; i.e., it is possible to indicate what kind of evidence, if actually encountered, would disconfirm the hypothesis. In the light of this remark, consider now a simple "hypothesis" from arithmetic: $3 + 2 = 5$. If this is actually an empirical generalization of past experiences, then it must be possible to state what kind of evidence would oblige us to concede the hypothesis was not generally true after all. If any disconfirming evidence for the given proposition can be thought of, the following illustration might well be typical of it: We place some microbes on a slide, putting down first three of them and then another two. Afterwards we count all the microbes to test whether in this instance 3 and 2 actually added up to 5. Suppose now that we counted 6 microbes altogether. Would we consider this as an empirical disconfirmation of the given proposition, or at least as a proof that it does not apply to microbes?

Clearly not; rather, we would assume we had made a mistake in counting or that one of the microbes had split in two between the first and the second count. But under no circumstances could the phenomenon just described invalidate the arithmetical proposition in question; for the latter asserts nothing whatever about the behavior of microbes; it merely states that any set consisting of 3 + 2 objects may also be said to consist of 5 objects. And this is so because the symbols "3 + 2" and "5" denote the same number: they are synonymous by virtue of the fact that the symbols "2," "3," "5," and "+" are *defined* (or tacitly understood) in such a way that the above identity holds as a consequence of the meaning attached to the concepts involved in it.

4. THE ANALYTIC CHARACTER OF MATHEMATICAL PROPOSITIONS

The statement that $3 + 2 = 5$, then, is true for similar reasons as, say, the assertion that no sexagenarian is 45 years of age. Both are true simply by virtue of definitions or of similar stipulations which determine the meaning of the key terms involved. Statements of this kind share certain important characteristics: Their validation naturally requires no empirical evidence; they can be shown to be true by a mere analysis of the meaning attached to the terms which occur in them. In the language of logic, sentences of this kind are called analytic or true *a priori*, which is to indicate that their truth is logically independent of, or logically prior to, any experiential evidence.[1] And while the statements of empirical science, which are synthetic and can be validated only *a posteriori*, are constantly subject to revision in the light of new evidence, the truth of an analytic statement can be established definitely, once and for all. However, this characteristic "theoretical certainty" of analytic propositions has to be paid for at a high price: An analytic statement conveys no factual information. Our statement about sexagenarians, for example, asserts nothing that could possibly conflict with any factual evidence: it has no factual implications, no empirical content; and it is precisely for this reason that the statement can be validated without recourse to empirical evidence.

Let us illustrate this view of the nature of mathematical propositions by reference to another, frequently cited, example of a mathematical—or

---

[1] The objection is sometimes raised that without certain types of experience, such as encountering several objects of the same kind, the integers and the arithmetical operations with them would never have been invented, and that therefore the propositions of arithmetic do have an empirical basis. This type of argument, however, involves a confusion of the logical and the psychological meaning of the term "basis." It may very well be the case that certain experiences occasion psychologically the formation of arithmetical ideas and in this sense form an empirical "basis" for them; but this point is entirely irrelevant for the logical questions as to the *grounds* on which the propositions of arithmetic may be accepted as true. The point made above is that no empirical "basis" or evidence whatever is needed to establish the truth of the propositions of arithmetic.

rather logical—truth, namely the proposition that whenever $a = b$ and $b = c$ then $a = c$. On what grounds can this so-called "transitivity of identity" be asserted? Is it of an empirical nature and hence at least theoretically disconfirmable by empirical evidence? Suppose, for example, that $a, b, c,$ are certain shades of green, and that as far as we can see, $a = b$ and $b = c$, but clearly $a \neq c$. This phenomenon actually occurs under certain conditions; do we consider it as disconfirming evidence for the proposition under consideration? Undoubtedly not; we would argue that if $a \neq c$, it is impossible that $a = b$ and also $b = c$; between the terms of at least one of these latter pairs, there must obtain a difference, though perhaps only a subliminal one. And we would dismiss the possibility of empirical disconfirmation, and indeed the idea that an empirical test should be relevant here, on the grounds that identity is a transitive relation by virtue of its definition or by virtue of the basic postulates governing it.[2] Hence, the principle in question is true *a priori*.

5. MATHEMATICS AS AN AXIOMATIZED DEDUCTIVE SYSTEM

I have argued so far that the validity of mathematics rests neither on its alleged self-evidential character nor on any empirical basis, but derives from the stipulations which determine the meaning of the mathematical concepts, and that the propositions of mathematics are therefore essentially "true by definition." This latter statement, however, is obviously oversimplified and needs restatement and a more careful justification.

For the rigorous development of a mathematical theory proceeds not simply from a set of definitions but rather from a set of non-definitional propositions which are not proved within the theory; these are the postulates or axioms of the theory.[3] They are formulated in terms of certain basic or primitive concepts for which no definitions are provided within the theory. It is sometimes asserted that the postulates themselves represent "implicit definitions" of the primitive terms. Such a characterization of the postulates, however, is misleading. For while the postulates do limit, in a specific sense, the meanings that can possibly be ascribed to the primitives, any self-consistent postulate system admits, nevertheless, many different interpretations of the primitive terms (this will soon be illustrated), whereas a set of definitions in the strict sense of the word determines the meanings of the definienda in a unique fashion.

Once the primitive terms and the postulates have been laid down, the entire theory is completely determined; it is derivable from its postulational basis in the following sense: Every term of the theory is definable

[2] A precise account of the definition and the essential characteristics of the identity relation may be found in A. Tarski, *Introduction to Logic*, New York, 1941, ch. III.
[3] For a lucid and concise account of the axiomatic method, see A. Tarski, *loc. cit.*, ch. VI.

in terms of the primitives, and every proposition of the theory is deducible from the postulates. To be entirely precise, it is necessary to specify the principles of logic which are to be used in the proof of the propositions, i.e., in their deduction from the postulates. These principles can be stated quite explicitly. They fall into two groups: Primitive sentences, or postulates, of logic (such as: If *p* and *q* is the case, then *p* is the case), and rules of deduction or inference (including, for example, the familiar *modus ponens* rule and the rules of substitution which make it possible to infer, from a general proposition, any one of its substitution instances). A more detailed discussion of the structure and content of logic would, however, lead too far afield in the context of this article.

6. PEANO'S AXIOM SYSTEM AS A BASIS FOR MATHEMATICS

Let us now consider a postulate system from which the entire arithmetic of the natural numbers can be derived. This system was devised by the Italian mathematician and logician G. Peano (1858–1932). The primitives of this system are the terms "0" "number," and "successor." While, of course, no definition of these terms is given within the theory, the symbol "0" is intended to designate the number 0 in its usual meaning, while the term "number" is meant to refer to the natural numbers 0, 1, 2, 3 . . . exclusively. By the successor of a natural number *n*, which will sometimes briefly be called *n'*, is meant the natural number immediately following *n* in the natural order. Peano's system contains the following 5 postulates:

P1. 0 is a number

P2. The successor of any number is a number

P3. No two numbers have the same successor

P4. 0 is not the successor of any number

P5. If *P* is a property such that (a) 0 has the property *P*, and (b) whenever a number *n* has the property *P*, then the successor of *n* also has the property *P*, then every number has the property *P*.

The last postulate embodies the principle of mathematical induction and illustrates in a very obvious manner the enforcement of a mathematical "truth" by stipulation. The construction of elementary arithmetic on this basis begins with the definition of the various natural numbers. 1 is defined as the successor of 0, or briefly as 0'; 2 as 1', 3 as 2', and so on. By virtue of P2, this process can be continued indefinitely; because of P3 (in combination with P5), it never leads back to one of the numbers previously defined, and in view of P4, it does not lead back to 0 either.

As the next step, we can set up a definition of addition which expresses in a precise form the idea that the addition of any natural number to some given number may be considered as a repeated addition of 1; the latter operation is readily expressible by means of the successor relation. This definition of addition runs as follows:

D1. (a) $n + 0 = n$; (b) $n + k' = (n + k)'$.

The two stipulations of this recursive definition completely determine the sum of any two integers. Consider, for example, the sum $3 + 2$. According to the definitions of the numbers 2 and 1, we have $3 + 2 = 3 + 1' = 3 + (0')'$; by D1 (b), $3 + (0')' = (3 + 0')' = ((3 + 0)')'$; but by D1 (a), and by the definitions of the numbers 4 and 5, $((3 + 0)')' = (3')' = 4' = 5$. This proof also renders more explicit and precise the comments made earlier in this paper on the truth of the proposition that $3 + 2 = 5$: Within the Peano system of arithmetic, its truth flows not merely from the definition of the concepts involved, but also from the postulates that govern these various concepts. (In our specific example, the postulates P1 and P2 are presupposed to guarantee that 1, 2, 3, 4, 5, are numbers in Peano's system; the general proof that D1 determines the sum of any two numbers also makes use of P5.) If we call the postulates and definitions of an axiomatized theory the "stipulations" concerning the concepts of that theory, then we may say now that the propositions of the arithmetic of the natural numbers are true by virtue of the stipulations which have been laid down initially for the arithmetical concepts. (Note, incidentally, that our proof of the formula "$3 + 2 = 5$" repeatedly made use of the transitivity of identity; the latter is accepted here as one of the rules of logic which may be used in the proof of any arithmetical theorem; it is, therefore, included among Peano's postulates no more than any other principle of logic.)

Now, the multiplication of natural numbers may be defined by means of the following recursive definition, which expresses in a rigorous form the idea that a product nk of two integers may be considered as the sum of k terms each of which equals n.

D2. (a) $n \cdot 0 = 0$; (b) $n \cdot k' = n \cdot k + n$.

It now is possible to prove the familiar general laws governing addition and multiplication, such as the commutative, associative, and distributive laws ($n + k = k + n$; $n \cdot k = k \cdot n$; $n + (k + l) = (n + k) + l$; $n \cdot (k \cdot l) = (n \cdot k) \cdot l$; $n \cdot (k + l) = (n \cdot k) + (n \cdot l)$). In terms of addition and multiplication, the inverse operations of subtraction and division can then be defined. But it turns out that these "cannot always be performed"; i.e., in contradistinction to the sum and the product, the difference and the quotient are not defined for every couple of numbers; for example, $7 - 10$ and $7 \div 10$ are undefined. This situation suggests an enlargement of the number system by the introduction of negative and of rational numbers.

It is sometimes held that in order to effect this enlargement, we have to "assume" or else to "postulate" the existence of the desired additional kinds of numbers with properties that make them fit to fill the gaps of subtraction and division. This method of simply postulating what we want has its advantages; but, as Bertrand Russell [4] puts it, they are the same as the advantages of theft over honest toil; and it is a remarkable fact that the negative as well as the rational numbers can be obtained from Peano's primitives by the honest toil of constructing explicit definitions for them, without the introduction of any new postulates or assumptions whatsoever. Every positive and negative integer (in contradistinction to a natural number which has no sign) is definable as a certain set of ordered couples of natural numbers; thus, the integer $+2$ is definable as the set of all ordered couples (m, n) of natural numbers where $m = n + 2$; the integer -2 is the set of all ordered couples (m, n) of natural numbers with $n = m + 2$. (Similarly, rational numbers are defined as classes of ordered couples of integers.) The various arithmetical operations can then be defined with reference to these new types of numbers, and the validity of all the arithmetical laws governing these operations can be proved by virtue of nothing more than Peano's postulates and the definitions of the various arithmetical concepts involved.

The much broader system thus obtained is still incomplete in the sense that not every number in it has a square root, and more generally, not every algebraic equation whose coefficients are all numbers of the system has a solution in the system. This suggests further expansions of the number system by the introduction of real and finally of complex numbers. Again, this enormous extension can be effected by mere definition, without the introduction of a single new postulate.[5] On the basis thus obtained, the various arithmetical and algebraic operations can be defined for the numbers of the new system, the concepts of function, of limit, of derivative and integral can be introduced, and the familiar theorems pertaining to these concepts can be proved, so that finally the huge system of mathematics as here delimited rests on the narrow basis of Peano's system: Every concept of mathematics can be defined by means of Peano's three primitives, and every proposition of mathematics can be deduced from the five postulates enriched by the definitions of the non-primitive

[4] Bertrand Russell, *Introduction to Mathematical Philosophy*, New York and London, 1919, p. 71.
[5] For a more detailed account of the construction of the number system on Peano's basis, cf. Bertrand Russell, *loc. cit.*, esp. chs. I and VII. A rigorous and concise presentation of that construction, beginning, however, with the set of all integers rather than that of the natural numbers, may be found in G. Birkhoff and S. MacLane, *A Survey of Modern Algebra*, New York, 1941, chs. I, II, III, V. For a general survey of the construction of the number system, cf. also J. W. Young, *Lectures on the Fundamental Concepts of Algebra and Geometry*, New York, 1911, esp. lectures X, XI, XII.

terms.[6] These deductions can be carried out, in most cases, by means of nothing more than the principles of formal logic; the proof of some theorems concerning real numbers, however, requires one assumption which is not usually included among the latter. This is the so-called axiom of choice. It asserts that given a class of mutually exclusive classes, none of which is empty, there exists at least one class which has exactly one element in common with each of the given classes. By virtue of this principle and the rules of formal logic, the content of all of mathematics can thus be derived from Peano's modest system—a remarkable achievement in systematizing the content of mathematics and clarifying the foundations of its validity.

7. INTERPRETATIONS OF PEANO'S PRIMITIVES

As a consequence of this result, the whole system of mathematics might be said to be true by virtue of mere definitions (namely, of the non-primitive mathematical terms) provided that the five Peano postulates are true. However, strictly speaking, we cannot, at this juncture, refer to the Peano postulates as propositions which are either true or false, for they contain three primitive terms which have not been assigned any specific meaning. All we can assert so far is that any specific interpretation of the primitives which satisfies the five postulates—i.e., turns them into true statements—will also satisfy all the theorems deduced from them. But for Peano's system, there are several—indeed, infinitely many—interpretations which will do this. For example, let us understand by 0 the origin of a half-line, by the successor of a point on that half-line the point 1 cm. behind it, counting from the origin, and by a number any point which is either the origin or can be reached from it by a finite succession of steps each of which leads from one point to its successor. It can then readily be seen that all the Peano postulates as well as the ensuing theo-

---

[6] As a result of very deep-reaching investigations carried out by K. Gödel it is known that arithmetic, and *a fortiori* mathematics, is an incomplete theory in the following sense: While all those propositions which belong to the classical systems of arithmetic, algebra, and analysis can indeed be derived, in the sense characterized above, from the Peano postulates, there exist nevertheless other propositions which can be expressed in purely arithmetical terms, and which are true, but which cannot be derived from the Peano system. And more generally: For any postulate system of arithmetic (or of mathematics for that matter) which is not self-contradictory, there exist propositions which are true, and which can be stated in purely arithmetical terms, but which cannot be derived from that postulate system. In other words, it is impossible to construct a postulate system which is not self-contradictory, and which contains among its consequences all true propositions which can be formulated within the language of arithmetic.

This fact does not, however, affect the result outlined above, namely, that it is possible to deduce, from the Peano postulates and the additional definitions of non-primitive terms, all those propositions which constitute the classical theory of arithmetic, algebra, and analysis; and it is to these propositions that I refer above and subsequently as the propositions of mathematics.

rems turn into true propositions, although the interpretation given to the primitives is certainly not the customary one, which was mentioned earlier. More generally, it can be shown that every progression of elements of any kind provides a true interpretation, or a "model," of the Peano system. This example illustrates our earlier observation that a postulate system cannot be regarded as a set of "implicit definitions" for the primitive terms: The Peano system permits of many different interpretations, whereas in everyday as well as in scientific language, we attach one specific meaning to the concepts of arithmetic. Thus, e.g., in scientific and in everyday discourse, the concept 2 is understood in such a way that from the statement "Mr. Brown as well as Mr. Cope, but no one else is in the office, and Mr. Brown is not the same person as Mr. Cope," the conclusion "Exactly two persons are in the office" may be validly inferred. But the stipulations laid down in Peano's system for the natural numbers, and for the number 2 in particular, do not enable us to draw this conclusion; they do not "implicitly determine" the customary meaning of the concept 2 or of the other arithmetical concepts. And the mathematician cannot acquiesce in this deficiency by arguing that he is not concerned with the customary meaning of the mathematical concepts; for in proving, say, that every positive real number has exactly two real square roots, he is himself using the concept 2 in its customary meaning, and his very theorem cannot be proved unless we presuppose more about the number 2 than is stipulated in the Peano system.

If therefore mathematics is to be a correct theory of the mathematical concepts in their intended meaning, it is not sufficient for its validation to have shown that the entire system is derivable from the Peano postulates plus suitable definitions; rather, we have to inquire further whether the Peano postulates are actually true when the primitives are understood in their customary meaning. This question, of course, can be answered only after the customary meaning of the terms "0," "natural number," and "successor" has been clearly defined. To this task we now turn.

8. DEFINITION OF THE CUSTOMARY MEANING OF THE CONCEPTS OF ARITHMETIC IN PURELY LOGICAL TERMS

At first blush, it might seem a hopeless undertaking to try to define these basic arithmetical concepts without presupposing other terms of arithmetic, which would involve us in a circular procedure. However, quite rigorous definitions of the desired kind can indeed be formulated, and it can be shown that for the concepts so defined, all Peano postulates turn into true statements. This important result is due to the research of the German logician G. Frege (1848–1925) and to the subsequent systematic and detailed work of the contemporary English logicians and philosophers B. Russell and A. N. Whitehead. Let us consider briefly the basic ideas underlying these definitions.[7]

A natural number—or, in Peano's term, a number—in its customary meaning can be considered as characteristic of certain *classes* of objects. Thus, e.g., the class of the apostles has the number 12, the class of the Dionne quintuplets the number 5, any couple the number 2, and so on. Let us now express precisely the meaning of the assertion that a certain class C has the number 2, or briefly, that $n(C) = 2$. Brief reflection will show that the following definiens is adequate in the sense of the customary meaning of the concept 2: There is some object x and some object y such that (1) $x \epsilon C$ (i.e., x is an element of C) and $y \epsilon C$, (2) $x \neq y$, and (3) if z is any object such that $z \epsilon C$, then either $z = x$ or $z = y$. (Note that on the basis of this definition it becomes indeed possible to infer the statement "The number of persons in the office is 2" from "Mr. Brown as well as Mr. Cope, but no one else is in the office, and Mr. Brown is not identical with Mr. Cope"; C is here the class of persons in the office.) Analogously, the meaning of the statement that $n(C) = 1$ can be defined thus: There is some x such that $x \epsilon C$, and any object y such that $y \epsilon C$, is identical with x. Similarly, the customary meaning of the statement that $n(C) = 0$ is this: There is no object such that $x \epsilon C$.

The general pattern of these definitions clearly lends itself to the definition of any natural number. Let us note especially that in the definitions thus obtained, the definiens never contains any arithmetical term, but merely expressions taken from the field of formal logic, including the signs of identity and difference. So far, we have defined only the meaning of such phrases as "$n(C) = 2$," but we have given no definition for the numbers 0, 1, 2, . . . apart from this context. This desideratum can be met on the basis of the consideration that 2 is that property which is common to all couples, i.e., to all classes C such that $n(C) = 2$. This common property may be conceptually represented by the class of all those classes which share this property. Thus we arrive at the definition: 2 is the class of all couples, i.e., the class of all classes C for which $n(C) = 2$. This definition is by no means circular because the concept of couple—in other words, the meaning of "$n(C) = 2$"—has been previously defined without any reference to the number 2. Analogously, 1 is the class of all unit classes, i.e., the class of all classes C for which $n(C) = 1$. Finally,

[7] For a more detailed discussion, cf. Russell, *loc. cit.*, chs. II, III, IV. A complete technical development of the idea can be found in the great standard work in mathematical logic, A. N. Whitehead and B. Russell, *Principia Mathematica*, Cambridge, England, 1910–1913. For a very precise development of the theory, see W. V. O. Quine, *Mathematical Logic*, New York, 1940. A specific discussion of the Peano system and its interpretations from the viewpoint of semantics is included in R. Carnap, *Foundations of Logic and Mathematics*, International Encyclopedia of Unified Science, Vol. I, no. 3, Chicago, 1939; especially sections 14, 17, 18.

0 is the class of all null classes, i.e., the class of all classes without elements. And as there is only one such class, 0 is simply the class whose only element is the null class. Clearly, the customary meaning of any given natural number can be defined in this fashion.[8] In order to characterize the intended interpretation of Peano's primitives, we actually need, of all the definitions here referred to, only that of the number 0. It remains to define the terms "successor" and "integer."

The definition of "successor," whose precise formulation involves too many niceties to be stated here, is a careful expression of a simple idea which is illustrated by the following example: Consider the number 5, i.e., the class of all quintuplets. Let us select an arbitrary one of these quintuplets and add to it an object which is not yet one of its members. 5′, the successor of 5, may then be defined as the number applying to the set thus obtained (which, of course, is a sextuplet). Finally, it is possible to formulate a definition of the customary meaning of the concept of natural number; this definition, which again cannot be given here, expresses, in a rigorous form, the idea that the class of the natural numbers consists of the number 0, its successor, the successor of that successor, and so on.

If the definitions here characterized are carefully written out—this is one of the cases where the techniques of symbolic, or mathematical, logic prove indispensable—it is seen that the definiens of every one of them contains exclusively terms from the field of pure logic. In fact, it is possible to state the customary interpretation of Peano's primitives, and thus also the meaning of every concept definable by means of them—and that includes every concept of mathematics—in terms of the following seven expressions (in addition to variables such as "x" and "C"): *not, and, if— then; for every object x it is the case that . . .; there is some object x such that . . . ; x is an element of class C; the class of all things x such that . . .* And it is even possible to reduce the number of logical concepts needed to a mere four: The first three of the concepts just mentioned are all definable in terms of "*neither—nor,*" and the fifth is definable by means of the fourth and "*neither—nor.*" Thus, all the concepts of mathematics prove definable in terms of four concepts of pure logic. (The

---

[8] The assertion that the definitions given above state the "customary" meaning of the arithmetical terms involved is to be understood in the logical, not the psychological sense of the term "meaning." It would obviously be absurd to claim that the above definitions express "what everybody has in mind" when talking about numbers and the various operations that can be performed with them. What is achieved by those definitions is rather a "logical reconstruction" of the concepts of arithmetic in the sense that if the definitions are accepted, then those statements in science and everyday discourse which involve arithmetical terms can be interpreted coherently and systematically in such a manner that they are capable of objective validation. The statement about the two persons in the office provides a very elementary illustration of what is meant here.

definition of one of the more complex concepts of mathematics in terms of the four primitives just mentioned may well fill hundreds or even thousands of pages; but clearly this affects in no way the theoretical importance of the result just obtained; it does, however, show the great convenience and indeed practical indispensabiilty for mathematics of having a large system of highly complex defined concepts available.)

9. THE TRUTH OF PEANO'S POSTULATES IN THEIR CUSTOMARY INTERPRETATION

The definitions characterized in the preceding section may be said to render precise and explicit the customary meaning of the concepts of arithmetic. Moreover—and this is crucial for the question of the validity of mathematics—it can be shown that the Peano postulates all turn into true propositions if the primitives are construed in accordance with the definitions just considered.

Thus, P1 (0 is a number) is true because the class of all numbers—i.e., natural numbers—was defined as consisting of 0 and all its successors. The truth of P2 (the successor of any number is a number) follows from the same definition. This is true also of P5, the principle of mathematical induction. To prove this, however, we would have to resort to the precise definition of "integer" rather than the loose description given of that definition above. P4 (0 is not the successor of any number) is seen to be true as follows: By virtue of the definition of "successor," a number which is a successor of some number can apply only to classes which contain at least one element; but the number 0, by definition, applies to a class if and only if that class is empty. While the truth of P1, P2, P4, P5 can be inferred from the above definitions simply by means of the principles of logic, the proof of P3 (no two numbers have the same successor) presents a certain difficulty. As was mentioned in the preceding section, the definition of the successor of a number n is based on the process of adding, to a class of n elements, one element not yet contained in that class. Now if there should exist only a finite number of things altogether then this process could not be continued indefinitely, and P3, which (in conjunction with P1 and P2) implies that the integers form an infinite set, would be false. This difficulty can be met by the introduction of a special "axiom of infinity" [9] which asserts, in effect, the existence of infinitely many objects, and thus makes P3 demonstrable. The axiom of infinity does not belong to the generally recognized laws of logic; but it is capable of expression in purely logical terms and may be considered as an additional postulate of modern logical theory.

[9] Cf. Bertrand Russell, *loc. cit.*, p. 24 and ch. XIII.

10. MATHEMATICS AS A BRANCH OF LOGIC

As was pointed out earlier, all the theorems of arithmetic, algebra, and analysis can be deduced from the Peano postulates and the definitions of those mathematical terms which are not primitives in Peano's system. This deduction requires only the principles of logic plus, in certain cases, the axiom of choice. By combining this result with what has just been said about the Peano system, the following conclusion is obtained, which is also known as *the thesis of logicism concerning the nature of mathematics*:

Mathematics is a branch of logic. It can be derived from logic in the following sense:

a. All the concepts of mathematics, i.e., of arithmetic, algebra, and analysis, can be defined in terms of four concepts of pure logic.

b. All the theorems of mathematics can be deduced from those definitions by means of the principles of logic (including the axioms of infinity and choice).[10]

In this sense it can be said that the propositions of the system of mathematics as here delimited are true by virtue of the definitions of the mathematical concepts involved, or that they make explicit certain characteristics with which we have endowed our mathematical concepts by definition. The propositions of mathematics have, therefore, the same unquestionable certainty which is typical of such propositions as "All bachelors are unmarried," but they also share the complete lack of empirical content which is associated with that certainty: The propositions of mathematics are devoid of all factual content; they convey no information whatever on any empirical subject matter.

11. ON THE APPLICABILITY OF MATHEMATICS TO EMPIRICAL SUBJECT MATTER

This result seems to be irreconcilable with the fact that after all mathematics has proved to be eminently applicable to empirical subject matter, and that indeed the greater part of present-day scientific knowledge has been reached only through continual reliance on and application of the propositions of mathematics. Let us try to clarify this apparent paradox by reference to some examples.

Suppose that we are examining a certain amount of some gas, whose

---

[10] The principles of logic developed in modern systems of formal logic embody certain restrictions as compared with those logical rules which had been rather generally accepted as sound until about the turn of the 20th century. At that time, the discovery of the famous paradoxes of logic, especially of Russell's paradox (cf. Russell, *loc. cit.*, ch. XIII), revealed the fact that the logical principles implicit in customary mathematical reasoning involved contradictions and therefore had to be curtailed in one manner or another.

volume v, at a certain fixed temperature, is found to be 9 cubic feet when the pressure p is 4 atmospheres. And let us assume further that the volume of the gas for the same temperature and $p = 6$ *at.*, is predicted by means of Boyle's law. Using elementary arithmetic we reason thus: For corresponding values of v and p, $vp = c$, and $v = 9$ when $p = 4$; hence $c = 36$: Therefore, when $p = 6$, then $v = 6$. Suppose that this prediction is borne out by subsequent test. Does that show that the arithmetic used has a predictive power of its own, that its propositions have factual implications? Certainly not. All the predictive power here deployed, all the empirical content exhibited stems from the initial data and from Boyle's law, which asserts that $vp = c$ for *any* two corresponding values of v and p, hence also for $v = 9$, $p = 4$, and for $p = 6$ and the corresponding value of v.[11] The function of the mathematics here applied is not predictive at all; rather, it is analytic or explicative: it renders explicit certain assumptions or assertions which are included in the content of the premises of the argument (in our case, these consist of Boyle's law plus the additional data); mathematical reasoning reveals that those premises contain—hidden in them, as it were,—an assertion about the case as yet unobserved. In accepting our premises—so arithmetic reveals—we have—knowingly or unknowingly—already accepted the implication that the p-value in question is 6. Mathematical as well as logical reasoning is a conceptual technique of making explicit what is implicitly contained in a set of premises. The conclusions to which this technique leads assert nothing that is *theoretically new* in the sense of not being contained in the content of the premises. But the results obtained may well be *psychologically new*: we may not have been aware, before using the techniques of logic and mathematics, what we committed ourselves to in accepting a certain set of assumptions or assertions.

A similar analysis is possible in all other cases of applied mathematics, including those involving, say, the calculus. Consider, for example, the hypothesis that a certain object, moving in a specified electric field, will undergo a constant acceleration of 5 feet/sec^2. For the purpose of testing this hypothesis, we might derive from it, by means of two successive integrations, the prediction that if the object is at rest at the beginning of the motion, then the distance covered by it at any time t is $\frac{5}{2}t^2$ feet. This conclusion may clearly be psychologically new to a person not acquainted with the subject, but it is not theoretically new; the content of the conclusion is already contained in that of the hypothesis about the constant acceleration. And indeed, here as well as in the case of the compression of a gas, a failure of the prediction to come true would be considered as indicative of the factual incorrectness of at least one of the premises in-

---

[11] Note that we may say "hence" by virtue of the rule of substitution, which is one of the rules of logical inference.

volved (*f.ex.*, of Boyle's law in its application to the particular gas), but never as a sign that the logical and mathematical principles involved might be unsound.

Thus, in the establishment of empirical knowledge, mathematics (as well as logic) has, so to speak, the function of a theoretical juice extractor: the techniques of mathematical and logical theory can produce no more juice of factual information than is contained in the assumptions to which they are applied; but they may produce a great deal more juice of this kind than might have been anticipated upon a first intuitive inspection of those assumptions which form the raw material for the extractor.

At this point, it may be well to consider briefly the status of those mathematical disciplines which are not outgrowths of arithmetic and thus of logic; these include in particular topology, geometry, and the various branches of abstract algebra, such as the theory of groups, lattices, fields, etc. Each of these disciplines can be developed as a purely deductive system on the basis of a suitable set of postulates. If P be the conjunction of the postulates for a given theory, then the proof of a proposition T of that theory consists in deducing T from P by means of the principles of formal logic. What is established by the proof is therefore not the truth of T, but rather the fact that T is true provided that the postulates are. But since both P and T contain certain primitive terms of the theory, to which no specific meaning is assigned, it is not strictly possible to speak of the truth of either P or T; it is therefore more adequate to state the point as follows: If a proposition T is logically deduced from P, then every specific interpretation of the primitives which turns all the postulates of P into true statements, will also render T a true statement. Up to this point, the analysis is exactly analogous to that of arithmetic as based on Peano's set of postulates. In the case of arithmetic, however, it proved possible to go a step further, namely to define the customary meanings of the primitives in terms of purely logical concepts and to show that the postulates—and therefore also the theorems—of arithmetic are unconditionally true by virtue of these definitions. An analogous procedure is not applicable to those disciplines which are not outgrowths of arithmetic: The primitives of the various branches of abstract algebra have no specific "customary meaning"; and if geometry in its customary interpretation is thought of as a theory of the structure of physical space, then its primitives have to be construed as referring to certain types of physical entities, and the question of the truth of a geometrical theory in this interpretation turns into an *empirical* problem.[12] For the purpose of applying any one of these non-arithmetical disciplines to some specific field of mathematics or empirical science, it is therefore necessary first to assign to the primitives some specific meaning and then to ascertain whether in this interpretation

[12] For a more detailed discussion of this point, cf. the next article in this collection.

the postulates turn into true statements. If this is the case, then we can be sure that all the theorems are true statements too, because they are logically derived from the postulates and thus simply explicate the content of the latter in the given interpretation. In their application to empirical subject matter, therefore, these mathematical theories no less than those which grow out of arithmetic and ultimately out of pure logic, have the function of an analytic tool, which brings to light the implications of a given set of assumptions but adds nothing to their content.

But while mathematics in no case contributes anything to the content of our knowledge of empirical matters, it is entirely indispensable as an instrument for the validation and even for the linguistic expression of such knowledge: The majority of the more far-reaching theories in empirical science—including those which lend themselves most eminently to prediction or to practical application—are stated with the help of mathematical concepts; the formulation of these theories makes use, in particular, of the number system, and of functional relationships among different metrical variables. Furthermore, the scientific test of these theories, the establishment of predictions by means of them, and finally their practical application, all require the deduction, from the general theory, of certain specific consequences; and such deduction would be entirely impossible without the techniques of mathematics which reveal what the given general theory implicitly asserts about a certain special case.

Thus, the analysis outlined on these pages exhibits the system of mathematics as a vast and ingenious conceptual structure without empirical content and yet an indispensable and powerful theoretical instrument for the scientific understanding and mastery of the world of our experience.

2 Geometry and Empirical Science

By CARL G. HEMPEL

1. INTRODUCTION

THE most distinctive characteristic which differentiates mathematics from
the various branches of empirical science, and which accounts for its fame
as the queen of the sciences, is no doubt the peculiar certainty and necessity of its results. No proposition in even the most advanced parts of
empirical science can ever attain this status; a hypothesis concerning
"matters of empirical fact" can at best acquire what is loosely called a
high probability or a high degree of confirmation on the basis of the relevant evidence available; but however well it may have been confirmed by
careful tests, the possibility can never be precluded that it will have to be
discarded later in the light of new and disconfirming evidence. Thus, all
the theories and hypotheses of empirical science share this provisional
character of being established and accepted "until further notice," whereas
a mathematical theorem, once proved, is established once and for all; it
holds with that particular certainty which no subsequent empirical discoveries, however unexpected and extraordinary, can ever affect to the
slightest extent. It is the purpose of this paper to examine the nature of
that proverbial "mathematical certainty" with special reference to geometry, in an attempt to shed some light on the question as to the validity
of geometrical theories, and their significance for our knowledge of the
structure of physical space.

The nature of mathematical truth can be understood through an analysis
of the method by means of which it is established. On this point I can be
very brief: it is the method of mathematical demonstration, which consists in the logical deduction of the proposition to be proved from other
propositions, previously established. Clearly, this procedure would involve
an infinite regress unless some propositions were accepted without proof;
such propositions are indeed found in every mathematical discipline which
is rigorously developed; they are the *axioms* or *postulates* (we shall use
these terms interchangeably) of the theory. Geometry provides the historically first example of the axiomatic presentation of a mathematical
discipline. The classical set of postulates, however, on which Euclid based
his system, has proved insufficient for the deduction of the well-known

theorems of so-called euclidean geometry; it has therefore been revised and supplemented in modern times, and at present various adequate systems of postulates for euclidean geometry are available; the one most closely related to Euclid's system is probably that of Hilbert.

2. THE INADEQUACY OF EUCLID'S POSTULATES

The inadequacy of Euclid's own set of postulates illustrates a point which is crucial for the axiomatic method in modern mathematics: Once the postulates for a theory have been laid down, every further proposition of the theory must be proved exclusively by logical deduction from the postulates; any appeal, explicit or implicit, to a feeling of self-evidence, or to the characteristics of geometrical figures, or to our experiences concerning the behavior of rigid bodies in physical space, or the like, is strictly prohibited; such devices may have a heuristic value in guiding our efforts to find a strict proof for a theorem, but the proof itself must contain absolutely no reference to such aids. This is particularly important in geometry, where our so-called intuition of geometrical relationships, supported by reference to figures or to previous physical experiences, may induce us tacitly to make use of assumptions which are neither formulated in our postulates nor provable by means of them. Consider, for example, the theorem that in a triangle the three medians bisecting the sides intersect in one point which divides each of them in the ratio of $1:2$. To prove this theorem, one shows first that in any triangle ABC (see figure) the line segment MN which connects the centers of AB and AC is parallel to BC and therefore half as long as the latter side. Then the lines BN and CM are drawn, and an examination of the triangles MON and BOC leads to the proof of the theorem. In this procedure, it is usually taken for granted that BN and CM intersect in a point O which lies between B and N as well as between C and M. This assumption is based on geometrical

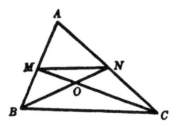

intuition, and indeed, it cannot be deduced from Euclid's postulates; to make it strictly demonstrable and independent of any reference to intuition, a special group of postulates has been added to those of Euclid; they are the postulates of order. One of these—to give an example—asserts that if A, B, C are points on a straight line l, and if B lies between A and C, then B also lies between C and A. Not even as "trivial" an assumption as this may be taken for granted; the system of postulates has to be made so complete that all the required propositions can be deduced from it by purely logical means.

Another illustration of the point under consideration is provided by the proposition that triangles which agree in two sides and the enclosed angle, are congruent. In Euclid's Elements, this proposition is presented as a theorem; the alleged proof, however, makes use of the ideas of motion and superimposition of figures and thus involves tacit assumptions which are based on our geometric intuition and on experiences with rigid bodies, but which are definitely not warranted by—i.e., deducible from—Euclid's postulates. In Hilbert's system, therefore, this proposition (more precisely: part of it) is explicitly included among the postulates.

3. MATHEMATICAL CERTAINTY

It is this purely deductive character of mathematical proof which forms the basis of mathematical certainty: What the rigorous proof of a theorem —say the proposition about the sum of the angles in a triangle—establishes is not the truth of the proposition in question but rather a conditional insight to the effect that that proposition is certainly true *provided that* the postulates are true; in other words, the proof of a mathematical proposition establishes the fact that the latter is logically implied by the postulates of the theory in question. Thus, each mathematical theorem can be cast into the form

$$(P_1 \cdot P_2 \cdot P_3 \cdot \; \ldots \; . P_n) \to T$$

where the expression on the left is the conjunction (joint assertion) of all the postulates, the symbol on the right represents the theorem in its customary formulation, and the arrow expresses the relation of logical implication or entailment. Precisely this character of mathematical theorems is the reason for their peculiar certainty and necessity, as I shall now attempt to show.

It is typical of any purely logical deduction that the conclusion to which it leads simply re-asserts (a proper or improper) part of what has already been stated in the premises. Thus, to illustrate this point by a very elementary example, from the premise, "This figure is a right triangle," we can deduce the conclusion, "This figure is a triangle"; but this conclusion clearly reiterates part of the information already contained in the premise. Again, from the premises, "All primes different from 2 are odd" and "n is a prime different from 2," we can infer logically that n is odd; but this consequence merely repeats part (indeed a relatively small part) of the information contained in the premises. The same situation prevails in all other cases of logical deduction; and we may, therefore, say that logical deduction—which is the one and only method of mathematical proof—is a technique of conceptual analysis: it discloses what assertions are concealed in a given set of premises, and it makes us realize to what we committed ourselves in accepting those premises; but none of the results obtained by this technique ever goes by one iota beyond the information already contained in the initial assumptions.

Since all mathematical proofs rest exclusively on logical deductions from certain postulates, it follows that a mathematical theorem, such as the Pythagorean theorem in geometry, asserts nothing that is *objectively* or *theoretically new* as compared with the postulates from which it is derived, although its content may well be *psychologically new* in the sense that we were not aware of its being implicitly contained in the postulates.

The nature of the peculiar certainty of mathematics is now clear: A mathematical theorem is certain *relatively* to the set of postulates from which it is derived; i.e., it is necessarily true *if* those postulates are true; and this is so because the theorem, if rigorously proved, simply re-asserts part of what has been stipulated in the postulates. A truth of this conditional type obviously implies no assertions about matters of empirical fact and can, therefore, never get into conflict with any empirical findings, even of the most unexpected kind; consequently, unlike the hypotheses and theories of empirical science, it can never suffer the fate of being disconfirmed by new evidence: A mathematical truth is irrefutably certain just because it is devoid of factual, or empirical content. Any theorem of geometry, therefore, when cast into the conditional form described earlier, is analytic in the technical sense of logic, and thus true *a priori*; i.e., its truth can be established by means of the formal machinery of logic alone, without any reference to empirical data.

4. POSTULATES AND TRUTH

Now it might be felt that our analysis of geometrical truth so far tells only half of the relevant story. For while a geometrical proof no doubt enables us to assert a proposition conditionally—namely on condition that the postulates are accepted—is it not correct to add that geometry also unconditionally asserts the truth of its postulates and thus, by virtue of the deductive relationship between postulates and theorems, enables us unconditionally to assert the truth of its theorems? Is it not an unconditional assertion of geometry that two points determine one and only one straight line that connects them, or that in any triangle, the sum of the angles equals two right angles? That this is definitely not the case, is evidenced by two important aspects of the axiomatic treatment of geometry which will now be briefly considered.

The first of these features is the well-known fact that in the more recent development of mathematics, several systems of geometry have been constructed which are incompatible with euclidean geometry, and in which, for example, the two propositions just mentioned do not necessarily hold. Let us briefly recollect some of the basic facts concerning these *non-euclidean geometries*. The postulates on which euclidean geometry rests include the famous postulate of the parallels, which, in the case of plane geometry, asserts in effect that through every point *P* not on a given line *l* there exists exactly one parallel to *l*, i.e., one straight line which does not meet *l*. As this postulate is considerably less simple than the others, and as it was also felt to be intuitively less plausible than the latter, many efforts were made in the history of geometry to prove that this proposition need not be accepted as an axiom, but that it can be deduced as a theorem from the remaining body of postulates. All attempts in this direction failed, however; and finally it was conclusively demonstrated that a proof of the parallel principle on the basis of the other postulates of euclidean geometry (even in its modern, completed form) is impossible. This was shown by proving that a perfectly self-consistent geometrical theory is obtained if the postulate of the parallels is replaced by the assumption that through any point *P* not on a given straight line *l* there exist at least two parallels to *l*. This postulate obviously contradicts the euclidean postulate of the parallels, and if the latter were actually a consequence of the other postulates of euclidean geometry, then the new set of postulates would clearly involve a contradiction, which can be shown not to be the case. This first non-euclidean type of geometry, which is called hyperbolic geometry, was discovered in the early 20's of the last century almost simultaneously, but independently by the Russian N. I. Lobatschefski, and by the Hungarian J. Bolyai. Later, Riemann developed an alternative geometry, known as elliptical geometry, in which the axiom of the parallels is replaced by the postulate that no line has any parallels. (The acceptance of this postulate, however, in contradistinction to that of hyperbolic geometry, requires the modification of some further axioms of euclidean geometry, if a consistent new theory is to result.) As is to be expected, many of the theorems of these non-euclidean geometries are at variance with those of euclidean theory; thus, e.g., in the hyperbolic geometry of two dimensions, there exist, for each straight line *l*, through any point *P* not on *l*, infinitely many straight lines which do not meet *l*; also, the sum of the angles in any triangle is less than two right angles. In elliptic geometry, this angle sum is always greater than two right angles; no two straight lines are parallel; and while two different points usually determine exactly one straight line connecting them (as they always do in euclidean geometry), there are certain pairs of points which are connected by infinitely many different straight lines. An illustration of this latter type of geometry is provided by the geometrical structure of that curved two-dimensional space which is represented by the surface of a sphere, when the concept of straight line is interpreted by that of great circle on the sphere. In this space, there are no parallel lines since any two great circles intersect; the endpoints of any diameter of the sphere are points connected by infinitely many different "straight lines," and the sum of the angles in a triangle is always in excess of two right angles. Also, in this space, the ratio between the circumference and the diameter of a circle (not necessarily a great circle) is always less than 2π.

Elliptic and hyperbolic geometry are not the only types of non-euclidean geometry; various other types have been developed; we shall later have occasion to refer to a much more general form of non-euclidean geometry which was likewise devised by Riemann.

The fact that these different types of geometry have been developed in modern mathematics shows clearly that mathematics cannot be said to assert the truth of any particular set of geometrical postulates; all that pure mathematics is interested in, and all that it can establish, is the deductive consequences of given sets of postulates and thus the necessary truth of the ensuing theorems relatively to the postulates under consideration.

A second observation which likewise shows that mathematics does not assert the truth of any particular set of postulates refers to *the status of the concepts in geometry*. There exists, in every axiomatized theory, a close parallelism between the treatment of the propositions and that of the concepts of the system. As we have seen, the propositions fall into two classes: the postulates, for which no proof is given, and the theorems, each of which has to be derived from the postulates. Analogously, the concepts fall into two classes: the primitive or basic concepts, for which no definition is given, and the others, each of which has to be precisely defined in terms of the primitives. (The admission of some undefined concepts is clearly necessary if an infinite regress in definition is to be avoided.) The analogy goes farther: Just as there exists an infinity of theoretically suitable axiom systems for one and the same theory—say, euclidean geometry —so there also exists an infinity of theoretically possible choices for the primitive terms of that theory; very often—but not always—different axiomatizations of the same theory involve not only different postulates, but also different sets of primitives. Hilbert's axiomatization of plane geometry contains six primitives: point, straight line, incidence (of a point on a line), betweenness (as a relation of three points on a straight line), congruence for line segments, and congruence for angles. (Solid geometry, in Hilbert's axiomatization, requires two further primitives, that of plane and that of incidence of a point on a plane.) All other concepts of geometry, such as those of angle, triangle, circle, etc., are defined in terms of these basic concepts.

But if the primitives are not defined within geometrical theory, what meaning are we to assign to them? The answer is that it is entirely unnecessary to connect any particular meaning with them. True, the words "point," "straight line," etc., carry definite connotations with them which relate to the familiar geometrical figures, but the validity of the propositions is completely independent of these connotations. Indeed, suppose that in axiomatized euclidean geometry, we replace the over-suggestive terms "point," "straight line," "incidence," "betweenness," etc., by the neutral terms "object of kind 1," "object of kind 2," "relation No. 1," "relation No. 2," etc., and suppose that we present this modified wording of geometry to a competent mathematician or logician who, however, knows nothing of the customary connotations of the primitive terms. For this logician, all proofs would clearly remain valid, for as we saw before, a rigorous proof in geometry rests on deduction from the axioms alone without any reference to the customary interpretation of the various geometrical concepts used. We see therefore that indeed no specific meaning has to be attached to the primitive terms of an axiomatized theory; and in a precise logical presentation of axiomatized geometry the primitive concepts are accordingly treated as so-called logical variables.

As a consequence, geometry cannot be said to assert the truth of its postulates, since the latter are formulated in terms of concepts without any specific meaning; indeed, for this very reason, the postulates themselves do not make any specific assertion which could possibly be called true or false! In the terminology of modern logic, the postulates are not sentences, but sentential functions with the primitive concepts as variable arguments. This point also shows that the postulates of geometry cannot be considered as "self-evident truths," because where no assertion is made, no self-evidence can be claimed.

5. PURE AND PHYSICAL GEOMETRY

Geometry thus construed is a purely formal discipline; we shall refer to it also as *pure geometry*. A pure geometry, then,—no matter whether it is of the euclidean or of a non-euclidean variety—deals with no specific subject-matter; in particular, it asserts nothing about physical space. All its theorems are analytic and thus true with certainty precisely because they are devoid of factual content. Thus, to characterize the import of pure geometry, we might use the standard form of a movie-disclaimer: No portrayal of the characteristics of geometrical figures or of the spatial properties or relationships of actual physical bodies is intended, and any similarities between the primitive concepts and their customary geometrical connotations are purely coincidental.

But just as in the case of some motion pictures, so in the case at least of euclidean geometry, the disclaimer does not sound quite convincing: Historically speaking, at least, euclidean geometry has its origin in the generalization and systematization of certain empirical discoveries which were made in connection with the measurement of areas and volumes, the practice of surveying, and the development of astronomy. Thus understood, geometry has factual import; it is an empirical science which might be called, in very general terms, the theory of the structure of physical space, or briefly, *physical geometry*. What is the relation between pure and physical geometry?

When the physicist uses the concepts of point, straight line, incidence, etc., in statements about physical objects, he obviously connects with each of them a more or less definite physical meaning. Thus, the term "point" serves to designate physical points, i.e., objects of the kind illustrated by pin-points, cross hairs, etc. Similarly, the term "straight line" refers to straight lines in the sense of physics, such as illustrated by taut strings or by the path of light rays in a homogeneous medium. Analogously, each of the other geometrical concepts has a concrete physical meaning in the statements of physical geometry. In view of this situation, we can say that physical geometry is obtained by what is called, in contemporary logic, a semantical interpretation of pure geometry. Generally speaking, a semantical interpretation of a pure mathematical theory, whose primitives are not assigned any specific meaning, consists in giving each primitive (and thus, indirectly, each defined term) a specific meaning or designatum. In the case of physical geometry, this meaning is physical in the sense just illustrated; it is possible, however, to assign a purely arithmetical meaning to each concept of geometry; the possibility of such an arithmetical interpretation of geometry is of great importance in the study of the consistency and other logical characteristics of geometry, but it falls outside the scope of the present discussion.

By virtue of the physical interpretation of the originally uninterpreted primitives of a geometrical theory, physical meaning is indirectly assigned also to every defined concept of the theory; and if every geometrical term is now taken in its physical interpretation, then every postulate and every theorem of the theory under consideration turns into a statement of physics, with respect to which the question as to truth or falsity may meaningfully be raised—a circumstance which clearly contradistinguishes the propositions of physical geometry from those of the corresponding uninterpreted pure theory. Consider, for example, the following postulate of pure euclidean geometry: For any two objects x, y of kind 1, there exists exactly one object l of kind 2 such that both x and y stand in relation No. 1 to *l*. As long as the three primitives occurring in this postulate are uninterpreted, it is obviously meaningless to ask whether the postulate is true. But by virtue of the above physical interpretation, the postulate turns into the following statement: For any two physical points *x*, *y* there exists exactly one physical straight line *l* such that both *x* and *y* lie on *l*. But this is a physical hypothesis, and we may now meaningfully ask whether it is true or false. Similarly, the theorem about the sum of the angles in a triangle turns into the assertion that the sum of the angles (in the physical sense) of a figure bounded by the paths of three light rays equals two right angles.

Thus, the physical interpretation transforms a given pure geometrical theory—euclidean or non-euclidean—into a system of physical hypotheses which, if true, might be said to constitute a theory of the structure of physical space. But the question whether a given geometrical theory in physical interpretation is factually correct represents a problem not of pure mathematics but of empirical science; it has to be settled on the basis of suitable experiments or systematic observations. The only assertion the mathematician can make in this context is this: If all the postulates of a given geometry, in their physical interpretation, are true, then all the theorems of that geometry, in their physical interpretation, are necessarily true, too, since they are logically deducible from the postulates. It might seem, therefore, that in order to decide whether physical space is euclidean or non-euclidean in structure, all that we have to do is to test the respective postulates in their physical interpretation. However, this is not directly feasible; here, as in the case of any other physical theory, the basic hypotheses are largely incapable of a direct experimental test; in geometry, this is particularly obvious for such postulates as the parallel axiom or Cantor's axiom of continuity in Hilbert's system of euclidean geometry, which makes an assertion about certain infinite sets of points on a straight line. Thus, the empirical test of a physical geometry no less than that of any other scientific theory has to proceed indirectly; namely, by deducing from the basic hypotheses of the theory certain consequences, or predictions, which are amenable to an experimental test. If a test bears out a prediction, then it constitutes confirming evidence (though, of course, no conclusive proof) for the theory; otherwise, it disconfirms the theory. If an adequate amount of confirming evidence for a theory has been established, and if no disconfirming evidence has been found, then the theory may be accepted by the scientist "until further notice."

It is in the context of this indirect procedure that pure mathematics and logic acquire their inestimable importance for empirical science: While formal logic and pure mathematics do not in themselves establish any assertions about matters of empirical fact, they provide an efficient and entirely indispensable machinery for deducing, from abstract theoretical assumptions, such as the laws of Newtonian mechanics or the postulates of euclidean geometry in physical interpretation, consequences concrete and specific enough to be accessible to direct experimental test. Thus, e.g., pure euclidean geometry shows that from its postulates there may be deduced the theorem about the sum of the angles in a triangle, and that this deduction is possible no matter how the basic concepts of geometry are interpreted; hence also in the case of the physical interpretaion of euclidean geometry. This theorem, in its physical interpretation, is accessible to experimental test; and since the postulates of elliptic and of hyperbolic geometry imply values different from two right angles for the angle sum of a triangle, this particular proposition seems to afford a good opportunity for a crucial experiment. And no less a mathematician than Gauss did indeed perform this test; by means of optical methods—and thus using the interpretation of physical straight lines as paths of light rays—he ascertained the angle sum of a large triangle determined by three mountain tops. Within the limits of experimental error, he found it equal to two right angles.

6. ON POINCARÉ'S CONVENTIONALISM CONCERNING GEOMETRY

But suppose that Gauss had found a noticeable deviation from this value; would that have meant a refutation of euclidean geometry in its physical interpretation, or, in other words, of the hypothesis that physical space is euclidean in structure? Not necessarily; for the deviation might have been accounted for by a hypothesis to the effect that the paths of the light rays involved in the sighting process were bent by some disturbing force and thus were not actually straight lines. The same kind of reference to deforming forces could also be used if, say, the euclidean theorems of congruence for plane figures were tested in their physical interpretation by means of experiments involving rigid bodies, and if any violations of the theorems were found. This point is by no means trivial; Henri Poincaré, the great French mathematician and theoretical physicist, based on considerations of this type his famous *conventionalism concerning geometry*. It was his opinion that no empirical test, whatever its outcome, can conclusively invalidate the euclidean conception of physical space; in other words, the validity of euclidean geometry in physical science can always be preserved—if necessary, by suitable changes in the theories of physics, such as the introduction of new hypotheses concerning deforming or deflecting forces. Thus, the question as to whether physical space has a euclidean or a non-euclidean structure would become a matter of convention, and the decision to preserve euclidean geometry at all costs would recommend itself, according to Poincaré, by the greater simplicity of euclidean as compared with non-euclidean geometrical theory.

It appears, however, that Poincaré's account is an oversimplification. It rightly calls attention to the fact that the test of a physical geometry G always presupposes a certain body P of non-geometrical physical hypotheses (including the physical theory of the instruments of measurement and observation used in the test), and that the so-called test of G actually bears on the combined theoretical system $G \cdot P$ rather than on G alone. Now, if predictions derived from $G \cdot P$ are contradicted by experimental findings, then a change in the theoretical structure becomes necessary. In classical physics, G always was euclidean geometry in its physical interpretation, GE; and when experimental evidence required a modification of the theory, it was P rather than GE which was changed. But Poincaré's assertion that this procedure would always be distinguished by its greater simplicity is not entirely correct; for what has to be taken into consideration is the simplicity of the total system $G \cdot P$, and not just that of its geometrical part. And here it is clearly conceivable that a simpler total theory in accordance with all the relevant empirical evidence is obtainable by going over to a non-euclidean form of geometry rather than by preserving the euclidean structure of physical space and making adjustments only in part P.

And indeed, just this situation has arisen in physics in connection with the development of the general theory of relativity: If the primitive terms of geometry are given physical interpretations along the lines indicated before, then certain findings in astronomy represent good evidence in favor of a total physical theory with a non-euclidean geometry as part G. According to this theory, the physical universe at large is a three-dimensional curved space of a very complex geometrical structure; it is finite in volume and yet unbounded in all directions. However, in comparatively small areas, such as those involved in Gauss' experiment, euclidean geometry can serve as a good approximative account of the geometrical structure of space. The kind of structure ascribed to physical space in this theory may be illustrated by an analogue in two dimensions; namely, the surface of a sphere. The geometrical structure of the latter, as was pointed out before, can be described by means of elliptic geometry, if the primitive term "straight line" is interpreted as meaning "great circle," and if the other primitives are given analogous interpretations. In this sense, the surface of a sphere is a two-dimensional curved space of non-euclidean structure, whereas the plane is a two-dimensional space of euclidean structure. While the plane is unbounded in all directions, and infinite in size, the spherical surface is finite in size and yet unbounded in all directions: a two-dimensional physicist, travelling along "straight lines" of that space would never encounter any boundaries of his space; instead, he would finally return to his point of departure, provided that his life span and his technical facilities were sufficient for such a trip in consideration of the size of his "universe." It is interesting to note that the physicists of that world, even if they lacked any intuition of a three-dimensional space, could empirically ascertain the fact that their two-dimensional space was curved. This might be done by means of the method of travelling along straight lines; another, simpler test would consist in determining the angle sum in a triangle; again another in determining, by means of measuring tapes, the ratio of the circumference of a circle (not necessarily a great circle) to its diameter; this ratio would turn out to be less than π.

The geometrical structure which relativity physics ascribes to physical space is a three-dimensional analogue to that of the surface of a sphere, or, to be more exact, to that of the closed and finite surface of a potato, whose curvature varies from point to point. In our physical universe, the curvature of space at a given point is determined by the distribution of masses in its neighborhood; near large masses such as the sun, space is strongly curved, while in regions of low mass-density, the structure of the universe is approximately euclidean. The hypothesis stating the connection between the mass distribution and the curvature of space at a point has been approximately confirmed by astronomical observations concerning the paths of light rays in the gravitational field of the sun.

The geometrical theory which is used to describe the structure of the physical universe is of a type that may be characterized as a generalization of elliptic geometry. It was originally constructed by Riemann as a purely mathematical theory, without any concrete possibility of practical application at hand. When Einstein, in developing his general theory of relativity, looked for an appropriate mathematical theory to deal with the structure of physical space, he found in Riemann's abstract system the conceptual tool he needed. This fact throws an interesting sidelight on the importance for scientific progress of that type of investigation which the "practical-minded" man in the street tends to dismiss as useless, abstract mathematical speculation.

Of course, a geometrical theory in physical interpretation can never be validated with mathematical certainty, no matter how extensive the experimental tests to which it is subjected; like any other theory of empirical science, it can acquire only a more or less high degree of confirmation. Indeed, the considerations presented in this article show that the demand for mathematical certainty in empirical matters is misguided and unreasonable; for, as we saw, mathematical certainty of knowledge can be attained only at the price of analyticity and thus of complete lack of factual content. Let me summarize this insight in Einstein's words:

"As far as the laws of mathematics refer to reality, they are not certain; and as far as they are certain, they do not refer to reality."

> The errors of definitions multiply themselves according as the reckoning
> proceeds; and lead men into absurdities, which at last they see but cannot
> avoid, without reckoning anew from the beginning.   —THOMAS HOBBES

> Mathematicians are like lovers . . . Grant a mathematician the least prin-
> ciple, and he will draw from it a consequence which you must also grant
> him, and from this consequence another.   —FONTENELLE

3   The Axiomatic Method

By RAYMOND L. WILDER

I

1. EVOLUTION OF THE METHOD

IF the reader has at hand a copy of an elementary plane geometry, of a
type frequently used in high schools, he may find two groupings of fundamental assumptions, one entitled "Axioms," the other entitled "Postulates." The intent of this grouping may be explained by such accompanying remarks as: "An *axiom* is a *self-evident truth*." "A *postulate* is a *geometrical fact* so simple and obvious that its validity may be assumed." The
"axioms" themselves may contain such statements as: "The whole is
greater than any of its parts." "The whole is the sum of its parts." "Things
equal to the same thing are equal to one another." "Equals added to
equals yield equals." It will be noted that such geometric terms as "point"
or "line" do not occur in these statements; in some sense the axioms are
intended to transcend geometry—to be "universal truths." In contrast, the
"postulates" probably contain such statements as: "Through two distinct
points one and only one straight line can be drawn." "A line can be extended indefinitely." "If L is a line and P is a point not on L, then through
P there can be drawn one and only one line parallel to L." (Some so-called
"definitions" of terms usually precede these statements.)

This grouping into "axioms" and "postulates" has its roots in antiquity.
Thus we find in Aristotle (384–321 B.C.) the following viewpoint:[1]

Every demonstrative science must start from indemonstrable principles;
otherwise, the steps of demonstration would be endless. Of these indemonstrable principles some are (a) common to all sciences, others are
(b) particular, or peculiar to the particular science; (a) the common principles are the *axioms*, most commonly illustrated by the axiom that, if
equals be subtracted from equals, the remainders are equal. In (b) we
have first the *genus* or subject-matter, the *existence* of which must be
assumed.

[1] As summarized by T. L. Heath, *The Thirteen Books of Euclid's Elements*, Cambridge (England), 1908, p. 119. The reader is referred to this book for citations from
Aristotle, Proclus, et al.

1.1. In Euclid's *Elements* (written about 300 B.C.), the two groups occur, respectively labeled "Common notions" and "Postulates." From these and a collection of definitions, Euclid deduced 465 propositions in a logical chain. Although the actual background for Euclid's work is not clear, apparently he did not originate this method of deducing logically from certain unproved propositions, given at the start, all the remaining propositions. As we have just noted, Aristotle, and probably other scholars of the period, had a well-conceived notion of the nature of a demonstrative science; and the logical deduction of mathematical propositions was common in Plato's Academy and perhaps among the Pythagoreans. Nevertheless, the influence of Euclid's work has been tremendous; probably no other document has had a greater influence on scientific thought. For example, modern high school geometries are usually modeled after Euclid's famous work (in England, Euclid is still used as a textbook), thus explaining the still common grouping into "axioms" and "postulates." Also the use in non-mathematical writings of such phrases as "It is axiomatic that . . . ," and "It is a fundamental postulate of . . . ," in the sense of something being "universal" or beyond opposition, is explained by this traditional use of the terms in mathematics.

The method featured in Euclid's work was employed by Archimedes (287–212 B.C.) in his two books which provided a foundation for the science of theoretical mechanics; in Book I of this treatise Archimedes proved 15 propositions from 7 postulates. Newton's famous *Principia*, first published in 1686, is organized as a deductive system in which the well-known laws of motion appear as unproved propositions, or postulates, given at the start. The treatment of analytic mechanics published by Lagrange in 1788 has been considered a masterpiece of logical perfection, moving from explicitly stated primary propositions to the other propositions of the system.

1.2. There exists a large literature devoted to the discussion of the *nature* of axioms and postulates and their philosophical background. Most of this is influenced by the fact that only within comparatively recent years have axioms and postulates been very generally employed in parts of mathematics other than geometry. Even though the method popularized by Euclid is acknowledged now as a fundamental part of the scientific method in every realm of human endeavor, our modern understanding of axioms and postulates, as well as our comprehension of deductive methods in general, has resulted to a great extent from studies in the field of geometry. And, since geometry was conceived to be an attempt to describe the actual physical space in which we live, there arose a conviction that axioms and postulates possessed a character of *logical necessity*. For example, Euclid's fifth postulate (the "parallel postulate") was "Let the following be postulated that, if a straight line falling on two straight lines make the interior angles on the same side less than two right angles, the two straight lines, if produced indefinitely, meet on that side on which are the angles less than the two right angles." [2] Proclus (A.D. 410–485) described vividly in his writings the controversy that was taking place in connection with this postulate even in his time; in fact, he argued in favor of the elimination "from our body of doctrine this merely plausible and unreasoned statement." [2] With the renewal of interest in Greek learning during the Renaissance, controversy in regard to the fifth postulate was renewed. Attempts were made to prove the "parallel postulate," often from logical—non-geometrical—principles alone. Surely if a statement is a "logical necessity" the assumption of its invalidity should lead to contradiction—such was the motivation of much of the work on the postulates of geometry. With the invention of non-euclidean geometries the futility of such attempts became clear.

1.3. The development of the non-euclidean geometries was evidence of a growing recognition of the independent nature of the fifth postulate; that is, this postulate cannot be demonstrated as a logical consequence of the other axioms and postulates in the euclidean system. By a suitable replacement of the fifth postulate, one may obtain the alternative and logically consistent geometry of Bolyai, Lobachevski, and Gauss in which the fifth postulate of Euclid fails to hold. In it appears, for example, the proposition that the sum of the interior angles of a triangle is less than two right angles. Riemann in 1854 developed another non-euclidean geometry, likewise composed of a non-contradictory collection of propositions, in which all lines are of finite length and the sum of the interior angles of a triangle is greater than two right angles.

The invention of the non-euclidean geometries was only part of the rapidly moving developments of the nineteenth century that were to lead to the acceptance of formal geometries apart from those that might be regarded as constituting definitive sciences of space or extension. Grassmann's *Ausdehnungslehre*, published in 1844 and a critical landmark during this era of changing ideas, was described by its author in these terms: "My *Ausdehnungslehre* is the abstract foundation for the doctrine of space, i.e., it is free from all spatial intuition, and is a purely mathematical discipline whose application to space yields the science of space. This latter science, since it refers to something given in nature (i.e., space), is no branch of mathematics, but is an application of mathematics to nature." [3] In explanation of Grassmann's concept of a formal science, Nagel writes: "Formal sciences are characterized by the fact that their sole principles of procedure are the rules of logic as well as by the further

---

[2] Quoted from T. L. Heath, op. cit., pp. 154–155, 203.
[3] As quoted by E. Nagel, "The formation of modern conceptions of formal logic in the development of geometry," *Osiris*, vol. 8 (1939), pp. 142–222, pp. 169, 172.

fact that their theorems are not 'about' some phase of the existing world but are 'about' whatever is *postulated* by thought." [3]

1.4. The idea expressed by Grassmann is essentially the one held at the present time; that is, a mathematical system called "geometry" is not necessarily a description of actual space. One must distinguish, of course, between the origin of a theory and the form to which it evolves. Geometry, like arithmetic, originated in things "practical," but to assert that any particular type of geometry is a description of physical space is to make a *physical* assertion, not a mathematical statement. In short, the modern viewpoint is that one must distinguish between mathematics and *applications* of mathematics.

A natural consequence of this change in viewpoint on the significance of a mathematical system was a re-examination of the nature of the basic, unproved propositions. It became clear, for instance, that the euclidean "common notion" that "the whole is greater than the part" has no more of an absolute character than the "parallel postulate" but is contingent upon the meaning of "greater than"; in fact, the proposition may even fail to hold, as in the theory of the infinite. Although there was much discussion as to whether the parallel postulate should be listed as a "postulate" or as a "common notion" (axiom), it was finally realized that neither had any more universality than the other and the distinction might as well be deleted.[4] Accordingly we find in the classical work of Hilbert on the foundations of geometry,[5] published in 1899, that only one name, "axioms," is applied to the fundamental statements or assumptions, and that certain basic terms such as "point" and "line" are left completely undefined. To be sure, Hilbert made a grouping of his axioms—into five groups—but this pertained only to the technical character of the statements, and not to their relative status of "trueness" or "commonness."

1.5. Although this work of Hilbert has come to be regarded by many as the first to display the axiomatic method in its modern form, it should be recognized that similar ideas were appearing in works of his contemporaries. . . .

1.6. . . . Such studies as those of Pasch, Peano, Hilbert, and Pieri in euclidean geometry provided a tremendous impetus for investigations of possible formal organizations of the subject matter of this old discipline; these considerations, in turn, provided new understanding of mathematical systems in general and were partly responsible for the remarkable mathematical advances of the twentieth century. . . .

    [4] For an excellent non-technical description of this "revolution" in thought, see E. T. Bell, *The Search for Truth*, Baltimore, 1934, chap. XIV.
    [5] Hilbert, *Grundlagen der Geometrie*, Leipzig, 1899 (published in *Festschrift zur Feier der Enthüllung des Gauss-Weber-Denkmals in Göttingen*); *The Foundations of Geometry*, Chicago, 1902. See also the seventh edition of *Grundlagen der Geometrie*, Leipzig, 1930.

It is noteworthy that these early studies in the field of geometry were revealing the great generality that was inherent in formal mathematical systems. Mathematics was evolving in a direction that was to compel the development of a method which could *encompass in a single framework* of undefined terms and basic statements concepts like *group* and *abstract space* that were appearing in seemingly unrelated branches of mathematics. . . . The economy of effort so achieved is one of the characteristic features of modern mathematics.

2. DESCRIPTION OF THE METHOD; THE UNDEFINED TERMS AND AXIOMS

As commonly used in mathematics today, the axiomatic method consists in setting forth certain basic statements about the concept (such as the geometry of the plane) to be studied, using certain undefined technical terms as well as the terms of classical logic. Usually no description of the meanings of the logical terms is given, and no rules are stated about their use or the methods allowable for proving theorems; perhaps these omissions form a weakness of the method.[6] The basic statements are called *axioms* (or, synonymously, *postulates*). It is assumed that in proving theorems from the axioms the rules of classical logic regarding contradictions and "excluded middle" may be employed; hence the "reductio ad absurdum" type of proof is in common use. The statements of both the axioms and the theorems proved from them are said to be *implied by* or *deduced from* the axioms. An example might be instructive:

2.1. Let us consider again the subject of plane geometry. It will be unnecessary to recall many details. We may perhaps assume, however, that the reader recalls from his high school course that *points* and *straight lines*, and such notions as that of *parallel* lines, were fundamental. Now, if we were going to set forth an axiomatic system for plane geometry in rigorous modern form, we would first of all select certain basic terms that we would leave undefined; perhaps "point" and "line" would be included here (the adjective "straight" can be omitted, since the *undefined* character of the term "line" enables us to choose to *mean* "straight line" in our thinking as well as in the later selection of statements for the axioms). Next we would scan the propositions of geometry and try to select certain basic ones with an eye to both their simplicity and their adequacy for proving the ones not selected; these we would call our *primary propositions* or *axioms*, to be left unproved in our system.

2.2. To be more explicit, let us proceed as though we were actually carry-

[6] We are not here describing the method as used in modern mathematical logic or the formalistic treatises of Hilbert and his followers, where the rules for operations with the basic symbols and formulas are (of necessity) set forth in the language of ordinary discourse.

ing out the above procedure; although we do not intend to give a complete system of axioms, a miniature sample of what the axioms and *secondary propositions* or *theorems* might be like, together with sample proofs of the latter, follows:

Undefined terms: Point; line.

Axiom 1. Every line is a collection of points.

Axiom 2. There exist at least two points.

Axiom 3. If p and q are points, then there exists one and only one line containing p and q.

Axiom 4. If L is a line, then there exists a point not on L.

Axiom 5. If L is a line, and p is a point not on L, then there exists one and only one line containing p that is parallel to L.

These axioms would not by any means suffice as a basis for proof of all the theorems of plane geometry, but they will be sufficient to prove a certain number of the theorems found in any organization of plane geometry. Their selection is motivated as follows: In the first place, the undefined terms "point" and "line" are to play a role like that of the variables in algebra. Thus, in the expression

$$x^2 - y^2 = (x - y)(x + y)$$

the x and y are undefined, in the sense that they may represent any individual numbers in a certain domain of numbers (as for instance the domain of ordinary integers). In the present instance, "point" may be any individual in a domain sufficiently delimited as to satisfy the statements set forth in the axioms. On the other hand, "line," as indicated in Axiom 1, has a range of values (= meanings) limited to certain collections of the individuals that are selected as "points." Thus Axiom 1 is designed to set up a *relationship* between the undefined entities *point* and *line*. It is *not* a *definition* of line, since (if the study is carried through) there will be other collections of points (circles, angles, etc.) that are not lines. Furthermore, it enables us, as we shall see presently, to define certain terms needed in the statements of the later axioms. Axiom 2 is the first step toward introducing lines into our geometry, and this is actually accomplished by adding Axiom 3. Before the latter can have meaning, however, we need the following formal definition:

2.3. Definition. If a point p is an element of the collection of points which constitutes a line L (cf. Axiom 1), then we say, variously, that L *contains p, p is on L,* or L is a line containing p.

Having stated Axioms 2 and 3, we would have that there exists a line in our geometry, but in order to have *plane* geometry and not merely a *line* or "one-dimensional" geometry we would have to say something to insure that not all points lie on a single line; Axiom 4 is designed to accomplish this. We would now imagine, intuitively (since we have a line L, a point p not on that line, and also a line through p and each point q of L), that we have practically a plane; however, so far as euclidean geometry is concerned, we have not provided, in Axioms 1–4, for the *parallel* to L through p until we have stated Axiom 5. And of course Axiom 5 is not significant until we have the definition:

2.4. Definition. Two lines L_1 and L_2 are called *parallel* if there is no point which is on both L_1 and L_2. (We may also call L_1 "parallel to" L_2, or conversely.)

2.5. Let us denote the above set of five axioms, together with the undefined terms point, line, by Γ and call it the *axiom system* Γ.

(We shall also frequently use the term "axiom system" in a broader sense to include the theorems, etc., implied by the axioms.)

For future purposes we note two aspects of Γ, but we shall not go into these fully at this point: (1) In addition to the *geometrical* ("technical") undefined terms *point, line*, we have used *logical* ("universal") undefined terms such as *collection, there exist, one, every*, and *not*. (2) That Γ is far from being a set of axioms adequate for plane geometry may be shown as follows: Since *point* and *line* are left undefined, we are at liberty to consider possible meanings for them, subject of course to the restriction that we take into account the statements made in the axioms. If we have been educated in the American or English school systems, our reactions to these terms will no doubt immediately be specialized, our geometric experience in the schools having the upper hand in our response. But let us imagine that the terms are entirely unfamiliar, although the logical terms used in the axioms are not unfamiliar, so that we may consider other possible meanings for point and line. Unquestionably this will involve considerable experimentation before suitable meanings are found. For example, we might first try letting "point" mean book and "line" mean library; we know from the statement in Axiom 1 that a line is a collection of points, and libraries form one of the most familiar collections in our daily experience. We can imagine that we live in a city, C, which has two distinct libraries, and that by library we mean either one of the libraries of C, and by book any one of the books in these two libraries. Axiom 2 becomes a valid statement: "There exist at least two books." However, Axiom 3 fails, since, if p and q designate books in different libraries, then there is no library that contains p *and* q. However, before trying other meanings for point and line, we notice that Axioms 4 and 5 are valid, becoming, respectively, "If L is a library, then there exists a book not on (i.e., in) L," and "If L is a library and p a book not on (i.e., in) L, then there exists one and only one library containing p that is parallel to (has no books in common with) L." [7]

---

[7] In parentheses we have placed the terms commonly employed in connection with libraries and books that are indicated by our definition of "on" and "parallel."

Now, impressed by our failure to satisfy Axiom 3 on our first attempt at meanings for "point" and "line," we may, with an eye on Axiom 3, try to imagine a community, which we denote by Z, of people in which everyone belongs to some club, but in such a manner that, if p and q are two persons in Z, then there is one and only one club of which p and q are both members. In other words, we may try letting "point" mean a *person in Z* and "line" mean a *club in Z*, and imagine that the club situation in Z is such that the statement just made is valid, so that Axiom 3 is satisfied. We would then have no difficulty in seeing that Axioms 1, 2, and 4 are satisfied: "A club in Z is a collection of people in Z"; "There exist at least two people in Z"; etc. However, Axiom 5 becomes (with suitable change of wording to suit the new meanings): "If L is a club in Z, and p is a person in Z not in the club L, then there exists one and only one club in Z of which p is a member and which has no members in common with L." This is a statement which apparently makes a rather strong convention regarding the club situation in Z, and which may conceivably fail to apply; in any case, the stipulation that only one club have a given pair of persons as members can hardly be expected to suffice for Axiom 5! To clinch the matter suppose that Z is a "ghost" community, there being only three persons, whom we shall designate by a, b and c respectively, living in Z; and that as a result of certain circumstances each pair ab, bc, and ac shares a secret from the third member of the community, so that we may consider this bond between each two as forming them into a club ("secret society") excluding the third member. Now, with the meaning of point and line as before, we see that Axioms 1–4 hold but Axiom 5 does not hold.

Before rejecting the latter attempt as impossible, however, let us imagine that Z has *four* citizens: a, b, c, and d. And suppose that *each pair* of these people forms a club excluding the other two members of the community; that is, there are six club consisting of ab, ac, ad, bc, bd, and cd. Now *all* the axioms of Γ are satisfied with the meanings *person in Z* for "point" and *club in Z* for "line"! And we may then notice that we could arrive at a similar example by taking *any* collection Z of *four* things a, b, c, and d, and, by letting "point" mean a member or *element* of the collection Z, and "line" mean any pair of elements of Z, satisfy the statements embodied in the axioms of Γ.

2.6. Although we may experience no particular thrill at this discovery—may, rather, begin to feel that it is a rather trivial game we are playing in toying with possible meanings for the system Γ—we might conceivably be beguiled into seeking an answer to questions such as: How many "points" must a collection have in order to serve as the basis for an example satisfying the statements in Γ? For a given collection at hand, how many "points" must a "line" have in order to satisfy Γ? (For example, a "line" above could not have consisted of *three* persons in Z in the case where Z has exactly four citizens.) Furthermore, if we have already a general knowledge of, or experience with, plane geometry, the above example shows us that Γ is far from being a sufficient basis for euclidean geometry; certainly an adequate set of axioms for plane geometry would exclude the possibility of the geometry permitting a set of only *four* points satisfying all the axioms.

Before proceeding any further with this general discussion, however, let us notice how theorems would be proved from such a system as Γ.

3. DESCRIPTION OF THE METHOD; THE PROVING OF THEOREMS

Having set down a system, such as Γ for instance, we then proceed to see what statements are *implied*, or can be *proved* or *deduced from* the system. Contrary to the manner in which we proceeded in high school, when we brought in all kinds of propositions and assumptions not included in the fundamental terms and axioms (such as "breadth"; "a line has no breadth"), and even drew diagrams and pictures embodying properties that we promptly accepted as part of our equipment,[8] we take care to use only points and lines, and those relations and properties of points and lines that are given in the axioms. (Of course, after we have proved a statement, we may use it in later proofs instead of going back to the axioms and proving it all over again.) There is no objection to drawing diagrams, provided they are used only to aid in the reasoning process and do not trick us into making assumptions not implied by the axioms; indeed, the professional mathematician uses them constantly. . . .

3.1. Consider the following formal theorem and proof:

Theorem 1. *Every point is on at least two distinct lines.*

Proof. Let *p* denote any point. Since by Axiom 2 there exist at least two points, there must exist a point *q* distinct from *p*. And by Axiom 3 there exists a line *L* containing *p* and *q*. Furthermore, by Axiom 4 there exists a point *r* not on *L*, and (again by Axiom 3) a line *K* containing *p* and *r*.

Now by Axiom 1 every line is a collection of points. Hence, for two lines to be distinct (i.e., different), the two collections which constitute them must be different; or, what amounts to the same thing, one of them must contain a point that is not on the other. The lines *L* and *K* are distinct, then, because *K* contains the point *r* which is not on *L*. As *p* is on both *L* and *K*, the theorem is proved.

3.2. Now it will be noticed that we have used Axioms 1–4 in the proof,

---

[8] A classical example may be found in the well-known "proof" that all triangles are isosceles, which is based on a diagram that deceives the eye by placing a certain point *within* an angle instead of outside, where rigorous reasoning about the situation would place it. This may be found in J. W. Young, *Lectures on Fundamental Concepts of Algebra and Geometry*, New York, 1916, pp. 143–145.

but not Axiom 5. We could, then, go back to the example of the community Z, let "point" mean *person in Z* and "line" mean *pair of persons in Z*, rephrase Axioms 1–4 in these terms, and carry through the proof of Theorem 1 in these terms. That is, Theorem 1 is a "true" statement about any example, such as Z, which satisfies the statements embodied in Axioms 1–4 of Γ. In proving Theorem 1, then, *we have in one step proved many different statements about many different examples*, namely, the statements corresponding to Theorem 1 as they appear in the different examples that satisfy Axioms 1–4 of Γ. This [is an important] aspect of the "economy" achieved in using the axiomatic method. . . . If, because of some diagram or other aid to thought, we had used some property of point or line not stated in Axioms 1–4, we could not expect to make the above assertions, and the "economy" cited would be lost!

Note, too, that Theorem 1 will remain valid in any axiom system (such as Γ) that *contains* the undefined terms point and line as well as Axioms 1–4. In particular, it is valid for euclidean plane geometry, which is only one of the possible geometries embodying these four axioms, and which, as we stated before, would require many more axioms than those stated above.

3.3. Now consider the following statement, which we call a *corollary* of Theorem 1:

Corollary. *Every line contains at least one point.*

3.4. Before considering a proof, we hasten to meet an objection which the "uninitiated" might make at this point; to wit, since Axiom 1 explicitly states that a line is a collection of points, *of course* every line contains at least one point, so why should this be repeated as a corollary of Theorem 1? This is not a trivial matter, and it leads directly to a question which causes considerable concern in modern mathematics, namely, what is meant by *collection?* We said above that "collection" is an undefined logical term, and as such we took it for granted that its use is universally understood and employed, just as the word "the" is universally understood and used by anyone familiar with the English language. But now we find ourselves almost immediately in need of explaining the use of the term in the above corollary.

However, there is nothing so very astonishing about this if we reflect that, whenever we try to make very precise a term in ordinary use, it is usually necessary to adopt certain conventions. For example, such terms as *vegetable, fruit, animal* are commonly "understood" and used by anyone who habitually uses the English language, but, when we come to apply them to certain special objects, it is frequently necessary to agree on some convention; as, for example, that a certain type of living substance shall be called "animal" rather than "fish" (e.g., *whale*). So, for instance, we may want to make the convention that, if person A wishes to talk about "the collection of all coins in B's pockets," he may do so even though person B is literally penniless! In other words, no matter whether there actually are coins in B's pockets or not, the collection of all such coins is to be regarded as an existing entity; we call the collection *empty* if B has no coins. (In case B is penniless, we may also talk about "the coin in B's pocket," but in this case there is no existing entity to which the phrase refers.) And this is the convention that is generally agreed on throughout mathematics and logic, namely, that a collection may "exist," as in the case of the collection of all coins in B's pockets, even though it is empty. . . .

3.5. Proof of corollary to Theorem 1. There exists a point p by Axiom 2, and by Theorem 1 there exist two distinct lines L_1 and L_2 containing p.

Now, if there exists a line L that contains no points, then both L_1 and L_2 are parallel to L (by definition). As this would stand in contradiction to Axiom 5, it follows that there cannot exist such a line L.

3.6. A statement "stronger" than the above corollary is embodied in the next theorem:

Theorem 2. *Every line contains at least two points.*

Proof. Let L be any line. By the above corollary, L contains a point p. To show p not the only point on L, we shall use a "proof by contradiction." Suppose p is the only point that L contains. By Theorem 1 there is another line L_1 containing p. Now L_1 must contain at least one other point, q; for otherwise L and L_1 would each contain only p, hence be the same collection of points and ergo the same line (Axiom 1). By Axiom 4 there is a point x which is not on L_1, and by Axiom 5 there is a line L_2 containing x and parallel to L_1. But both L and L_1 are lines containing p and parallel to L_2, in violation of Axiom 5. We must conclude, then, that the supposition that p is the only point on L cannot hold and hence that L contains at least two points.

Now, since by Theorem 2 every line contains at least two points, and since by Axiom 3 two given points can lie simultaneously on only one line, we can state:

Corollary (to Theorem 2). *Every line is completely determined by any two of its points that are distinct.*

3.7. Theorem 3. *There exist at least four distinct points.*

Proof. By Axiom 2 there exist at least two distinct points p and q. By Axiom 3 there exists a line L containing p and q, and by Axiom 4 there exists a point x not on L. By Axiom 5 there exists a line L_1 containing x and parallel to L, and by Theorem 2 L_1 contains at least two distinct points (cf. Definition 2.4).

3.8. Theorem 4. *There exist at least six distinct lines.*

Before proving Theorem 4, we perhaps need to make sure that the meaning of another one of our "common" terms is agreed upon, namely the word "distinct." As we are using the term, two collections are distinct if they are not the *same*. Thus, the lines L and L_1 which figure in the proof of Theorem 2 are distinct, although under the supposition made there L_1 *contains* L, for they are not the *same* line (L_1 contains q and L does not).

Proof of Theorem 4. We proceed, as in the proof of Theorem 3, to obtain the line L containing the points p and q, and the line L_1 parallel to L containing two distinct points (Theorem 2) x and y. By Axiom 3 there exist lines K and K_1 determined respectively by the pairs (p, x), (q, y). Now the point q is not on K, else by Axiom 3 K and L would be the same line (which is impossible since x is not on L). Also, y is not on K, else K and L_1 would be the same line. Similarly, p is not on K_1 and x is not on K_1. Now there also exist lines M and M_1 determined respectively by the pairs (p, y), (q, x); and we can show that q is not on M, x is not on M, p is not on M_1, and y is not on M_1. It follows that no two of the lines L, L_1, K, K_1, M, M_1 are the same.

4. COMMENT ON THE ABOVE THEOREMS AND PROOFS

If the reader has followed the proofs given above, he has probably resorted to the use of figures by this time! This would be quite natural, since in high school geometry he used figures; and they help to keep the various symbols (L, p, q, $\cdots$) and their significance in mind. However, as we stated above, no special meanings have been assigned to "point" and "line," and consequently the above proofs should, and do, hold just as well if the reader uses coins for "points" and pairs of coins for "lines." As a matter of fact, if *any* collection of four objects is employed, "point" meaning any object of the collection and "line" any pair of the objects, then the reader may follow the above proofs with these meanings in mind.

Of course, the theorems we have stated in the preceding sections are not by any means all the theorems that we might state. For example, we can show that any collection of objects satisfying the axioms of the system Γ must, if not infinite as in ordinary geometry, satisfy certain conditions regarding the number of points (there cannot be just 5 points in the collection, for instance), and that there must be a relation between the number of lines and the number of points in the collection. In fact, we can continue the above study to a surprising extent; we could hardly expect to reach a point where we could confidently assert that no more theorems could be proved. It is not our intention to extend the number of theorems, however, since we believe that we have already obtained enough theorems and proofs to serve as specimens for our later purposes.

4.1. As a useful terminology in what follows, let us agree that, when we use the term "statement" in connection with an axiom system Σ, we shall mean a sentence phrased, or phrasable, in the undefined terms and universal terms of Σ; such a statement may be called a Σ-*statement*. Thus the axioms of Γ are Γ-statements (Axiom 5 contains the word "parallel," but this is "phrasable" in the undefined terms and universal terms), as are also the theorems.

4.2. In conformance with the conventions made in Section 2, we shall say that an axiom system Σ *implies* a statement S if S follows by logical argument, such as used above, from Σ. In particular, each axiom is itself implied, trivially. We shall also say S is *logically deducible* from Σ if Σ implies S.

4.3. In the course of our work above we had to pause in two instances to explain the conventions we were making in regard to the use of two words commonly used in ordinary discourse, namely "collection" and "distinct." These words were left undefined, to be sure, in the sense that they are supposedly universally understood non-technical terms; but, as we discovered, not so "universal" but that it was felt advisable to give some conventions we were making in regard to their use here! On the other hand, the words "point" and "line" we left strictly undefined, saying that any meaning whatsoever could be assigned to them as long as these meanings were consistent with the statements embodied in the axioms. We saw that the "collection = library," "point = book" meanings were not permissible, but that, if C is any collection of four objects, then "point = object of C," "line = pair of objects of C" are permissible meanings. The terms "point," "line," "parallel," etc., we may call *technical* terms of the system, the terms "point" and "line" being the *undefined technical terms*. The terms "collection," "distinct" might be called *universal terms* or *logical terms*.

Other examples of universal terms in Γ are "exist" (in Axiom 2), "one" (Axiom 3), "two" (Theorem 1), "four" (Theorem 3), "six" (Theorem 4), "and" (Axiom 3), "or" (Definition 1), "not" (Axiom 4), and "every" (Theorem 1). However, if we were setting up an axiom system for the elementary arithmetic of integers ($1 + 2 = 3$, $2 \times 2 = 4$, etc.), we might use a term like "one" as an undefined technical term. Thus the same term may have different roles in different axiom systems! . . . As the term "exist" is used above, it is chiefly *permissive* so far as proofs are concerned, and *stipulative* for examples; thus in the proof of Theorem 3 we were permitted to introduce the line L_1 by virtue of Axiom 5, and the example of the "ghost community" containing only three persons failed because it could not meet the stipulation concerning the existence of a certain line parallel to another line which is made in Axiom 5.

5. SOURCE OF THE AXIOMS

Let us consider more fully the *source of the statements* embodied in the axioms. We chose axioms for *geometry* in our example Γ since we felt that we could assume the reader had studied some elementary geometry in high school. That is, we were careful to pick a subject already familiar! The undefined technical terms "point" and "line" already have a meaning of some sort for us. And . . . this is the usual way in which axioms are obtained; *they are statements about some concept with which we already have some familiarity*. Thus, if we are already familiar with arithmetic, we might begin to set down axioms for arithmetic. Of course the method is not restricted to mathematics. If we are familiar with some field such as physics, philosophy, chemistry, zoology, economics, for instance, we might choose to set down some axioms for it, or a portion of it, and see what theorems we might logically deduce from them.[9] We may say, then, that an *axiom*, as used in the modern way, is *a statement which seems to hold for an underlying concept*, an *axiom system* being *a collection of such statements about the concept*.

Thus, in practice, the concept comes first, the axioms later. Theoretically this is not necessary, of course. Thus, we may say "Let us take as undefined terms *aba* and *daba*, and set down some axioms in these and universal logical terms." With no concept in mind, it is difficult to think of anything to say! That is, unless we first *give some meanings* to "aba" and "daba"—that is, introduce some *concept* to talk about—it is difficult to find anything to say at all. And, if we finally do make some statements without first fitting a suitable concept to "aba" and "daba," we shall, very likely, make statements that contradict one another! The underlying concept is not only a source of the axioms, but it also guides us to *consistency* (about which we shall speak directly).

Thus, we select the concept; then we select the terms that are to be left undefined and the statements that are to form our axioms; and finally we prove theorems as we did above. This is a simplification of the process, to be sure, but in a general way it describes the method. It is to be noted how the procedure, as so formulated, differs from the classical use of the method. In the classical use the axioms were regarded as absolute truths—absolutely true statements about material space—and as having a certain character of necessity. To have stated the parallel axiom, Axiom 5 above, was to have stated something "obviously true," something one took for granted if one had thought about the character of the space in which he lived. It would have been inconceivable before the nineteenth century to state an axiom such as "If L is a line and p a point not on L, then there

---

[9] As an example in genetics and embryology, see J. H. Woodger, *The Axiomatic Method in Biology*, Cambridge (England), 1937.

exist at least two distinct lines containing p and parallel to L." To have in mathematics, simultaneously, two axiom systems Γ_1 and Γ_2 with axioms in Γ_1 denying axioms in Γ_2 as is the case in mathematics today with the euclidean and non-euclidean geometries, would also have been inconceivable! But, if we take the point of view that an axiom is only a statement about some concept,[10] so that axioms contradicting one another in different systems only express basic differences in the concepts from which they were derived, we see that no fundamental difficulty exists. What is important is that axioms in the *same* system should not contradict one another. This brings us to the point where we should discuss consistency and other characteristics of an axiom system.

5.1. REMARK

The derivation of an axiom system for non-euclidean geometry from axioms for euclidean geometry, using the device of replacing the parallel axiom by one of its denials, is an example of another manner in which new axiom systems may be obtained. In general, we may select a given axiom system and change one or more of the axioms therein in suitable manner to derive a new axiom system.

II

ANALYSIS OF THE AXIOMATIC METHOD

[When the undefined terms and primary propositions or axioms of a system have been selected at least three relevant questions suggest themselves: (1) Is the system suited to the purposes for which it was set up? (2) Are the axioms truly independent, i.e., are any of them provable from the others (in which case they should perhaps be deleted from the system and transferred to the body of theorems to be proved)? (3) Does the system imply any contradictory theorems (if so, this defect must be eliminated if the theorems are to be relied on)? Of these questions, the third, relating to contradictoriness, is by far the most fundamental and critical. In the selection below, a continuation of the preceding discussion, I have excerpted Wilder's analysis of the consistency and independence of an axiom system. ED.]

1. CONSISTENCY OF AN AXIOM SYSTEM

From a logical point of view we can make the following definition:
1.1. Definition. An axiom system Σ is called *consistent* if contradictory statements are not implied by Σ.

Now this definition gives rise to certain questions and criticisms. In the first place, given an axiom system Σ, *how are we going to tell whether*

---

[10] It is only in this sense—that an axiom is a statement *true* of some concept—that the word "true" can be used of an axiom.

it is consistent or not? Conceivably we might prove two theorems from Σ which contradict one another, and hence conclude that Σ is *not* consistent.

For example, if we added to the system Γ, discussed above, the new axiom, "There exist at most three points," it would become apparent, as soon as Theorem 3 of Γ was proved, that the new system of axioms is not consistent.

But, supposing that this does not happen, are we going to conclude that Σ is consistent? How can we tell that, if we continued stating and proving theorems, we might not ultimately arrive at contradictory statements and hence *inconsistency*? We remarked about the system Γ that we could hardly expect to reach a point where we could say with confidence that no more theorems could be stated. And, unless we could have all possible theorems in front of our eyes, capable of being scanned for contradictions, how could we assert that the system is consistent? We are immediately faced with the problem: Is there any *procedure for proving a system of axioms consistent?* And, if so, on what basis does the proof rest, since the proof cannot be conducted *within* the system as in the case of the theorems of the system?

Another difficulty would arise from the fact that it might be very hard to recognize that a contradiction is implied, when such is the case. There are examples in mathematical literature of cases where considerable material was published concerning systems which later were found to be inconsistent. Until someone suspected the inconsistency and set out to prove it, or (as in some cases) stumbled upon it by chance, the systems seemed quite valid and worth while. It can also happen, for example, that the theorems become so numerous and complicated that we fail to detect a pair of contradictory ones. For example, although two theorems might really be of the form "S" and "not S" respectively, because of the manner in which they are stated it might escape our attention that they contradict one another. In short, the usefulness of the above definition is limited by our ability to recognize a contradiction even when it is staring us in the face, so to speak.

The former objection, that hinging upon the probable impossibility of setting down all theorems implied by the system, is the more serious from the point of view of the working mathematician. And as a consequence the mathematician usually resorts to the procedure described below:

Let us make the definition:

1.2. Definition. If Σ is an axiom system, then an *interpretation* of Σ is the assignment of meanings to the undefined technical terms of Σ in such a way that the axioms become true statements for all values of the variables (such as p and q of Axiom 3, I 2.2, for instance).

This definition requires some explanation. First, as an example we can cite the system Γ above and the meanings "point" = any one of a collection of four coins and "line" = any pair of coins in this collection. The axioms now become statements about the collection of coins and are easily seen to be true thereof. Hence this assignment of meanings is an interpretation of Γ. As the axioms stand, with "point" and "line" having no assigned meanings, they cannot be called either true or not true. (Similarly, we cannot speak of the expression "$x^2 - y^2 = (x - y)(x + y)$" as being either true or false until *meanings*, such as "x and y are integers," are assigned.) But, with the meanings assigned above, they are true statements about a "meaningful" concept. As a rule, we shall use the word "model" to denote the *result* of the assignment of meanings to the undefined terms. Thus the collection of four coins, considered a collection of points and lines according to the meanings assigned above, is a model of Γ. Generally, if an interpretation I is made of an axiom system, we shall denote the model resulting from I by 𝔐(I).

For some models of an axiom system Σ, certain axioms of Σ may be *vacuously satisfied*. That is, axioms of the form "If . . . , then . . . ," such as Axiom 3 of Γ, which we might call "conditional axioms," may be true as interpreted only because the conditional "If . . ." part is not fulfilled by the model.

Suppose, for example, we delete Axioms 2 and 4 from Γ and denote the resulting system by Γ'. Then a collection of coins containing just one coin is a model for Γ', if we interpret "point" to mean coin and "line" to mean a collection containing just one coin. For in this model the "If . . ." parts of Axioms 3 and 5 are not fulfilled. (Note that, for Axiom 3 to be false of a model 𝔐, there must be *two* points p and q in 𝔐 such that either no line of 𝔐 contains p and q or more than one line of 𝔐 contains p and q.)

This may be better illustrated, perhaps, by the following digression: Suppose boy A tells girl B, "If it happens that the sun shines Sunday, then I will take you boating." And let us suppose that on Sunday it rains all day, the sun not once peeping out between the clouds. Then, no matter whether A takes B boating or not, it cannot be asserted that he made her a false promise. For his promise to have been false, (1) the sun must have shone Sunday and (2) A must not have taken B boating. And, in general, for a statement of the form "If . . . , then . . ." to be false, the "If . . ." condition must be fulfilled and the "then . . ." not be fulfilled.

Now we did not have in mind a collection of four coins when we set down the axioms of Γ. We were thinking of something entirely different, namely euclidean geometry as we knew it in high school. "Point" had for us then an entirely different meaning—something "without length, breadth, or thickness"; and "line" meant a "straight" line that had "length, but no breadth or thickness." Do not these meanings also yield a model of Γ—what we might perhaps call an "ideal" model? We may admit that this is so, and we resort frequently in mathematics to such ideal models—always, of course, when it happens that every collection of objects serving as a model must of necessity be infinite in number. (Such is the case, for instance, when we have enough axioms in a geometry to insure an infinite number of lines.) We return to this discussion later (see 2.3); at present, let us go on to the so-called "working definition" of consistency:

1.3. Definition. An axiom system Σ is *satisfiable* if there exists an interpretation of Σ.

Now what is the relation between the two definitions 1.1 and 1.3? What we actually want of any axiom system is that it be consistent in the sense of 1.1. But we saw that 1.1 was not a practicable definition except in cases where contradictory statements are actually found to be implied by the system and inconsistency is thus recognized. Where a system is consistent, we are usually unable to tell the fact from 1.1. But, as in the case of the four-coin interpretation of Γ above, we have a very simple test showing "satisfiability" in the sense of 1.3. Does this imply consistency in the sense of 1.1? The mathematician and the logician take the point of view that it does, and, in order to explain why, we have to go into the domain of logic for a few moments.

2. THE PROOF OF CONSISTENCY OF AN AXIOM SYSTEM

The Law of Contradiction and the Law of the Excluded Middle.

First let us recall two basic "laws" of classical (i.e., Aristotelian) logic, namely the Law of Contradiction and the Law of the Excluded Middle; the latter is also called the Law of the Excluded Third ("tertium non datur"). These are frequently, and loosely, described as follows: If S is any statement, then the Law of Contradiction states that S and a contradiction (i.e., any denial) of S cannot both hold. And the Law of the Excluded Middle states that either S holds or a denial of S holds. For example, let S be the statement "Today is Tuesday." The Law of Contradiction certainly holds here, for today cannot be *both* Tuesday and Wednesday, for example. And the Law of the Excluded Middle states that either today is Tuesday or it is not Tuesday.

But "things are not so simple as they seem" here. Unless one limits himself to a specified point on the earth (or parallel of longitude), it *can* be both Tuesday and Wednesday at the same time! And, without including in S such a geographical provision, the statement "Either today is Tuesday or it is not Tuesday" can hardly be accepted. As a matter of fact, whenever such statements are made there usually exists a tacit understanding between speaker and listener that their locale at the time is the place being referred to.

Or consider the statement "The king of the United States wears bow ties." Does the Law of the Excluded Middle hold here? Or let S be the statement "All triangles are green."

The upshot of this is merely that, although these "laws" are called "universally valid," some sort of qualifications have to be made with regard to their applicability in order for them to have validity. In so far as axiomatic systems are concerned, the problem is not so great, since we can restrict our use of the term "statement" to the convention already made in section I 4.1 ("Σ-statement"). And this will be our understanding from now on.

2.2. As soon as an interpretation of a system Σ is made, the statements of the system become statements about the resulting model. Let us assume the following, which may be considered *basic principles of applied logic*.

2.2.1. All statements implied by an axiom system Σ hold true for all models of Σ;

2.2.2. The Law of Contradiction holds for all statements about a model of an axiom system Σ, provided they are Σ-statements whose technical terms have the meanings given in the interpretation. We can make this clearer and more precise by introducing the notion of an I-Σ-statement:

2.2.3. If Σ is an axiom system and I denotes an interpretation of Σ, then the result of assigning to the technical terms in a Σ-statement their meanings in I will be called an I-Σ-statement.

Then 2.2.1 and 2.2.2 become respectively:

2.2.1. Every I-Σ-statement, such that the corresponding Σ-statement is implied by Σ, holds true for $\mathfrak{M}(I)$ (cf. 1.2);

2.2.2. Contradictory I-Σ-statements cannot both hold true of $\mathfrak{M}(I)$.

Under the assumption that 2.2.1 and 2.2.2 hold, satisfiability implies consistency. For, if an axiom system Σ implies two contradictory Σ-statements, then by 2.2.1 these statements as I-Σ-statements hold true for the model $\mathfrak{M}(I)$; but the latter is impossible by 2.2.2. Hence we must conclude that, if 2.2.1 and 2.2.2 are valid, then the existence of an interpretation for an axiom system Σ guarantees the consistency of Σ in the sense of 1.1. And this is the basis for the "working definition" 1.3. For example, the existence of the "four-coin interpretation" of the system Γ above guarantees the consistency of Γ if we grant 2.2.1 and 2.2.2.

The reader will have noticed that we have not proved that consistency in the sense of 1.1 implies satisfiability. To go into this question would be impractical, since it would necessitate going into detail concerning formal logical systems and is too complicated to describe here.

2.3. In Section 1 we used the term "ideal" model, by way of contrast to such models as that of the four coins for Γ; the latter might be termed a "concrete" model. It was pointed out that, whenever an axiom system Σ requires an infinite collection in each of its models, then of necessity the models are "ideal."

This raises not only the question as to how reliable are "ideal" models, but also the question as to what constitutes an *allowable* model. What we should like, of course, is a criterion which would allow only models that satisfy assumptions 2.2.1, 2.2.2, and especially the latter. If there is any danger that an ideal model may require such a degree of abstraction that it harbors contradictions in violation of 2.2.2, then clearly the use of models is no general guarantee of consistency in spite of what we have said above.

Further light can be shed on this matter by a consideration of well-known examples. It is not an uncommon practice, for instance, to obtain a model of an axiom system Σ in another branch of mathematics—even in a branch of mathematics that is, in its turn, based on an axiom system Σ'. How valid are such models? Do they necessarily satisfy 2.2.2? For example, to establish the consistency of a non-euclidean geometry we give a model of it in euclidean geometry. (See Richardson [*Fundamentals of Mathematics*, New York, 1941, pp. 418–19] for instance.) But suppose that the euclidean geometry harbors contradictions; what then? Evidently all we can conclude here is that, *if* euclidean geometry is consistent, then so is the non-euclidean geometry whose model we have set up in the euclidean framework.

We are forced to admit that in such cases we have no absolute test for consistency, but only what we may call a *relative consistency proof*. The axiom system Σ' may be one in whose consistency we have great confidence, and then we may feel that we achieve a high degree of plausibility for consistency, but in the final analysis we have to admit that we are not *sure* of it.[11] . . .

3. INDEPENDENCE OF AXIOMS

Earlier we mentioned "independence" of axioms. By "independence" we mean essentially that we are "not saying too much" in stating our axioms. For example, if to the five axioms of the system Γ (I 2.2) we added a sixth axiom stating "There exist at least four points," we would provide no new information inasmuch as the axiom is already implied by Γ (see Theorem 3 of I 3.7). Of course the addition of such an axiom would not destroy the property of consistency inherent in Γ.

3.1. In order to state a formal definition of independence, let Σ denote an axiom system and let A denote one of the axioms of Σ. Let us denote some denial of A by $\sim$A, and let $\Sigma - $A denote the system Σ with A deleted. If S is any Σ-statement, $\Sigma + $ S will mean the axiom system con-

---

[11] In one well-known case, the system Σ' is a *subsystem* of Σ; viz., the Gödel proof (Gödel, *The Consistency of the Axiom of Choice and the Generalized Continuum Hypothesis with the Axioms of Set Theory*, Princeton, 1940) of the relative consistency of the axiom of choice when adjoined to the set theory axioms.

taining the axioms of Σ and the statement S as a new axiom. Then we define:

3.1.1. Definition. If Σ is an axiom system and A is one of the axioms of Σ, then A is called *independent* in Σ, or an independent axiom of Σ, if both Σ and the axiom system $(\Sigma - A) + \sim A$ are satisfiable.

3.2. Just which of the many forms of $\sim A$ is used is immaterial. Thus Axiom 5 is independent in Γ (I 2.2) if Γ is satisfiable and if the first four axioms of Γ togther with a "non-euclidean" form of the axiom constitute a satisfiable system. For example, for a denial of Axiom 5 take the statement: "There exist a line L and point p not on L, such that there does not exist a line containing p and parallel to L." To show that the system Γ with Axiom 5 replaced by this statement forms a satisfiable system, let us take a collection of *three* coins, let "point" mean a coin of this collection, and "line" mean any pair of points of this collection. Then we have an interpretation of the new system, showing it to be satisfiable. We have already ascertained (2.2) that Γ is satisfiable, and so we conclude that Axiom 5 is independent in Γ.

3.3. The reader will probably gather by this time that the reason for specifying the satisfiability of Σ, in Definition 3.1.1, is to insure that some $\sim A$ is not a *necessary* consequence of the axioms of $\Sigma - A$; for, if it were, we would not wish to call A "independent." And, as the definition is phrased, it insures that neither A nor any denial, $\sim A$, of A is implied by the system $\Sigma - A$, so that the addition of A to $\Sigma - A$ is really the supplying of new information.

3.4. Actually, however, we do not place the same emphasis on independence as we do on consistency. Consistency is *always* desired, but there may be cases where independence is *not* desired. . . . Generally speaking, of course, it is preferable to have all axioms independent; but, if some axiom turns out not to be independent, the system is not invalidated.

As a matter of fact, some well-known and important axiom systems, when first published, contained axioms that were not independent (a fact unknown at the time to the authors, of course). An example of this is the original formulation of the set of axioms for geometry given by Hilbert in 1899. This set of axioms contained two axioms which were later discovered to be implied by the other axioms.[12] This in no way invalidated the system; it was only necessary to change the axioms to theorems (supplying the proofs of the latter, of course). . . .

[12] See, for example, E. H. Moore, "On the projective axioms of geometry," *Trans. Amer. Math. Soc.*, vol. 3 (1902), pp. 142–158.

> *There can* never *be surprises in logic.* —Ludwig Wittgenstein

> *Do I contradict myself?*
> *Very well then I contradict myself.*
> *(I am large, I contain multitudes.)* —Walt Whitman

> *Thus, be it understood, to demonstrate a theorem, it is neither necessary*
> *nor even advantageous to know what it means. The geometer might be*
> *replaced by the "logic piano" imagined by Stanley Jevons; or, if you choose,*
> *a machine might be imagined where the assumptions were put in at one*
> *end, while the theorems came out at the other, like the legendary Chicago*
> *machine where the pigs go in alive and come out transformed into hams*
> *and sausages. No more than these machines need the mathematician know*
> *what he does.* —Henri Poincaré

4   Goedel's Proof

By ERNEST NAGEL and JAMES R. NEWMAN

IN 1931 there appeared in a German scientific periodical an exceptionally difficult and brilliant paper entitled *"Ueber formal unentscheidbare Saetze der Principia Mathematica und verwandter Systeme"* ("On Formally Undecidable Propositions of Principia Mathematica and Related Systems"). The author of the paper was Kurt Goedel, then a young mathematician of 25 at the University of Vienna, now a member of the Institute for Advanced Study at Princeton. When at a convocation in 1952 Harvard University awarded Goedel an honorary degree, the citation described his achievement as the most important advance in mathematical logic in a quarter century.

"On Formally Undecidable Propositions of Principia Mathematica and Related Systems" is a milestone in the history of modern logic and mathematics, yet probably neither its title nor its contents were at the time of its appearance intelligible to the great majority of professional mathematicians. This is not surprising. The term "undecidable propositions" may for the moment be briefly identified as the name of propositions which can be neither proved nor disproved within a given system; the *Principia Mathematica*, to which the paper referred, is the monumental three-volume treatise by Alfred North Whitehead and Bertrand Russell on mathematical logic and the foundations of mathematics. Now familiarity with the thesis and the techniques of the *Principia*, let alone with some of the questions it raised, was not in 1931 (and is not now) a prerequisite to successful research in most branches of mathematics. There were, to be sure, a number of mathematicians, chiefly under the influence of the outstanding German mathematician David Hilbert, who were profoundly interested in these matters; but the group was small. Logico-mathematical problems have never attracted a wide audience even among those who are partial to abstract reasoning. On the other hand, to those who were able to read Goedel's paper with understanding, its conclusions came as an astounding and a melancholy revelation. For the central theorems which it demonstrated challenged deeply rooted preconceptions concerning mathematical method, and put an end to one great hope that motivated decades of research on the foundations of mathematics. Goedel showed that the axiomatic method, which mathematicians had been exploiting with increasing power and rigor since the days of Euclid, possesses certain inherent limitations when it is applied to sufficiently complex systems—indeed, when it is applied even to relatively simple systems such as the familiar arithmetic of cardinal numbers. He also proved, in effect, that it is impossible to demonstrate the internal consistency (non-contradictoriness) of such systems, except by employing principles of inference so complex that their internal consistency is as open to doubt as that of the systems themselves. Goedel's paper was not, however, exclusively negative in import. It introduced a novel technique of analysis into the foundations of mathematics that is comparable in fertility with the power of the algebraic method which Descartes introduced into the study of geometry. It suggested and initiated new problems and branches of logico-mathematical research. It provoked a critical reappraisal, not yet completed, of widely held philosophies of knowledge in general, and of philosophies of mathematics in particular.

Despite the novelty of the techniques Goedel introduced, and the complexity of the details in his demonstrations, the major conclusions of his epoch-making paper can be made intelligible to readers with even limited mathematical preparation. The aim of the present article is to make the substance of Goedel's findings generally understandable. This aim will perhaps be most easily achieved if the reader is first briefly reminded of certain relevant developments in the history of mathematics and modern formal logic.

I

The nineteenth century witnessed a tremendous expansion and intensification of mathematical research. Many fundamental problems that had long withstood the best efforts of earlier thinkers received definitive solutions; new areas of mathematical study were created; and the foundations for various branches of the discipline were either newly laid, or were recast with the help of more rigorous techniques of analysis. In particular, the development of the non-Euclidean geometries stimulated the revision and completion of the axiomatic basis for many mathematical systems;

and axiomatic foundations were supplied for fields of inquiry which hitherto had been cultivated in a more or less intuitive manner. One important conclusion that emerged from this critical examination of the foundations of mathematics was that the traditional conception of mathematics as the "science of quantity" is both inadequate and misleading. For it became evident that mathematics is the discipline *par excellence* which draws necessary conclusions from any given set of axioms (or postulates), and that the validity of the inferences drawn does not depend upon any particular interpretation which may be assigned to the postulates. Mathematics was thus recognized to be much more "abstract" and "formal" than had been traditionally supposed. The postulates of any branch of demonstrative mathematics are not inherently "about" space, quantity, or anything else; and any special meaning which may be associated with the "descriptive" terms (or predicates) in the postulates plays no essential role in the process of deriving theorems. The sole question which confronts the pure mathematician (as distinct from the scientist who employs mathematics in investigating a special subject matter) is not whether the postulates he assumes or the conclusions he deduces from them are *true,* but only whether the alleged conclusions are in fact the *necessary logical consequences* of the initial assumptions. For example, among the undefined terms employed by Hilbert in his famous axiomatization of geometry are the following: "point," "line," "plane," "lies on," and "between." The customary meanings attributed to these (predicate) expressions undoubtedly promote the cause of discovery and learning. That is, because of the very familiarity of these notions, they not only motivate and facilitate the formulation of axioms, but they also suggest the goals of inquiry, i.e., the statements one wishes to establish as theorems. Nevertheless, as Hilbert states explicitly, for mathematical purposes familiar connotations are to be banished and the "meanings" of the expressions are to be taken as completely described by the axioms into which they enter. In more technical language the expressions are "implicitly defined" by the axioms and whatever is not embraced by the implicit definitions is irrelevant to the demonstration of theorems. The procedure recalls Russell's famous epigram: pure mathematics is the subject in which we do not know what we are talking about, nor whether what we are saying is true.

This land of rigorous abstraction, empty of all familiar landmarks, was certainly not easy to get around in. But it offered compensations in the form of a new freedom of movement and fresh vistas. The intensified formalization of mathematics emancipated men's minds from the restrictions which the standard interpretation of expressions placed on the construction of novel systems of postulates. As the meaning of certain terms became more general, less explicit, their use became broader, the inferences to be drawn from them less confined. Formalization led in fact to a great variety of axiomatized deductive systems of considerable mathematical interest and value. Some of these systems, it must be admitted, did not lend themselves to an intuitively obvious interpretation, but this fact caused no alarm. Intuition, for one thing, is an elastic faculty; our children will have no difficulty in accepting as intuitively obvious the paradoxes of relativity, just as we do not boggle at ideas which were regarded as wholly unintuitive a couple of generations ago. Moreover, intuition, as we all know, is not a dependable guide: it cannot be used safely as a criterion of either truth or fruitfulness in scientific explorations.

A more serious problem, however, was raised by the increased abstractness of mathematics. This turned on the question whether a given set of postulates underlying a new system was internally consistent, so that no mutually contradictory theorems could be deduced from the set. The problem does not seem pressing when a set of axioms is taken to be "about" a definite and familiar domain of objects; for then it is not only significant to ask, but it may be possible to ascertain, whether the axioms are indeed true of these objects. Thus, since the Euclidean axioms were generally supposed to be true statements about space (or objects in space), apparently no mathematician prior to the nineteenth century ever entertained the question whether a pair of contradictory theorems might not some day be deduced from the axioms. The basis for this confidence in the consistency of Euclidean geometry was the sound principle that logically incompatible statements cannot be simultaneously true; accordingly, if a set of statements are true (and this was generally assumed to be the case for the Euclidean axioms), they are also mutually consistent.

But the non-Euclidean geometrics were clearly in a different case. For since their axioms were initially regarded as being plainly false of space, and, for that matter, doubtfully true of anything, the problem of establishing the internal consistency of non-Euclidean systems was recognized to be both substantial and serious. In Riemannian geometry, for example, the famous parallel postulate of Euclid (which is equivalent to the assumption that through a given point in a plane just one parallel can be drawn to a given line in the plane) is replaced by the assumption that through a given point in a plane *no* parallel can be drawn to a given line in the plane. Now suppose the question: is the Riemannian set of postulates consistent? They are evidently not true of the ordinary space of our experience. How then is their consistency to be tested? How can one prove they will not lead to contradictory theorems?

A general method was devised for solving this problem. The underlying idea is to find a "model" (or interpretation) for the postulates so that each postulate is converted into a true statement about the model. In the case of Euclidean geometry, as we have seen, the model was ordinary space. Now the method was extended to find other models, the elements of which could be used as crutches for the abstractions of the postulates. The procedure goes something like this. Suppose the following set of postulates is given concerning two classes K and L, whose special nature is left undetermined except as "implicitly" defined in the postulates:

(1) Any two members of K are contained in just one member of L.

(2) No member of K is contained in more than two members of L.

(3) The members of K are not all contained in a single member of L.

(4) Any two members of L contain just one member of K.

(5) No member of L contains more than two members of K.

From this little set, using customary rules of inference, theorems can be derived. For example, it can be shown that K contains just three members. But is the set a consistent one, so that mutually contradictory theorems can never be derived from it? The fact that no one has as yet deduced such theorems does not settle the question, because this does not prove that contradictory theorems may not eventually be deduced. The question is readily resolved, however, with the help of the following model. Let K be the class of points constituting the vertices of a triangle, and L the class of lines constituting its sides; and let us understand the phrase "a member of K is contained in a member of L" to mean that a point which is a vertex lies on a line which is a side. Each of the five abstract postulates is then converted into a true statement—for example, the first asserts that any two points which are vertices of the triangle lie on just one line which is a side. Thereby the set is proved to be consistent.

In a similar fashion the consistency of plane Riemannian geometry can be established. Let us interpret the expression "plane" in the Riemannian postulates to signify the surface of a Euclidean sphere, the expession "point" to signify a point on this surface, the expression "straight line" to signify an arc of a great circle on this surface, and so on. Each Riemannian postulate is then converted into a truth of Euclid. For example, on this interpretation the Riemannian parallel postulate reads as follows: Through a point on the surface of a sphere, no arc of a great circle can be drawn parallel to a given arc of a great circle.

All this is very tidy, no doubt, but we must not become complacent. For as any sharp eye will have seen by now we are not so much answering the problem as removing it to familiar ground. We seek to settle the question of Riemannian consistency by appealing, in effect, to the authority of Euclid. But what about his system of geometry—are its axioms consistent? To say that they are "self-evidently true," and therefore consistent, is today no longer regarded as an acceptable reply. To describe the axioms as inductive generalizations from experience would be to claim for them only some degree of probable truth. A great mass of evidence might be adduced to support them, yet a single contrary item would destroy their title of universality. Induction therefore will not suffice to establish the consistency of Euclid's geometry as logically certain. A different approach was tried by David Hilbert. He undertook to interpret the Euclidean postulates in a manner made familiar by Cartesian co-ordinate geometry, so that they are transformed into algebraic truths. Thus, in the axioms for plane geometry, construe the expression "point" to signify a pair of real numbers, the expression "straight line" to signify the relation between real numbers which is expressed by a first degree equation with two unknowns, the expression "circle" to signify the relation between numbers expressed by a quadratic equation of a certain form, and so on. The geometric statement that two distinct points uniquely determine a straight line is then transformed into the algebraic truth that two pairs of real numbers uniquely determine a linear form; the geometric theorem that a straight line intersects a circle in at most two points, is transformed into the algebraic theorem that a linear form and a quadratic form of a certain type determine at most two pairs of real numbers; and so on. In brief, the consistency of the Euclidean postulates is established by showing that they are satisfied by an algebraic model.

This method for establishing consistency is powerful and effective. Yet it too remains vulnerable to the objections set forth above. In other words the problem has again been solved in one domain only by transferring it to another. Hilbert's proof of the consistency of his postulates simply shows that *if* algebra is consistent, then so is his geometric system. The proof is merely *relative* to the assumed consistency of some other system and is not an "absolute" proof.

In attempting to solve the problem of consistency one notices a recurrent source of difficulty. It is encountered whenever a non-finite model is invoked for purposes of interpretation. It is evident that in making generalizations about space only a very limited portion—that which is accessible to our senses—serves as the basis of grand inferences; we extrapolate from the small to the universal. But where the model has a finite number of elements the difficulty is minimized, if it does not completely vanish. The vertex-triangle model used above to show the consistency of the five abstract K and L class postulates is finite; it was therefore comparatively simple to determine by actual inspection whether all the elements in the model actually satisfied the postulates. If this condition is fulfilled they are "true" and hence consistent. To illustrate: by examining in turn all the vertices of the model triangle one can learn whether any two of them lie on one side—so that the first postulate is established as true. Unfortunately, however, most of the postulate systems that constitute the foundations of important branches of mathematics cannot be mirrored in finite models and can be satisfied only by non-finite ones. One of the postulates, for example, in a well known axiomatization of elementary arithmetic asserts that *every* integer has an immediate successor which differs from any integer preceding it in the progression. It is evident that the set of postulates containing this one cannot be interpreted by means of a finite model; the model itself will have to mirror the infinity of elements postulated by the axioms. The truth (and so the consistency) of the set cannot therefore be established by inspection and enumeration. Apparently then we have reached an impasse. Finite models suffice to establish the consistency of certain sets of postulates, but these are of lesser importance. Non-finite models, necessary for the interpretation of most postulate systems, can be described only in general terms, and we are not warranted in concluding as a matter of course that the descriptions themselves are free from a concealed contradiction.

It may be tempting to suggest at this point that we can be assured of the consistency of descriptions which postulate non-finite models, if the basic notions employed in such descriptions are transparently "clear" and "certain." But the history of thought has not dealt kindly with the doctrine of intuitive knowledge which is implicit in the suggestion. In certain areas of mathematical research, in which assumptions about infinite domains play central roles, radical contradictions (or "antinomies") have turned up, despite the "intuitive" clarity of the notions involved in the assumptions, and despite the seemingly consistent character of the intellectual constructions performed. Such antinomies have emerged in the theory of transfinite numbers developed by Georg Cantor in the nineteenth century; and the occurrence of these contradictions has made plain that the apparent clarity of even such an elementary notion as that of *class*, does not guarantee the consistency of the system built on it. Now the theory of classes (or aggregates) is often made the foundation for other branches of mathematics, and in particular for elementary arithmetic. It is therefore pertinent to ask whether antinomies similar to those encountered in the theory of transfinite numbers may not infect other parts of mathematics.

In point of fact, Russell constructed a contradiction within the framework of elementary logic itself, a contradiction which is the precise analogue of the antinomy first developed in the Cantorian theory of transfinite numbers. Russell's antinomy can be stated as follows: Classes may be divided in two groups: those which do not, and those which do contain themselves as members. A class will be called "normal" if, and only if, it does not contain itself as a member. Otherwise it is "non-normal." An example of a normal class is the class of mathematicians, for patently the class itself is not a mathematician and is therefore not a member of itself. An example of a non-normal class is the class of all thinkable things; for the class of all thinkable things is itself a thinkable thing and is therefore a member of itself. Now let "N" by definition stand for the class of *all* normal classes. We ask whether N itself is a normal class. If N is normal, it is a member of itself for, by definition of "N," N is to include all normal classes; but in that case also N is non-normal because by definition of "non-normal," non-normal classes are those which contain themselves as members. On the other hand, if N is non-normal, then again it is a member of itself by definition of "non-normal," but then also it is normal because it belongs to N which is defined as normal. N, in other words, is normal if and only if N is non-normal. It follows that the statement "N is normal" is both true and false. This fatal contradiction results from an uncritical use of the apparently pellucid notion of class.

Moreover, additional antinomies were found subsequently, each of them constructed by means of familiar and seemingly cogent modes of reasoning. But the intellectual construction and formulation of non-finite models generally involves the use of possibly inconsistent sets of postulates. Accordingly, although the classical method for establishing the consistency of axioms continues to be an invaluable mathematical tool, that method does not supply a final answer to the problem it was designed to resolve.

II

The inadequacies of the model method of demonstrating consistency, and the growing apprehension, based on the discovery of the antinomies, that established mathematical systems were infected by contradictions, led to new attacks upon the problem. An alternative to relative proofs of consistency was proposed by Hilbert. He sought to construct so-called "absolute" proofs of freedom from contradiction. These we must explain briefly as a further preparation for discussing Goedel's proof.

The first requirement of an absolute proof as Hilbert conceived it is the *complete formalization* of the system. This, the reader will recall, means draining the expressions occurring within the system of any meaning whatever; they are to be regarded simply as empty, formal signs. How these signs are to be manipulated is then to be set forth explicitly in a set of rules. The purpose of this procedure is to construct a calculus which conceals nothing, which has in it only that which we intended to put in it. When theorems of this calculus are derived from the postulates by the combination and transformation of its meaningless signs in accord with precisely stated rules of operation, the danger is eliminated of the use of any unavowed principles of reasoning. Formalization is a difficult and tricky business, but it serves a valuable purpose. It reveals structure and function in naked clarity as does a cut-away working model of a machine. When a system has been formalized the logical relations between mathematical propositions are exposed to view; one is able to see the structures of configurations of certain "strings" (or sequences) of "meaningless" signs, how they hang together, are syntactically combined, nest in one another and so on.

A page covered with the "meaningless" marks of such a formalized mathematics does not *assert* anything—it is simply an abstract design or mosaic possessing a determinate structure. But suppose we as observers wish to make statements *about* a given configuration in the calculus, for example, that one "string" is longer than another or that one "string" is made up of three others. Such statements are evidently meaningful, and are expressed in a language belonging not to the calculus (or to mathematics) but to what Hilbert called "meta-mathematics" (or the language about mathematics). Meta-mathematical statements are statements *about* the signs in a calculus. They describe the kinds and arrangements of such signs when they are combined to form longer strings of marks called "formulas," and the relations between formulas in consequence of the rules of manipulation that have been specified for them. The following table illustrates some of the differences between expressions within arithmetic (mathematics) and statements about such expressions (meta-mathematics).

Mathematics	*Meta-mathematics*
For every x, if x is a prime and $x > 2$, then x is odd.	'x' is a numerical variable. '2' is a numerical constant. 'prime' is a predicate expression. '>' is a binary predicate.
$2 + 3 = 5$	If the sign '$=$' occurs in an expression which is a formula of arithmetic, the sign must be flanked on both its left and right sides by numerical expressions. '$2 + 3 = 5$' is a formula.
$x = x$ $0 = 0$	The formula '$0 = 0$' is derivable from the formula '$x = x$' by substituting the numeral '0' for the numerical variable 'x'.
$0 \neq 0$	'$0 \neq 0$' is not a theorem. Arithmetic is consistent—that is, it is not possible to derive from the axioms of arithmetic both the formula '$0 = 0$' and the formula '$0 \neq 0$'.

It is worth observing that, despite appearances to the contrary, the meta-mathematical statements in the right-hand column do not actually contain any of the mathematical expressions listed in the left-hand column. The right-hand column contains only the *names* of some of the arithmetical expressions in the left-hand column. This is so, because the rules of English grammar require that no English sentence shall contain the *objects* to which it refers, but only their *names*. The rule is enforced in the above table through the convention of enclosing an expression within single quotation marks in order to obtain a name for that expression. In consonance with this convention, it is correct to say that $2 + 3$ is identical with 5, but it is false to say that '$2 + 3$' is identical with '5'.

The importance of the division between the mathematical and the meta-mathematical language cannot be overemphasized. By erecting a separate, formal calculus whose symbols are free of all hidden assumptions and intuitive associations, and each of whose operations are precisely and rigidly defined, we have an instrument which exposes to plain view the nature of mathematical reasoning. But as human beings who wish to analyze this stark symbolism and to communicate our findings, we must construct another language which will enable us to describe, discuss, explain and theorize about the more formal system. Thus we separate the theory of the thing from the thing itself and devise the discourse of meta-mathematics.

It was by the application of this meta-mathematical language that Hilbert hoped to prove the consistency of the formalized calculus itself. Specifically, he sought to develop a theory of proof (*Beweistheorie*) that would yield demonstrations of consistency by an analysis of the purely structural features of expressions in uninterpreted calculi. Such an analysis consists exclusively of noting the kinds and arrangements of signs in formulas, and of showing whether a given combination of signs can be obtained from others in accordance with the explicitly stated rules of operation. An essential requirement for demonstrations of consistency, as propounded in the original version of Hilbert's program, is that they employ only *finitary* notions, and make no reference either to an infinite number of formulas or to an infinite number of operations upon them. A proof of the consistency of a set of postulates which conforms to these requirements is called "absolute." Such a proof achieves its objective by means of a bare minimum of inferential principles, without assuming the consistency of some other set of axioms. An absolute proof of the consistency of arithmetic, if one could be devised, would afford a demonstration, by finitary meta-mathematical means, that two "contradictory" formulas, such as '$(0 = 0)$' and '$\sim (0 = 0)$'—where the sign '$\sim$', called a tilde, signifies negation—are not both derivable from the axioms or initial formulas of the system, when the derivations conform to the stated rules of inference.

It may be useful, by way of illustration, to compare meta-mathematics as a theory of proof with the theory of some game, such as chess. Chess is a game played with 32 pieces of specified design on a square board containing 64 square subdivisions, where the pieces may be moved in accordance with fixed rules. The game can obviously be played without assigning any "interpretation" to the pieces or to their various positions on the board, although it is clear that such interpretations could be supplied if desired. There is thus an analogy between the game and a formalized mathematical calculus. The pieces and the squares of the board correspond to the elementary signs of the calculus; the permitted configurations

of pieces on the board correspond to the formulas of the calculus; the
initial positions of pieces on the board correspond to the axioms or initial
formulas of the calculus; the subsequent configurations of pieces on the
board correspond to formulas derived from the axioms (i.e., to the
theorems); and the rules of the game correspond to the rules of derivation
for the calculus. Again, although configurations of pieces on the board,
like the formulas of the calculus, are "meaningless," statements about
these configurations, like meta-mathematical statements about formulas,
are quite meaningful. A meta-chess statement may assert, for example,
that there are 20 possible opening moves for White, or that, given a certain configuration of pieces on the board with White to move, Black is
mate in three moves. It is pertinent to note, moreover, that general meta-
chess theorems can be established, whose proof involves the consideration
of only a finite number of permissible configurations on the board. The
meta-chess theorem about the number of possible opening moves for
White can be established in this way; and so can the meta-chess theorem
that if White has only two Knights and the King, and Black only his King,
it is impossible for White to force a mate against Black. These and other
meta-chess theorems can thus be proved by finitary methods of reasoning,
consisting in the examination in turn of each of a finite number of configurations that can occur under stated conditions. The aim of Hilbert's
theory of proof, similarly, was to demonstrate by such finitary methods
the impossibility of deriving certain formulas in a calculus.

III

There are two more bridges to cross before entering upon Goedel's
proof itself. Something needs be said about how and why the *Principia
Mathematica* came into being; also we must give a short illustration of
the formalization of a deductive system—we shall take a fragment of
Principia—and how its consistency can be established.

Ordinarily, even when mathematical proofs conform to accepted standards of professional rigor, they suffer from one important omission. They
employ principles (or rules) of inference which are not explicitly formulated, and of which mathematicians are frequently unaware. Take as
example Euclid's proof that there is no greatest prime number. This is
cast in the form of a *reductio ad absurdum* argument and runs as follows.
Suppose there is a greatest prime x. Then:

(1) x is the greatest prime number.

(2) Form the product of all primes less than or equal to x and add 1
to the product. This yields a new number y, where $y = (2 \times 3 \times 5 \times 7$
$\ldots \times x) + 1$.

(3) Now if y is itself a prime, then x is not the greatest prime, for y is
greater than x.

(4) But suppose y is composite, i.e., not a prime; then again x is not the greatest prime. For if y is composite, it must have a prime divisor z, which is different from each of the primes 2, 3, 5, 7 . . . x; hence z itself is a prime greater than x.

(5) But y is either prime or composite, and in either case x is not the greatest prime.

(6) Hence, since x is not the greatest prime, and x can be *any* prime number, there is no greatest prime.

We have shown the essential steps of this proof, and we could show also—though we cannot here take the time—that a number of elementary rules of inference are essential to its development, (e.g., the "Rule of Substitution," the "Rule of Detachment") and even rules and theorems belonging to more advanced parts of logical theory (e.g., the theory of "quantification," having to do with the proper use of expressions such as "all," "every," "some" and their synonyms). It has been pointed out that the use of these rules and theorems is an all but unconscious process; however, even more noteworthy is the fact that the analysis of Euclid's proof which uncovers the use of these logical props depends upon advances in the theory of logic which have occurred only within the past century. Like Molière's M. Jourdain, who spoke prose without knowing it, mathematicians have been reasoning without knowing their reasons. Modern students have had to show them the real nature of the tools of their craft.

For almost 2,000 years Aristotle's codification of valid forms of deduction was widely regarded as complete and as incapable of essential improvement. As late as 1787, the German philosopher Immanuel Kant was able to say that since Aristotle, formal logic "has not been able to advance a single step, and is to all appearances a closed and completed body of doctrine." But the fact is that the traditional logic is seriously incomplete and fails to give an account of many principles of inference employed in even quite elementary mathematical reasoning, such as the above proof of Euclid. In any event, a renaissance of logical studies in modern times began with the publication in 1847 of George Boole's *The Mathematical Analysis of Logic*. The primary concern of Boole and his immediate successors was to develop a non-numerical algebra of logic, which would provide a precise algorithm for handling more general and more varied types of deductions than were covered by traditional logical principles.

Another line of inquiry, intimately related to the work of 19th-century mathematicians on the foundations of analysis, became associated eventually with the Boolean program. This new development sought to exhibit all of pure mathematics as simply a chapter of formal logic; and it received its classical embodiment in the *Principia Mathematica* of

Whitehead and Russell in 1910. Mathematicians of the 19th century succeeded in "arithmetizing" algebra and the so-called "infinitesimal calculus," by showing that the various notions employed in mathematical analysis are definable exclusively in arithmetical terms (i.e., in terms of the integers and the arithmetical operations upon them). What Russell (and, before him, the German mathematician Gottlob Frege) sought to show was that all arithmetical notions are in turn definable in terms of purely logical ideas, and that, furthermore, the axioms of arithmetic are all deducible from a small number of basic propositions certifiable as purely logical truths. For instance, the notion of *class* belongs to general logic. Two classes are defined to be "similar," if there is a one-to-one correspondence between their members, the notion of such a correspondence being specifiable in terms of other logical ideas. A class which has no members (e.g., the class of satellites of the planet Venus) is said to be "empty." Then the cardinal number 0 can be defined as the class of all classes which are similar to an empty class. Again, a class which has a single member is said to be a "unit" class (e.g., the class of satellites of the planet Earth); and the cardinal number 1 can be defined as the class of all classes similar to a unit class. Analogous definitions can be given of the other cardinal numbers, and the various arithmetical operations can also be defined in terms of the notions of formal logic. An arithmetical statement, e.g., $1 + 1 = 2$, can then be exhibited as a condensed transcription of a statement containing only expressions belonging to general logic; and such purely logical statements can be shown to be deducible from certain logical axioms, some of which will be mentioned presently.

Principia Mathematica thus appeared to advance the final solution of the problem of consistency of mathematical systems, and of arithmetic in particular, by reducing that problem to the question of the consistency of formal logic. For if the axioms of arithmetic are simply transcriptions of theorems in logic, then the question whether these axioms are consistent is immediately transposed into the problem whether the fundamental axioms of logic are consistent.

The Frege-Russell thesis that mathematics is but a chapter of logic has not won universal acceptance from mathematicians, for various reasons of detail. Moreover, as we pointed out earlier, the antinomies of the Cantorian theory of transfinite numbers can be duplicated within logic itself, unless special measures are taken to prevent such an outcome. But are the measures adopted in *Principia Mathematica* to outflank these antinomies sufficient to exclude *all* forms of self-contradictory constructions? This cannot be asserted as a matter of course. It follows that the Frege-Russell reduction of arithmetic to logic does not provide a final answer to the consistency problem—indeed, the problem simply emerges in a more general form. On the other hand, irrespective of the validity of the Frege-Russell thesis, two features of *Principia* have proved to be of inestimable value for the further study of the problem. *Principia* supplies an inclusive system of notation, with the help of which all statements of pure mathematics can be codified in a standard manner; and *Principia* makes explicit most of the rules of formal inference (eventually these rules were made more precise and complete) which are employed in mathematical demonstrations. In short, *Principia* provides the essential instrument for investigating the entire system of formal logic as an uninterpreted calculus, whose formulas are combined and transformed in accordance with explicitly stated rules of operation.

We turn now to the formalization of a small portion of *Principia*, namely, the elementary logic of propositions. The task is to convert this fragment into a "meaningless" calculus of uninterpreted signs and to show how its freedom from contradiction can be proved.

Four steps are involved. First a complete catalogue is presented of the signs to be employed in the calculus. These are its vocabulary. Second, the "Formation Rules" are laid down. These indicate the permissible combinations of the elementary signs which are acceptable as formulas (or sentences). The rules may be said to constitute the grammar of the system. Third, the "Transformation Rules" are specified. They describe the precise structure of formulas from which some other formula is derivable. Finally, certain formulas are selected as axioms (or as "primitive formulas"). They serve as foundation for the entire system. By the expression "theorems of the system" we denote all the formulas, including the axioms, which can be derived from the axioms by successively applying the Transformation Rules. By "proof" we mean a finite sequence of legitimate formulas, each of which is either an axiom or is derivable from preceding formulas in the sequence by the Transformation Rules.

For the elementary logic of propositions (often also called the "sentential calculus") the vocabulary is extremely simple. It consists of sentential variables (which stand for sentences) and are written

$$'p', \ 'q', \ 'r', \ \text{etc.,}$$

of sentential connectives

'~' is short for 'not'
'v' is short for 'or'
'⊃' is short for 'if . . . then'
'.' is short for 'and'

and of parentheses, used as signs of punctuation. It is convenient to define the last two connectives in terms of the first two, so that expressions containing '⊃' or '.' can be replaced by expressions containing only 'v'

and '$\sim$'. For example '$p \supset q$' is defined as being simply shorthand for
the slightly longer expression '$\sim p \vee q$'.[1]

The Formation Rules are so laid down that combinations of the elementary signs which would normally be called "sentences" are designated
as "formulas." Accordingly, each sentential variable will count as a
formula. Moreover, if S is a formula, so is its negation $\sim$ (S); and if S_1
and S_2 are formulas, so is $(S_1) \vee (S_2)$, with similar conventions for the
other connectives. Two Transformation Rules are adopted. One of them,
the Rule of Substitution, says that if a sentence containing sentential variables has been accepted as logically true, any formulas may be uniformly
substituted for these variables, whereupon the new sentence will also be
logically true. The other rule, that of Detachment, simply says that if we
have two logically true sentences of the form S_1, and $S_1 \supset S_2$, we may
also accept as logically true the sentence S_2.

The axioms of the calculus (essentially those of *Principia*) are the
following:

1. $(p \vee p) \supset p$
2. $p \supset (p \vee q)$
3. $(p \vee q) \supset (q \vee p)$
4. $(p \supset q) \supset [(r \vee p) \supset (r \vee q)]$

Their meaning is easily understood. The second, for instance, says that
a proposition (or sentence) implies that either it or some other proposition (or sentence) is true.

Our purpose is to show that this set of axioms is not contradictory; in
other words that, by using the stated Transformation Rules, it is impossible to derive from the axioms any formula S together with its negation
$\sim$ S.

Now it happens that '$p \supset (\sim p \supset q)$' is a theorem in the calculus. (We
shall simply accept this as a fact without exhibiting the derivation.) Suppose, then, that some formula S, as well as $\sim$ S were deducible from the
axioms. (The reader will recognize the *reductio ad absurdum* approach of
Euclid's proof.) By substituting S for 'p' in the theorem (as permitted by
the Rule of Substitution), and applying the Rule of Detachment twice,
the formula 'q' would be deducible. But this immediately has the consequence that by substituting any formula whatsoever for 'q', any formula
whatsoever would be deducible from the axioms. It is thus clear that if
both some formula S and its contradictory $\sim$ S were deducible from the
axioms, then *any* formula would be deducible. In short, if the calculus is

---

[1] That is, "if *p* then *q*" is defined as short for "either not–*p* or *q*." In view of this
definition, the statement "If Galileo played the lute then Galileo was a musician" is
simply a slightly more compact way of rendering what is expressed by the statement
"Either Galileo did not play the lute or Galileo was a musician."

not consistent, *every* formula is a theorem. And likewise, if *not* every formula is a theorem (i.e., if there is at least one formula which is not derivable from the axioms), then the calculus *is* consistent. The task, therefore, is to exhibit some formula which cannot be derived from the axioms.

The way this is done is to employ meta-mathematical reasoning upon the system to be tested. We place ourselves, so to speak, *outside* the calculus and consider how theorems are generated *within* it. The actual procedure is pretty. (1) We try to find a characteristic common to all four axioms; (2) we try to show that this characteristic is "hereditary" under the Transformation Rules—i.e., that if all the axioms have this characteristic, any formula derived from them by the rules (which is to say, any *theorem*) also has it; (3) we try to exhibit a formula that does not have this characteristic. If we succeed in this triple task, we will have an absolute proof of consistency. For if the common characteristic exists and is hereditary, so that it is transmitted to all properly derived formulas, then any array of symbols which satisfies the requirements for being a formula but nevertheless does not possess the characteristic in question cannot be a theorem. That is to say, structurally it may be a formula, yet not one which could have been derived from the axioms; or to put it yet another way, since the suspected offspring (formula) lacks an invariably inherited trait of the forebears (axioms) it cannot in fact be their descendant (theorem). Furthermore, if we can find such a formula we will have established the consistency of the calculus; because, as we noted a moment ago, if the calculus were not consistent, *every* formula could be derived from the axioms, i.e., every formula would possess the characteristic and therefore be a theorem.

Let us specify a common characteristic. The trait we have in mind is that of being a *tautology*. In common parlance *tautology* is defined as the saying of a thing twice over in different words, e.g., "John is the father of Charles and Charles is the son of John." In logic, however, a tautology is defined as a statement that excludes no logical possibilities, e.g., "Either it is raining or it is not raining." The essence of a tautology is that it is "true in all possible worlds," whence it is a *truth of logic*. Now it can be shown (though we shall not turn aside to give the demonstration) with the aid of an ingenious device known as a "truth-table," that each of the four axioms of our little set is a tautology. That is to say, if each axiom is regarded as a formula made up of simpler formulas (e.g., the compound formula or sentence $p \supset (p \vee q)$ is constituted of the simple formulas 'p' and 'q'), it must be accepted as true *irrespective* of the truth or falsity of its elementary constituents. Even the skeptical reader will have no difficulty accepting the fact, for example, that axiom 1: $(p \vee p) \supset p$ is "true in all possible worlds," if he substitutes the elementary sentence "2 is a prime number" for the sentential variable *p* and derives the sentence "If 2 is a prime number or 2 is a prime number, then 2 is a prime number."

It is also possible to show that the characteristic of being a tautology is hereditary under the Transformation Rules. In sum, if the axioms are tautologous, so are all the formulas derivable from them.

Having performed these two steps, we are ready for the third. We must look for a formula which from the standpoint of its vocabulary (sentential variables) and structure (the use of the connectives) belongs to our system; yet, because it does not possess the characteristic of being a tautology, cannot be a theorem (i.e., be derivable from the axioms) and therefore cannot belong to the system. We do not have to look very hard; it is easy to exhibit such a formula. For example '*p* v *q*' fits the requirements. It purports to be a gosling but is in fact a duckling; it does not belong to the family; it is a formula but it is not a theorem. There can be no doubt that it is not a tautology. Any correct interpretation shows this at once. As an illustration, we obtain by substitution for the variables in

$$\text{'}p \text{ v } q\text{'}$$

the sentence

"Either John is a philosopher or Charles reads *Scientific American*."

Clearly this is not a truth of logic; which is to say, it is not a sentence that is true irrespective of the truth or falsity of its elementary constituents.

We have, therefore, achieved our goal. At least one formula has been found which is not a theorem. It follows, for reasons already explained, that it is not possible to derive from the axioms of the sentential calculus both a formula and its negation. We have constructed an absolute proof of the consistency of the system.

One final point must be mentioned. It has been shown that every theorem of the sentential calculus is a tautology, a truth of logic. It is natural to ask whether, conversely, every logical truth which is expressible in the vocabulary of the calculus (i.e., every tautology) is also a theorem. The answer is yes, though the proof is too long to be shown here. Since the axioms of the calculus are sufficient for generating all logical truths expressible in the system, the axioms are said to be "complete." It is frequently of paramount interest to determine whether an axiomatized system is complete. Indeed, a powerful motive for axiomatizing various branches of mathematics has been the desire to specify a sufficient set of initial assumptions from which all the true statements in some field of analysis are deducible. Thus, when Euclid axiomatized elementary geometry, he apparently selected his axioms so as to make it possible to deduce from them all geometric truths which were already established, as well as those still to be discovered. (Euclid's inclusion of his famous parallel postulate in his list of axioms showed remarkable insight. For as was subsequently proved, this postulate is not derivable from the others of the set, so that without the parallel postulate the remaining axioms are surely incomplete.) A similar objective controlled the axiomatization of elementary arithmetic toward the close of the nineteenth century. Until recently, it was assumed as a matter of course that a complete set of axioms for any given branch of mathematics could always be specified. In particular, it seems to have been generally believed that the axioms proposed for arithmetic by nineteenth-century mathematicians were in fact complete, or at worst could be made complete by the addition of a finite number of further axioms. The discovery that this is not so, is one of Goedel's achievements.

IV

The sentential calculus is an example of a mathematical system for which the objectives of Hilbert's theory of proof are fully realized. As we have pointed out, however, this calculus codifies only a fragment of formal logic, and its vocabulary and formal apparatus do not suffice to develop within its framework even elementary arithmetic. On the other hand, Hilbert's program has been successfully carried out for more inclusive systems, which have been shown to be both consistent and complete by meta-mathematical reasoning. For example, an absolute proof of consistency has been given for a system of arithmetic which allows for the *addition*, though not for the *multiplication*, of cardinal numbers. But can a system such as *Principia*, in which the whole and not merely a fragment of arithmetic is expressible, be proved consistent in the sense of Hilbert's program? Repeated attempts at constructing such a proof were unsuccessful; and the publication of Goedel's paper in 1931 showed, finally, that all such efforts are doomed to failure.

What did Goedel establish, and how did he prove his results? His main conclusions are two-fold. In the first place, he showed that no meta-mathematical proof is possible for the formal consistency of a system comprehensive enough to contain the *whole* of arithmetic; unless, that is, the meta-mathematical proof employs rules of inference whose consistency is as doubtful as is the consistency of the Transformation Rules used in deriving theorems *within* the system. But thus one dragon is slain only to create another.

Goedel's second main conclusion is even more surprising and revolutionary in its import, for it makes evident a fundamental limitation in the power of the axiomatic method. Goedel showed that *Principia*, or any other system within which arithmetic can be developed, is *essentially incomplete*. In other words, given *any* consistent set of arithmetical axioms, there are true arithmetical statements which are not derivable from the set. This essential point deserves illustration. Mathematics abounds in statements which seem self-evident, to which no exceptions have been found, which nevertheless have thwarted all attempts at proof. A simple example is Goldbach's "theorem" which states that every even number is the sum of two primes; yet no one has succeeded in finding a proof valid for *all* even numbers. Goldbach's conjecture presents us with a statement that may be true, but may not be derivable from the axioms of arithmetic. Now it may be suggested that the axioms should be modified or augmented to take care of this and related theorems by making them derivable. But Goedel has shown that this approach promises no final cure. That is, even if the given set of axioms is augmented by the addition of any finite number of arithmetical postulates, there will always be *further* arithmetic truths which are not formally derivable from the augmented set. Such further truths may, to be sure, be established by some form of meta-mathematical reasoning *about* an arithmetical system; but this procedure does not fit the requirement that the calculus must so to speak be self-contained, that the logical truths in question must be exhibited as the formal consequences of the specified axioms *within* the system. There is, it seems, an inherent limitation in the axiomatic method as a way of systematizing the whole of arithmetic.

How did Goedel prove his conclusions? Up to a point, the structure of his demonstration is modeled, as he himself noted, on the reasoning involved in one of the logical antinomies known as the "Richard Paradox," first propounded by the French mathematician, Jules Richard, in 1905. The paradox can be stated as follows. Assume some definite language (e.g., English) in which the various purely arithmetical properties of the integers can be expressed; and consider the definitions of these properties which can be formulated in the notation of that language. Thus, the property of being a prime number may be defined by: "not divisible by any integer other than one and itself"; the property of being a perfect square may be defined by: "being equal to the product of some integer by that integer"; and so on. It is easily seen that each such definition will contain a finite number of words and therefore a finite number of letters of the alphabet. This being so, the definitions can be placed in serial order, according to the number of letters they contain (definitions with the same number of letters can be arranged alphabetically under their serial tag). To each definition there will then correspond a unique integer—for example, the definition with the smallest number of letters will correspond to the number 1, the next definition in the ordered series will correspond to 2, and so on. We come now to an odd little point. Since each definition has an integer attached to it, it may happen in certain cases that an integer possesses the very property designated by the definition to which the integer is serially attached. (This is the same sort of thing that would happen if we prefixed to each of a list of English words the descriptive tags "short" or "long," and the word "short" itself appeared in the list.

"Short" itself would of course have the tag "short" attached to it.) Suppose, for instance, the defining expression "not divisible by any integer other than one and itself" happens to be correlated with the number 17; then 17 has the property designated by that expression. On the other hand, suppose that the defining expression "being equal to the product of some integer by that integer" turns out to be correlated with the number 20; then 20 does not have the property designated by that expression. We shall now say that (in the second example) the number 20 has the property of being "Richardian," while (in the first example) the number 17 does not have the property of being "Richardian." More generally, we define the property of being Richardian as follows: "not having the property designated by the defining expression with which an integer is correlated in the serially ordered set of definitions." But observe that this last expression itself defines a numerical property, so that this expression also must belong to the above series of definitions. This being so, it must correspond to some number, say n, giving its position in the series. The question may then be posed, reminiscent of Russell's antinomy, whether the number n itself is Richardian. Almost at once we can see the fatal contradiction looming. For n is Richardian if, and only if, it does *not* possess the property designated by the definition with which n is correlated; and it is easy to see that therefore n is Richardian if and only if n is not Richardian. Accordingly, the statement "n is Richardian" is both true and false.

The contradiction can be avoided if we notice that in constructing it we have not played the game quite fairly. Pursuant to our initial stipulations, we were invited to consider the definitions expressible in a language about numbers and their arithmetical properties. However, it was not intended to consider definitions involving reference to the *notation* used in *formulating* numerical properties. In other words, the series of definitions in the above construction was supposed to include only expressions which refer exclusively to such notions as arithmetical addition, multiplication, and the like. Accordingly, the definition of being Richardian does not belong to this series, for this definition makes reference to such *meta-mathematical* notions as the number of *letters* occurring in *expressions*. The Richard Paradox can therefore be outflanked by distinguishing carefully between statements *within* arithmetic (which make no reference to any system of notation) and statements *about* the language in which arithmetic is codified.

The reasoning in the Richard Paradox is evidently fallacious. Its construction nevertheless suggests that it might be possible to "map" (or "mirror") meta-mathematical statements *about* a sufficiently comprehensive formal system *into* the system itself. If this were possible, then meta-mathematical statements about a system would be *represented* by statements within the system. Thereby one could achieve the desirable end of getting the formal system to speak about itself—a most valuable form of self-consciousness. The idea of such mapping is a familiar one in mathematics. It is employed in coordinate geometry, which translates geometric statements into algebraic ones, so that geometric relations are mapped onto algebraic ones. The idea is manifestly used in the construction of ordinary maps, since the construction consists in projecting configurations on the surface of a sphere onto a plane, so that relations between plane figures can mirror the relations between the spherical ones. The idea also plays a role in mathematical physics when, for example, relations between properties of electric currents are represented in the language of hydrodynamics. The basic fact which underlies all these mapping procedures is that an abstract structure of relations embodied in one domain of "objects" is exhibited to hold between "objects" in some other domain. In consequence, deductive relations between statements about the first domain can be established by exploring (often more conveniently and easily) the deductive relations between statements about their counterparts. For example, complicated geometrical relations between surfaces in space are usually more readily studied by way of the algebraic formulas for such surfaces. Similarly, questions about complicated logical relations between assertions may be more readily handled *via* the arithmetical representatives of those assertions.

In any event, the exploitation of this notion of mapping is the key to the argument in Goedel's revolutionary paper. In a manner suggested by the Richard Paradox, but without falling victim to the fallacy involved in its construction, Goedel showed that meta-mathematical statements *about* a formalized arithmetical calculus can indeed be represented *within* that calculus. In fact, he found a method of representation such that neither the arithmetical formula corresponding to a certain true meta-mathematical statement about the formula, nor the arithmetical formula corresponding to the denial of the statement, is demonstrable within the calculus. Since one of these arithmetical formulas must codify an arithmetical truth, but neither is derivable from the axioms, the axioms are incomplete. As we shall see, this incompleteness is incurable. Moreover, Goedel indicated how to construct an arithmetical formula to represent the meta-mathematical statement "The calculus is consistent," and he showed that this formula is not demonstrable within the calculus. Accordingly, the consistency of arithmetic cannot be established except by using rules of inference whose consistency is at least as doubtful as is the consistency of arithmetic itself.

Goedel's paper, as we said at the outset, is difficult. Forty-six preliminary definitions together with several important lemmas must be mastered before the main results are reached. We shall take a much easier road; nevertheless we hope at least to offer glimpses of the argument.

Goedel first showed that a formalized system of arithmetic can be set up in which it is possible to associate with each elementary sign, each formula (or sequence of signs) and each proof (that is, each finite sequence of formulas) a *unique integer*. This integer, a distinctive label, is called the "Goedel number" of the sign, formula or proof.

As an illustration take the following correspondence. The top row lists part of the basic vocabulary with the help of which the whole of arithmetic can be formulated. The second row lists under each of these basic signs its corresponding Goedel number.

'~'	'v'	'⊃'	'∃'	(a qualifying symbol meaning "there
1	2	3	4	is" or "there are," so that "$(\exists x)$"
				means "There is an x" or "For
				every x")

'='	'0'	's'	(for representing the immediate suc-
5	6	7	cessor of a number)

'(' left-hand   ')' right-hand   ',' (the comma)
8   parenthesis)   9   parenthesis)   10

In addition to these basic signs we also require sentential variables ('p', 'q', etc.) for which sentences may be substituted, individual variables ('x', 'y', etc.) for which numerals and numerical expressions may be substituted, and predicate variables ('P', 'Q', etc.) for which predicates may be substituted. These are assigned Goedel numbers in accordance with the following rules.

Associate with each sentential variable an integer greater than 10 but divisible by 3; with each individual variable, an integer greater than 10 which leaves a remainder of 1 on division by 3; and with each predicate variable, an integer greater than 10 which leaves a remainder of 2 on division by 3. This done, each elementary sign of the system is now associated with a unique number.

Consider next a formula of the system, for example, '$(\exists x)(x = sy)$' (when literally translated, it reads: 'There is an x, such that x is the immediate successor of y,' and in effect says that every number has an immediate successor). The numbers associated with its ten constituent elementary signs are, respectively, 8, 4, 13, 9, 8, 13, 5, 7, 16, 9. We now agree to associate with the formula itself the number which is the product of the first ten primes in order of magnitude, each prime being raised to a power equal to the Goedel number of the corresponding elementary sign. By this convention the formula is associated with the number $2^8 \times 3^4 \times 5^{13} \times 7^9 \times 11^8 \times 13^{13} \times 17^5 \times 19^7 \times 23^{16} \times 29^9$; let us refer to the number as m. In a similar fashion, every formula can be made to correspond to a unique number.

Consider, finally, a sequence of formulas as may occur in some proof, for example, the sequence:

$$(\exists x)(x = sy)$$
$$(\exists x)(x = s0)$$

This second formula when translated reads '0 has an immediate successor'; it is derivable from the first formula by substituting '0' for the "free" variable 'y'. (A variable is said to be "free" in a formula if it is not preceded in that formula by a quantifier containing this variable. Thus, the variable 'x' is *not* free in either of these formulas.) We have already determined the Goedel number of the first formula: it is m; and suppose that n is the Goedel number of the second formula. We now agree to associate with the sequence of two formulas the number which is the product of the first two primes in order of magnitude, each prime being raised to a power equal to the Goedel number of the corresponding formula in the sequence. The above sequence is accordingly associated with the number $k = 2^m \times 3^n$. In like manner, a number is associated with each sequence of formulas. It is easy to see that following this procedure, every expression in the system can be tagged with a unique Goedel number.

What has been done so far is to establish a method for completely arithmetizing a formal system. The method is essentially a set of directions for establishing a correspondence between certain integers and the various elements or combinations of elements of the system. Once an expression is given, it can be uniquely numbered. But more than that, once a Goedel number is given, the expression it represents can be exactly analyzed or "retrieved," since the number itself, having been arrived at as a product of prime numbers, can be factored into these primes (as we know from a classic theorem of arithmetic) in only one way. In other words, we can take the number apart like a machine, see how it was constructed and what went into it; which is to say we can dissect an expression, a proof, in the same way.

This leads to the next step. We have already spoken of "mapping." Now we can extend the process with the help of the Goedel numbers so that meta-mathematical statements can be completely mirrored within the calculus; so that meta-mathematics itself becomes completely "arithmetized." In particular, every *meta-mathematical characterization* of the structure of expressions in the system, everything we say about them, is mapped into an arithmetical function of integers; and *every meta-mathematical statement about relations between formulas* is mapped into an arithmetical relation between integers. (By an *arithmetical function* is meant an expression such as $2 + 3$, $(7 \times 5) + 8$, and so on: that is, a function of an integer in itself an integer. By an *arithmetical relation* is meant a proposition—which may be true or false—such as $5 = 3$, $7 > 4$, and so on.) The importance of this arithmetization of meta-mathematics stems from the fact that, since each of its statements can be uniquely represented in the formal system by an expression tagged with a Goedel number, relations of logical dependence between meta-mathematical statements can be explored by examining relations between integers and their factors. To take a trivial analogue: if customers in a supermarket are given tickets with numbers determining the order in which they are to be waited on when buying meat, it is a simple matter, merely by scrutinizing the numbers themselves to discover (a) how many persons have been served, (b) how many are waiting, (c) who precedes whom and by how many customers, etc.

Consider the meta-mathematical statement: 'The sequence of formulas whose Goedel number is x is a demonstration for the formula whose Goedel number is z.' This statement is represented (mirrored) by a definite formula *in* the arithmetical calculus, a formula which expresses a purely arithmetical relation between x and z. (In the above example of assigning the Goedel number k to a demonstration, we found that $k = 2^m \times 3^n$; and a little reflection shows that there is a definite though complex arithmetical relation between k, the Goedel number of the proof, and n, the Goedel number of the conclusion.) We write this arithmetical relation between x and z as the formula 'Dem (x,z)', to remind ourselves of the meta-mathematical statement to which it corresponds. Similarly, the meta-mathematical statement: "The sequence of formulas with Goedel number x is *not* a demonstration for the formula with the Goedel number z," is also represented by a definite formula in the arithmetical formalism. This formula we shall write as '$\sim$ Dem (x,z)'.

We shall need one additional bit of special notation for stating the crux of Goedel's argument. Begin with an example. The formula '$(\exists x)(x = sy)$' has the Goedel number m, and the variable 'y' has the Goedel number 16. Substitute in this formula for the variable with Goedel number 16 (i.e., for 'y') the numeral for m. We then obtain the formula '$(\exists x)(x = sm)$'. This latter formula obviously also has a Goedel number—a number which can be actually calculated, and which, in fact, is a certain complex arithmetical function of the two numbers m and 16. However, instead of calculating this Goedel number, we can give an unambiguous meta-mathematical characterization for it: it is the Goedel number of the formula which is obtained from the formula with Goedel number m, by substituting for the variable with Goedel number 16 the numeral for m. Accordingly, this meta-mathematical characterization corresponds to a definite arithmetical function of the numbers m and 16, a function which can be expressed within the arithmetical calculus. We shall write this function as 'sub $(m, 16, m)$', to remind ourselves of the meta-mathematical description which it represents. More generally, the expression 'sub $(y, 16, y)$' is the mirror-image *within* the arithmetical formalism of the meta-mathematical characterization: "the Goedel number of the formula which is obtained from the formula with Goedel number y, by substituting for the variable with Goedel number 16 the numeral for y." It should be noted that, when a definite numeral is substituted for 'y' in 'sub $(y, 16, y)$', sub $(y, 16, y)$ is a definite integer which is the Goedel number of a certain formula.

We are now equipped to follow in outline Goedel's argument. Consider the formula '$(x) \sim \text{Dem}\ (x,z)$'. This represents, in the arithmetical calculus, the meta-mathematical statement "For every x, where x is the Goedel number of a demonstration, x is not the number of a demonstration for the formula whose Goedel number is z." This formula may therefore be regarded as a formal paraphrase of the statement "The formula with Goedel number z is not demonstrable." What Goedel was able to show was that a certain special case of this formula itself is in fact not formally demonstrable. To construct this special case we start with a formula which we shall display as line (1):

(1) $(x) \sim \text{Dem}\ (x, \text{sub}\ (y, 16, y))$

It corresponds to the meta-mathematical statement that the formula with the Goedel number sub $(y, 16, y)$ is not demonstrable. Moreover, since line (1) is a formula within the arithmetical calculus, it has its own Goedel number, say n. Let us now obtain another formula from the one on line (1) by substituting the numeral for n for the variable with Goedel number 16 (i.e., for 'y'). We thus arrive at the special case we wished to construct, and display it as line (2):

(2) $(x) \sim \text{Dem}\ (x, \text{sub}\ (n, 16, n))$

Since this last formula occurs within the arithmetical calculus, it must have a Goedel number. What is its Goedel number? A little reflection shows that it is sub $(n, 16, n)$. To see this, we must recall that sub $(n, 16, n)$ is the Goedel number of the formula which is obtained from the formula with Goedel number n, by substituting for the variable with Goedel number 16 (i.e., for 'y') the numeral for n. But the formula (2) has indeed been obtained from the formula with Goedel number n (i.e., from the formula on line (1)) by substituting for the variable 'y' the numeral for n. Let us also remind ourselves, however, that the formula '$(x) \sim \text{Dem}\ (x, \text{sub}\ (n, 16, n))$' is the mirror-image *within* the arithmetical calculus of the meta-mathematical statement: "The formula whose Goedel number is sub $(n, 16, n)$ is not demonstrable." It follows that the *arithmetical formula* '$(x) \sim \text{Dem}\ (x, \text{sub}\ (n, 16, n))$' *represents the meta-mathematical statement*: "The formula '$(x) \sim \text{Dem}\ (x, \text{sub}\ (n, 16, n))$' is not demonstrable." In a sense, therefore, this arithmetical formula can be construed as saying that it itself is not demonstrable.

Goedel is now able to show, in a manner reminiscent of the Richard Paradox, but free from the fallacious reasoning involved in that puzzle, that this arithmetical formula is indeed not demonstrable. The argument from this point on is relatively simple and straightforward. He shows that *if* the formula were demonstrable, then its formal contradictory (i.e.,

'~ (x) ~ Dem (x, sub (n, 16, n)))', which in effect says that the formula *is* in fact demonstrable) would also be demonstrable; and conversely, if the formal contradictory of the formula were demonstrable, the formula itself would also be demonstrable. But as was noted earlier, if a formula as well as its contradictory can both be derived from a set of axioms, the axioms are not consistent. Accordingly, if the axioms are consistent, neither the formula nor its contradictory is demonstrable. In short, if the axioms are consistent, the formula is "undecidable"—neither the formula nor its contradictory can be formally deduced from the axioms.

Very well. Yet there is a surprise coming. For although the formula is undecidable if the axioms are consistent, it can nevertheless be shown by meta-mathematical reasoning to be true. That is to say, the formula is a true arithmetical statement which expresses a complex but definite numerical property of integers—just as the formula '$(x) \sim (x + 3 = 2)$' (in words, "There is no positive integer which when added to 3 will equal 2") expresses another but much simpler property of integers. The reasoning that shows the truth of the undecidable formula is rather simple. In the first place, on the assumption that arithmetic is consistent, we have already established the meta-mathematical statement: "The formula '$(x) \sim$ Dem $(x, \text{sub} (n, 16, n))$'" is not demonstrable." It must be accepted, then, that this meta-mathematical statement is true. Secondly, the statement is *represented* within arithmetic by that very formula itself. Third, we recall that meta-mathematical statements have been mapped upon the arithmetical formalism in such a way that true-mathematical statements always correspond to true arithmetical formulas. (Indeed, this is the whole point of the mapping procedure—just as in analytic geometry, geometric statements are mapped onto algebra in such a way that true geometric statements always correspond to true algebraic ones.) Accordingly, the formula in question must be true. We have thus established an arithmetical truth, not by deducing it formally from the axioms of arithmetic, but by a meta-mathematical argument.

When we were discussing the sentential calculus, we explained that the axioms of that system are "complete," since all the logical truths expressible in the system are formally derivable from the axioms. More generally, we can say that the axioms of any formalized system are "complete" if every true statement expressible in the system is formally deducible from the axioms. A set of axioms is therefore "incomplete" if not every true statement expressible in the system is formally derivable from them. It follows, since we have now established as true an arithmetical formula which is not derivable from the axioms of arithmetic, that the system is incomplete. Moreover, the system is *essentially incomplete*, which means that even if we added this true but undemonstrable formula to the axioms as a further axiom, the augmented system would still not suffice to yield formally all arithmetical truths: another true arithmetical formula could be constructed, such that neither the formula nor its contradictory would be demonstrable within the enlarged system. This remarkable conclusion would hold, no matter how often we enlarged the system by adding further axioms to it.

We come then to the coda of Goedel's amazing and profound intellectual symphony. It can be shown that the meta-mathematical statement just established, namely, "If arithmetic is consistent, then it is incomplete," itself corresponds to a demonstrable formula in the arithmetical system. But the antecedent clause of this formula (the one corresponding to the meta-mathematical statement "arithmetic is consistent") is not demonstrable within the system. For if it were, the consequent clause of the formula (the one corresponding to the statement "arithmetic is incomplete," and which in fact turns out to be our old friend '$(x) \sim$ Dem $(x$, sub $(n, 16, n))$') would also be demonstrable. This conclusion would, however, be incompatible with the previously obtained result that the latter formula is not demonstrable. The grand final step is now before us: we must conclude that the consistency of arithmetic cannot be established by any meta-mathematical reasoning which can be represented within the formalism of arithmetic!

A meta-mathematical proof of the consistency of arithmetic is not excluded by this capital result of Goedel's analysis. In point of fact, meta-mathematical proofs of the consistency of arithmetic have been constructed, notably by Gerhard Gentzen, a member of the Hilbert school, in 1936. But such proofs are in a sense pointless if, as can be demonstrated, they employ rules of inference whose *own* internal consistency is as much open to doubt as is the formal consistency of arithmetic itself. Thus, Gentzen used the so-called "principle of transfinite mathematical induction" in his proof. But the principle in effect stipulates that a formula is derivable from an *infinite* class of premises. Its use therefore requires the employment of nonfinitistic meta-mathematical notions, and so raises once more the question which Hilbert's original program was intended to resolve.

The import of Goedel's conclusions is far-reaching, though it has not yet been fully fathomed. They seem to show that the hope of finding an absolute proof of consistency for any deductive system in which the whole of arithmetic is expressible cannot be realized, if such a proof must satisfy the finitistic requirements of Hilbert's original program. They also show that there is an endless number of true arithmetical statements which cannot be formally deduced from any specified set of axioms in accordance with a closed set of rules of inference. It follows, therefore, that an axiomatic approach to number theory, for example, cannot exhaust the domain of arithmetic truth, and that mathematical proof does not coincide with the exploitation of a formalized axiomatic method. Just in what way a general notion of mathematical or logical truth is to be defined which is adequate to the fact here stated, and whether, as Goedel himself appears to believe, only a thoroughgoing Platonic realism can supply such a definition, are problems still under debate and too difficult for more than mention here.

Goedel's conclusions also have a bearing on the question whether calculating machines can be constructed which would be substitutes for a living mathematical intelligence. Such machines, as currently constructed and planned, operate in obedience to a fixed set of directives built in, and they involve mechanisms which proceed in a step-by-step manner. But in the light of Goedel's incompleteness theorem, there is an endless set of problems in elementary number theory for which such machines are inherently incapable of supplying answers, however complex their built-in mechanisms may be and however rapid their operations. It may very well be the case that the human brain is itself a "machine" with built-in limitations of its own, and that there are mathematical problems which it is incapable of solving. Even so, the human brain appears to embody a structure of rules of operation which is far more powerful than the structure of currently conceived artificial machines. There is no immediate prospect of replacing the human mind by robots.

None of this is to be construed, however, as an invitation to despair, or as an excuse for mystery mongering. The discovery that there are formally indemonstrable arithmetic truths does not mean that there are truths which are forever incapable of becoming known, or that a mystic intuition must replace cogent proof. It does mean that the resources of the human intellect have not been, and cannot be, fully formalized, and that new principles of demonstration forever await invention and discovery. We have seen that mathematical propositions which cannot be established by formal deduction from a given set of axioms, may nevertheless be established by "informal" meta-mathematical reasoning. It would be an altogether irresponsible claim to maintain that the formally indemonstrable truths Goedel established by meta-mathematical arguments are asserted in the absence of any proof or by appeals simply to an uncontrolled intuition. Nor do the inherent limitations of calculating machines constitute a basis for valid inferences concerning the impossibility of physico-chemical explanations of living matter and human reason. The possibility of such explanations is neither precluded nor affirmed by Goedel's incompleteness theorem. The theorem does indicate that in structure and power the human brain is far more complex and subtle than any nonliving machine yet envisaged. Goedel's own work is a remarkable example of such complexity and subtlety. It is an occasion not for dejection because of the limitations of formal deduction but for a renewed appreciation of the powers of creative reason.

Mathematics is the science which draws necessary conclusions.
—BENJAMIN PEIRCE

5   A Mathematical Science

By OSWALD VEBLEN
and JOHN WESLEY YOUNG

1. UNDEFINED ELEMENTS AND UNPROVED PROPOSITIONS

GEOMETRY deals with the properties of figures in space. Every such figure is made up of various elements (points, lines, curves, planes, surfaces, etc.), and these elements bear certain relations to each other (a point lies on a line, a line passes through a point, two planes intersect, etc.). The propositions stating these properties are logically interdependent, and it is the object of geometry to discover such propositions and to exhibit their logical interdependence.

Some of the elements and relations, by virtue of their greater simplicity, are chosen as fundamental, and all other elements and relations are defined in terms of them. Since any defined element or relation must be defined in terms of other elements and relations, it is necessary that one or more of the elements and one or more of the relations between them remain entirely *undefined*; otherwise a vicious circle is unavoidable. Likewise certain of the propositions are regarded as fundamental, in the sense that all other propositions are derivable, as logical consequences, from these fundamental ones. But here again it is a logical necessity that one or more of the propositions remain entirely *unproved*; otherwise a vicious circle is again inevitable.

The starting point of any strictly logical treatment of geometry (and indeed of any branch of mathematics) must then be a set of undefined elements and relations, and a set of unproved propositions involving them; and from these all other propositions (theorems) are to be derived by the methods of formal logic. Moreover, since we assumed the point of view of formal (i.e., symbolic) logic, the undefined elements are to be regarded as mere symbols devoid of content, except as implied by the fundamental propositions. Since it is manifestly absurd to speak of a proposition involving these symbols as self-evident, the unproved propositions referred to above must be regarded as mere *assumptions*. It is customary to refer to these fundamental propositions as axioms or postulates, but we prefer to retain the term *assumption* as more expressive of their real logical character.

We understand the term *a mathematical science* to mean *any set of propositions arranged according to a sequence of logical deduction.* From the point of view developed above such a science is purely *abstract.* If any concrete system of things may be regarded as satisfying the fundamental assumptions, this system is a *concrete application* or *representation* of the abstract science. The practical importance or triviality of such a science depends simply on the importance or triviality of its possible applications. These ideas will be illustrated and further discussed in the next section, where it will appear that an abstract treatment has many advantages quite apart from that of logical rigor.

2. CONSISTENCY, CATEGORICALNESS, INDEPENDENCE. EXAMPLE OF A MATHEMATICAL SCIENCE

The notion of a *class* [1] of objects is fundamental in logic and therefore in any mathematical science. The objects which make up the class are called the *elements* of the class. The notion of a class, moreover, and the relation of *belonging to a class* (being included in a class, being an element of a class, etc.) are primitive notions of logic, the meaning of which is not here called in question.[2]

The developments of the preceding section may now be illustrated and other important conceptions introduced by considering a simple example of a mathematical science. To this end let S be a class, the elements of which we will denote by $A, B, C, \ldots$ Further, let there be certain undefined subclasses [3] of S, any one of which we will call an *m-class*. Concerning the elements of S and the m-classes we now make the following

Assumptions:

I. *If A and B are distinct elements of S, there is at least one m-class containing both A and B.*

II. *If A and B are distinct elements of S, there is not more than one m-class containing both A and B.*

III. *Any two m-classes have at least one element of S in common.*

IV. *There exists at least one m-class.*

V. *Every m-class contains at least three elements of S.*

VI. *All the elements of S do not belong to the same m-class.*

VII. *No m-class contains more than three elements of S.*

The reader will observe that in this set of assumptions we have just two undefined terms, viz., *element of S* and *m-class,* and one undefined rela-

---

[1] Synonyms for *class* are *set, aggregate, assemblage, totality*; in German, *Menge*; in French, *ensemble.*

[2] Cf. B. Russell, *The Principles of Mathematics,* Cambridge, 1903; and L. Couturat, *Les principes des mathématiques,* Paris, 1905.

[3] A class S' is said to be a *subclass* of another class S, if every element of S' is an element of S.

tion, *belonging to a class.* The undefined terms, moreover, are entirely devoid of content except such as is implied in the assumptions.

Now the first question to ask regarding a set of assumptions is: *Are they logically consistent?* In the example above, of a set of assumptions, the reader will find that the assumptions are all true statements, if the class S is interpreted to mean the digits 0, 1, 2, 3, 4, 5, 6 and the *m*-classes to mean the columns in the following table:

$$
(1) \quad
\begin{array}{ccccccc}
0 & 1 & 2 & 3 & 4 & 5 & 6 \\
1 & 2 & 3 & 4 & 5 & 6 & 0 \\
3 & 4 & 5 & 6 & 0 & 1 & 2
\end{array}
$$

This interpretation is a concrete representation of our assumptions. Every proposition derived from the assumptions must be true of this system of triples. Hence none of the assumptions can be logically inconsistent with the rest; otherwise contradictory statements would be true of this system of triples.

Thus, in general, *a set of assumptions is said to be consistent if a single concrete representation of the assumptions can be given.*[4]

Knowing our assumptions to be consistent, we may proceed to derive some of the *theorems* of the mathematical science of which they are the basis:

Any two distinct elements of S *determine one and only one m-class containing both these elements* (Assumptions I, II).

The *m*-class containing the elements *A* and *B* may conveniently be denoted by the symbol *AB.*

Any two m-classes have one and only one element of S *in common* (Assumptions II, III).

There exist three elements of S *which are not all in the same m-class* (Assumptions IV, V, VI).

In accordance with the last theorem, let *A, B, C* be three elements of S not in the same *m*-class. By Assumption V there must be a third element in each of the *m*-classes *AB, BC, CA,* and by Assumption II these elements must be distinct from each other and from *A, B,* and *C.* Let the new elements be *D, E, G,* so that each of the triples *ABD, BCE, CAG* belongs to the same *m*-class. By Assumption III the *m*-classes *AE* and *BG,* which are distinct from all the *m*-classes thus far obtained, have an element of S in common, which, by Assumption II, is distinct from those hitherto mentioned; let it be denoted by *F,* so that each of the triples *AEF* and *BFG* belong to the same *m*-class. No use has as yet been made of Assumption VII. We have, then, the theorem:

---

[4] It will be noted that this test for the consistency of a set of assumptions merely shifts the difficulty from one domain to another. It is, however, at present the only test known. On the question as to the possibility of an absolute test of consistency, cf. Hilbert, *Grundlagen der Geometrie,* 2d ed., Leipzig (1903), p. 18, and *Verhandlungen d. III. intern. math. Kongresses zu Heidelberg,* Leipzig (1904), p. 174; Padoa, *L'Enseignement mathématique,* Vol. V (1903), p. 85.

Any class S *subject to Assumptions I–VI contains at least seven elements.*

Now, making use of Assumption VII, we find that the *m*-classes thus far obtained contain only the elements mentioned. The *m*-classes *CD* and *AEF* have an element in common (by Assumption III) which cannot be *A* or *E*, and must therefore (by Assumption VII) be *F*. Similarly, *ACG* and the *m*-class *DE* have the element *G* in common. The seven elements *A, B, C, D, E, F, G* have now been arranged into *m*-classes according to the table

(1')
$$\begin{array}{ccccccc}
A & B & C & D & E & F & G \\
B & C & D & E & F & G & A \\
D & E & F & G & A & B & C
\end{array}$$

in which the columns denote *m*-classes. The reader may note at once that this table is, except for the substitution of letters for digits, entirely equivalent to Table (1); indeed (1') is obtained from (1) by replacing 0 by *A*, 1 by *B*, 2 by *C*, etc. We can show, furthermore, that S can contain no other elements than *A, B, C, D, E, F, G*. For suppose there were another element, *T*. Then, by Assumption III, the *m*-classes *TA* and *BFG* would have an element in common. This element cannot be *B*, for then *ABTD* would belong to the same *m*-class; it cannot be *F*, for then *AFTE* would all belong to the same *m*-class; and it cannot be *G*, for then *AGTC* would all belong to the same *m*-class. These three possibilities all contradict Assumption VII. Hence the existence of *T* would imply the existence of four elements in the *m*-class *BFG*, which is likewise contrary to Assumption VII.

The properties of the class S and its *m*-classes may also be represented vividly by the accompanying figure (Figure 1). Here we have represented the elements of S by points (or spots) in a plane, and have joined by a

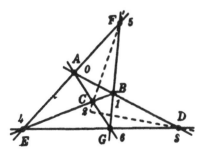

FIGURE 1

line every triple of these points which form an *m*-class. It is seen that the points may be so chosen that all but one of these lines is a straight line. This suggests at once a similarity to ordinary plane geometry. Suppose we interpret the elements of S to be the points of a plane, and interpret the *m*-classes to be the straight lines of the plane, and let us reread our assumptions with this interpretation. Assumption VII is false, but all the others are true with the exception of Assumption III, which is also true except when the lines are parallel. How this exception can be removed we will discuss in the next section, so that we may also regard the ordinary plane geometry as a representation of Assumptions I–VI.

Returning to our miniature mathematical science of triples, we are now in a position to answer another important question: *To what extent do Assumptions I–VII characterize the class* S *and the m-classes?* We have just seen that any class S satisfying these assumptions may be represented by Table (1′) merely by properly labeling the elements of S. In other words, if S_1 and S_2 are two classes S subject to these assumptions, every element of S_1 may be made to correspond [5] to a unique element of S_2, in such a way that every element of S_2 is the correspondent of a unique element of S_1, and that to every *m*-class of S_1 there corresponds an *m*-class of S_2. The two classes are then said to be in *one-to-one reciprocal correspondence*, or to be *simply isomorphic*.[6] Two classes S are then abstractly equivalent; i.e., there exists essentially only one class S satisfying Assumptions I–VII. This leads to the following fundamental notion:

A set of assumptions is said to be categorical, if there is essentially only one system, for which the assumptions are valid; i.e., if any two such systems may be made simply isomorphic.

We have just seen that the set of Assumptions I–VII is categorical. If, however, Assumption VII be omitted, the remaining set of six assumptions is not categorical. We have already observed the possibility of satisfying Assumptions I–VI by ordinary plane geometry. Since Assumption III, however, occupies as yet a doubtful position in this interpretation, we give another, which, by virtue of its simplicity, is peculiarly adapted to make clear the distinction between categorical and noncategorical. The reader will find, namely, that each of the first six assumptions is satisfied by interpreting the class S to consist of the digits 0, 1, 2, $\cdots$, 12, arranged according to the following table of *m*-classes, every column constituting one *m*-class:

$$
(2) \quad
\begin{array}{ccccccccccccc}
0 & 1 & 2 & 3 & 4 & 5 & 6 & 7 & 8 & 9 & 10 & 11 & 12 \\
1 & 2 & 3 & 4 & 5 & 6 & 7 & 8 & 9 & 10 & 11 & 12 & 0 \\
3 & 4 & 5 & 6 & 7 & 8 & 9 & 10 & 11 & 12 & 0 & 1 & 2 \\
9 & 10 & 11 & 12 & 0 & 1 & 2 & 3 & 4 & 5 & 6 & 7 & 8
\end{array}
$$

Hence Assumptions I–VI are not sufficient to characterize completely the class S, for it is evident that Systems (1) and (2) cannot be made iso-

---

[5] The notion of *correspondence* is another primitive notion which we take over without discussion from the general logic of classes.

[6] The isomorphism of Systems (1) and (1′) is clearly exhibited in Figure 1, where each point is labeled both with a digit and with a letter. This isomorphism may, moreover, be established in $7 \cdot 6 \cdot 4$ different ways.

morphic. On the other hand, it should be noted that all theorems derivable from Assumptions I–VI are valid for both (1) and (2). These two systems are two essentially different concrete representations of the same mathematical science.

This brings us to a third question regarding our assumptions: *Are they independent?* That is, can any one of them be derived as a logical consequence of the others? Table (2) is an example which shows that Assumption VII is independent of the others, because it shows that they can all be true of a system in which Assumption VII is false. Again, if the class S is taken to mean the three letters A, B, C, and the m-classes to consist of the pairs AB, BC, CA, then it is clear that Assumptions I, II, III, IV, VI, VII are true of this class S, and therefore that any logical consequence of them is true with this interpretation. Assumption V, however, is false for this class, and cannot, therefore, be a logical consequence of the other assumptions. In like manner, other examples can be constructed to show that each of the Assumptions I–VII is independent of the remaining ones.

3. IDEAL ELEMENTS IN GEOMETRY

The miniature mathematical science which we have just been studying suggests what we must do on a larger scale in a geometry which describes our ordinary space. We must first choose a set of undefined elements and a set of fundamental assumptions. This choice is in no way prescribed *a priori*, but, on the contrary, is very arbitrary. It is necessary only that the undefined symbols be such that all other elements and relations that occur are definable in terms of them; and the fundamental assumptions must satisfy the prime requirement of logical consistency, and be such that all other propositions are derivable from them by formal logic. It is desirable, further, that the assumptions be independent [7] and that certain sets of assumptions be categorical. There is, further, the desideratum of utmost symmetry and generality in the whole body of theorems. The latter means that the applicability of a theorem shall be as wide as possible. This has relation to the arrangement of the assumptions, and can be attained by using in the proof of each theorem a minimum of assumptions.[8]

Symmetry can frequently be obtained by a judicious choice of terminology. This is well illustrated by the concept of "points at infinity" which is fundamental in any treatment of projective geometry. Let us note first the reciprocal character of the relation expressed by the two statements:

---

[7] This is obviously necessary for the precise distinction between an assumption and a theorem.

[8] If the set of assumptions used in the proof of a theorem is not categorical, the applicability of the theorem is evidently wider than in the contrary case. Cf. example of preceding section.

A point lies on a line.   A line passes through a point.

To exhibit clearly this reciprocal character, we agree to use the phrases

A point is *on* a line;   A line is *on* a point

to express this relation. Let us now consider the following two propositions:

1. *Any two distinct points of a*   1'. *Any two distinct lines of a*
plane are on one and only one line.[9] *plane are on one and only one point.*

Either of these propositions is obtained from the other by simply interchanging the words *point* and *line*. The first of these propositions we recognize as true without exception in the ordinary Euclidean geometry. The second, however, has an exception when the two lines are parallel. In view of the symmetry of these two propositions it would clearly add much to the symmetry and generality of all propositions derivable from these two, if we could regard them both as true without exception. This can be accomplished by *attributing to two parallel lines a point of intersection.* Such a point is not, of course, a point in the ordinary sense; it is to be regarded as an *ideal point*, which we suppose two parallel lines to have in common. Its introduction amounts merely to a change in the ordinary terminology. Such an ideal point we call a *point at infinity*; and we suppose one such point to exist on every line.[10]

The use of this new term leads to a change in the statement, though not in the meaning, of many familiar propositions, and makes us modify the way in which we think of points, lines, etc. Two non-parallel lines cannot have in common a point at infinity without doing violence to propositions 1 and 1'; and since each of them has a point at infinity, there must be at least two such points. Proposition 1, then, requires that we attach a meaning to the notion of a line on two points at infinity. Such a line we call a *line at infinity*, and think of it as consisting of all the points at infinity in a plane. In like manner, if we do not confine ourselves to the points of a single plane, it is found desirable to introduce the notion of a plane through three points at infinity which are not all on the same line at infinity. Such a plane we call a *plane at infinity*, and we think of it as consisting of all the points at infinity in space. Every ordinary plane is supposed to contain just one line at infinity; every system of parallel planes in space is supposed to have a line at infinity in common with the plane at infinity, etc.

---

[9] By *line* throughout we mean *straight line*.
[10] It should be noted that (since we are taking the point of view of Euclid) we do not think of a line as containing more than one point at infinity; for the supposition that a line contains two such points would imply either that two parallels can be drawn through a given point to a given line, or that two distinct lines can have more than one point in common.

The fact that we have difficulty in presenting to our imagination the notions of a point at infinity on a line, the line at infinity in a plane, and the plane at infinity in space, need not disturb us in this connection, provided we can satisfy ourselves that the new terminology is self-consistent and cannot lead to contradictions. The latter condition amounts, in the treatment that follows, simply to the condition that the assumptions on which we build the subsequent theory be consistent. That they are consistent will be shown at the time they are introduced. The use of the new terminology may, however, be justified on the basis of ordinary analytic geometry. This we do in the next section, the developments of which will, moreover, be used frequently in the sequel for proving the consistency of the assumptions there made.

4. CONSISTENCY OF THE NOTION OF POINTS, LINES, AND PLANE AT INFINITY

We will now reduce the question of the consistency of our new terminology to that of the consistency of an algebraic system. For this purpose we presuppose a knowledge of the elements of analytic geometry of three dimensions. In this geometry a point is equivalent to a set of three numbers (x, y, z). The totality of all such sets of numbers constitutes the analytic space of three dimensions. If the numbers are all real numbers, we are dealing with the ordinary "real" space; if they are any complex numbers, we are dealing with the ordinary "complex" space of three dimensions. The following discussion applies primarily to the real case.

A *plane* is the set of all points (number triads) which satisfy a single linear equation

$$ax + by + cz + d = 0.$$

A *line* is the set of all points which satisfy two linear equations,

$$a_1x + b_1y + c_1z + d_1 = 0,$$
$$a_2x + b_2y + c_2z + d_2 = 0,$$

provided the relations

$$\frac{a_1}{a_2} = \frac{b_1}{b_2} = \frac{c_1}{c_2}$$

do not hold.[11]

Now the points (x, y, z), with the exception of $(0, 0, 0)$, may also be

---

[11] It should be noted that we are not yet, in this section, supposing anything known regarding points, lines, etc., at infinity, but are placing ourselves on the basis of elementary geometry.

denoted by the direction cosines of the line joining the point to the origin of coördinates and the distance of the point from the origin; say by

$$\left(l, m, n, \frac{1}{d} \right),$$

where $d = \sqrt{x^2 + y^2 + z^2}$, and $l = \dfrac{x}{d}, m = \dfrac{y}{d}, n = \dfrac{z}{d}$. The origin itself may be denoted by $(0, 0, 0, k)$, where k is arbitrary. Moreover, any four numbers (x_1, x_2, x_3, x_4) $(x_4 \neq 0)$, proportional respectively to

$$\left(l, m, n, \frac{1}{d} \right),$$

will serve equally well to represent the point (x, y, z), provided we agree that (x_1, x_2, x_3, x_4) and (cx_1, cx_2, cx_3, cx_4) represent the same point for all values of c different from 0. For a point (x, y, z) determines

$$x_1 = \frac{cx}{\sqrt{x^2 + y^2 + z^2}} = cl, \qquad x_2 = \frac{cy}{\sqrt{x^2 + y^2 + z^2}} = cm,$$

$$x_3 = \frac{cz}{\sqrt{x^2 + y^2 + z^2}} = cn, \qquad x_4 = \frac{c}{\sqrt{x^2 + y^2 + z^2}} = \frac{c}{d},$$

where c is arbitrary $(c \neq 0)$, and (x_1, x_2, x_3, x_4) determines

$$(1) \qquad\qquad x = \frac{x_1}{x_4}, \qquad y = \frac{x_2}{x_4}, \qquad z = \frac{x_3}{x_4},$$

provided $x_4 \neq 0$.

We have not assigned a meaning to (x_1, x_2, x_3, x_4) when $x_4 = 0$, but it is evident that if the point $\left(cl, cm, cn, \dfrac{c}{d} \right)$ moves away from the origin an unlimited distance on the line whose direction cosines are l, m, n, its coördinates approach $(cl, cm, cn, 0)$. A little consideration will show that as a point moves on any other line with direction cosines l, m, n, so that its distance from the origin increases indefinitely, its coördinates also approach $(cl, cm, cn, 0)$. Furthermore, these values are approached, no matter in which of the two opposite directions the point moves away from the origin. We now *define* $(x_1, x_2, x_3, 0)$ as a *point at infinity* or an *ideal point*. We have thus associated with every set of four numbers (x_1, x_2, x_3, x_4) a point, ordinary or ideal, with the exception of the set $(0, 0, 0, 0)$, which we exclude entirely from the discussion. The ordinary points are those for which x_4 is not zero; their ordinary Cartesian coördinates are given by the equations (1). The ideal points are those for which $x_4 = 0$. The numbers (x_1, x_2, x_3, x_4) we call the *homogeneous* coördinates of the point.

We now define a *plane* to be the set of all points (x_1, x_2, x_3, x_4) which satisfy a linear homogeneous equation:

$$ax_1 + bx_2 + cx_3 + dx_4 = 0.$$

It is at once clear from the preceding discussion that as far as all ordinary points are concerned, this definition is equivalent to the one given at the beginning of this section. However, according to this definition all the ideal points constitute a plane $x_4 = 0$. This plane we call the *plane at infinity*. In like manner, we define a line to consist of all points (x_1, x_2, x_3, x_4) which satisfy two distinct linear homogeneous equations:

$$a_1x_1 + b_1x_2 + c_1x_3 + d_1x_4 = 0,$$
$$a_2x_1 + b_2x_2 + c_2x_3 + d_2x_4 = 0.$$

Since these expressions are to be distinct, the corresponding coefficients throughout must not be proportional. According to this definition the points common to any plane (not the plane at infinity) and the plane $x_4 = 0$ constitute a line. Such a line we call a *line at infinity*, and there is one such in every ordinary plane. Finally, the line defined above by two equations contains one and only one point with coördinates $(x_1, x_2, x_3, 0)$; that is, an ordinary line contains one and only one point at infinity. It is readily seen, moreover, that with the above definitions two parallel lines have their points at infinity in common.

Our discussion has now led us to an analytic definition of what may be called, for the present, an analytic *projective space* of three dimensions. It may be defined, in a way which allows it to be either real or complex, as consisting of:

Points: All sets of four numbers (x_1, x_2, x_3, x_4), except the set $(0, 0, 0, 0)$, where (cx_1, cx_2, cx_3, cx_4) is regarded as identical with (x_1, x_2, x_3, x_4), provided c is not zero.

Planes: All sets of points satisfying one linear homogeneous equation.

Lines: All sets of points satisfying two distinct linear homogeneous equations.

Such a projective space cannot involve contradictions unless our ordinary system of real or complex algebra is inconsistent. The definitions here made of points, lines, and the plane at infinity are, however, precisely equivalent to the corresponding notions of the preceding section. We may therefore use these notions precisely in the same way that we consider ordinary points, lines, and planes. Indeed, the fact that no exceptional properties attach to our ideal elements follows at once from the symmetry of the analytic formulation; the coördinate x_4, whose vanishing gives rise to the ideal points, occupies no exceptional position in the algebra of the homogeneous equations. The ideal points, then, are not to be regarded as different from the ordinary points.

All the assumptions we shall make in our treatment of projective geometry will be found to be satisfied by the above analytic creation, which therefore constitutes a proof of the consistency of the assumptions in question. . . .

5. PROJECTIVE AND METRIC GEOMETRY

In projective geometry no distinction is made between ordinary points and points at infinity, and it is evident by a reference forward that our assumptions provide for no such distinction. We proceed to explain this a little more fully, and will at the same time indicate in a general way the difference between *projective* and the ordinary Euclidean *metric* geometry.

Confining ourselves first to the plane, let m and m' be two distinct lines, and P a point not on either of the two lines. Then the points of m may be made to correspond to the points of m' as follows: To every point A on m let correspond that point A' on m' in which m' meets the line joining A to P (Figure 2). In this way every point on either line is assigned a unique corresponding point on the other line. This type of correspondence is called *perspective*, and the points on one line are said to be transformed into the points of the other by a *perspective transformation with center P*. If the points of a line m be transformed into the points of a line m' by a perspective transformation with center P, and then the points of m' be transformed into the points of a third line m'' by a perspective transformation with a new center Q; and if this be continued any finite number of times, ultimately the points of the line m will have been brought into correspondence with the points of a line $m^{(n)}$, say, in such

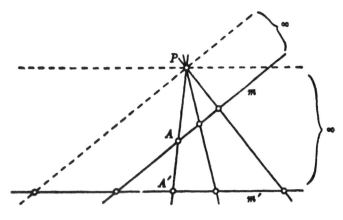

FIGURE 2

a way that every point of m corresponds to a unique point of $m^{(n)}$. A correspondence obtained in this way is called *projective*, and the points of m are said to have been transformed into the points of $m^{(n)}$ by a *projective transformation.*

Similarly, in three-dimensional space, if lines are drawn joining every point of a plane figure to a fixed point P not in the plane π of the figure, then the points in which this totality of lines meets another plane π' will form a new figure, such that to every point of π will correspond a unique point of π', and to every line of π will correspond a unique line of π'. We say that the figure in π has been transformed into the figure in π' by a *perspective transformation with center P*. If a plane figure be subjected to a succession of such perspective transformations with different centers, the final figure will still be such that its points and lines correspond uniquely to the points and lines of the original figure. Such a transformation is again called a *projective transformation*. In projective geometry two figures that may be made to correspond to each other by means of a projective transformation are not regarded as different. In other words, *projective geometry is concerned with those properties of figures that are left unchanged when the figures are subjected to a projective transformation.*

It is evident that no properties that involve essentially the notion of measurement can have any place in projective geometry as such; [12] hence the term *projective*, to distinguish it from the ordinary geometry, which is almost exclusively concerned with properties involving the idea of measurement. In case of a plane figure, a perspective transformation is clearly equivalent to the change brought about in the aspect of a figure by looking at it from a different angle, the observer's eye being the center of the perspective transformation. The properties of the aspect of a figure that remain unaltered when the observer changes his position will then be properties with which projective geometry concerns itself. For this reason von Staudt called this science *Geometrie der Lage*.

In regard to the points and lines at infinity, we can now see why they cannot be treated as in any way different from the ordinary points and lines of a figure. For, in the example given of a perspective transformation between lines, it is clear that to the point at infinity on m corresponds in general an ordinary point on m', and·conversely. And in the example given of a perspective transformation between planes we see that to the line at infinity in one plane corresponds in general an ordinary line in the other. In projective geometry, then, there can be no distinction between the ordinary and the ideal elements of space.

---

[12] The theorems of metric geometry may however be regarded as special oases of projective theorems.

The observation of phenomena cannot tell us anything more than that the mathematical equations are correct: the same equations might equally well represent the behavior of some other material system. For example, the vibrations of a membrane which has the shape of an ellipse can be calculated by means of a differential equation known as Mathieu's equation: but this same equation is also arrived at when we study the dynamics of a circus performer who holds an assistant balanced on a pole while he himself stands on a spherical ball rolling on the ground.

If now we imagine an observer who discovers that the future course of a certain phenomenon can be predicted by Mathieu's equation, but who is unable for some reason to perceive the system which generated the phenomenon, then evidently he would be unable to tell whether the system in question is an elliptic membrane or a variety artist.

—Sir Edmund T. Whittaker

6 Mathematics and the World

By DOUGLAS GASKING

MY OBJECT is to try to elucidate the nature of mathematical propositions, and to explain their relation to the everyday world of counting and measurement—of clocks, and yards of material, and income-tax forms. I should like to be able to summarize my views in a few short phrases, and then go on to defend them. Unfortunately I cannot do this, for, as I shall try to demonstrate, I do not think any short statement will do to express the truth of the matter with any precision. So I shall proceed by approximations—I shall discuss several different views in the hope that in showing what is right and what is wrong with them, clarification will come about.

The opinions of philosophers about the nature of mathematical propositions can be divided, as can their opinions about so many things, into two main classes. There are those who try to analyse mathematical propositions away—who say that they are *really* something else (like those writers on ethics who say that goodness is really only pleasure, or those metaphysicians who say that chairs and tables are really groups of sensations, or colonies of souls). I shall call such 'analysing-away' theories 'radical' theories. On the other hand there are those who insist that mathematical propositions are *sui generis*, that they cannot be analysed into anything else, that they give information about an aspect of reality totally different from any other (compare those philosophers who maintain, e.g., that goodness is a simple unanalysable quality, or those realists who maintain that a chair is a chair, an external material substance, known, perhaps, by means of sensations, but not to be confused with those sensations). For convenience, I shall call these types of theory which oppose any analysing-away, 'conservative.' I should maintain that in general what I call 'conservative' opinions in philosophy are perfectly correct, but rather unsatisfactory and unilluminating, whereas opinions of the 'radical' type are untrue, but interesting and illuminating.

I shall start by considering the 'radical' theories about the nature of mathematics. Those I know of fall into two main types. (1) Some people maintain that a proposition of mathematics is *really* a particularly well-founded empirical generalization of a certain type, or that it is logically on the same footing as a very well-established scientific law. Mill's theory was of this type, and many scientists I have talked to have tended to have similar opinions. Let us call these 'empirical' theories about mathematics. (2) Then, on the other hand, there is a great variety of theories usually called 'conventionalist,' which analyse away mathematical propositions into propositions about the use of symbols. Examples: 'By a mathematical proposition the speaker or writer merely expresses his intention of manipulating symbols in a certain way, and recommends or commands that others should do likewise.' 'A mathematical proposition is really an empirical proposition describing how educated people commonly use certain symbols.' 'A mathematical proposition is really a rule for the manipulation of symbols.' (Ayer, for example, and C. I. Lewis have expressed opinions of this general type.)

First for the 'empirical' theories. According to these a mathematical proposition just expresses a particularly well-founded empirical generalization or law about the properties and behaviour of objects, obtained by examining a large number of instances and seeing that they conform without exception to a single general pattern. The proposition '7 + 5 = 12,' for instance, just expresses (on one version of this theory) the fact of experience that if we count up seven objects of any sort, and then five more objects, and then count up the whole lot, we always get the number twelve. Or again, it might be maintained that the geometrical proposition 'Equilateral triangles are equiangular' just expresses the fact that wherever, by measurement, we find the sides of a triangle to be equal, we will find, on measuring the angles with the same degree of accuracy, that the angles are equal too. It is contended that such propositions are essentially like, for example, Boyle's Law of gases, only much better founded.

But '7 + 5 = 12' does not mean the same as the proposition about what you get on counting groups. For it is true that 7 + 5 does equal 12, but it is not true that on counting seven objects and then five others, and then counting the whole, you will always get twelve. People sometimes miscount, and sometimes the objects counted melt away (if they are wax) or coalesce (if they are globules of mercury). Similarly the geometrical proposition that equilateral triangles are equiangular does not mean the same as the proposition that any triangle which is equilateral by measurement will be found to be equiangular when measured. The former is true; the latter false. We sometimes make mistakes with our rulers and protractors.

To this it might be objected that this shows that the empirical proposition offered as a translation of the mathematical one is not a correct translation, but that it has not been demonstrated that it is impossible to find an empirical proposition about counting and measurement, which is a correct translation. Let us try some alternatives, then. It might be suggested that '7 + 5 = 12' means 'If you count *carefully and with attention*, you will get such and such a result.' But, even with the greatest care in counting, mistakes sometimes happen at any rate with large numbers. Shall we then say: '7 + 5 = 12' means 'If you count *correctly* you will get such and such results'? But, in the first place, even if you count objects correctly, you do not always get a group of seven objects and a group of five adding up to twelve. It sometimes happens that a person correctly counts seven objects, then correctly counts five, and then correctly counts the total and gets eleven. Sometimes one of the objects does disappear in the course of counting, or coalesces with another. And even if this were not so, the suggested translation would not give you a simple empirical proposition about what happened when people counted, as a translation of 7 + 5 = 12, but would give you a mere tautology. For what is the criterion of correctness in counting? Surely that when you add seven and five you should get twelve. 'Correctness' has no meaning, in this context, independent of the mathematical proposition. So our suggested analysis of the meaning of '7 + 5 = 12' runs, when suitably expanded: '7 + 5 = 12' means 'If you count objects *correctly* (i.e., in such a way as to get 12 on adding 7 and 5) you will, on adding 7 to 5, get 12.'

No doubt there *are* important connections between mathematical propositions, and propositions about what results people will usually get on counting and measuring. But it will not do to say that a mathematical proposition means the same as, or is equivalent to, any such empirical proposition, for this reason: A mathematical proposition is 'incorrigible,' whereas an empirical proposition is 'corrigible.'

The difference between 'corrigible' and 'incorrigible' propositions can best be explained by examples. Most everyday assertions that we make, such as that 'Mr. Smith has gone away for the day,' are corrigible. By this I mean simply that, whenever we make such an assertion, however strong our grounds for making it, we should always freely withdraw it and admit it to have been false, *if* certain things were to happen. Thus my assertion, that Smith is away for the day, is corrigible, because (although I may have the excellent grounds for making it that when I met him in the street this morning he said he was on his way to the railway-station) if, for example, I were to go to his room now and find him sitting there, I should withdraw my assertion that he was away and admit it to have been false. I should take certain events as proving, if they happened, that my assertion was untrue.

A mathematical proposition such as '7 + 5 = 12,' on the other hand, is incorrigible, because no future happenings whatsoever would ever prove the proposition false, or cause anyone to withdraw it. You can imagine any sort of fantastic chain of events you like, but nothing you can think of would ever, if it happened, disprove '7 + 5 = 12.' Thus, if I counted out 7 matches, and then 5 more, and then on counting the whole lot, got 11, this would not have the slightest tendency to make anyone withdraw the proposition that 7 + 5 = 12 and say it was untrue. And even if this constantly happened, both to me and to everyone else, and not only with matches, but with books, umbrellas and every sort of object —surely even this would not make us withdraw the proposition. Surely in such a case we should not say: 'the proposition "7 + 5 = 12" has been empirically disproved; it has been found that 7 + 5 really equals 11.' There are plenty of alternative explanations to choose from. We might try a psychological hypothesis, such as this: we might say that it had been discovered by experiment that everyone had a curious psychological kink, which led him, whenever he performed counting operations of a certain sort, always to miss out one of the objects in his final count (like the subject in some experiments on hypnosis who, under suggestion, fails to see any 't's on a printed page). Or we might prefer a physical hypothesis and say: a curious physical law of the universe has been experimentally established, namely, that whenever 5 objects are added to 7 objects, this process of addition causes one of them to disappear, or to coalesce with another object. The one thing we should *never* say, whatever happened, would be that the proposition that 7 + 5 = 12 had been experimentally disproved. If curious things happened, we should alter our physics, but not our mathematics.

This rather sweeping assertion that mathematical propositions are completely incorrigible is, I think, an over-simplification, and needs qualifying. I shall mention the qualifications later, rather than now, for simplicity of exposition. So if you will accept it for the moment as very nearly true, I should like to draw your attention to certain of its consequences. A *corrigible* proposition gives you some information about the world—a completely *incorrigible* proposition tells you nothing. A corrigible proposition is one that you would withdraw and admit to be false if certain things happened in the world. It therefore gives you the information that *those* things (i.e., those things which would make you withdraw your proposition *if* they happened) will *not* happen. An incorrigible proposition is one which you would never admit to be false *whatever* happens: it therefore does not tell you *what* happens. The truth, for example, of the corrigible proposition that Smith is away for the day, is compatible with certain things happening (e.g., your going to his room and finding it empty) and is not compatible with certain other happenings (e.g., your going to his room and finding him there). It therefore tells you what sort of thing will happen (you will find his room empty) and what sort of thing will not happen (you will not find him in). The truth of an incorrigible proposition, on the other hand, is compatible with any and every conceivable state of affairs. (For example: whatever is your experience on counting, it is still true that $7 + 5 = 12$.) It therefore does not tell you which events will take place and which will not. That is: the proposition '$7 + 5 = 12$' tells you nothing about the world.

If such a proposition tells you nothing about the world, what, then, is the point of it—what does it do? I think that in a sense it is true to say that it prescribes what you are to *say*—it tells you *how to describe* certain happenings. Thus the proposition '$7 + 5 = 12$' does not tell you that on counting $7 + 5$ you will not get 11. (This, as we have seen, is false, for you sometimes do get 11.) But it does *lay it down*, so to speak, that *if* on counting $7 + 5$ you do get 11, you are to describe what has happened in some such way as this: *Either* 'I have made a mistake in my counting' *or* 'Someone has played a practical joke and abstracted one of the objects when I was not looking' *or* 'Two of the objects have coalesced' *or* 'One of the objects has disappeared,' etc.

This, I think, is the truth that is in the various 'conventionalist' theories of mathematics. Only, unfortunately, the formulae expressing such theories are usually misleading and incorrect. For example, to say that: 'a mathematical proposition merely expresses the speaker's or writer's determination to use symbols in a certain way,' is obviously untrue. For if it were true, and if I decided to use the symbol '+' in such a way that $5 + 7 = 35$, I would then be speaking truly if I said '$5 + 7 = 35$.' But this proposition is not true. The truth of any mathematical proposition does not depend on my decision or determination. It is independent of my will. This formula neglects the 'public' or 'over-individual' character of mathematics.

Or, consider the formula: 'A mathematical proposition is really an empirical statement describing the way people commonly use certain symbols.' This, I think, is nearer. But it is open to the following obvious objection: If '$7 + 5 = 12$' were really an assertion about the common usage of symbols, then it would follow that $7 + 5$ would not equal 12 if people had a different symbolic convention. But even if people did use symbols in a way quite different from the present one, the fact which we now express by '$7 + 5 = 12$' would still be true. No change in our language-habits would ever make this false.

This objection is, I think, sufficient to show that the suggested formula is untrue, as it stands. But we should be blind to its merits if we did not see *why* it is that no change in our language-habits would make the proposition '7 + 5 = 12' untrue. The reason is this: As we use symbols at present, this proposition is incorrigible—one which we maintain to be true whatever happens in the world, and never admit to be false under any circumstances. Imagine a world where the symbolic conventions are totally different—say on Mars. How shall we *translate* our incorrigible proposition into the Martian symbols? If our translation is to be correct—if the proposition in the Martian language is to mean the same as our '7 + 5 = 12,' it *too* must be incorrigible—otherwise we should not call it a correct translation. Thus a correct Martian translation of our '7 + 5 = 12' must be a proposition which the Martians maintain to be true whatever happens. Thus '7 + 5 = 12,' and any correct translation into any other symbolic convention will be incorrigible, i.e., true whatever happens. So its truth does, in a sense, depend on the empirical fact that people use symbols in certain ways. But it is an inaccurate way of stating this fact to say that it describes how people use symbols.

A better formulation is: 'A mathematical proposition really expresses a rule for the manipulation of symbols.' But this, too, is unsatisfactory, and for the following reason: To say that it is a 'rule for the manipulation of symbols' suggests that it is entirely arbitrary. A symbolic rule is something which we can decide to use or not, just as we wish. (We could easily use 'hice' as the plural of 'house,' and get on as well as we do now.) But, it seems, we cannot just change our mathematical propositions at will, without getting into difficulties. An engineer, building a bridge, has to use the standard multiplication tables and no others, or else the bridge will collapse. Thus which mathematical system we use does not seem to be entirely arbitrary—one system works in practice, and another does not. Which system we are to use seems to depend in some way not on our decision, but on the nature of the world. To say that '7 + 5 = 12' really expresses a rule for the use of symbols, suggests that this proposition is just like ' "house" forms its plural by adding "s".' But there *is* a difference between the two, and so the formula is misleading.

I want to conclude this paper by considering in some detail the objection that you cannot build bridges with any mathematics, and that therefore mathematics does depend on the nature of reality. Before doing so, however, I should like to mention the type of theory I called 'conservative.' We saw that the (radical) theory, that mathematical propositions are 'really' empirical propositions about the results of counting, is untrue. But there is a close connection between the two sorts of proposition, and therefore the 'empirical' theory, although untrue, has a point. It emphasizes the connection between mathematical propositions and our everyday practice of counting and calculation; thus it serves as a useful corrective to that type of theory which would make mathematics too abstract and pure—a matter of pure intellect and Platonic 'Forms,' far from the mundane counting of change. Similarly the various 'conventionalist' theories are also, strictly speaking, untrue, but they too have their point. Mathematical propositions in certain respects are *like* rules for the use of symbols, *like* empirical propositions about how symbols are used, *like* statements of intention to use symbols in certain ways. But conventionalist formulae are untrue because mathematical propositions are not *identical* with any of these. They are what they are; they function in the way they do, and not exactly like any other sort of proposition.

And this it is which makes that sort of theories I have called 'conservative' perfectly correct. Mathematical propositions are *sui generis*. But merely to say: 'They are what they are' is not very helpful. Nor is it any better if this is dressed up in learned language: e.g., 'Mathematical propositions state very general facts about the structure of reality; about the necessary and synthetic relations between the universals number, shape, size, and so on.' If you are inclined to think that such answers as this, to the question 'What are mathematical propositions about?', are informative and illuminating, ask yourself: 'How does any hearer come to understand the meaning of such phrases as "structure of reality," "necessary relations between universals," and so on? How were these phrases explained to him in the first place?' Surely he was told what was meant by 'necessary relation between universals,' by being told, for example, that colour, shape, size, number, etc., are universals, and that an example of a necessary relation between universals would be 'everything that has shape has size,' '2 + 2 = 4,' 'two angles of an isosceles triangle are equal,' and so on. These phrases, such as 'necessary relation between universals,' are *introduced* into his language *via* or *by means of* such phrases as '2 + 2 = 4'; they are introduced *via* mathematical propositions, among others. To use an expression of John Wisdom's,[1] they are 'made to measure.' So to tell someone that mathematical propositions are 'so-and-so' does not help, if, in explaining what is meant by 'so-and-so,' you have to introduce mathematical propositions, among others, as illustrative examples. Compare giving a 'conservative' answer to the question 'What are mathematical propositions?' with the following example: A child learns the meaning of the words 'see,' 'can't see,' 'blindfolded,' etc., before he learns the meaning of the word 'blind.' The latter word is then introduced into his vocabulary by the explanation: 'A blind man is one who can't see in broad daylight even when not blindfolded.' If the child then asks of a blind man 'Why can't he see in broad daylight even when not blindfolded?', it is not much use answering 'Because he is blind.' Like the 'con-

---

[1] My debt to the lectures of Wisdom and Wittgenstein, in writing this paper, is very great.

servative' answer in philosophy, it may serve to stop any further questions, but it usually leaves a feeling of dissatisfaction.

Then what sort of answer *can* be given to one who is puzzled about the nature of mathematics? Mathematical propositions are what they are, so any radical answer equating them with something else, such as symbolic rules, or statements of the results of counting and measurement, or of common symbolic usage, will be untrue. Such answers will be untrue, because the two sides of the equation will have different meanings. Similarly conservative answers will be unhelpful, because the two sides of the equation will have the same meaning. The definiens will be useless, because it will contain terms which are introduced into the language *via* the definiendum, and can only be explained in terms of it. It is 'made to measure.' No simple formula will do. The only way of removing the puzzle is to describe the use and function of mathematical propositions in detail and with examples. I shall now try to do this, to some extent, in considering the natural objection to the strictly untrue but illuminating theory: 'Mathematical propositions express rules for the manipulation of symbols.' The objection is that symbolic rules are essentially arbitrary, whereas mathematics does, to some extent at least, depend not on our choice of symbolic conventions, but on the nature of reality, because only our present system gives useful results when applied to the practical tasks of the world. Against this, I shall maintain that we could use *any* mathematical rules we liked, and still get on perfectly well in the business of life.

Example 1. 6 × 4, according to our current multiplication table, equals 24. You might argue: this cannot be merely a conventional rule for our use of symbols, for if it were we could use any other rule we liked, e.g., 6 × 4 = 12, and still get satisfactory results. But if you tried this alternative rule, you would, in fact, find your practical affairs going all wrong. A builder, for example, having measured a room to be paved with tiles, each one yard square, and having found the length of the sides to be 6 yards and 4 yards, could not use the alternative table. He could not say to himself: 'The room is 6 by 4; now 6 × 4 = 12, so I shall have to get 12 tiles for this job.' For, if he did, he would find he had not enough tiles to cover his floor.

But the builder could quite easily have used an arithmetic in which 6 × 4 = 12, and by measuring and counting could have paved his room perfectly well, with exactly the right number of tiles to cover the floor. How does he do it? Well, he:

(1) Measures the sides, and writes down '4' and '6.'
(2) Multiplies 4 by 6 according to a 'queer' multiplication table which gives 4 × 6 = 12.
(3) Counts out 12 tiles, lays them on the floor. And they fit perfectly.

The 'queer' multiplication table he uses gives $2 \times 2 = 4$, $2 \times 4 = 6$, $2 \times 8 = 10$, $4 \times 4 = 9$, $4 \times 6 = 12$, etc. The number found by multiplying a by b according to *his* table, is that which in *our* arithmetic we should get by the formula:

$$(a + 2)(b + 2)/4$$

And he could pave any other size of floor, using the queer multiplication table described, and still always get the right number of tiles to cover it.

How is this possible? He measures the sides of the room with a yardstick as follows: He lays his yardstick along the longer side, with the '0' mark of the yardstick in the corner, and the other end of the stick, marked '36 inches,' some distance along the stick. As he does this, he counts 'one.' He then pivots the yardstick on the 36 inches mark, and swings it round through two right angles, till it is once more lying along the side of the room—this time with the '36 inches' mark nearer to the corner from which he started, and the '0' mark further along the side. As he does this, he counts 'two.' But now the direction of the stick has been reversed, and it is the convention for measuring that lengths should always be measured in the same direction. So he pivots the stick about its middle and swings it round so that the '36' mark is now where the '0' mark was, and vice versa. As he does this, he counts 'three.' He then swings the stick round through two right angles, pivoting on the '36' mark, counting 'four.' He then reverses its direction, as before, counting 'five.' He swings it over again, counting 'six.' It now lies with its end in the corner, so he writes down the length of the side as 'six yards.' (If we had measured it in our way, we should have written its length down as four yards.) He then measures the shorter side in the same way, and finds the length (using his measuring technique) to be four yards. (We should have made it three.) He then multiplies 4 by 6, according to his table, making it 12, counts out 12 tiles, and lays them down. So long as he uses the technique described for measuring lengths, he will always get the right number of tiles for any room with his 'queer' multiplication table.

This example shows you that we use the method we do for multiplying lengths to get areas, because we use a certain method of measuring lengths. Our technique of calculating areas is relative to our technique of measuring lengths.

Here you might say: Admitting that this is true, it is still the case that mathematics is not arbitrary, for you could not use the method of measuring we do, *and* a different multiplication table, and *still* get the right number of tiles for our room. Can't we? Let's see.

Example 2. Suppose our 'queer' multiplication table gave $3 \times 4 = 24$. The builder measures the sides of a room exactly as we do, and finds that they are 3 yards and 4 yards, respectively. He then 'multiplies' 3 by 4, and gets 24. He counts out 24 tiles, places them on the floor, and they fit perfectly, with none over. How does he do it?

He measures the sides as we do, and writes down '3' and '4.' He 'multiplies' and gets 24. He then counts out 24 tiles as follows: He picks up a tile from his store, and counts 'one.' He puts the tile on to his truck and counts 'two.' He picks up another tile and counts 'three.' He puts it on his truck and counts 'four.' He goes on in this way until he reaches a count of 'twenty-four.' He then takes his 'twenty-four' tiles and paves the room, and they fit perfectly.

This example shows that our technique of calculating areas is relative both to a certain technique of measurement, *and* to a certain technique of counting.

At this stage you might make a further objection. You might say: Mathematics *does* tell you something about the world, and is not an arbitrary rule of symbolic usage. It tells you that *if* you both count and measure lengths in the way we do, you will not get the right number of tiles for a room unless you multiply the lengths according to our present table. It is not arbitrary, because if, for example, you measure the sides of a room as we do, and find them to be 4 and 3, and if you count tiles as we do, you would get the wrong number of tiles to pave your room if you used some other multiplication table—say one in which $3 \times 4 = 24$. I maintain, on the contrary, that we could quite well use such a 'queer' table, and count and measure as at present, and still get the right number of tiles. To help us to see what is involved here, let us consider a rather analogous case.

Example 3. Imagine that the following extraordinary thing happened. You measure a room normally, and find the sides to be 6 and 4. You multiply normally and get 24. You then count out 24 tiles in the normal way. (Each tile is 1×1.) But when you come to try and lay the tiles in the room, you find that you can only get 12 such tiles on to the floor of the room, and there are 12 tiles over. What should we say if this happened?

The first thing we should say would be: 'You must have made a mistake in your measuring' or 'You must have made a slip in multiplying' or 'You must have counted your tiles wrongly, somehow.' And we should immediately check again the measurements, calculations and counting. But suppose that, after the most careful checking and re-checking, by large numbers of highly qualified persons, *no* mistake at all of this sort can be found anywhere. Suppose, morever, that this happened to everyone constantly, with all sorts of rooms and tiles. What should we say then? There are still a number of ways in which we might explain this curious phenomenon. I shall mention two conceivable hypotheses:

(1) Measuring rods do not, as we supposed, stay a constant length wherever they are put. They stay the same size when put against things the same length as themselves, and also when put against things larger than themselves running from north to south. But when put against things larger than themselves running east-west, they always contract to half their previous length (and this contraction happens so smoothly that we never notice it). Thus the room is in fact 6 by 2 yards, i.e., 12 square yards, and twelve tiles are needed. When the measuring rod is put along the north-south wall of six yards' length, it stays a yard long, and so we get a measurement of 6. When, however, it is put along the shorter east-west wall it contracts to half a yard in length, and can be put four times along the two-yard wall. If you now say the dimensions are 6 and 4, and multiply to get 24, you are overestimating the real area.

(2) An alternative hypothesis: When we measure the room our yardstick always stays a constant length, and thus the area of the room is really 24 square yards. But since we can only get 12 tiles in it, each tile being 1 yard square, it follows that the tiles must *expand*, on being put into the room, to double their area. It is just a curious *physical* law that objects put into a room double their area instantaneously. We do not see this expansion because it is instantaneous. And we can never measure it, by measuring the tiles, first out of the room and then inside, because our yardstick itself expands proportionately on being taken into the room.

This example (which might easily have been put in much more detail with *ad hoc* hypotheses to cover every discrepancy) shows that, however much the practical predictions of builders' requirements are upset when we use our present multiplication table, this need never cause us to alter our present rules for multiplication. Anomalies are accounted for by saying our knowledge of the relevant *physical* laws is defective, not by saying that the multiplication table is 'untrue.' If, when working things out in the usual way, we found that we had constantly 12 tiles too many, we should not say that we had been wrong in thinking that $6 \times 4 = 24$. We should rather say that we had been wrong in thinking that physical objects did not expand and contract in certain ways. If things go wrong, we always change our physics rather than our mathematics.

If we see, from Example 3, what we should do if things went wrong when we used our present arithmetic, we can now answer the objection it was intended to throw light on. The objection was this:

'It is wrong to say that we could use any arithmetic we liked and still get on perfectly well in our practical affairs. Mathematics is not a collection of arbitrary symbolic rules, therefore, and does tell us something about, and does depend on, the nature of reality. For if you *both* count and measure as we do, *and* use a "queer" multiplication table, you won't get the right number of tiles to pave a room. Thus the proposition "3 × 4

= 12" tells you that for a room 3 yards by 4, measured normally, you need neither more nor less than 12 tiles, counted normally. Its truth depends on this fact about the world.'

But I deny this. I say we could have

(1) used our present technique of counting and measurement,
(2) multiplied according to the rule 3 × 4 = 24 (for example),
(3) and still have got exactly the right number of tiles to pave our room.

I therefore say that 3 × 4 = 12 depends on *no* fact about the world, other than some fact about the usage of symbols.

Example 4. Imagine that we did use a 'queer' arithmetic, in which 3 × 4 = 24. If this was our universally accepted and standard arithmetic, we should treat the proposition '3 × 4 = 24' *exactly* as we now treat the proposition '3 × 4 = 12' of our present standard arithmetic. That is to say, if we did use this queer system, we should stick to the proposition '3 × 4 = 24' no matter *what* happened, and ascribe any failure of prediction of builders' requirements, and so on, *always* to a mistaken view of the physical laws that apply to the world, and *never* to the untruth of the formula '3 × 4 = 24.' This latter proposition, if it *were* part of our mathematical system, would be *incorrigible*, exactly as '3 × 4 = 12' is to us now.

In Example 3 we saw what would be done and said if things went wrong in using '3 × 4 = 12.' Now *if* 3 × 4 = 24 were our rule, and incorrigible, and *if* in using it we found ourselves getting the wrong practical results, we should do and say exactly the same sort of thing as we did in Example 3. Thus, assuming that our rule is 3 × 4 = 24, a builder measures his floor normally and writes down '3' and '4.' He multiplies according to his table and gets 24. He counts out 24 tiles normally and tries to put them in the room. He finds that he can only get 12 tiles in. What does he say? He *does not* say 'I have proved by experiment that 3 × 4 does not equal 24,' for his proposition '3 × 4 = 24' is *incorrigible*, and no event in the world, however extraordinary, will ever lead him to deny it, or be counted as relevant to its truth or falsity. What he does say is something like this: 'The area of the room is *really* 24 square yards. Since I can only get 12 yard square tiles into it, it follows that the tiles must expand to double their area on being put into the room.' (As we have seen, he might use other hypotheses, e.g., about the behaviour of measuring rods. But this is probably the most convenient.)

Thus we could easily have counted and measured as at present, *and* used an arithmetic in which 3 × 4 = 24, *and* have got perfectly satisfactory results. Only, of course, to get satisfactory practical results, we should use a physics different in some respects from our present one. Thus a builder having found the area of a room to be 24 square yards would never attempt to put 24 tiles in it, for he would have learnt in his physics lessons at school that tiles put in a room double in area. He would there-fore argue: 'Since the tiles double in area, I must put half of 24 tiles, or 12 tiles, in the room.' He would count out 12 tiles and pave the room perfectly with them.

But even here an obstinate objector might admit all this, and still main-tain that mathematics was not an arbitrary convention; that it did depend on certain facts about the world. He might say ' "3 × 4 = 12" is true, and it is true because of this fact about the world, namely that *if* tiles and rulers do not expand and contract (except slightly with changes in temperature), and if we measure and count normally, we need exactly 12 tiles, no more and no less, to pave a room that is 3 by 4. And "3 × 4 = 24" is false, because of the "brute fact" that *if* tiles, etc., don't ex-pand, and *if* you measure and count normally, 24 tiles are too many to pave a room that is 3 by 4.'

The point that is, I think, missed by this objection could be brought out by asking: 'How do we *find out* whether a tile or a yardstick has or has not expanded or contracted?' We normally have two ways of doing so. We can *watch* it growing bigger or smaller. Or we can *measure* it before and after.

Now in the case described in Example 4, where our queer arithmetic gives 3 × 4 = 24, and things double in area on being put in a room, how do we find out that the things do expand? Not by watching them grow— *ex hypothesi* we do not observe this. Nor by measuring them before and after. For, since we assume that a measuring rod *also* expands on being taken into the room, the dimensions of the tile as measured by a yardstick outside the room are the same as its dimensions as measured by the same (now expanded) yardstick inside the room. In this case we find out that the tiles expand by *measuring, counting and calculating in a certain way* —by finding that the tiles each measure 1 × 1, that the room measures 3 × 4, or 24 square yards, and that we can only get 12 tiles in it. This is our sole *criterion* for saying that the tiles expand. That the tiles expand *follows from* our queer arithmetic. Similarly, as we do things at present, our criterion for saying that tiles do not expand, is that when 12 tiles measuring 1 × 1 are put into a room 3 × 4, or 12 square yards, they fit exactly. From our present arithmetic, it follows that tiles do not expand.

In Example 4, where we have a 'queer' arithmetic in which 3 × 4 = 24, and a 'queer' physics, it is a 'law of nature' that tiles expand on being put into a room. But it is not a 'law of nature' which describes what happens in the world. Rather is it a law 'by convention,' analogous to that law of our present physics which says that when a body falls on the floor with a certain downward force, the floor itself exerts an equal force in the opposite direction. It is just put into the system to balance our calculations, not to describe anything that happens.

This last objection might have been put in a slightly different form. It might have been said: ' "3 × 4 = 12" does describe and depend on the nature of reality, because it entails a certain purely empirical proposition about what does and does not happen, namely the complex proposition. "It is not the case *both* that tiles do not expand *and* that we need less than 12 tiles to pave a floor measuring 3 by 4".' But I should maintain that this complex proposition (of the form 'Not both p and q') is not empirical; that it does not describe anything that happens in the world, because it is incorrigible. Nothing whatsoever that we could imagine happening would ever lead us, if it happened, to deny this complex proposition. Therefore it does not tell us what happens in the world. The simple propositions which are elements in this complex one—the propositions 'Tiles do not expand' and 'We need less than 12 tiles to pave a 3 by 4 floor'—are both of them corrigible, and both describe the world (one of them falsely). But the complex proposition that they are not both true is incorrigible, and therefore, for the reasons given earlier, does not describe or depend on the nature of the world. There is nothing out of the ordinary about this. The propositions 'My curtains are now red over their whole surface,' and 'My curtains are now green all over' are both of them corrigible propositions, descriptive of the world. (One is true, the other false, as a matter of fact.) But the complex proposition 'My curtains are not both red and green over their whole surface' is incorrigible, because nothing would ever make me deny it, and it is therefore not descriptive of the world.

I have talked, throughout the paper, as if mathematical propositions were completely incorrigible, in the sense that *whatever* queer things happened, we should *never* alter our mathematics, and always prefer to change our physics. This was a convenient over-simplification that must now be qualified. I maintain that we *need* never alter our mathematics. But it might happen that we found our physical laws getting very complicated indeed, and might discover that, by changing our mathematical system, we could effect a very great simplification in our physics. In such a case we might decide to use a different mathematical system. (So far as I can understand, this seems to be what has actually happened in certain branches of contemporary physics.) And mathematics does depend on and reflect the nature of the world at least to this extent, that we would find certain systems enormously inconvenient and difficult to use, and certain others relatively simple and handy. Using one sort of arithmetic or geometry, for example, we might find that our physics could be reduced to a logically neat and simple system, which is intellectually satisfying, whereas using different arithmetics and geometries, we should find our physics full of very complicated *ad hoc* hypotheses. But what we find neat, simple, easy, and intellectually satisfying surely depends rather on our psychological make-up, than on the behaviour of measuring rods, solids and fluids, electrical charges—the 'external world.'

If we take in our hand any volume; of divinity or school metaphysics, for instance; let us ask, "Does it contain any abstract reasoning concerning quantity or number?" No. "Does it contain any experimental reasoning concerning matter of fact and existence?" No. Commit it then to the flames: for it can contain nothing but sophistry and illusion.   —DAVID HUME

7   Mathematical Postulates and Human Understanding

By RICHARD VON MISES

AXIOMATICS

1. *High-School Axiomatics.* Early in our high-school education all of us had occasion to learn certain "axioms" of geometry and arithmetic. They were presented as irrefutable truths and soon came to haunt our memories like nightmares. Here are familiar examples of such propositions: every quantity is equal to itself; the whole is bigger than any of its parts; all right angles are equal to each other; and so on. Two things were asserted of these axioms: in the first place, that they are clearly self-evident, and in the second place, that all mathematical theorems follow from them in a strictly logical way.

The student, normally, does not feel any apprehension toward the assertion of obviousness. For how, indeed, could he imagine that a quantity is *not* equal to itself? The situation is not quite the same concerning the claim that all mathematical propositions are derived from the axioms, and exclusively from them. Here, the intelligent student soon senses that in the customary derivations many other "self-evident" concepts besides the explicitly stated axioms have been used. From the modern investigations of the foundations of geometry we know that no geometry can be built up from the few basic propositions that are listed as axioms in the school books. Hilbert's axiomatics, for example, comprises five groups of axioms, among which are propositions such as the so-called axiom of continuity, which is certainly a far cry from the simple statements of the high-school textbook.

Nevertheless, it may be argued that the simple high-school axiomatics is useful for training the student in the method of logical deduction. But this practice has serious deficiencies too. For in such a sentence as: the whole is larger than any of its parts, even the meanings of the words themselves are rather obscure. The statement presupposes that the student to whom it is addressed is, from his everyday language experience, acquainted with the two relations, part to whole and larger to smaller. The axiom asserts that these two relations, in a certain sense, are in each instance simultaneously present or not present, and hence that one can discuss independently the presence of the one as well as of the other in every case. In some examples this is evidently more or less true. For if one says that France is part of Europe, it is generally admitted that France must be smaller than Europe, although, even in this instance, some acquaintance with the concept of area has to be presupposed. However, consider the following: according to common usage it is quite legitimate to say that sound sleep is a part of one's well-being; here, if we really consider it as an independent property and not simply as one aspect of the relation part to whole, the question of bigger and smaller breaks down. One might object that such cases are not meant by the axiom; but this would mean that the given proposition needs a preceding explanation and thus cannot stand at the base of a logically constructed system. From an ambiguous premise one cannot draw unambiguous conclusions. No proposition that presupposes complicated experiences and appeals to a necessarily vague use of colloquial language can be fit to serve as the starting point of a rigorously systematized branch of science.

There have been frequent objections to the so-called axiomatic method of instructing the beginner. However, most of them were made from a point of view quite different from ours. In general, such a procedure has been criticized as "too formal," appealing too little to intuition and thus "apart from life." These are clearly considerations of a pedagogic nature, with which we need not concern ourselves here. Our critique is directed from the purely logical point of view.

The formulation of axioms found in high-school textbooks, being based on uncertain and imprecise customs of language and therefore unsuited for drawing unambiguous conclusions, is a failure.

2. *Classical Axiomatics.* Historically, we find the origin of axiomatics in the tradition of Euclid's *Elements* of geometry. However, the presently existing Greek text begins with a number of definitions, which are followed by five postulates and nine common notions. In the older Latin translations, the latter (among which is, for example, the addition of equals results in equals) are named "communes notiones sive axiomata." According to our contemporary terminology, the "postulates," e.g., the parallel postulate, would also count among the axioms.

The remainder of Euclid's first book consists of theorems and solutions of problems for which the proofs are always explicitly derived from the preceding definitions, postulates, and common notions. The subsequent books contain further definitions (of the circle and the tangent to the circle, etc.), from which additional theorems are derived by means of the original axioms. Here the relation of the various propositions is clear:

at first definitions are given, then unproved statements are made about the defined things, and finally other statements are derived from these by ordinary deductive methods.

We find the same schema in the work which about two thousand years later became the foundation of mathematical physics—Newton's mechanics in the *Philosophiae naturalis principia mathematica*. It begins with a series of eight definitions for the concepts of mass, force, and so on. Next, we find the assertion that words such as time, space, location, and motion do not need any definition—only certain refinements. After these follow the "axiomata sive leges motus," while the consequences of the laws of motion form the bulk of the work. We shall not discuss the question here of how far it is really possible to derive rigorously the laws of motion of a rigid body (for example, the physical pendulum) without any premises but those expressed in Newton's three laws of motion.

However, the Newtonian mechanics is above all useful in offering us instructive insight into the construction of an axiomatic system, and the relation of definitions and axioms. Without a doubt the principal part of the Newtonian system is contained in the Second Law of Motion (which includes the first one, the so-called Law of Inertia): "Mutationem motus proportionalem esse vi motrici impressae . . ."; "the change of motion (of momentum, as we should say today) is proportional to the impressed force . . ." Hence, if no impressed force is present, the motion remains unchanged (which is the Law of Inertia). One may compare this with the preceding definition (IV): "Vis impressa est actio in corpus exercita, ad mutandum eius statum vel quiescendi vel movendi uniformiter in directum"; "the impressed force is the action on a body that changes its state of rest or of uniform rectilinear motion." As one can see, the axiom is anticipated by the definition. For, if an impressed force is defined as that which changes the state of motion, it follows that in the absence of impressed forces the uniform rectilinear motion remains unchanged. Thus we see that definitions and axioms are not independent of each other at all, and one recognizes the naïveté of the notion that the axioms by themselves state anything *"about"* the defined concepts. The relation between definitions and postulates in Euclid is quite the same, though not so immediately apparent.

The classical axiomatics of Euclid and Newton, which for a long time were taken as the model for the construction of every branch of the exact sciences, are characterized by a subtle confusion of apparent definitions and explicit postulates, which in fact cannot be regarded as independent of each other.

3. *Mach's Reform.* In the area of Newtonian mechanics the confusion arising from an insufficient differentiation between definitions and axioms was remedied by Ernst Mach. It was in Mach's *Mechanics*, published in

1883, that for the first time the principle was established which today is so generally accepted and which forms the essence of modern axiomatics. It can be briefly stated by saying that the fundamental concepts are *defined by the axioms*, that is, that, apart from the introduction of new terms, there are no definitions in addition to the axioms at the basis of a deductive science. We shall elucidate this by the example of mechanics.

There are two new basic concepts that enter into the construction of Newtonian mechanics—those of force and of mass. Newton explains mass in his first definition as "quantity of matter." One notices immediately that this definition is completely empty and in no way helps us to gain an understanding of the phenomena of motion. All one has to do is to reflect that it would be possible to substitute the words "quantity of matter" for the word "mass" wherever it appears in a contemporary text, i.e., not to have the sequence of letters m-a-s-s occur at all, and then one would be able to dispense with the first definition completely without anything in mechanics being changed. If we disregard the fact that perhaps in the age of Newton one phrasing was more colloquial than the other, the definition serves, in fact, only as the equation of two expressions, both of which are equally in need of an explanation.

On the other hand, we have already shown that Newton's definition of force largely anticipates the content of the first two laws of motion: the force neither changes the location of a body, nor determines its velocity, but rather *changes its velocity*. Thus force is first defined as something that changes the velocity, and then the law is stated that velocities are changed by forces. This manner of inference has been well put by Molière; the poppy seed is soporiferous; why? because it has the power of soporiferousness.

But scoffing here is ill-advised, for Newton's *Principia* expresses one of the most far-reaching and original discoveries ever made in physics. One can conveniently describe it by two statements: first, the circumstances in which a body is at a given time (its position with respect to other bodies, and other observable properties) determine the instantaneous *change* of velocity (or the acceleration) but not its velocity; and second, for different bodies under the same circumstances the observed accelerations differ by a numerical factor which is proper to the body in question, and hence is a constant for *each body considered*. These are the two discoveries of Newton. Once they are found, it is easy enough to add that the constant associated with each body shall be called "mass" (or if one wishes, "quantity of matter") and the circumstances determining the acceleration, "impressed force."

Thus definitions are reduced to explanations of words, to the (in principle, dispensable) introduction of abbreviations. Everything essential is contained in the axioms themselves. They delineate the concepts, for which verbal denotations can then be chosen as the need arises. We can summarize briefly:

As Mach showed, the Newtonian foundations of mechanics can be remodeled in such a way that one does not begin with definitions, but with assertions (axioms), which also suffice to define the fundamental concepts of force and mass. Then, all one need add to the axioms is explanations of words serving as verbal abbreviations.

4. *Hilbert's Geometry.* It was David Hilbert who for the first time, in the year 1899, created the new form of axiomatics for geometry. This was the starting point for the extension of the axiomatic method to many branches of the exact sciences. As had become apparent from the works of Euclid and several of Hilbert's predecessors (Pasch, Veronese, and others), geometry is particularly well suited for such a treatment. The properties of space, described by geometry, are the simplest physical phenomena known. They do not refer back to experiences of a different kind as, for instance, mechanics has to refer back to geometry even in its fundamental concepts of motion. Only *one* realm of general experience is used in geometry from the start—that of counting. In the construction of geometry the foundations of arithmetic are taken for granted.

Hilbert's system, as remarked before, differs from Euclid's in that it does not start with definitions of space elements, but considers these as defined by the axioms:

We consider three different systems of things: the things of the first system we call points . . . ; those of the second, straight lines . . . ; those of the third, planes . . . We think of the points, straight lines, planes in certain relations to each other and denote these relations by words as "lying," 'between,' 'parallel,' 'congruent,' 'continuous.' The precise and, for mathematical purposes, complete description of these relations is contained in the *axioms of geometry.*

Hilbert's first group of axioms contains the axioms of connection, examples of which are: two noncoincident points determine a straight line. . . ; there are at least four points that do not lie in one plane. (Notice here that the knowledge of counting is taken for granted.) The second group, the axioms of order, mainly delineate the concept "between"; for instance: among any three points of a straight line there is always one and only one that lies between the other two. These axioms are followed by the third group, the axioms of congruence, which explain the concept of equality. Here is the legitimate place for the proposition, every straight line is equal to itself. For this serves, together with other similar propositions, to determine precisely and completely the way in which the word "congruent" shall be used thereafter.

The main interest of geometric axiomatics is concentrated upon the parallel axiom, which forms Hilbert's fourth group. From the axioms of connection one can conclude only that two straight lines in a plane have either *one* point in common or none. Now the "parallel axiom" is added: through a point outside a given straight line there is at most one straight line that lies in the same plane and does not intersect the first one. It is a result (already found earlier) of far-reaching consequences that such an explicit assumption is necessary in order to draw the conclusions that form the contents of ordinary geometry. If one gives up this assumption, or replaces it by a different one, one can arrive at new geometric theories which contradict certain theorems of Euclidean geometry. Quite apart from the practical importance which these theories have acquired in modern times for the description of certain physical phenomena, they are highly instructive as regards the position of geometry, and hence also of the other exact sciences, in the total realm of our experience. We shall return to the contents of the so-called non-Euclidean geometry in the next section.

The last group of Hilbert's axioms contains the axioms of continuity which are missing in Euclid and whose necessity follows only from a more profound analysis of geometric propositions. Hilbert states the essential hypothesis here in the form of a "completeness" theorem: it is impossible to add to the system of points, straight lines, and planes another system of things such that in the resulting total system all the previously stated axioms are still valid. Hilbert shows how one can construct a manifold of points, straight lines, and planes in which all the remaining axioms are valid, but which does not satisfy the demand of completeness. This manifold does not yield a useful picture of the well-known geometric phenomena.

Hilbert's axiomatics of geometry rigorously carries through the principle that the elementary concepts are to be defined by the axioms themselves; and, in particular, it demonstrates the role of the parallel axiom and the axiom of continuity.

5. *Non-Euclidean Geometry.* The best-known and in many respects most important result of axiomatic investigations is the invention of the so-called non-Euclidean geometries. We know today that Gauss had much of the essential knowledge (some of it since the year 1792), but refrained from publishing it because he was afraid of the "Geschrei der Böotier." This goes to show how strong the influence of traditional scholastic opinions may become and how necessary an education toward a free outlook and unprejudiced judgment is for the progress of science. The history of non-Euclidean geometry starts with the publication of a work by the Russian mathematician Lobachevski about 1840.

The essential point is to establish the following matter of fact: Dropping Euclid's parallel axiom (previously quoted in Hilbert's form) and substituting for it another suitable hypothesis but retaining all the others, it is possible, by the customary rules of deduction, to derive from this new set of axioms a new geometry. Such a geometric system *does not contain in itself any contradictions*, although it contradicts certain theorems of Euclidean geometry. This fact refutes the allegation that the theorems of geometry taught in our schools are *imposed by the ,laws of thought*, and absolutely assured truths, independent of all experience. In the philosophical system of Kant's *Critique of Pure Reason* this assertion plays a decisive role. Kant says,

For geometric principles are always apodictic, i.e., united with the consciousness of their necessity . . . ; theorems of this kind cannot be empirical judgments or conclusions from them.

There is no doubt that Kant wanted to establish with these words a fundamental and profound difference between the theorems of geometry and those of other natural sciences, such as optics or mechanics, for instance.

One often tries to make clear how a non-Euclidean geometry is possible by pointing out spatial phenomena that are governed by laws similar in structure to a non-Euclidean geometry. Imagine, for instance, the spatial situation on the surface of a sphere of very large diameter with creatures whose entire existence is limited to this two-dimensional surface. As long as they know only a limited portion of their world, i.e., only a piece of the spherical surface small compared to the diameter, they would have a geometry different in nothing from our Euclidean plane geometry. Their straight lines would be, in fact, arcs of great circles, these being the lines of shortest distance between two points (and the shapes of stretched strings) on the surface of the sphere. As soon, however, as the surface inhabitants extend their experience beyond the immediate neighborhood, they will be forced to change their hypotheses. There are no "parallel" lines on the spherical surface; all great circles intersect each other in two points. Now, it is indeed possible—by means which cannot be explained here in any detail—to describe the situation on the surface of a sphere in such a way that all statements become equivalent to those of a plane geometry in which the parallel axiom does not hold. But the same relation as that between a plane and a very large spherical surface exists also between the three-dimensional space described by Euclidean geometry and a "curved" space not satisfying the parallel postulate. Assume that the space we live in has an extraordinarily large "radius," i.e., an extremely slight "curvature"; then our measuring instruments would not be sufficiently accurate to determine practically whether the Euclidean or a modified geometry is in better agreement with the facts. But the decision would be of no practical interest, anyway, as long as the differences are not noticeable in some way. We may leave open the question whether today there is enough evidence for such a decision. The notion of "apodictic certainty" of the geometric theorems, however, has to be abandoned definitely.

The study of axiomatics of geometry has proved that a consistent sys-
tem of geometry not obeying the parallel axiom and hence not agreeing
with the Euclidean geometry can be constructed; thus the assertion that
the customary geometry is logically inevitable, apodictically certain, and
independent of any experience, is disproved.

6. *Applications of the Axiomatic Method.* Contemporary axiomaticists
are mainly concerned with two general problems: the questions of *con-*
sistency and of *independence* of the axioms. By a "consistency proof" is
meant the demonstration that from an assumed set of axioms, using the
recognized customary rules of inference, one can never deduce both the
statements *A* and non-*A* (the contradictory opposite). Such a proof has so
far, strictly speaking, never been given for any axiomatic system of general interest. All one has succeeded in doing is to reduce the consistency
of, e.g., geometry (Euclidean as well as non-Euclidean) to the consistency
of arithmetic. In other words, it has been shown that an inconsistency in
the set of axioms of the particular geometry would imply an inconsistency
in the structure of elementary arithmetic. It is only because one credits
arithmetic with consistency that one believes this to be proved in the other
cases. We shall return later to the problem in arithmetic itself.[1]

The question of independence is somehow related to that of consistency.
The (consistent) group of axioms *A* is said to be independent of the (consistent) group of axioms *B*, if the group *B* can be exchanged for a different group *B'* such that *A* and *B'* together again form a consistent system. It is, for instance, proved that the parallel postulate is independent
of the axioms of connection, etc. In each case the consistency is measured
by that of arithmetic. If an axiom proves to be dependent upon others,
one can try to restrict it in such a way that the remaining weaker axiom
becomes independent of them.

The usual theoretical discussion of axiomatic systems suggests a question
which belongs to epistemology and which seems of importance to us. Are
consistency, we ask, and perhaps independence, *all* one can demand of a
set of axioms? Is every system that satisfies these conditions permissible,
and are all these systems equally justifiable objects of scientific research?
As long as this question is not formulated in another way it cannot be
answered reasonably, for the meaning of what is "permissible" or a "justifiable object of scientific research" is not clear. Human activities, of which
research is but an example, can only be described, i.e., studied and classified as to their relations to each other and to other facts. A "valuation"
would be possible only after an arbitrary standard of values had first been
adopted. If we limit ourselves to an objective description of the situation,
we find that there are two extreme cases. On one end we find axiomatic
systems more or less similar to the rules of the game of chess. They satisfy

---

[1] [See p. 1744 (section on Foundations of Mathematics). ED.]

all logical postulates but are far from being applicable to more vital phenomena. On the opposite end there are investigations that explicitly aim to influence practical action in certain areas of life, by, e.g., answering directly or indirectly questions arising from technology. All axiomatic systems fall somewhere between these two extremes.

In all cases, however, even in the case of the axioms of chess, there is but one way to establish a connection between the system of axioms and their consequences on the one hand and observable facts on the other: The words and phrases used for the basic or derived concepts of the axiomatic system must be given a more or less definite interpretation by reducing them to *protocol sentences*.[2] We may agree with Hilbert that the words "point," "straight line," "plane" within the axiomatic system are nothing but arbitrary signs. Yet, if one substitutes for them the words "shoes" and "ships" and "sealing wax," attaching to these words their customary (incidental) meanings, one will find that the relations expressed in the axioms and theorems either are not reducible to protocol sentences or contradict certain protocol sentences. Two points determine a straight line: this is not only an axiom, but with the usual meaning of the words it is also the approximate expression of a matter of fact. That analogy breaks down if one asserts as an axiom that two shoes determine a ship. This situation can be described by saying that the axiomatic system itself remained intact, but it could not be applied. One may also adopt the formulation that an axiomatic system that with no choice of names (i.e., with no coördination of the arbitrary symbols to protocol statements) leads to useful applications in the indicated sense is "worthless." All systems of axioms considered in science are formulated in such a way that they are paralleled by some interpretation in terms of observable facts. As far as the non-Euclidean geometries are concerned, we have mentioned already that their application to observable space phenomena, with the assumption of a slight space curvature, seems possible. In modern relativity theory this is an indispensable tool for the description of physical facts.

The connection of a system of axioms and their consequent theorems with reality, i.e., the meaning of its statements, has to be determined in the same way as that of all other statements, by coördinating the words and idioms used in them to elementary experiences by means of reduction

---

[2] [In Von Mises' terminology the expression "protocol sentence" is meant to indicate that "one is dealing with statements having immediate, present events as their subject and which are instantly written down (or otherwise recorded), as is the case with protocols in the juristic sense. According to the conception of Carnap, one might imagine schematically that all pertinent experiences, sensations, feelings and thoughts are recorded in the form of a written (or mentally fixed) protocol and then become the basis of further study. . . . If we say that any statement can be reduced to protocol sentences we mean that the statements can be supplemented by further (explanatory) sentences in such a way that for those who understand the protocol language a correspondence between these statements and experienceable events is established." ED.]

*to protocol statements. On this basis one can differentiate between axio-
matic systems for which consistency and independence have already been
proved, according to whether or not they can be used for the description
of observable facts.*

7. *Axiomatization in General.* The axiomatic method, having had so
great a success in geometry, has more recently been extended to various
other parts of the natural sciences. We have mentioned previously the
clarification of the foundations of mechanics due to Mach, an investigation that must be considered as a forerunner of modern axiomatics. Later,
the model of geometry was used in a more formal manner to set up,
among others, axioms of mechanics by G. Hamel, axioms of set theory
by E. Zermelo, axioms of thermodynamics and axioms of the special
theory of relativity by C. Carathéodory, etc. Attempts at extending the
axiomatic method to many other branches of the natural sciences are in
various stages of development.

The followers of modern axiomatics see in that method the highest form
of scientific theory, the form toward which all scientific endeavor strives.
Thus Hilbert said,

I think that everything that can be an object of scientific thought at all,
as soon as it is ripe for the formation of a theory, falls into the lap of the
axiomatic method and thereby indirectly of mathematics. Under the flag
of the axiomatic method mathematics seems to be destined for a leading
role in science.

If we disregard the reference to the place of mathematics (it is not at all
obvious why an axiomatics of biology should subordinate biology to
mathematics), there still remains the claim that every theory in a certain
stage of its development takes the form of axiomatics. This form was defined by Hermann Weyl as "the complete collection of the basic concepts
and facts from which all concepts and theorems of a science can be derived by definitions or deductions." In the last analysis, everybody who
wants to exhibit a branch of science systematically intends to proceed
according to this scheme, i.e., to collect all the essential facts in the fundamental theorems; it is only the manner in which this is done, its completeness and precision, that will not always be as Hilbert meant it. Nothing
can be said against Hilbert's claim as long as the following two points are
kept in mind.

In the first place, the axiomatic method is only a *form* of description,
hence something that can be taken into consideration only after the essential content of what is to be described is known. It was the discovery of
Newton, prepared by Galileo, that external circumstances (to a body) determine its momentary *change* of velocity (acceleration), and not the
velocity itself. Once this discovery (or, if one prefers, the "invention")
was made, it became another problem to put its content into a set of clean

formulas that do not say too much or too little. To take another example, a long time ago it was found that in the mathematical concept of probability one deals with the limit of the relative frequency in an unlimited number of trials, where a certain irregularity in the sequence of the results is essential. Anyone who accepts this basic idea is faced with the problem of expressing the facts in a set of precisely formulated axioms. In geometry we find the same situation, only there the essential discoveries date so far back that we are not conscious of them any more. The axiomatization is always a secondary activity which follows the actual discovery of the pertinent relations and puts them in a precise form.

The second reservation one has to keep in mind against the exaggerated claims of the axiomaticists concerns the concept of absolute rigor. All followers of the axiomatic method and most mathematicians think that there is some such thing as an *absolute* "mathematical rigor" which has to be satisfied by any deduction if it is to be valid. The history of mathematics shows that this is not the case, that, on the contrary, every generation is surpassed in rigor again and again by its successors. All classical mathematicians, Gauss included, as any student today can show, have been guilty, on some occasion or other, of a faulty lack of rigor. Certain new developments of mathematics, of which we shall speak,[3] will show that some things that were taken as quite rigorous thirty or forty years ago have to be doubted today. There is by no means an eternally valid agreement about the admissible methods of logical deduction. Thus we have to regard the task of axiomatization of a science as a relative one, subject to change with time. One tries to give to the basic theorems of a theory such a form that they satisfy all *present* requirements of logical rigor.

The axiomatic formulation of a discipline may be regarded in each period as the highest level of scientific presentation, if one keeps in mind, first, that it is merely a form, which can be taken into consideration only after the essential relations are known; and second, that the logical requirements which the axioms have to satisfy are themselves subject to evolution.

LOGISTIC

1. *Tautologies and Factual Statements.* "Mathematical logic" or "logistic" is a modern name for the fundamental system of logical rules, presented in the form of concise formulas, and the study of them. Before we discuss this matter, we have to find out what is meant here by "logical."

The present stage in the development of positivism is governed by an idea first enunciated by L. Wittgenstein, though in a slightly different form, and advanced with particular emphasis by the so-called Vienna Circle. It is the conception according to which all meaningful statements

[3] [See p. 1744 (section on Foundations of Mathematics). ED.]

—or, as we prefer to say, all connectible statements [4]—have to be divided
into two groups: those expressing a state of fact which can be tested by
experience, and those which, independently of all experience, are true or
false by virtue of their wording. Statements of the second kind are called
"tautological" in the first instance and "contradictory" in the second.
Tautological sentences form the content of logic, of pure mathematics,
and of all other axiomatically formulated scientific theories.

At first sight it might seem that here Kant's a priori is revived, including
the "synthetic judgments a priori" which are supposed to form the content
of mathematics. However, the difference is quite considerable. A sentence
is tautological if it is true independently of all experience, *because* it does
not say anything about reality at all and is nothing but a reformulation or
recasting of arbitrarily fixed linguistic rules (definitions). Consider for
instance, the statement, the sum of two natural numbers is itself a natural
number; everybody can see that this is a statement of a different kind from
this: alcohol boils at a temperature of 78° C. For if we try to explain
what we mean by the sum of two natural numbers, i.e., if we want to define the word "sum," or to give the rules according to which it is to be
used, then necessarily these rules include the result that the sum is itself
a natural number. The tautology is a little more difficult to recognize in
the statement, the sum of two prime numbers, neither of which is 2, is an
even number. Here one needs, in order to demonstrate the tautological
and noncontradictory character of the statement, a "proof." But according
to our fundamental conception every mathematical proof is the tautological transformation of definitions and other linguistic rules and in any case
has nothing to do with observation and experience. Finally, as a third example we might take the assertion, every even number is the sum of two
prime numbers. Here we have to remark that no living person today could
say whether this is tautological or contradictory (true or false). But we
know that this decision does not depend upon some sort of physical experiment or observation of facts, but can be made solely on the grounds
of a study of definitions and a carrying out of calculations and similar
transformations.

Thus, the theorems of logic and mathematics are in the framework of
this conception "tautological," but neither synthetic nor a priori. They are
not synthetic, because they say nothing about reality, and not a priori,
because they do not come from a superempirical "source" but are the
result of arbitrary definitions introduced by us. Furthermore, tautological

[4] [Von Mises defines a "connectible statement" as follows: "We shall call a group
of words (sentences and sequences of sentences) 'connectible' if they are compatible
with a system of statements which, it is assumed, regulate the use of language—connectible, that is, with respect to this system. Strictly speaking, one would always have
to say 'connectible with . . .'; this addition can be omitted only if the system of
reference is supposed to be known." ED.]

character is by no means confined to mathematics; it is possessed by any system of sentences that is constructed according to the axiomatic method and that serves as the theoretical exposition of an area of facts, e.g., a chapter of biology.

Theorems of logic and pure mathematics and of any axiomatically formulated theory are, according to the conception of L. Wittgenstein, "tautological," i.e., they do not say anything about reality and are nothing but transformations of arbitrarily agreed upon linguistic rules. The notion of tautology by no means coincides with that of Kant's a priori synthetic judgments.

2. *Tautological Systems.* Against the thesis according to which logical, mathematical, and similar theorems are "tautologies," various objections have been raised. Some of them—and these can most easily be settled—originate in the rather unhappy choice of the name "tautology." Obviously Wittgenstein, and the members of the Vienna Circle (who like to call themselves today "logical empiricists"), have created a new concept by making the distinction between two kinds of statements, and for its denotation—even though in analogy with linguistic usage—they arbitrarily drew upon a word belonging to the existing language. But previously the word "tautology" had been used almost entirely in a derogatory sense, to designate empty or superfluous talk; by "tautology" one meant something like a sentence which, without any loss to the reader or listener, could just as well be omitted. The new use of the word is *burdened* with this old meaning and even people who are used to abstract thinking and who are quite familiar with the process of nomenclature in the exact sciences cannot always free themselves from this influence. Thus, a beautiful book about *Numbers and Figures*, having two well-known contemporary mathematicians as authors, ends with a remark which shows how much the authors feel hurt by the thesis that "mathematics is fundamentally but a chain of tautologies." This is a psychologically understandable resistance to a new terminology.

It might be suggested that we replace the word "tautological" by "analytic," but that would result in much worse misunderstandings, since it would be reminiscent of Kant's concept of analytic judgments. We have shown above that tautologies in our sense are by no means Kant's a priori synthetic judgments. But neither are they his analytic judgments. According to Kant's theory and that of idealistic philosophy in general, there exist concepts, independently of all human influence and all linguistic conventions, whose delineation we can more or less precisely discover through pure thought. A statement which expresses the result of such acts of thinking is called an analytic judgment. We, however, mean by a tautology a sentence that is derived by *arbitrarily* fixed transformations of *arbitrarily* chosen basic assumptions. In this chapter we shall have to show first of all what such a system of tautological statements looks like and this will bring us back to the discussion of the axiomatic method. On the other hand, since the aim of all language, in everyday life as well as in science, is in our opinion to assert something about observable reality, i.e., to find statements verifiable in experience, we shall have to explain further the role of tautological systems in this respect.

By no means can we accept a view that would assign to tautological sentences something like dogmatic or absolute validity. At most one could call them "apodictic," as long as one keeps in mind the frame of reference that determines their range of validity. The starting theorems (definitions) as well as the rules of transformation (methods of deduction) are, as previously mentioned, to a certain degree arbitrary conventions. Even though at the present time among logicians there is, on most counts, agreement about the basic assumptions, there are still differences of opinion upon certain specific points, e.g., the axiom of the excluded middle.[5] A decision about whether one or the other of the logical systems is "right" is impossible. It is only the usefulness of a specific system for the representation of observable phenomena which can prove it to be more or less *expedient*.

What contemporary logical empiricism calls tautologies must not be confused with the customary meaning of the word in ordinary language, nor with Kant's "analytic judgments." Tautologies have no absolute validity but are valid only within a specific system of basic logical concepts. The application of such a system as a means of representation of a part of reality determines its usefulness.

3. *Basic Logical Relations.* The system of logistic (or, as it is often called today, of mathematical or theoretical logic) starts with the setting of the basic relations that can exist between statements. These fundamenal connectives, of which there are four, are denoted in ordinary language by the expressions "and," "or," "not," "if . . . then." We call them, in this order, conjunction (also logical product), disjunction, negation, and implication (or conditional). The sign "and," for which we write simply a comma in the formulas, means, if put between two statements, that both statements are posited or assumed. The sign "or," whose abbreviation is "v" (*vel*), means that of the two statements between which it stands at least one (but not one and only one) is assumed. In order to indicate negation, i.e., the contradictory opposite, we use the sign "∼" put before or above the statement sign to which it applies. Implication may be expressed by an arrow, "→"; it signifies that if the left-hand statement is posited, the right-hand one is assumed to be posited too: the first one implies the second; the second is implied by the first.

To these basic concepts one can add others, e.g., that of equivalence; but one recognizes at once that this is dispensable, for by the equivalence

[5] [See p. 1748 (section on Foundations of Mathematics). ED.]

of two statements one means that the first follows from the second and the second from the first. Hence equivalence is expressible by the two relations of implication and conjunction. If we want to use the formula way of writing, we must say that "*A* equ *B*" can be replaced by "*A→B, B→A*".

These remarks suggest the question whether all four of these relations are independent of one another or whether some of them can be derived from the others. The latter is, indeed, the case. It suffices, for instance, as one easily sees, to introduce the signs "v" and "∼" as undefinable basic signs, or, as Brentano has shown, only the concepts "and" and "not." We shall show how implication and disjunction can be expressed by the two signs "," and "∼".

That *B* follows from *A* means the same as that *A* and non-*B* cannot exist simultaneously. Hence, in order to express the implication, one has only to signify that the conjunction of *A* and non-*B* is not true. The sign "*A→B*" can therefore be replaced by "∼[*A*,∼*B*]", which has to be read: *A* does not exist together with non-*B*. It is equally simple to express disjunction by means of conjunction and negation. That of the two statements *A* and *B* at least one is true can be expressed by stating that the statements non-*A* and non-*B* do not hold simultaneously. The compound of signs "*A*v*B*" is thus replaceable by "∼[∼*A*,∼*B*]", which is to be read: non-*A* and non-*B* do not exist together.

A similar consideration shows that conjunction and implication are reducible to disjunction and negation. On the other hand, it is impossible to construct the other relations by means of negation and conjunction. Of course, it cannot be the purpose of such investigations to express all logical deductions, whenever they occur, by means of the two basic relations previously chosen. On the contrary, once we have seen that signs like "equ", "→", "v" can be replaced by suitable combinations of "," and "∼", we know that the new signs can be used as abbreviations. In the same way one seeks to form larger and more efficient units and by their means to reach more complicated deductive patterns. The whole of mathematics is a construction of this kind.

The first steps at the basis of mathematical logic consist of showing how the various simplest relations—conjunction, negation, disjunction, and implication—are connected with one another and in a certain way can be reduced to each other.

4. *Further Formalization.* The foregoing argument can be further formalized and then yields definite rules for the derivation of new tautologies. That is the essential aim of theoretical logic since Leibnitz, who must be considered the founder of the entire discipline. We start with two symbols, "F" and "T" (which stand for the words "false" and "true"), without giving any definition for them. To every statement we assign one of these two letters; that is, only such statements are admitted as premises as are

accompanied by one of the symbols "T" or "F". Then one defines operations on statements (truth functions) by directing in what way the assignment of T or F to the objects of the operation transfers to the result. The only operation acting on a *single* statement is the negation. It is defined by the rule that, if the statement A has the truth value F, $\sim A$ has the truth value T, and if A is assigned T, $\sim A$ is to be assigned F.

For every truth function that connects *two* independent statements, e.g., the implication "$A \rightarrow B$", there are four possible combinations of the assignments of T and F to the two parts. The implication "$A \rightarrow B$" is defined thus: the sign "$A \rightarrow B$" has the truth value F, if A has the value T and B the value F; in the other three cases "$A \rightarrow B$" has the value T. On the other hand, the definition of the disjunction "$A \vee B$" is this: the sign "$A \vee B$" has the value F if F is the value of both A and B; otherwise its truth value is T. Once a number of symbols have been defined in this manner, one also knows the distribution of the T and F values in formulas that result from combinations. For instance, the combination "$A \rightarrow \sim B$" has the value F whenever both A and B have the value T; in all other cases, its value is T.

Of special interest are those formulas which have under all circumstances the value T. These are the actual propositions of mathematical logic. For example, according to the above conventions about the operations "$\sim$" and "$\vee$" (negation and disjunction), the combination "$A \vee \sim A$" has always the value T. For A and $\sim A$ cannot, according to the definition of "$\sim$", both have the value F, and according to the definition of "$\vee$" the combination could have the value F only if both parts, the one before as well as the one after the sign "$\vee$", have the value F. The formula "$A \vee \sim A$" can be called the "theorem of the excluded middle" (in the simplest case). Other "always true" formulas or theorems in truth-function theory (propositional calculus) are: "$[(A \rightarrow B), (B \rightarrow C)] \rightarrow (A \rightarrow C)$" and "$(A \rightarrow \sim A) \rightarrow \sim A$". The first one is, in words, the fact that A implies C, or more briefly (but not exactly correctly), if B follows from A and C follows from B, then it follows that C follows from A. The second formula is the basis of the so-called indirect proof; for it says in words (approximately), if a statement implies its own contradiction, it follows that the statement is false.

If one writes down any formula consisting of complexes of the four signs we have introduced above and any letters $A, B, C, \ldots$, then one can determine purely mechanically the truth conditions of the formula, and in this fashion it is possible to derive theorems. Further pursuit along these lines forms the first chapter of any textbook of mathematical logic, the so-called truth-function theory or propositional calculus, which has a certain analogy to elementary algebra.

As in algebra, where according to fixed rules algebraic formulas and theorems are derived, truth-function theory arrives at formal laws determining the connection between the basic operations.

5. *Difficulty of Deduction.* From the foregoing examples of theorems of truth-function theory, which of course can only give a faint idea of the first beginnings of mathematical logic, there are many interesting things to be learned. In the first place, they show how vague and unreliable ordinary linguistic usage is in such simple cases. For the sign "∨", introduced above, corresponds to the English word "or" only when the latter is used in the sense of the Latin "vel", but not in the case of the *exclusive* "or" (aut). If "or" in this latter sense is to be expressed one has to write "$(A \lor B), \sim (A,B)$", i.e., "A or B and not at the same time A and B". The sign "→" of implication, too, does not always coincide with that which is customarily expressed by the words "it follows". According to our definition, "$A \rightarrow B$" is always true when A is false, whatever B may be, i.e., a false statement implies any arbitrary true or false statement. If, e.g., to the correct theorems of arithmetic the theorem "$2 \times 2 = 5$" is added, then any true or false result can be derived from the premises, in particular, of course, the theorem "$2 \times 2 = 4$". In the customary way of speaking it is not so; there the conception—which is not exactly definable —is used that from false antecedents only some conclusions, namely false ones, can be drawn.

On the other hand, our examples show how logical theorems can be constructed by applying arbitrarily adopted rules, as in a game. Such theorems represent "eternal truths" only as long as one takes the rules for granted. They do not become statements about reality, verifiable in experience, before the language of formulas is translated by a more or less vague correspondence into the language of everyday life. Analogously, in chess it is "unshakably true" that a player has lost the game when his king is open and without defense against the attack of his opponent. If one takes this theorem and gives the word "player," "king," "attack," "opponent," etc., the meanings customary in everyday language, it becomes vague and doubtful. There is here another analogy with the relation between logic and reality in that the theorem does not become meaningless, but must be considered false or true, according to experiential circumstances. We shall later return to this point.[6]

The comparison with the game of chess can also serve to illustrate another important point. Thinking of certain situations in practical life where decisions are to be made, one might be inclined to believe that the difficulties arising here are due exclusively to vagueness, indefinite and flexible conditions, and the impossibility of exact description by language. If

[6] [See p. 1742.]

everything were as clear and unambiguous as in logic, one may think, then decisions would be easy, and almost obvious. The chess-game analogy refutes this idea. Taking, say, an advanced point in the development of the game as a start, in which, say, black has the next move, the set of all possible continuations of the game is completely given by the rules of chess. Thus it is uniquely "determined" which next moves of black will lead to certain loss, which will lead to certain victory, and finally, which leave open both possibilities. (Incidentally, one or another of these three classes of next moves may be empty; also, certain well-defined assumptions about the manner of playing of the opponent may be included in the premises.) But something that is determined theoretically or in principle is by no means actually *known*. The complete enumeration of all possible continuations of the game goes beyond our capabilities. Under certain conditions such an enumeration could be replaced by "theorems," i.e., by tautological transformations of the chess rules, perhaps of this form: For certain starting positions (x) and next moves (y), only a checkmate of black can result. To find such theorems, i.e., to derive them from the original rules, seems in the case of the chess game extraordinarily difficult. A type of game in which the players have too much knowledge of this kind at their disposal is of no interest for them; the game of nim is appreciated only by children.

Theoretical logic, including mathematics, is based upon a system of rules which is immensely more abundant, diversified, and complicated than that of chess; moreover, it is subject to change in time. Of all the theorems derivable in the system of rules, only an extremely small part has been found so far and further progress often presents extraordinary difficulty. However highly one may rate the difficulties of *practical* decisions that cannot be reduced to theoretical questions, there is no reason to underestimate the intricacies of decisions in logic and mathematics.

The theorems of truth-function theory are additions to language indispensable for the purpose of higher precision. They do not contain anything that is not implied somehow in the basic assumptions (the rules of the game). Nevertheless, it remains a very hard problem, and one that can never be exhausted, to find explicitly all theorems that can be produced in this way.

6. *Russell's Theory of Types.* We want to mention here another important chapter of logistic which shows more clearly than truth-function theory how much present-day logistic theory differs from the classical logic, of which Kant said that since Aristotle it had made neither a step forward nor a step back. We mean the so-called theory of types of B. Russell, which proved to be indispensable in getting rid of certain otherwise unsolvable contradictions.

In introducing the concept of connectibility we spoke about what is meant by a sentential function or propositional function.[7] An important part of all linguistic rules has the purpose of delineating the area of applicability of predicates. When we say "x bears green fruit," this sentence is meaningful only if we put for "x" the name of a plant, but not if we put for it, e.g., "subway" or "circle sector." The totality of the objects x for which the individual statement is meaningful—no matter whether it is true or false—forms a "class," namely, the class belonging to the particular predicate. In our example, the class belonging to the predicate "bears green fruit" consists of the totality of plants, at least as long as one gives the words only their customary, nonsymbolic meaning, corresponding to usual parlance. Each single plant is then called an individual member of the class.

Now, undoubtedly statements can be made whose subjects are classes themselves; or better, that which appears as a class in one case may in another case be an individual of another class. When we say, e.g., "x possesses 23 vertebrae," this is a sentential function defined for a vertebrate, i.e., the corresponding class is formed by the totality of vertebrates. But this totality is an individual member of the class "classes of animals," of which one could, for example, make the statement "x contains more than 200 species." Russell was the first to point out clearly that the two classes, "classes of animals" and "vertebrates" stand in a certain relation of subordination; they belong to different types or rank. No statement can be meaningful for classes of different types at the same time.

This concept of types must not be confused with a distinction of smaller and larger classes. The sentential function "x has four legs" is defined for a bigger class of animals than that of the vertebrates, perhaps for the totality of all animals. But this totality is not of a higher type than that of the vertebrates; it is only bigger, more comprehensive, the vertebrates being a subset of the set of all animals. On the other hand, the concept of the class of animals, i.e., the class of which one could say "x contains many species," is from the point of view of the theory of types a higher concept than that of a vertebrate, an insect, or any individual animal, etc.

The theory of types, or the doctrine of the "hierarchy of types," served originally to resolve certain contradictions that appeared in the theory of sets, a branch of pure mathematics. No satisfactory solution had existed previously. In a work of great scope, *Principia Mathematica*, Whitehead and Russell showed how to reduce all mathematical concepts to the simplest logical operations. But far beyond its original purpose, the theory of types serves as an important reference for the constitution of a general conceptual system and for the logical construction of a scientific language comprehending broad areas of experience. Today we are still a long way

---

[7] [For the meaning of sentential or propositional function, see the selections on logic by Ernest Nagel and Alfred Tarski. ED.]

from being able to show decisive practical results. But considering that for centuries philosophical discussions were governed by the controversy between so-called universalists and nominalists, i.e., the question whether classes are as real as individuals or, on the contrary, of higher reality, or are merely abstract names, one may harbor a faint hope that today's mathematical logic, one of whose pillars is Russell's theory of types, constitutes the first step toward a useful conceptual structure which does not admit of such absurd questions.

Russell's theory of types forms an important part of theoretical logic which increases its efficiency as compared to classical forms of logic. It constitutes the first step toward a general rational conceptual structure and discards old pseudo problems, like the controversy between the universalists and the nominalists.

7. *Universal Physics.* In the framework of the present efforts, going on in different countries along different paths, to arrive at a "philosophie scientifique," at a conception of the world free of metaphysics, theoretical logic certainly plays an important role. But we cannot object too strenuously to the opinion occasionally voiced by representatives of the Viennese and the Polish schools (Łukasiewicz, Ajdukiewicz, Kotarbinski, Tarski, etc.) that *all* one need do is to develop mathematical logic further.

Certainly mathematical logic, i.e., the formal construction of symbolic systems and their tautological transformations, is an indispensable means for a useful description of reality, at least as indispensable as elementary and nonelementary mathematics is for the correct presentation of physical phenomena, say the propagation of light. Occasionally one finds also mathematicians who are of the opinion that physics is reducible to mathematics; they hold, for instance, that electrodynamics has become a "part of geometry" through the theory of relativity. Such utterances are logical misconceptions and go ill with the critical subtlety which the mathematician otherwise often exhibits.

By the mere manipulation of signs according to chosen rules one can indeed learn nothing about the external world. All the knowledge we gain through mathematics about reality depends upon the fact that the signs as well as the rules of transformation are in some wise made to correspond to certain observable phenomena. This correspondence is *not a part of logic* or mathematics and has no place within its tautological construction. To explore how the correspondence works is a big problem. For in order to describe it in any concrete case, one cannot use the filtered, sharp, and more or less formularized language of logistic; rather one has to resort to ordinary speech which is imperfectly built up and not sufficiently reducible to element statements. Between the exact theories and reality there lies an area of vagueness and "unspeakability" in the literal sense.

The question whether logic rests upon experience or originates in the human mind independently of all experience is wrongly posed. We observe that there are things which are black and things which are not black. But this distinction is not strict, the disjunction is incomplete, and we cannot even circumscribe precisely the range of objects to which it applies. But it is this observation, duplicated again and again, that led us to create the logical relation of *A* to non-*A*, which after suitable agreement about the use of the signs is free from uncertainty and vagueness. The same is true for the other relations of logic—the disjunction, conjunction, etc. They are abstracted from experiences in which they are realized only *approximately*. The beginning of this schematization (using a more general but not very expedient expression, one could say rationalization) goes back, as we saw in the first chapter, to the creation of ordinary language, hence to the distant past of the origins of human history. Its latest stage is the systematic synthesis of a logical language of formulas. Of course, one may ask the question, which qualities enable man—man alone or also some higher animals—to carry on such intelligent activity. But even this question would not lead us out of the domain of experience and into the realm of the transcendental or of metaphysics; it merely opens up another discipline of (biological) research, which is to be attacked by the usual means of empirical science.

However, it is not this historical question of the genesis of the basic logical concepts that is essential for us here; rather it is the present relation between logical insights and reality. Logic does not float freely in mid-air, without connection to the world of observation. What we have said above about axiomatics also holds for logistic. The words and idioms appearing in it—that they are presented in other forms than by customary letters does not make any difference in this respect—correspond in an unexact and never precisely determinable way to elements of observable matters of fact. It is only because logical formulas, after having been interpreted in ordinary language, according to this correspondence, yield a useful description of experienceable relations, that the study of theoretical logic is regarded as a part of scientific activity. We agree with the mathematician F. Gonseth—in a certain deviation from the "logical empiricism" of the Vienna Circle—who in a theory which he calls "idoneism" conceives of logic as of a "physique de l'objet quelconque" (physics of all things) and says that

the intuitive rules of logic and common sense are nothing else than an abstract schema drawn from the world of concrete objects.

Remembering the explanation of the concept of axiomatization given in the previous section, we may say that theoretical logic is in axiomatic form the doctrine of the most general and the simplest *typical relations* that are observed among objects of any kind.

Theoretical logic—like the axiomatic of a specific discipline—is a tauto-logical symbolic system developed according to accepted rules; its impor-tance for science lies merely in the fact that if the single signs and basic formulas are made to correspond to certain very general factual situations, any derived formulas correspond to certain observable matters of fact. It is, therefore, just as false to say that the logical inferences rest upon ex-perience, as that logic has nothing whatever to do with experience.

THE FOUNDATIONS OF MATHEMATICS

1. *Tautological Part.* When anyone, on any occasion, wants to give an example of an absolutely certain and indubitable truth, he does not hesi-tate to cite some mathematical theorem known to him, perhaps the Pythagorean theorem or even the formula of multiplication, "2 times 2 equals 4." If thus the *results* of mathematical rules are taken as completely and unshakably true, one should think that this is even more the case with the general foundations from which such results are derived. Nothing is more astonishing for the layman than to hear that among mathematicians there are differences of opinion as to the basic principles of mathematics— their meaning, their applicability, and their content. If, moreover, an emi-nent contemporary mathematician declares that the present uncertainties by no means concern merely questions on the frontier of mathematical knowledge but go directly to its core, then the general naïve faith in the "most absolute" of all sciences must be completely shaken.

After all that has been said in the previous chapters it is not difficult for us to understand what one has to think of the certainty or uncertainty of mathematical theories. "Pure" mathematics in the sense of customary parlance is only a system of tautologies, i.e., of conventions about signs and transformations according to accepted rules, or, we could say, a system of deductions. From the point of view of the extreme formalists (which we do not share), all assumptions about symbols and rules are *completely* arbitrary, such that without any further justification one may construct different mathematical systems which are not in agreement with one another. From our point of view this is true only with the essential reser-vation that all tautological systems that play any role within science rest in the last analysis upon a certain coördination of the symbols with other observable experiences. Only if one disregards these coördinations com-pletely, which is, e.g., possible by excluding all use of colloquial expres-sions, can one regard different systems that are complete and consistent within themselves as equivalent to each other. This relation of mathematics to reality, which is so often neglected, is particularly well pointed out by F. Gonseth in his interesting book *Les Mathématiques et la Réalité.*

After all, it is not a factual question but rather one of nomenclature whether one wants to define the concept "pure mathematics" in such a way that it is sharply distinguished from all other natural sciences. The theory of each domain of experience has its tautological part which appears in its most complete form as an axiomatic structure. Now, one may single out a specific realm of experience and call its treatment "mathematics"; then mathematics is in principle not different from other areas of knowledge and consists of a tautological and a nontautological part. This point of view was taken by M. Pasch, who for half a century tried to clarify the basic concepts of mathematics in a number of valuable works. However, it corresponds better to present usage to reserve the name "mathematics" for the purely tautological parts within the different branches. As far as geometry, which is usually considered as a part of mathematics, is concerned, the interplay of tautological and empirical questions was clearly described earlier by various scholars like Helmholtz and Mach. It would be a grave error to think that the situation in arithmetic is fundamentally different. If we say that the number 17 is a prime number (which is a tautological proposition) we have also in mind a simple empirical fact, which can be observed with 17 apples or coins, that a division into "equal" groups is impossible. A result which is not so trivial, e.g., that the number 681,199 does not admit of any factorization but 727 × 937 can be verified by trying to divide a corresponding number of objects into equal parts. A shepherd who has to care for 30 sheep can check experimentally that ⅓ plus ⅙ is more than a half. All these examples belong to the domain of experience the theory of which has arithmetic as its tautological part. It is only because the transition between the facts and the axiomatic concepts seems so extremely simple, familiar, and clear here that one usually pays no attention to it and forms from the tautological considerations alone the concept of a science called arithmetic. As far as geometry is concerned, the customary linguistic usage is a little vague in this respect. In mechanics or other parts of physics, it occurs to nobody any more to call their theorems, as far as they refer to real facts, independent of experience and absolutely certain truths, i.e., to confuse the empirical and the tautological aspects.

There is no difference in principle between the disciplines of arithmetic, geometry, mechanics, thermodynamics, optics, electricity, etc. It is merely a habit (suggested by the actual situation) in arithmetic, and sometimes in geometry, to reserve the name "arithmetic," or "geometry," for the purely tautological part of the studies. Thus the foundations and basic assumptions of arithmetic are debatable in the same sense as those of any part of physics, i.e., on the one hand, as to the internal questions of tautological structure, and on the other hand, as to the relations with the world of experience.

2. *Mathematical Evidence.* In all discussions of the foundations of

mathematics, without distinction between tautological and factual questions, the word "evidence" appears at an early stage. The simplest mathematical statements, so it is asserted, are "self-evident" (manifest) and their absolute certainty is derived from the evidence for them. In fact, even the contemporary controversy between the so-called formalists (Hilbert) and the so-called intuitionists (Brouwer) is for a large part based upon questions of "mathematical evidence."

We can assume that everybody, or at least everybody beyond a certain stage of education, is familiar with an experience for which he uses the word "evidence" or "self-evident truth." If anyone claims that to him the truth of a statement is evident or immediately apparent, he is hardly inclined to resolve this impression of his into simpler parts, i.e., to analyze it and to exhibit its components. This could lead us to accept the words "This is evident to me" as a kind of element statement or protocol sentence. As in the case of the statement "This is red," or "Here I have the sensation red," the words should, then, not be analyzed further, but should be regarded as a piece of raw material that has to be used along with other material of the same kind.

But physical optics, whose significance is derived from the fact that there are primitive impressions like red and blue, light and dark, or element statements of the form "Here I see red," by no means is as a scientific theory *based* upon such concepts as blue and red, or light and dark. Surely no course in optics starts with an attempt to make clear what the meanings of blue and red are, or derives anything from the assumption that these meanings are a priori known. The connection between the simple protocol sentences and the physical theory is not so immediate and in any case not so simple.

For the moment we do not have to go deeper into the question what is the connection in other areas between element statements and the theories. The point here is that, even granted the existence of a not further analyzable elementary experience of "evidence," it does not follow that mathematics (or any theory) should be built upon the assumption of evidence as a self-explanatory and obvious basic concept. For this to be possible, one condition above all would have to be satisfied: there must be a certain agreement about what knowledge is evident, that is, not only among the mathematicians themselves, but also among those who are only starting to study mathematics and who therefore do not yet draw conclusions from the theory.

It does not require much speculation to convince oneself of how divided, in all domains, the opinions are as to which statements have the property of evidence. We may even exclude such matters as the spherical shape of the earth, the existence of antipodes, or the rotation of the earth about its axis. Let us limit ourselves to purely mathematical concepts.

To almost everybody the properties of the positive integers, as they are taught in school, are evident. Most educated people claim, moreover, that the so-called existence of an infinite series of numbers is evident to them. More serious discussions arise if one asks whether the continuum of points on a line segment or the continuum of numbers between 0 and 1 is immediately evident or not. Some people even assert that it is immediately evident that all properties of geometric figures remain unaltered if one changes the dimensions of the figure. The latter lend to Euclidean geometry the character of evidence, while the former, who take the continuum for granted, jump over a wide area of mathematical difficulties. There is no doubt about the lack of agreement as to what is evident "to us." For each person it seems to depend upon education and incidental experience what appears evident to him, not to speak of the vagueness of linguistic formulation of allegedly "evident" propositions. The famous dictum of the mathematician C. G. J. Jacobi, "Mathematics is the science of what is clear by itself" can in no way be maintained today.

It is impossible to accept as the basis of mathematics merely statements that seem self-evident, if only because there is no agreement as to what statements actually belong to this class.

3. *Intuitionism.* Among the various controversial schools of thought in the field of the foundations of mathematics, it is the intuitionists who— as their name indicates—place the greatest emphasis upon intuition, evidence, and immediate apprehension or immediate insight. According to L. E. J. Brouwer, the founder of the intuitionist school, the simplest mathematical ideas are implied in the customary lines of thought of everyday life and all sciences make use of them; the mathematician is distinguished by the fact that he is conscious of these ideas, points them out clearly, and completes them. The only source of mathematical knowledge is, in Brouwer's opinion, the intuition that makes us recognize certain concepts and conclusions as absolutely evident, clear, and indubitable. However, he does not assume that it is possible to list in a precise and complete way all basic fundamental concepts and elementary methods of deduction, which in this sense are to serve as a basis of mathematical derivations. It should always be possible to supplement the once fixed set of assumptions by accepting new ones, if a further intuition leads that way. A first intuition yields us the concept of "two" from which the concept of multiplicity is inferred. Originally Brouwer regarded the continuum too as immediately given by intuition; later he tried to comprehend it by a new concept, the "sequences of free formation," i.e., to reduce the continuum in some way, after all, to a sequence of numbers.

Disregarding certain rather mystic formulations that Brouwer gave to his doctrine, one recognizes his point of view as very close to a radical empiricism. The thesis that the fundamental assumptions of mathematics cannot be formulated in a definitely fixed and complete form, but are subject to continued examination and possible supplementation by intuition (we should prefer to say, by experience which changes the stock of what appears as "evident"), corresponds exactly to our conception. The opposite point of view is that of Kantian a-priorism, according to which the basic mathematical concepts are once and for all impressed upon the human race by the properties of its reasoning power. Such a prediction that a specific chapter of science will never change—and the "a priori" means just that—has no place in the concept of science that this book [8] represents.

On the other hand, we see no reason why at any point of time the set of assumptions momentarily appearing as necessary should not be assembled and, in the form of axioms and fundamental rules, be made the basis of the derivations. *Today's* mathematics, as we know, can be derived only to a small extent from assumptions regarded as intuitive or evident by Brouwer. There is nothing to do but consider the remaining premises for the time being as further hypotheses which perhaps later will prove to be dispensable or replaceable by "intuitive" ones. This is without importance for the inner working of mathematics; witness the fact that the content of the textbooks in the various mathematical fields did not change appreciably in recent years, in spite of the immense influence intuitionism exerted, and justifiably so, upon mathematical thinking.

In agreement with the empiristic conception of science, intuitionism holds that the source of mathematics is the insight which we intuitively comprehend from experience of the external world, but which cannot once and for all be collected in a closed system of axioms.

4. *The Excluded Middle.* Not only did intuitionism bring new life into the discussions on the foundations of mathematics, which seemed to have reached somehow a dead end, but it also, for the first time in centuries, opened up again problems in elementary logic. Some rumors have spread to nonmathematicians—and were accepted with justified suspicion—that the intuitionists deny the validity of the simple rule of the excluded middle. That is, besides the two statements "today is Tuesday" and "today is not Tuesday" this revolutionary theory is supposed to admit a *tertium quid.* The situation here is as follows.

In no case of extramathematical application is the validity of the "tertium exclusum" questioned. Within mathematics, too, it remains absolutely unshaken as long as one deals with *finite* sets. If, for example, "*A*" signifies any precisely defined property of a natural number such that it is possible to determine unambiguously whether a given natural number possesses property *A* or not, then the following alternative holds: among the natural numbers from 1 to ten million there exists a number which has

[8] [*Positivism: A Study in Human Understanding.*]

the property A, or there is no such number (but there is no third possibility).

Brouwer, however, noticed that the problem is not so simple once one treats an "infinite" sequence of numbers (such as all even numbers) instead of a finite set (such as the numbers between 1 and ten million). The proposition, "there is a number with the property A" still signifies in this case the same as before, namely, running through the sequence of even numbers, one will find a number possessing the property A. But if we say there is no such number, that does not mean that running through the sequence of all even numbers we never hit upon a number of the property A, because it makes no sense to speak of an examination of infinitely many numbers. The negative statement is rather an abbreviation for the following much more complicated assertion: by means of the axioms and deductive methods of mathematics it can be *proved* that between the property of being an even number and the property A there is a *contradiction*. This is, we think, the only meaning one can reasonably assign to the statement that there is no number having the property A in the infinite set.

After this explanation it should be clear that between the assertions, there is a number . . ., and, there is no number . . ., in the case of an infinite set of numbers, the relation of contradictory opposites no longer exists. There is no difficulty in imagining that besides the finding of a number and the provability of a contradiction there is still a third possibility, namely, that neither does one find, on running through the numbers, one that has the property A, nor is a contradiction between the definition of the numbers and the property A derivable by means of mathematics. It is only when one makes the additional *assumption* that every mathematical problem is solvable that the extension of the theorem of the excluded middle to all problems dealing with infinite sets becomes justified. The nonmathematician will hardly be inclined to regard such an assumption as "logically necessary" and thus all that seems so objectionable in the intuitionist thesis disappears.

Brouwer, the founder of intuitionist mathematics, has shown that in certain mathematical problems dealing with infinite sets of numbers the elementary rule of the excluded middle is not admissible, without an additional arbitrary assumption. Statements like: there is a number . . . and: there is no number . . ., in this case only seemingly, by virtue of their abbreviated linguistic formulations, have the form of contradictory opposites.

5. *New Logic.* We saw in the discussion of logistic that the theorem of the tertium exclusum is an immediate consequence of the basic formulas of logic if one assumes these according to the conception of classical logic. Therefore, if one wants to have a logic that satisfies the requirements of

intuitionist mathematics and that does not break down on application to infinite sets, it is advisable, instead of admitting "exceptions," to formulate the basic axioms a little differently. This problem was solved for the first time by A. Heyting, but we shall follow the later exposition by A. Kolmogoroff, which points out more clearly the fundamental idea.

Let the letters A, B, C . . . signify, instead of statements, as before, *problems* to be solved. We may think of mathematical problems, e.g., to construct a triangle under given stipulations, or to calculate a number defined in a particular way, say the root of an equation. We shall use the sign for negation in order to signify that the solution of a problem is impossible, i.e., leads to a contradiction. Letters connected by means of a comma—"A,B"—will be taken to mean both problems; those connected by "v"—"$A \vee B$"—to mean at least one of the two problems, A,B. Finally, the sign for implication, "$\rightarrow$", will be used in the sense that "$A \rightarrow B$" means to reduce the solution of B to that of A. Now, for the use of the four signs "v", "$\sim$", "$\rightarrow$", ",", one can prescribe those and only those rules which correspond to the situations in solving problems. For example, the formula "$[A,(A \rightarrow B)] \rightarrow B$" is always true; in words it means that the solution of the problem B is reducible to the solution of the two problems, A and the reduction of B to A. Similarly "$[(A \rightarrow B),(B \rightarrow C)] \rightarrow (A \rightarrow C)$"; in words, if B is reducible to A, and C to B, then the reduction of C to A is reducible to these two problems, namely, to the carrying out of the two reductions. In the ordinary propositional calculus the first of these two formulas would state that if A holds and B follows from A, then B also holds; and the second, if B follows from A, and C from B, then it follows that C follows from A.

Thus in these two examples there is no essential difference between the propositional calculus and the new problem calculus. But while in the former the formula "$\sim A \vee A$" is valid, i.e., always, either non-A or A holds, we have no reason to admit the generality of the theorem: one of the two properties of problem A must be true, either A is solvable or A is recognizable as contradictory. One can see here that with a suitable agreement about symbols there will result an algorithm that agrees on the whole with ordinary truth-function theory, but that does not contain the formula which expresses the theorem of the excluded middle. This argument has the same significance as the acceptance of the logical independence of Euclid's parallel axiom, which led to the construction of the non-Euclidean geometries. We have arrived here at a special form of "non-Aristotelian logic," a form which is also called by the misleading name (since it is reminiscent of "intuitive") "intuitionist logic."

The main application of the intuitionist contribution consists in supplying us with a method of rejecting from all previous mathematical results those in whose derivation the tertium exclusum was used, including, in particular, all those theorems which rest upon so-called "indirect proofs." If, for instance, we are looking for an unknown quantity x and we can prove that the assumption that there is no such x leads to a contradiction, then, according to the new conception, the existence of x is not proved. Brouwer demands a "constructive" proof, i.e., the establishing of a method of calculating the number x. This way of looking at things proves to be fruitful in the study of the basic elements of mathematics, even though it leaves the actual work of mathematicians in most special branches almost unchanged. Mathematicians usually were satisfied with an indirect proof only if a constructive one could not be found. It makes no sense to argue about whether a quantity "really exists" if only an indirect proof can be given for its existence; the word "exist" cannot be defined independently of what one wants to admit as proofs.

The construction of a "problem calculus" in the sense of Heyting and Kolmogoroff yields a model of logic in which the theorem of the excluded middle does not appear among the basic formulas. The study of such a logic widens our insight into the basic elements of mathematics and, in particular, points out the special position of the so-called indirect proofs within mathematics.

6. *Formalism.* It is mainly Hilbert and his followers who have objected to the thesis of the intuitionists since it first became known. According to the ideas of Hilbert and the "formalistic" school led by him, mathematics in the narrower sense is replaceable by a purely mechanical method for deriving formulas, a method which has nothing to do with the meaning or interpretation of the symbols used. Certain aggregates of symbols are assumed as premises; these are the axioms, and from them further groups of signs are derived according to fixed rules and in a purely mechanical manner, i.e., without the use of conclusions drawn from their interpretation; the new groups are then the provable theorems. Thus, the entire content of mathematics is, according to Hilbert, transformable, in principle, into a system of symbolic formulas.

Besides this formal system, however, there is, as Hilbert states, still something else which serves as justification of the system of formulas and is called "metamathematics." It is not clearly stated whether the rules that govern the use of signs in the formal system and describe the methods of deriving new formulas are also considered as part of metamathematics. At any rate, it comprises all arguments that are supposed to lead to the proof of the consistency of the formal system. By a consistency proof is meant the proof that a certain "false" formula, e.g., the formula "$1 = 2$," cannot be derived; for, on the one hand, the appearance of this formula would make the system useless, and on the other hand, from any other "false" formula existing in the system the proposition "$1 = 2$" would follow.

Everything then depends upon how and with what means metamathe-

matics works. It uses meaningful deductions, i.e., it operates with words
and idioms whose meaning is somehow abstracted from linguistic usage.
According to Hilbert's original thesis, metamathematics should apply only
the most elementary and immediately evident logical premises and deductive methods, at any rate only the simplest inferences appearing in the
formal system of symbols under consideration, and should use those only
in a finite number of repetitions. Therefore, the theorem of the excluded
middle too can only be applied to a finite set. The essential idea was that
metamathematics, by using only finite means of elementary logic, should
be in a position to construct and support the structure of formal mathematics, which deals with infinite sets. That is exactly the point where
the questions posed by intuitionism start infiltrating into metamathematics.

One cannot say that the end which Hilbert posited for metamathematics
has been attained even in a single partial area or that there is any hope of
reaching it in the near future. On the contrary, the mathematician Gödel
recently showed that, in principle, in order to furnish a consistency proof
of a formal system one needs means that go beyond what is formalized in
the system. That does not necessarily mean a failure of Hilbert's efforts,
but it shows that metamathematics does not get around the questions
thrown into the discussion by the intuitionists. The opposition between the
formalists and the intuitionists, which was originally so violent and apparently irreconcilable, seems gradually to reduce to this: on the one hand,
Hilbert's formal mathematics comprises more than a formalization of
Brouwer's mathematics could yield, but on the other hand, the metamathematics, which is indispensable in Hilbert's total structure, has to incorpo-
rate essential ideas of Brouwer's intuitionism.

There is one point in which the opponents still seem irreconcilable.
While the formalists—in this they follow mainly Poincaré—regard a consistency proof, once it is given, as a complete and total justification of a
deductive system, the intuitionists, according to their rejection of the general principle of the excluded middle, do not consider consistency as something positive. According to our repeatedly stated view, formal (tautological) systems appear in science only because they can be coördinated with
certain sets of experiences or groups of phenomena, and from this coördination they derive, in the last analysis, their "justification." Inner consistency is certainly a necessary criterion of the usefulness of a system, but
for its applicability, i.e., for the possibility of coördination with the world
of experience, a sufficient "scientific proof" cannot be given at all.

*Dividing mathematics into a formal system, which progresses according
to mechanical rules, and a metamathematics, which is supposed to lead
to the justification of the formal system, does not exclude the difficulties
that intuitionism has pointed out. The coördination between mathematics*

(*its tautological side*) *and reality cannot be reached by a mathematicized doctrine and certainly cannot be settled by a consistency proof.*

7. *Logicism.* With the formalists and the intuitionists one frequently mentions the "logicists" as the third party in the controversy about the foundations of mathematics. Bertrand Russell attempted in his book *Principia Mathematica* (together with A. N. Whitehead), based upon essential preliminary studies by Frege and Peano, to construct completely the basic concepts used in mathematics, starting from their simplest and most plausible elements. The solution of this logical problem, to which the authors apply admirable ingenuity, will never be able to command universal acceptance. For it remains undecided, and depends on each individual, whether such concepts as sequence of numbers, cardinal numbers, etc., are less simple and immediate than successor, one-to-one coördination of elements, etc. The choice of the preferred starting point will always depend upon the experience of the individual. But about the relation to reality, to the world of experience, logicists do not *want* to say anything; they see their goal in the complete exhibition of the tautological relations, i.e., relations that are fixed by definitions and other linguistic rules. In this respect, their merits are indisputable; we mention here only one example.

In almost all branches of mathematics there appears a type of argument that always has been regarded as a "principle," characteristic of mathematics, and at times considered an inscrutable mystery: the process of so-called *complete induction.* If we divided 7 by 3, then the first digit after the decimal point of the quotient is a 3, and the remainder is 1. Furthermore it is easy to prove that in division by 3 a decimal which yields the remainder 1 must always be followed by another 3 with the remainder 1. From this one customarily "concludes" that all the "infinitely many" decimal digits of the quotient are 3's. This is the method of mathematical induction, and with respect to it the question was raised (e.g., by H. Poincaré), how is it possible to draw from such a small and in any case finite number of inferences an infinite number of conclusions (namely about all infinitely many decimals). The fact is that nothing at all is concluded here. The sentence "All decimal digits are 3's" is only a different linguistic expression for "The first decimal digit is 3, and as often as there appears a 3, it is followed by another 3." The word "all," in this context, i.e., applied to the infinite sequence of digits in the decimal fraction, has no other meaning than that determined by the concept of succession (by which the sequence of natural numbers is defined). There are other problems in mathematics in which the word "all," referring to infinite sets of a different kind, has another meaning and where the method of induction is not applicable at all. This situation is fully cleared up in Whitehead and Russell's foundation of mathematics. While the ordinary textbooks of mathematics have the purpose of developing chains of mathematical deductions ascending from simpler to more and more complicated forms, Russell drives the examination of the tautologies in the opposite direction, toward the origin.

We return now to what was said at the start of this chapter, that mathematics as a whole, like any other science, has a tautological and an empirical side; it differs from other sciences in that here the formal side is much more essential and decisive than anywhere else. Thus it becomes understandable that mathematics has often been identified completely with its tautological part, as witnessed by Goethe's well-known utterance,

Mathematics has the completely false reputation of yielding infallible conclusions. Its infallibility is nothing but identity. Two times two is not four, but it is just two times two, and that is what we call four for short. But four is nothing new at all. And thus it goes on and on in its conclusions, except that in the higher formulas the identity fades out of sight.

To this we have to say that four is not only two times two, but also three plus one, and fifteen minus eleven, and the cube root of 64, and so on. It is the task of mathematics in the narrower sense to describe the relations of these various "identities." But there is also an empirical side to mathematical doctrines, where there is no more talk about the "infallibility of conclusions," and which must not be neglected in an epistemological investigation.

The logicistic foundation of mathematics deals with an analysis of its tautological or deductive part, by attempting to reduce the basic mathematical concepts and methods to the simplest and most plausible elements. There is no reference to the relation with reality; the connection with the world of experience is hidden in the choice of the basic elements. Our final conclusion is this:

None of the three forms of the foundation of mathematics, the intuitionist, the formalistic, or the logistic, is capable of completely rationalizing the relation between tautological systems and (extramathematical) experiences, which is its very purpose, i.e., to make this relation a part of the mathematical system itself.

PART XII

The Mathematical Way of Thinking

An Eristic Controversy

THE famous address of J. J. Sylvester to the British Association in 1869 on the nature of mathematics was a genteel polemic fired at Thomas Henry Huxley. Huxley, himself a formidable controversialist, had contributed to *Macmillan's Magazine* an article on the sad shortcomings of the scientific training offered by contemporary British education, and to the celebrated *Fortnightly Review* a decapitating assault on the positivistic philosophy of Auguste Comte.[1] The Macmillan piece, enlivened by one of Huxley's stinging attacks on the clergy,[2] stressed the importance of scientific training—"in virtue of which it cannot be replaced by any other discipline whatsoever"—for bringing the mind "directly into contact with fact, and practicing the intellect in the completest form of induction." Mathematics, on the other hand (said Huxley), is almost purely deductive and cannot discipline the mind in this way. "The mathematician starts with a few propositions, the proof of which is so obvious that they are called self-evident, and the rest of his work consists of subtle deductions from them."

In the *Fortnightly Review*, Huxley quoted Comte: "C'est donc par l'étude des mathématiques, *et seulement* par elle, que l'on peut se faire une idée juste et approfondie de ce que c'est qu'une science." This placing of mathematics at the summit of knowledge, pointing to its method as the model and goal of all scientific enquiry, was the central principle of Comte's philosophy. To Huxley the principle implied an outrageous affront to the experimental method. He felt obliged to make this outrageous reply: "That is to say, the only study which can confer 'a just and comprehensive idea of what is meant by science,' and at the same time, furnish an exact conception of the general method of scientific investigation, is that which knows nothing of observation, nothing of experiment, nothing of induction, nothing of causation." Having thus disposed of mathematics, a subject which irritated him because he knew so little about it, he moved briskly to another front. The statement, however, created some excitement, and no one was better equipped than Sylvester "to speak for mathematics" and demonstrate the silliness of Huxley's eloquent tirade.

The address is typical of Sylvester.[3] It is learned, allusive, flowery, dis-

---

[1] "Scientific Education: Notes of an After-dinner Speech," *Macmillan's Magazine*, Vol. XX, 1869; *Fortnightly Review*, Vol. II, N.S. 5.

[2] For example: The clergy are "divisible in three sections: an immense body who are ignorant and speak out; a small proportion who know and are silent; and a minute minority who know and speak according to their knowledge."

[3] For a biographical essay on this eminent mathematician see p. 341.

cursive, amusing and long-winded. It is as richly stocked with excellent things as with irrelevancies. The crowning irrelevancy was the publication of the lecture, together with profuse annotations and ensuing correspondence provoked by an incidental reference to Kant's doctrine of space and time, as an appendix to Sylvester's *Laws of Verse*. The connection between the principles of sound versification and Huxley's, not to mention Kant's, critique is not obvious. Sylvester was never deterred by such considerations. It was his practice to give the public the benefit of his views on the subject uppermost in his mind, regardless of what topic was announced on the lecture program or title page. The story is told that he was scheduled one evening, while he was living in Baltimore, to give a reading of a 400-line poem he had written, all the lines rhyming with the name Rosalind. A large audience gathered to witness this improbable feat. "Professor Sylvester, as usual, had a number of footnotes appended to his production; and he announced that in order to save interruption in reading the poem itself, he would first read the footnotes. The reading of the footnotes suggested various digressions to his imagination; an hour had passed, still no poem; an hour and a half passed and the striking of the clock or the unrest of his audience reminded him of the promised poem. He was astonished to find how time had passed, excused all who had engagements, and proceeded to read the Rosalind poem." [4]

Huxley never replied to the British Association lecture. Yet it is doubtful that Sylvester disposed of the questions it raised as completely as he imagined. The extent to which abstract mathematical concepts are suggested by experience remains a topic of lively debate. Sylvester touches upon the problem, but in less than a paragraph rushes off in another direction. The effectiveness of the lecture lies not so much in its rebuttal of Huxley, as in its kaleidoscopic survey of nineteenth-century mathematical thought, and its display of Sylvester's amazing erudition and imagination.

[4] Alexander Macfarlane, *Lectures on Ten British Mathematicians of the Nineteenth Century*; New York, 1916, pp. 117–18. E. T Bell tells the same story in his essay on Cayley and Sylvester: see p. 341.

Having laid down fundamental principles of the wisdom of the Latins so far as they are found in language, mathematics and optics, I now wish to unfold the principles of experimental science, since without experience nothing can be sufficiently known. For there are two modes of acquiring knowledge, namely, by reasoning and experience. Reasoning draws a conclusion and makes us grant the conclusion, but does not make the conclusion certain, nor does it remove doubt so that the mind may rest on the intuition of truth, unless the mind discovers it by the path of experience; since many have the arguments relating to what can be known, but because they lack experience they neglect the arguments, and neither avoid what is harmful nor follow what is good. For if a man who has never seen a fire should prove by adequate reasoning that fire burns and injures things and destroys them, his mind would not be satisfied thereby, nor would he avoid fire, until he placed his hand or some combustible substance in the fire, so that he might prove by experience that which reasoning taught. But when he has actual experience of combustion his mind is made certain and rests in the full light of truth. Therefore, reasoning does not suffice, but experience does. —Roger Bacon

1 The Study That Knows Nothing of Observation

By JAMES JOSEPH SYLVESTER

EXCERPT OF ADDRESS TO BRITISH ASSOCIATION, 1869

. . . IT IS said of a great party leader and orator in the House of Lords that, when lately requested to make a speech at some religious or charitable (at all events a non-political) meeting, he declined to do so on the ground that he could not speak unless he saw an adversary before him—somebody to attack or reply to. In obedience to a somewhat similar combative instinct, I set to myself the task of considering certain recent utterances of a most distinguished member of this Assocation, one whom I no less respect for his honesty and public spirit than I admire for his genius and eloquence,[1] but from whose opinions on a subject which he has not studied I feel constrained to differ. Goethe has said—

> "Verständige Leute kannst du irren sehn
> In Sachen, nämlich, die sie nicht verstehn."

Understanding people you may see erring—in those things, to wit, which they do not understand.

I have no doubt that had my distinguished friend, the probable President-elect of the next Meeting of the Association, applied his uncommon powers of reasoning, induction, comparison, observation, and invention to

---

[1] Although no great lecture-goer, I have heard three lectures in my life which have left a lasting impression as masterpieces on my memory—Clifford on Mind, Huxley on Chalk, Dumas on Faraday.

the study of mathematical science, he would have become as great a mathematician as he is now a biologist; indeed he has given public evidence of his ability to grapple with the practical side of certain mathematical questions; but he has not made a study of mathematical science as such, and the eminence of his position and the weight justly attaching to his name render it only the more imperative that any assertions proceeding from such a quarter, which may appear to me erroneous, or so expressed as to be conducive to error, should not remain unchallenged or be passed over in silence.[2]

He says "mathematical training is almost purely deductive. The mathematician starts with a few simple propositions, the proof of which is so obvious that they are called self-evident, and the rest of his work consists of subtle deductions from them. The teaching of languages, at any rate as ordinarily practised, is of the same general nature—authority and tradition furnish the data, and the mental operations are deductive." It would seem from the above somewhat singularly juxtaposed paragraphs that, according to Prof. Huxley, the business of the mathematical student is from a limited number of propositions (bottled up and labelled ready for future use) to deduce any required result by a process of the same general nature as a student of language employs in declining and conjugating his nouns and verbs—that to make out a mathematical proposition and to construe or parse a sentence are equivalent or identical mental operations. Such an opinion scarcely seems to need serious refutation. The passage is taken from an article in *Macmillan's Magazine* for June last, entitled "Scientific Education—Notes of an After-dinner Speech," and I cannot but think would have been couched in more guarded terms by my distinguished friend had his speech been made *before* dinner instead of *after*.

The notion that mathematical truth rests on the narrow basis of a limited number of elementary propositions from which all others are to be derived by a process of logical inference and verbal deduction, has been stated still more strongly and explicitly by the same eminent writer in an article of even date with the preceding in the *Fortnightly Review*, where we are told that "Mathematics is that study which knows nothing of observation, nothing of experiment, nothing of induction, nothing of causation." I think no statement could have been made more opposite to the undoubted facts of the case, that mathematical analysis is constantly invoking the aid of new principles, new ideas, and new methods, not capable of being defined by any form of words, but springing direct from the inherent powers and activity of the human mind, and from continually renewed introspection of that inner world of thought of which the phenomena are as varied and require as close attention to discern as those of the outer

[2] In his *éloge* of Daubenton, Cuvier remarks, "Les savants jugent toujours comme le vulgaire les ouvrages qui ne sont pas de leur genre."

physical world (to which the inner one in each individual man may, I think, be conceived to stand in somewhat the same general relation of correspondence as a shadow to the object from which it is projected, or as the hollow palm of one hand to the closed fist which it grasps of the other), that it is unceasingly calling forth the faculties of observation and comparison, that one of its principal weapons is induction, that it has frequent recourse to experimental trial and verification, and that it affords a boundless scope for the exercise of the highest efforts of imagination and invention.

Lagrange, than whom no greater authority could be quoted, has expressed emphatically his belief in the importance to the mathematician of the faculty of observation; Gauss has called mathematics a science of the eye, and in conformity with this view always paid the most punctilious attention to preserve his text free from typographical errors; the ever to be lamented Riemann has written a thesis to show that the basis of our conception of space is purely empirical, and our knowledge of its laws the result of observation, that other kinds of space might be conceived to exist subject to laws different from those which govern the actual space in which we are immersed, and that there is no evidence of these laws extending to the ultimate infinitesimal elements of which space is composed. Like his master Gauss, Riemann refuses to accept Kant's doctrine of space and time being forms of intuition, and regards them as possessed of physical and objective reality. I may mention that Baron Sartorius von Waltershausen (a member of this Association) in his biography of Gauss ("Gauss zu Gedächtniss"), published shortly after his death, relates that this great man used to say that he had laid aside several questions which he had treated analytically, and hoped to apply to them geometrical methods in a future state of existence, when his conceptions of space should have become amplified and extended, for as we can conceive beings (like infinitely attenuated bookworms [3] in an infinitely thin sheet of paper) which possess only the notion of space of two dimensions, so we may imagine beings capable of realising space of four or a greater number of dimensions.[4] Our

[3] I have read or been told that eye of observer has never lighted on these depredators, living or dead. Nature has gifted me with eyes of exceptional microscopic power, and I can speak with some assurance of having repeatedly seen the creature wriggling on the learned page. On approaching it with breath or finger-nail it stiffens out into the semblance of a streak of dirt, and so eludes detection.

[4] It is well known to those who have gone into these views that the laws of motion accepted as a fact suffice to prove in a general way that the space we live in is a flat or level space (a "homaloid"), our existence therein being assimilable to the life of the bookworm in an *unrumpled page*: but what if the page should be undergoing a process of gradual bending into a curved form? Mr. W. K. Clifford has indulged in some remarkable speculations as to the possibility of our being able to infer, from certain unexplained phenomena of light and magnetism, the fact of our level space of three dimensions being in the act of undergoing in space of four dimensions (space as inconceivable to us as our space to the supposititious bookworm) a distortion analogous to the rumpling of the page to which that creature's powers of direct perception have been postulated to be limited.

Cayley, the central luminary, the Darwin of the English school of mathematicians, started and elaborated at an early age, and with happy consequences, the same bold hypothesis.

Most, if not all, of the great ideas of modern mathematics have had
their origin in observation. Take, for instance, the arithmetical theory of
forms, of which the foundation was laid in the diophantine theorems of
Fermat, left without proof by their author, which resisted all the efforts of
the myriad-minded Euler to reduce to demonstration, and only yielded up
their cause of being when turned over in the blowpipe flame of Gauss's
transcendent genius; or the doctrine of double periodicity, which resulted
from the observation by Jacobi of a purely analytical fact of transformation; or Legendre's law of reciprocity; or Sturm's theorem about the roots
of equations, which, as he informed me with his own lips, stared him in
the face in the midst of some mechanical investigations connected with the
motion of compound pendulums; or Huyghens' method of continued fractions, characterized by Lagrange as one of the principal discoveries of
"that great mathematician, and to which he appears to have been led by
the construction of his Planetary Automaton"; or the New Algebra, speaking of which one of my predecessors (Mr. Spottiswoode) has said, not
without just reason and authority, from this Chair, "that it reaches out and
indissolubly connects itself each year with fresh branches of mathematics,
that the theory of equations has almost become new through it, algebraic
geometry transfigured in its light, that the calculus of variations, molecular
physics, and mechanics" (he might, if speaking at the present moment, go
on to add the theory of elasticity and the highest developments of the
integral calculus) "have all felt its influence."

Now this gigantic outcome of modern analytical thought, itself, too,
only the precursor and progenitor of a future still more heaven-reaching
theory, which will comprise a complete study of the interoperation, the
actions and reactions, of algebraic forms (Analytical Morphology in its
absolute sense), how did this originate? In the accidental observation by
Eisenstein, some score or more years ago, of a single invariant (the Quad-
rinvariant of a Binary Quartic) which he met with in the course of certain
researches just as accidentally and unexpectedly as M. Du Chaillu might
meet a Gorilla in the country of the Fantees, or any one of us in London
a White Polar Bear escaped from the Zoological Gardens. Fortunately he
pounced down upon his prey and preserved it for the contemplation and
study of future mathematicians. It occupies only part of a page in his collected posthumous works. This single result of observation (as well entitled to be so called as the discovery of Globigerinæ in chalk or of the
Confoco-ellipsoidal structure of the shells of the Foraminifera), which remained unproductive in the hands of its distinguished author, has served
to set in motion a train of thought and to propagate an impulse which

have led to a complete revolution in the whole aspect of modern analysis, and whose consequences will continue to be felt until Mathematics are forgotten and British Associations meet no more.

I might go on, were it necessary, piling instance upon instance to prove the paramount importance of the faculty of observation to the process of mathematical discovery.[5] Were it not unbecoming to dilate on one's personal experience, I could tell a story of almost romantic interest about my own latest researches in a field where Geometry, Algebra, and the Theory of Numbers melt in a surprising manner into one another, like sunset tints or the colours of the dying dolphin, "the last still loveliest" (a sketch of which has just appeared in the *Proceedings of the London Mathematical Society* [6]), which would very strikingly illustrate how much observation, divination, induction, experimental trial, and verifica ion, causa on, too (if that means, as I suppose it must, mounting from phenomena to their reasons or causes of being), have to do with the work of the mathematician. In the face of these facts, which every analyst in this room or out of it can vouch for out of its own knowledge and personal experience, how can it be maintained, in the words of Professor Huxley, who, in this instance, is speaking of the sciences as they are in themselves and without any reference to scholastic discipline, that Mathematics "is that study which knows nothing of observation, nothing of induction, nothing of experiment, nothing of causation"?

I, of course, am not so absurd as to maintain that the habit of observation of external nature will be best or in any degree cultivated by the study of mathematics, at all events as that study is at present conducted; and no one can desire more earnestly than myself to see natural and experimental science introduced into our schools as a primary and indispensable branch of education: I think that that study and mathematical culture should go on hand in hand together, and that they would greatly influence each other for their mutual good. I should rejoice to see mathematics taught with that life and animation which the presence and example of her young and buoyant sister could not fail to impart, short roads preferred to long ones, Euclid honourably shelved or buried "deeper than did ever plummet sound" out of the schoolboy's reach, morphology introduced into the elements of Algebra—projection, correlation, and motion accepted as aids to geometry—the mind of the student quickened and elevated and his faith awakened by early initiation into the ruling ideas of polarity, continuity,

[5] Newton's Rule was to all appearance, and according to the more received opinion, obtained inductively by its author. My own reduction of Euler's problem of the Virgins (or rather one slightly more general than this) to the form of a question (or, to speak more exactly, a set of questions) in simple partitions was, strange to say, first obtained by myself inductively, the result communicated to Prof. Cayley, and proved subsequently by each of us independently, and by perfectly distinct methods.

[6] Under the title of "Outline Trace of the Theory of Reducible Cyclodes."

infinity, and familiarization with the doctrine of the imaginary and in-conceivable.

It is this living interest in the subject which is so wanting in our tradi-tional and mediaeval modes of teaching. In France, Germany, and Italy, everywhere where I have been on the Continent, mind acts direct on mind in a manner unknown to the frozen formality of our academic institutions; schools of thought and centres of real intellectual cooperation exist; the relation of master and pupil is acknowledged as a spiritual and a lifelong tie, connecting successive generations of great thinkers with each other in an unbroken chain, just in the same way as we read, in the catalogue of our French Exhibition, or of the Salon at Paris, of this man or that being the pupil of one great painter or sculptor and the master of another. When followed out in this spirit, there is no study in the world which brings into more harmonious action all the faculties of the mind than the one of which I stand here as the humble representative, there is none other which prepares so many agreeable surprises for its followers, more wonderful than the changes in the transformation-scene of a pantomime, or, like this, seems to raise them, by successive steps of initiation, to higher and higher states of conscious intellectual being.

This accounts, I believe, for the extraordinary longevity of all the greatest masters of the Analytical art, the Du Majores of the mathematical Pantheon. Leibnitz lived to the age of 70; Euler to 76; Lagrange to 77; Laplace to 78; Gauss to 78; Plato, the supposed inventor of the conic sec-tions, who made mathematics his study and delight, who called them the handles or aids to philosophy, the medicine of the soul, and is said never to have let a day go by without inventing some new theorems, lived to 82; Newton, the crown and glory of his race, to 85; Archimedes, the nearest akin, probably, to Newton in genius, was 75, and might have lived on to be 100, for aught we can guess to the contrary, when he was slain by the impatient and ill-mannered sergeant, sent to bring him before the Roman general, in the full vigour of his faculties, and in the very act of working out a problem; Pythagoras, in whose school, I believe, the word mathe-matician (used, however, in a somewhat wider than its present sense) originated, the second founder of geometry, the inventor of the matchless theorem which goes by his name, the precognizer of the undoubtedly mis-called Copernican theory, the discoverer of the regular solids and the musical canon, who stands at the very apex of this pyramid of fame, (if we may credit the tradition) after spending 22 years studying in Egypt, and 12 in Babylon, opened school when 56 or 57 years old in Magna Græcia, married a young wife when past 60, and died, carrying on his work with energy unspent to the last, at the age of 99. The mathematician lives long and lives young; the wings of his soul do not early drop off, nor do its pores become clogged with the earthy particles blown from the dusty highways of vulgar life.

Some people have been found to regard all mathematics, after the 47th proposition of Euclid, as a sort of morbid secretion, to be compared only with the pearl said to be generated in the diseased oyster, or, as I have heard it described, "une excroissance maladive de l'esprit humain." Others find its justification, its "raison d'être," in its being either the torch-bearer leading the way, or the handmaiden holding up the train of Physical Science; and a very clever writer in a recent magazine article, expresses his doubts whether it is, in itself, a more serious pursuit, or more worthy of interesting an intellectual human being, than the study of chess problems or Chinese puzzles. What is it to us, they say, if the three angles of a triangle are equal to two right angles, or if every even number is, or may be, the sum of two primes, or if every equation of an odd degree must have a real root. How dull, stale, flat, and unprofitable are such and such like announcements! Much more interesting to read an account of a marriage in high life, or the details of an international boat-race. But this is like judging of architecture from being shown some of the brick and mortar, or even a quarried stone of a public building, or of painting from the colours mixed on the palette, or of music by listening to the thin and screechy sounds produced by a bow passed haphazard over the strings of a violin. The world of ideas which it discloses or illuminates, the contemplation of divine beauty and order which it induces, the harmonious connexion of its parts, the infinite hierarchy and absolute evidence of the truths with which it is concerned, these, and such like, are the surest grounds of the title of mathematics to human regard, and would remain unimpeached and unimpaired were the plan of the universe unrolled like a map at our feet, and the mind of man qualified to take in the whole scheme of creation at a glance.

In conformity with general usage, I have used the word mathematics in the plural; but I think it would be desirable that this form of word should be reserved for the applications of the science, and that we should use mathematic in the singular number to denote the science itself, in the same way as we speak of logic, rhetoric, or (own sister to algebra [7]) music. Time was when all the parts of the subject were dissevered, when algebra, geometry, and arithmetic either lived apart or kept up cold relations of acquaintance confined to occasional calls upon one another; but that is now at an end; they are drawn together and are constantly becoming more and more intimately related and connected by a thousand fresh ties, and

[7] I have elsewhere (in my "Trilogy" published in the *Philosophical Transactions*) referred to the close connexion between these two cultures, not merely as having Arithmetic for their common parent, but as similar in their habits and affections. I have called "Music the Algebra of sense, Algebra, the Music of the reason; Music the dream, Algebra the waking life,—the soul of each the same!"

we may confidently look forward to a time when they shall form but one body with one soul. Geometry formerly was the chief borrower from arithmetic and algebra, but it has since repaid its obligations with abundant usury; and if I were asked to name, in one word, the pole-star round which the mathematical firmament revolves, the central idea which pervades as a hidden spirit the whole corpus of mathematical doctrine, I should point to Continuity as contained in our notions of space, and say, it is this, it is this! Space is the *Grand Continuum* from which, as from an inexhaustible reservoir, all the fertilizing ideas of modern analysis are derived; and as Brindley, the engineer, once allowed before a parliamentary committee that, in his opinion, rivers were made to feed navigable canals, I feel almost tempted to say that one principal reason for the existence of space, or at least one principal function which it discharges, is that of feeding mathematical invention. Everybody knows what a wonderful influence geometry has exercised in the hands of Cauchy, Puiseux, Riemann, and his followers Clebsch, Gordan, and others, over the very form and presentment of the modern calculus, and how it has come to pass that the tracing of curves, which was once to be regarded as a puerile amusement, or at best useful only to the architect or decorator, is now entitled to take rank as a high philosophical exercise, inasmuch as every new curve or surface, or other circumscription of space is capable of being regarded as the embodiment of some specific organized system of continuity.[8]

The early study of Euclid made me a hater of Geometry, which I hope may plead my excuse if I have shocked the opinions of any in this room (and I know there are some who rank Euclid as second in sacredness to the Bible alone, and as one of the advanced outposts of the British Constitution) by the tone in which I have previously alluded to it as a schoolbook; and yet, in spite of this repugnance, which had become a second nature in me, whenever I went far enough into any mathematical question, I found I touched, at last, a geometrical bottom: so it was, I may instance, in the purely arithmetical theory of partitions; so, again, in one of my more recent studies, the purely algebraical question of the invariantive criteria of the nature of the roots of an equation of the fifth degree: the first inquiry landed me in a new theory of polyhedra; the latter found its perfect and only possible complete solution in the construction of a surface of the ninth order and the subdivision of its infinite content into three distinct natural regions.

Having thus expressed myself at much greater length than I originally

---

[8] M. Camille Jordan's application of Dr. Salmon's Eikosi-heptagram to Abelian functions is one of the most recent instances of this reverse action of geometry on analysis. Mr. Crofton's admirable apparatus of a reticulation with infinitely fine meshes rotated successively through indefinitely small angles, which he applies to obtaining whole families of definite integrals, is another equally striking example of the same phenomenon.

intended on the subject, which, as standing first on the muster-roll of the Assocation, and as having been so recently and repeatedly arraigned before the bar of public opinion, is entitled to be heard in its defence (if anywhere) in this place,—having endeavoured to show what it is not, what it is, and what it is probably destined to become, I feel that I must enough and more than enough have trespassed on your forbearance, and shall proceed with the regular business of the Meeting.

COMMENTARY ON
CHARLES SANDERS PEIRCE

C HARLES SANDERS PEIRCE (1839–1914), the founder of pragmatism, was an able scientist, a creative philosopher, a great logician.
Today he is recognized as one of the most gifted and influential thinkers
this country has produced—an esteem in marked contrast to the neglect
he suffered during his life. Peirce's failure to win recognition was due
partly to outer circumstance, partly to his own prickly personality, partly
to a brilliance and breadth of vision that confused and frightened lesser
men. The present tendency is perhaps to overpraise him, to read into him
anticipations of the central ideas of some of the best-known philosophers
and logicians of the last quarter century.[1] But there can be no doubt that
he achieved much of lasting value. Whitehead judged him shrewdly: "The
essence of his thought was originality in every subject that he taught."

Peirce was born in Cambridge, Massachusetts, the second son of Benjamin Peirce, a Harvard professor and the foremost American mathematician of his time.[2] The elder Peirce, a forceful man, was bent on turning
his son into a thinking machine. He closely supervised every step of the
boy's education, bearing down hard on his evident mathematical talents,
but also putting him through somewhat unconventional exercises in the
"art of concentration." "From time to time they would play rapid games
of double dummy together, from ten in the evening until sunrise, the
father sharply criticizing every error." The results of this hair-raising
regimen were mixed. Charles was a rather poor student at Harvard; and
it is clear that by driving him, by making excessive demands, his father
aggravated, if indeed he did not instill, the unfortunate traits which contributed to Peirce's later misfortunes.[3] On the other hand Benjamin Peirce

[1] See Ernest Nagel, "Charles Peirce's Guesses at the Riddle," *The Journal of Philosophy*, Vols. XXX (July, 1933), XXI (1934), XXXIII (1936).
[2] The main source of the data for this sketch is the authoritative article on Peirce
by Paul Weiss in the *Dictionary of American Biography*, New York, 1928–1944. Many
of the unkeyed quotations are from this article. Other sources include Thomas A.
Goudge, *The Thought of C. S. Peirce*, Toronto, 1950; *Studies in the Philosophy of
Charles Sanders Peirce*, edited by Philip P. Wiener and Frederic H. Young, Cambridge (Mass.), 1952; *The Philosophy of Peirce, Selected Writings*, edited by Justus
Buchler, New York, 1950; *Chance, Love and Logic*, edited by Morris R. Cohen, New
York, 1949; W. B. Gallie, *Peirce and Pragmatism*, Penguin Books, Baltimore, 1952;
Ernest Nagel, "Charles Peirce's Guesses at the Riddle," *Journal of Philosophy*, July
16, 1933, pp. 365–386 (continued in Vols. XXXI and XXXIII); Ernest Nagel,
"Charles S. Peirce, Pioneer of Modern Empiricism," *Philosophy of Science*, Vol. 7,
no. 1, January, 1940, pp. 69–80. The most important collection of Peirce's writings,
a work which has contributed significantly to the revival of interest in his ideas, is
The Collected Papers of Charles Sanders Peirce, Vols. I–VI, edited by Charles Hartshorne and Paul Weiss, Cambridge (Mass.), 1931–35.
[3] Gallie, *op. cit.*, p. 35.

was a talented man of wide interests, and the training he gave his son
in experimental science, mathematics, logic and philosophy was invaluable.
"He educated me," Charles could justly say, "and if I do anything it will
be his work."

It was his father's wish that he become a scientist. Peirce was more inclined to pursue philosophy and found a way of following both profes-
sions. In 1861 he joined the United States Coast Survey, where he remained in various posts for thirty years. He served as a computer for the
nautical almanac, made pendulum investigations, was put in charge of
gravity research and wrote a number of solid scientific papers. Some of
these appeared in *Photometric Researches* (1878), the only book of his
published in his lifetime, and one that earned him "international recognition among contemporary astrophysicists." [4] The government post, though
obviously not a sinecure, left him with enough time to teach and to engage
in private researches in science, philosophy and logic. During the 1860s
he gave lecture courses at Harvard in the philosophy of science and logic;
for five years he taught logic at Johns Hopkins. In 1867 Peirce read before
the American Academy of Arts and Sciences a short paper on the work
of George Boole. This marked the beginning of a series of writings which
established him as "the greatest formal logician of his time, and the most
important single force in the period from Boole to Ernst Schröder." [5] His
technical logical papers are today regarded as "primarily of historical interest," but the weight of his contribution to the advancement of this
science is beyond dispute.[6] His main mathematical studies, published and
unpublished, dealt with foundation problems, associative algebra, the
theory of aggregates, transfinite arithmetic (in which he "anticipated or
ran parallel with" the work of Richard Dedekind and Georg Cantor),
analysis situs, and related topics.[7] He was one of the first proponents of
the frequency interpretation of probability. Though his writings in pure
mathematics were not extensive, they were of characteristic originality,
always suggestive, often prophetic; and his treatment of the logical and
philosophical aspects of mathematics was of high quality.

[4] Frederic Harold Young, "Charles Sanders Peirce: 1839–1914," in Wiener and
Young, *op. cit.*, p. 272.
[5] "He radically modified, extended and transformed the Boolean algebra, making
it applicable to propositions, relations, probability and arithmetic. Practically single-
handed, following De Morgan, Peirce laid the foundations of the logic of relations,
the instrument for the logical analysis of mathematics. He invented the copula of inclusion, the most important symbol in the logic of classes, two new logical algebras,
two new systems of logical graphs, discovered the link between the logic of classes
and the logic of propositions, was the first to give the fundamental principle for the
logical development of mathematics, and made exceedingly important contributions
to probability theory, induction, and the logic of scientific methodology." *Dictionary
of American Biography, loc. cit.*
[6] Ernest Nagel, *Philosophy of Science, op. cit.*, p. 72.
[7] Benjamin Peirce, attracted to associative algebra by his son's work in that field,
wrote the pioneer text, *Linear Associative Algebra*, which opens with the famous sentence: "Mathematics is the science which draws necessary conclusions."

Pragmatism is said to have had its origin in the discussions of a fortnightly "Metaphysical Club" (a name chosen, according to Peirce, "to alienate all whom it would alienate"), founded in Cambridge in the seventies, whose members included Oliver Wendell Holmes (the jurist), John Fiske and Francis E. Abbott. Others who played a significant part in the evolution of the concept were the mathematician and philosopher Chauncey Wright and, of course, William James, Peirce's "lifelong friend and benefactor." James was unusually persuasive in publicizing the pragmatic view but he gave it a "characteristic twist" which split it away from Peirce's doctrines.[8] Peirce offered the first definition of pragmatism in an article published in *Popular Science Monthly* (January 1878) under the title "How to Make Our Ideas Clear." To achieve this laudable goal he suggested that we "consider what effects, which might conceivably have practical bearings, we conceive the object of our conception to have. Then our conception of these effects is the whole of our conception of the object." The definition itself is not a promising beginning to the task of intellectual clarification, but while philosophers grumbled over and even ridiculed its awkwardness they obviously understood what it meant. It was a maxim proposed, in Ernest Nagel's words, as a "guiding principle of analysis. It was offered to philosophers in order to bring to an end disputes which no observation of facts could settle because they involved terms with no definite meaning." It was intended to "eliminate specious problems, and unmask mystification and obscurantism hiding under the cloak of apparent profundity. . . . Above all it pointed to the fact that the 'meaning' of terms and statements relevant in inquiry consist in their being *used* in determinate and *overt* ways." [9] Peirce's pragmatism is closely related to his "critical common-sensism" and "contrite fallibilism," two of his often-used expressions.[10] By "common-sensism" he meant that on a great many matters we have no sensible alternative to adopting vague but "indubitable beliefs" which "rest on the everyday experience of many generations of multitudinous populations." Examples of such beliefs are that fire burns, that incest is undesirable, that there is an element of order in the universe.[11] To be sure, these "instinctive beliefs" may change in time, or may in certain instances be proved false; but in the main it is insincere to pretend we can disregard them, start with a Cartesian clean

---

[8] James urged that pragmatism justified, in areas where proof was impossible, moral and religious issues, for example, the embracing of "faith" or the adoption of "unreasoned" beliefs if they were conducive to inward happiness or "beneficial" in other ways. Peirce described this doctrine of the "Will-to-Believe" as "suicidal." Gallie, *op. cit.*, pp. 25 *et seq.*

[9] Ernest Nagel, *Philosophy of Science, op. cit.*, p. 73.

[10] See, for example, Nagel, *op. cit.*, pp. 77–79; Gallie, *op. cit.*, pp. 106 *et seq.*; Roderick M. Chisholm, "Fallibilism and Belief," in Wiener and Young, *op. cit.*, pp. 93–110; Arthur F. S. Mullyan, "Some Implications of Common-Sensism," in Wiener and Young, *op. cit.*, pp. 111–120.

[11] See Thomas A. Goudge, *op. cit.*, pp. 16–17.

slate and make any real advances in knowledge.[12] Fallibilism is the twin to the tenet of common-sensism. All beliefs and all conclusions, however arrived at, are subject to error. The methods of science are more useful than old wives' gossip for achieving "stable beliefs and reliable conclusions," but even science offers no access to "perfect certitude or exactitude. We can never be absolutely sure of anything, nor can we with any probability ascertain the exact value of any measure or general ratio. This is my conclusion after many years study of the logic of science." [13]

I can do no more than mention Peirce's other labors in philosophy, which include a fairly explicit formulation of a cosmology, social theories of reality and logic, papers on epistemology and numerous writings on his difficult and obscure but valuable "theory of signs." Peirce's interest in logic, it should be observed, was a consequence of his initial concern with philosophical problems. But it soon came about that "he saw philosophy and other subjects almost entirely from a logic perspective." His importance in the history of thought, says Ernest Nagel, is due not only to his contributions to logic and mathematics, but to the stimulating effect on the study of scientific method of a mind at once philosophical and disciplined by first-hand knowledge of the sciences.

In ordinary social relations Peirce must have been hard to stomach. He was emotional, quarrelsome, vain, arrogant and snobbish. He was careless in money matters, gullible, impractical and slovenly in appearance. "I yield to no one," said James, "in admiration of his genius, but he is paradoxical and unsociable of intellect, and hates to *make connection* with anyone he is with." [14] He was erratic in the usual tiresome fashion—forgetting appointments, misplacing things, and so on; and a "queer being," as James described him, in more interesting ways: for example, he was not merely ambidextrous but could "write a question with one hand and the answer simultaneously with the other."

It is not surprising that in spite of strenuous efforts on his part and the enthusiastic support of influential admirers he could never get a permanent teaching job. His reputation as a heavy drinker (now known to be

[12] Peirce wrote: "A man may say 'I will content myself with common sense.' I, for one, am with him there, in the main. I shall show why I do not think there can be any *direct* profit in going behind common sense—meaning by common sense those ideas and beliefs that man's situation absolutely forces upon him. . . . I agree, for example, that it is better to recognize that some things are red and some others blue, in the teeth of what optical philosophers say, that it is merely that some things are resonant to shorter ether waves and some to longer ones." Hartshorne and Weiss, *op. cit.*, Vol. I, paragraph 129.

[13] Peirce's writings, Hartshorne and Weiss, *op. cit.*, Vol. I, paragraph 147. To the objection that the proposition "There is no absolute certainty" is itself inconsistent Peirce answered "If I must make any exception, let it be that the assertion that every assertion but this is fallible, is the only one that is absolutely infallible." Hartshorne and Weiss, *op. cit.*, Vol. 2, paragraph 75.

[14] These and other quotations from James are from Ralph Barton Perry, *The Thought and Character of William James*, 2 vols., Boston, 1935, and are quoted in Gallie, *op. cit.*

exaggerated) and as a "loose-living man"—mainly because he divorced his first wife—undoubtedly handicapped him in university circles; but his inability to get along with others, the independence and "violence" of his thought (James' word), the "sententiousness of his manner," were far greater obstacles to academic advancement.[15]

Yet Peirce was a much more attractive man than these reports of his personality (some of which are self-descriptions) would indicate. He could be charming and witty and gracious. He was "singularly free from academic jealousy." He was a fair-minded, indeed a most "chivalrous" opponent in controversy. There is no doubt he inspired love in the few men who knew him well, who recognized that he was a difficult but not an intractable child. James' formula for treating Peirce was to grasp him "after the famous 'nettle' receipt . . . contradict, push hard, make fun of him, and he is as pleasant as anyone." One of Peirce's warming attributes was his capacity for self-criticism. He knew himself and was brilliantly apt in his self-portrayals: "I insensibly put on a sort of swagger here, which is designed to say: 'You are a very good fellow in your way; who you are I don't know and I don't care, but I, you know, am Mr. Peirce, distinguished for my various scientific acquirements, but above all for my extreme modesty in which respect I challenge the world.'"[16] The self-critical faculty was relentlessly applied to his writings. It was not unusual for him to redraft a paper a dozen times "until it was as accurate and precisely worded as he could make it." Nevertheless his works are un-even in quality, not a few papers being obscure and fragmentary, and it cannot be said that he achieved anything approaching a systematic, unified expression of his conception of philosophy.

In 1887, having received a small legacy, Peirce retired to a house on the outskirts of Milford, Pennsylvania, though he continued to engage in researches for the Coast Survey until 1891. The legacy was insufficient for his needs and he hoped to eke it out with writing. He was a prodigious worker and regularly turned out, as he records, 2,000 words a day. Like Trollope, when he finished one paper he started with undiminished energy on the next. Since many of his manuscripts were not published, they piled up in his study and fell into disorder. The bulk of this work was on logic and philosophy, but he also wrote on mathematics, geodesy, religion, astronomy, chemistry, psychology, early English and classical Greek pro-nunciation, psychical research, criminology, the history of science, ancient history, Egyptology, and Napoleon. For the *Century Dictionary* he wrote

---

[15] A most interesting account of Peirce's brief career at Johns Hopkins, his rela-tions with the president, Daniel Coit Gilman, with J. J. Sylvester and other members of the faculty, his lectures, his library, his quarrels, and other personal items is un-folded in Max H. Fisch and Jackson I. Cope, "Peirce at the Johns Hopkins Uni-versity" in Wiener and Young, *op. cit.*, pp. 277–311.

[16] Ralph Barton Perry, *op. cit.*, Vol. I, p. 538.

all the definitions on more than half a dozen subjects; for the *Nation* he wrote many book reviews; he did translations, prepared a thesaurus and an editor's manual. Nonetheless he could earn barely enough to keep alive. "In his home he built an attic where he could work undisturbed, or, by pulling up the ladder, escape from his creditors." His last years were darkened by illness and the constant struggle to stave off poverty. In 1909, at the age of seventy, he was still working furiously, though he needed a grain of morphine a day to deaden the pain of cancer. For five more years he held on; on April 19, 1914, he died, "a frustrated, isolated man, still working on his logic, without a publisher, with scarcely a disciple, unknown to the public at large."

I should like to conclude this sketch with two quotations. The first is a poignant, remarkably revealing self-appraisal, a brief passage in which Peirce compares his own intellectual character with that of William James: "Who, for example, could be of a nature so different from his as I? He so concrete, so living; I a mere table of contents, so abstract, a very snarl of twine." [17] The second is an illuminating appraisal by Justus Buchler: "Even to the most unsympathetic, Peirce's thought cannot fail to convey something of lasting value. It has a peculiar property, like that of the Lernean hydra: discover a weak point, and two strong ones spring up beside it. Despite the elaborate architectonic planning of its creator, it is everywhere uncompleted, often distressingly so. There are many who have small regard for things uncompleted, and no doubt what they value is much to be valued. In his quest for magnificent array, in his design for a mighty temple that should house his ideas, Peirce failed. He succeeded only in advancing philosophy." [18]

[17] Hartshorne and Weiss, *op. cit.*, Vol. 6, paragraph 184, quoted in Gallie, *op. cit.*, pp. 57–58.
[18] Justus Buchler, *The Philosophy of Peirce, Selected Writings*, New York, 1940, XVI.

Certain characteristics of the subject are clear. To begin with, we do not, in this subject, deal with particular things or particular properties: we deal formally with what can be said about "any" thing or "any" property. We are prepared to say that one and one are two, but not that Socrates and Plato are two, because, in our capacity of logicians or pure mathematicians, we have never heard of Socrates or Plato. A world in which there were no such individuals would still be a world in which one and one are two. It is not open to us, as pure mathematicians or logicians, to mention anything at all, because, if we do so we introduce something irrelevant and not formal.
                                                                    —BERTRAND RUSSELL

2   The Essence of Mathematics

By CHARLES SANDERS PEIRCE

IT DOES not seem to me that mathematics depends in any way upon logic. It reasons, of course. But if the mathematician ever hesitates or errs in his reasoning, logic cannot come to his aid. He would be far more liable to commit similar as well as other errors there. On the contrary, I am persuaded that logic cannot possibly attain the solution of its problems without great use of mathematics. Indeed all formal logic is merely mathematics applied to logic.

It was Benjamin Peirce,[1] whose son I boast myself, that in 1870 first defined mathematics as "the science which draws necessary conclusions." This was a hard saying at the time; but today, students of the philosophy of mathematics generally acknowledge its substantial correctness.

The common definition, among such people as ordinary schoolmasters, still is that mathematics is the science of quantity. As this is inevitably understood in English, it seems to be a misunderstanding of a definition which may be very old,[2] the original meaning being that mathematics is the science of *quantities*, that is, forms possessing quantity. We perceive that Euclid was aware that a large branch of geometry had nothing to do with measurement (unless as an aid in demonstrating); and, therefore, a Greek geometer of his age (early in the third century B.C.) or later could not define mathematics as the science of that which the abstract noun quantity expresses. A line, however, was classed as a quantity, or *quantum*, by Aristotle and his followers; so that even perspective (which deals wholly with intersections and projections, not at all with lengths) could be said to be a science of quantities, "quantity" being taken in the con-

---

[1] "Linear Associative Algebra" (1870), sec. 1; see *American Journal of Mathematics*, vol. 4 (1881).
[2] From what is said by Proclus Diadochus, A.D. 485 (*Commentarii in Primum Euclidis Elementorum Librum*, Prologi pars prior, c. 12), it would seem that the Pythagoreans understood mathematics to be the answer to the two questions "how many?" and "how much?"

crete sense. That this was what was originally meant by the definition "Mathematics is the science of quantity," is sufficiently shown by the circumstance that those writers who first enunciate it, about A.D. 500, that is Ammonius Hermiæ and Boëthius, make astronomy and music branches of mathematics; and it is confirmed by the reasons they give for doing so.[3] Even Philo of Alexandria (100 B.C.), who defines mathematics as the science of ideas furnished by sensation and reflection in respect to their necessary consequences, since he includes under mathematics, besides its more essential parts, the theory of numbers and geometry, also the practical arithmetic of the Greeks, geodesy, mechanics, optics (or projective geometry), music, and astronomy, must be said to take the word 'mathematics' in a different sense from ours. That Aristotle did not regard mathematics as the science of quantity, in the modern abstract sense, is evidenced in various ways. The subjects of mathematics are, according to him, the how much and the continuous. He referred the continuous to his category of *quantum*; and therefore he did make *quantum*, in a broad sense, the one object of mathematics.

Plato, in the Sixth book of the *Republic*,[4] holds that the essential characteristic of mathematics lies in the peculiar kind and degree of its abstraction, greater than that of physics but less than that of what we now call philosophy; and Aristotle follows his master in this definition. It has ever since been the habit of metaphysicians to extol their own reasonings and conclusions as vastly more abstract and scientific than those of mathematics. It certainly would seem that problems about God, Freedom, and Immortality are more exalted than, for example, the question how many hours, minutes, and seconds would elapse before two couriers travelling under assumed conditions will come together; although I do not know that this has been proved. But that the methods of thought of the metaphysicians are, as a matter of historical fact, in any aspect, not far inferior to those of mathematics is simply an infatuation. One singular consequence of the notion which prevailed during the greater part of the history of philosophy, that metaphysical reasoning ought to be similar to that of mathematics, only more so, has been that sundry mathematicians have thought themselves, as mathematicians, qualified to discuss philosophy; and no worse metaphysics than theirs is to be found.

Kant regarded mathematical propositions as synthetical judgments *a priori*; wherein there is this much truth, that they are not, for the most part, what he called analytical judgments; that is, the predicate is not, in the sense he intended, contained in the definition of the subject. But if the propositions of arithmetic, for example, are true cognitions, or even forms

---

[3] I regret I have not noted the passage of Ammonius to which I refer. It is probably one of the excerpts given by Brandis. My MS. note states that he gives reasons showing this to be his meaning.

[4] 510C to the end; but in the *Laws* his notion is improved.

of cognition, this circumstance is quite aside from their mathematical truth. For all modern mathematicians agree with Plato and Aristotle that mathematics deals exclusively with hypothetical states of things, and asserts no matter of fact whatever; and further, that it is thus alone that the necessity of its conclusions is to be explained. This is the true essence of mathematics; and my father's definition is in so far correct that it is impossible to reason necessarily concerning anything else than a pure hypothesis. Of course, I do not mean that if such pure hypothesis happened to be true of an actual state of things, the reasoning would thereby cease to be necessary. Only, it never would be known apodictically to be true of an actual state of things. Suppose a state of things of a perfectly definite, general description. That is, there must be no room for doubt as to whether anything, itself determinate, would or would not come under that description. And suppose, further, that this description refers to nothing occult—nothing that cannot be summoned up fully into the imagination. Assume, then, a range of possibilities equally definite and equally subject to the imagination; so that, so far as the given description of the supposed state of things is general, the different ways in which it might be made determinate could never introduce doubtful or occult features. The assumption, for example, must not refer to any matter of fact. For questions of fact are not within the purview of the imagination. Nor must it be such that, for example, it could lead us to ask whether the vowel *OO* can be imagined to be sounded on as high a pitch as the vowel *EE*. Perhaps it would have to be restricted to pure spatial, temporal, and logical relations. Be that as it may, the question whether in such a state of things, a certain other similarly definite state of things, equally a matter of the imagination, could or could not, in the assumed range of possibility, ever occur, would be one in reference to which one of the two answers, *Yes* and *No*, would be true, but never both. But all pertinent facts would be within the beck and call of the imagination; and consequently nothing but the operation of thought would be necessary to render the true answer. Nor, supposing the answer to cover the whole range of possibility assumed, could this be rendered otherwise than by reasoning that would be apodictic, general, and exact. No knowledge of what actually is, no *positive* knowledge, as we say, could result. On the other hand, to assert that any source of information that is restricted to actual facts could afford us a necessary knowledge, that is, knowledge relating to a whole general range of possibility, would be a flat contradiction in terms.

Mathematics is the study of what is true of hypothetical states of things. That is its essence and definition. Everything in it, therefore, beyond the first precepts for the construction of the hypotheses, has to be of the nature of apodictic inference. No doubt, we may reason imperfectly and jump at a conclusion; still, the conclusion so guessed at is, after all, that in a certain supposed state of things something would necessarily be true. Conversely, too, every apodictic inference is, strictly speaking, mathematics. But mathematics, as a serious science, has, over and above its essential character of being hypothetical, an accidental characteristic peculiarity—a *proprium*, as the Aristotelians used to say—which is of the greatest logical interest. Namely, while all the "philosophers" follow Aristotle in holding no demonstration to be thoroughly satisfactory except what they call a "direct" demonstration, or a "demonstration why"—by which they mean a demonstration which employs only general concepts and concludes nothing but what would be an item of a definition if all its terms were themselves distinctly defined—the mathematicians, on the contrary, entertain a contempt for that style of reasoning, and glory in what the philosophers stigmatize as "mere" indirect demonstrations, or "demonstrations that." Those propositions which can be deduced from others by reasoning of the kind that the philosophers extol are set down by mathematicians as "corollaries." That is to say, they are like those geometrical truths which Euclid did not deem worthy of particular mention, and which his editors inserted with a garland, or corolla, against each in the margin, implying perhaps that it was to them that such honor as might attach to these insignificant remarks was due. In the theorems, or at least in all the major theorems, a different kind of reasoning is demanded. Here, it will not do to confine oneself to general terms. It is necessary to set down, or to imagine, some individual and definite schema, or diagram —in geometry, a figure composed of lines with letters attached; in algebra an array of letters of which some are repeated. This schema is constructed so as to conform to a hypothesis set forth in general terms in the thesis of the theorem. Pains are taken so to construct it that there would be something closely similar in every possible state of things to which the hypothetical description in the thesis would be applicable, and furthermore, to construct it so that it shall have no other characters which could influence the reasoning. How it can be that, although the reasoning is based upon the study of an individual schema, it is nevertheless necessary, that is, applicable, to all possible cases, is one of the questions we shall have to consider. Just now, I wish to point out that after the schema has been constructed according to the precept virtually contained in the thesis, the assertion of the theorem is not evidently true, even for the individual schema; nor will any amount of hard thinking of the philosophers' corollarial kind ever render it evident. Thinking in general terms is not enough. It is necessary that something should be DONE. In geometry, subsidiary lines are drawn. In algebra permissible transformations are made. Thereupon, the faculty of observation is called into play. Some relation between the parts of the schema is remarked. But would this relation subsist in every possible case? Mere corollarial reasoning will sometimes assure us of this. But, generally speaking, it may be necessary to draw distinct schemata to represent alternative possibilities. Theorematic reasoning invariably depends upon experimentation with individual schemata. We shall find that, in the last analysis, the same thing is true of the corollarial reasoning, too; even the Aristotelian "demonstration why." Only in this case, the very words serve as schemata. Accordingly, we may say that corollarial, or "philosophical" reasoning is reasoning with words; while theorematic, or mathematical reasoning proper, is reasoning with specially constructed schemata.

Another characteristic of mathematical thought is the extraordinary use it makes of abstractions. Abstractions have been a favorite butt of ridicule in modern times. Now it is very easy to laugh at the old physician who is represented as answering the question, why opium puts people to sleep, by saying that it is because it has a dormative virtue. It is an answer that no doubt carries vagueness to its last extreme. Yet, invented as the story was to show how little meaning there might be in an abstraction, nevertheless the physician's answer does contain a truth that modern philosophy has generally denied: it does assert that there really is in opium *something* which explains its always putting people to sleep. This has, I say, been denied by modern philosophers generally. Not, of course, explicitly; but when they say that the different events of people going to sleep after taking opium have really nothing in common, but only that the mind classes them together—and this is what they virtually do say in denying the reality of generals—they do implicitly deny that there is any true explanation of opium's generally putting people to sleep.

Look through the modern logical treatises, and you will find that they almost all fall into one or other of two errors, as I hold them to be; that of setting aside the doctrine of abstraction (in the sense in which an abstract noun marks an abstraction) as a grammatical topic with which the logician need not particularly concern himself; and that of confounding abstraction, in this sense, with that operation of the mind by which we pay attention to one feature of a percept to the disregard of others. The two things are entirely disconnected. The most ordinary fact of perception, such as "it is light," involves *precisive* abstraction, or *prescission*. But *hypostatic* abstraction, the abstraction which transforms "it is light" into "there is light here," which is the sense which I shall commonly attach to the word abstraction (since *prescission* will do for precisive abstraction) is a very special mode of thought. It consists in taking a feature of a percept or percepts (after it has already been prescinded from the other elements of the percept), so as to take propositional form in a judgment (indeed, it may operate upon any judgment whatsoever), and in conceiving this fact to consist in the relation between the subject of that judgment and another subject, which has a mode of being that merely consists in the truth of propositions of which the corresponding concrete term is the predicate. Thus, we transform the proposition, "honey is sweet," into "honey possesses sweetness." "Sweetness" might be called a fictitious thing, in one sense. But since the mode of being attributed to it *consists* in no more than the fact that some things are sweet, and it is not pretended, or imagined, that it has any other mode of being, there is, after all, no fiction. The only profession made is that we consider the fact of honey being sweet under the form of a relation; and so we really can. I have selected sweetness as an instance of one of the least useful of abstractions. Yet even this is convenient. It facilitates such thoughts as that the sweetness of honey is particularly cloying; that the sweetness of honey is something like the sweetness of a honeymoon; etc. Abstractions are particularly congenial to mathematics. Everyday life first, for example, found the need of that class of abstractions which we call *collections*. Instead of saying that some human beings are males and all the rest females, it was found convenient to say that *mankind* consists of the male *part* and the female *part*. The same thought makes classes of collections, such as pairs, leashes, quatrains, hands, weeks, dozens, baker's dozens, sonnets, scores, quires, hundreds, long hundreds, gross, reams, thousands, myriads, lacs, millions, milliards, milliasses, etc. These have suggested a great branch of mathematics.[5] Again, a point moves: it is by abstraction that the geometer says that it "describes a line." This line, though an abstraction, itself moves; and this is regarded as generating a surface; and so on. So likewise, when the analyst treats operations as themselves subjects of operations, a method whose utility will not be denied, this is another instance of abstraction. Maxwell's notion of a tension exercised upon lines of electrical force, transverse to them, is somewhat similar. These examples exhibit the great rolling billows of abstraction in the ocean of mathematical thought; but when we come to a minute examination of it, we shall find, in every department, incessant ripples of the same form of thought, of which the examples I have mentioned give no hint.

Another characteristic of mathematical thought is that it can have no success where it cannot generalize. One cannot, for example, deny that chess is mathematics, after a fashion; but, owing to the exceptions which everywhere confront the mathematician in this field—such as the limits of the board; the single steps of king, knight, and pawn; the finite number of squares; the peculiar mode of capture by pawns; the queening of pawns; castling—there results a mathematics whose wings are effectually clipped, and which can only run along the ground. Hence it is that a mathematician often finds what a chess-player might call a gambit to his

---

[5] Of course, the moment a collection is recognized as an abstraction we have to admit that even a percept is an abstraction or represents an abstraction, if matter has parts. It therefore becomes difficult to maintain that all abstractions are fictions.

advantage; exchanging a smaller problem that involves exceptions for a larger one free from them. Thus, rather than suppose that parallel lines, unlike all other pairs of straight lines in a plane, never meet, he supposes that they intersect at infinity. Rather than suppose that some equations have roots while others have not, he supplements real quantity by the infinitely greater realm of imaginary quantity. He tells us with ease how many inflexions a plane curve of any description has; but if we ask how many of these are real, and how many merely fictional, he is unable to say. He is perplexed by three-dimensional space, because not all pairs of straight lines intersect, and finds it to his advantage to use quaternions which represent a sort of four-fold continuum, in order to avoid the exception. It is because exceptions so hamper the mathematician that almost all the relations with which he chooses to deal are of the nature of correspondences; that is to say, such relations that for every relate there is the same number of correlates, and for every correlate the same number of relates.

Among the minor, yet striking characteristics of mathematics, may be mentioned the fleshless and skeletal build of its propositions; the peculiar difficulty, complication, and stress of its reasonings; the perfect exactitude of its results; their broad universality; their practical infallibility. It is easy to speak with precision upon a general theme. Only, one must commonly surrender all ambition to be certain. It is equally easy to be certain. One has only to be sufficiently vague. It is not so difficult to be pretty precise and fairly certain at once about a very narrow subject. But to reunite, like mathematics, perfect exactitude and practical infallibility with unrestricted universality, is remarkable. But it is not hard to see that all these, characters of mathematics are inevitable consequences of its being the study of hypothetical truth.

It is difficult to decide between the two definitions of mathematics; the one by its method, that of drawing necessary conclusions; the other by its aim and subject matter, as the study of hypothetical states of things. The former makes or seems to make the deduction of the consequences of hypotheses the sole business of the mathematician as such. But it cannot be denied that immense genius has been exercised in the mere framing of such general hypotheses as the field of imaginary quantity and the allied idea of Riemann's surface, in imagining non-Euclidian measurement, ideal numbers, the perfect liquid. Even the framing of the particular hypotheses of special problems almost always calls for good judgment and knowledge, and sometimes for great intellectual power, as in the case of Boole's logical algebra. Shall we exclude this work from the domain of mathematics? Perhaps the answer should be that, in the first place, whatever exercise of intellect may be called for in applying mathematics to a question not propounded in mathematical form [it] is certainly not pure mathe-

matical thought; and in the second place, that the mere creation of a
hypothesis may be a grand work of poietic genius, but cannot be said to
be scientific, inasmuch as that which it produces is neither true nor false,
and therefore is not knowledge. This reply suggests the further remark
that if mathematics is the study of purely imaginary states of things, poets
must be great mathematicians, especially that class of poets who write
novels of intricate and enigmatical plots. Even the reply, which is obvious,
that by *studying* imaginary states of things we mean *studying* what is true
of them, perhaps does not fully meet the objection. The article *Mathematics* in the ninth edition of the *Encyclopaedia Britannica* [6] makes
mathematics consist in the study of a particular sort of hypotheses, namely,
those that are exact, etc., as there set forth at some length. The article is
well worthy of consideration.

The philosophical mathematician, Dr. Richard Dedekind,[7] holds mathematics to be a branch of logic. This would not result from my father's
definition, which runs, not that mathematics is the science of *drawing*
necessary conclusions—which would be deductive logic—but that it is the
science which *draws* necessary conclusions. It is evident, and I know as
a fact, that he had this distinction in view. At the time when he thought
out this definition, he, a mathematician, and I, a logician, held daily discussions about a large subject which interested us both; and he was struck,
as I was, with the contrary nature of his interest and mine in the same
propositions. The logician does not care particularly about this or that
hypothesis or its consequences, except so far as these things may throw
a light upon the nature of reasoning. The mathematician is intensely interested in efficient methods of reasoning, with a view to their possible
extension to new problems; but he does not, qua mathematician, trouble
himself minutely to dissect those parts of this method whose correctness
is a matter of course. The different aspects which the algebra of logic will
assume for the two men is instructive in this respect. The mathematician
asks what value this algebra has as a calculus. Can it be applied to un-
ravelling a complicated question? Will it, at one stroke, produce a remote
consequence? The logician does not wish the algebra to have that character. On the contrary, the greater number of distinct logical steps, into
which the algebra breaks up an inference, will for him constitute a superiority of it over another which moves more swiftly to its conclusions. He
demands that the algebra shall analyze a reasoning into its last elementary
steps. Thus, that which is a merit in a logical algebra for one of these
students is a demerit in the eyes of the other. The one studies the science
of drawing conclusions, the other the science which draws necessary
conclusions.

[6] By George Chrystal.
[7] *Was sind und was sollen die Zahlen; Vorwort* (1888). [See p. 525 for discussion
of Dedekind. ED.]

But, indeed, the difference between the two sciences is far more than that between two points of view. Mathematics is purely hypothetical: it produces nothing but conditional propositions. Logic, on the contrary, is categorical in its assertions. True, it is not merely, or even mainly, a mere discovery of what really is, like metaphysics. It is a normative science. It thus has a strongly mathematical character, at least in its methodeutic division; for here it analyzes the problem of how, with given means, a required end is to be pursued. This is, at most, to say that it has to call in the aid of mathematics; that it has a mathematical branch. But so much may be said of every science. There is a mathematical logic, just as there is a mathematical optics and a mathematical economics. Mathematical logic is formal logic. Formal logic, however developed, is mathematics. Formal logic, however, is by no means the whole of logic, or even its principal part. It is hardly to be reckoned as a part of logic proper. Logic has to define its aim; and in doing so is even more dependent upon ethics, or the philosophy of aims, by far, than it is, in the methodeutic branch, upon mathematics. We shall soon come to understand how a student of ethics might well be tempted to make his science a branch of logic; as, indeed, it pretty nearly was in the mind of Socrates. But this would be no truer a view than the other. Logic depends upon mathematics; still more intimately upon ethics; but its proper concern is with truths beyond the purview of either.

There are two characters of mathematics which have not yet been mentioned, because they are not exclusive characteristics of it. One of these, which need not detain us, is that mathematics is distinguished from all other sciences except only ethics, in standing in no need of ethics. Every other science, even logic—logic, especially—is in its early stages in danger of evaporating into airy nothingness, degenerating, as the Germans say, into an arachnoid film, spun from the stuff that dreams are made of. There is no such danger for pure mathematics; for that is precisely what mathematics ought to be.

The other character—and of particular interest it is to us just now—is that mathematics, along with ethics and logic alone of the sciences, has no need of any appeal to logic. No doubt, some reader may exclaim in dissent to this, on first hearing it said. Mathematics, they may say, is preëminently a science of reasoning. So it is; preëminently a science that reasons. But just as it is not necessary, in order to talk, to understand the theory of the formation of vowel sounds, so it is not necessary, in order to reason, to be in possession of the theory of reasoning. Otherwise, plainly, the science of logic could never be developed. The contrary objection would have more excuse, that no science stands in need of logic, since our natural power of reason is enough. Make of logic what the majority of treatises in the past have made of it, and a very common class of English and French books still make of it—that is to say, mainly formal logic, and that formal logic represented as an art of reasoning— and in my opinion this objection is more than sound, for such logic is a great hindrance to right reasoning. It would, however, be aside from our present purpose to examine this objection minutely. I will content myself with saying that undoubtedly our natural power of reasoning is enough, in the same sense that it is enough, in order to obtain a wireless trans- atlantic telegraph, that men should be born. That is to say, it is bound to come sooner or later. But that does not make research into the nature of electricity needless for gaining such a telegraph. So likewise if the study of electricity had been pursued resolutely, even if no special attention had ever been paid to mathematics, the requisite mathematical ideas would surely have been evolved. Faraday, indeed, did evolve them without any acquaintance with mathematics. Still it would be far more economical to postpone electrical researches, to study mathematics by itself, and then to apply it to electricity, which was Maxwell's way. In this same manner, the various logical difficulties which arise in the course of every science except mathematics, ethics, and logic, will, no doubt, get worked out after a time, even though no special study of logic be made. But it would be far more economical to make first a systematic study of logic. If anybody should ask what are these logical difficulties which arise in all the sciences, he must have read the history of science very irreflectively. What was the famous controversy concerning the measure of force but a logical difficulty? What was the controversy between the uniformitarians and the catastrophists but a question of whether or not a given conclusion fol- lowed from acknowledged premisses? . . .

But it may be asked whether mathematics, ethics, and logic have not encountered similar difficulties. Are the doctrines of logic at all settled? Is the history of ethics anything but a history of controversy? Have no logical errors been committed by mathematicians? To that I reply, first, as to logic, that not only have the rank and file of writers on the subject been, as an eminent psychiatrist, Maudsley, declares, men of arrested brain-development, and not only have they generally lacked the most essential qualification for the study, namely mathematical training, but the main reason why logic is unsettled is that thirteen different opinions are current as to the true aim of the science. Now this is not a logical difficulty but an ethical difficulty; for ethics is the science of aims. Sec- ondly, it is true that pure ethics has been, and always must be, a theatre of discussion, for the reason that its study consists in the gradual develop- ment of a distinct recognition of a satisfactory aim. It is a science of subtleties, no doubt; but it is not logic, but the development of the ideal, which really creates and resolves the problems of ethics. Thirdly, in mathe- matics errors of reasoning have occurred, nay, have passed unchallenged for thousands of years. This, however, was simply because they escaped notice. Never, in the whole history of the science, has a question whether a given conclusion followed *mathematically* from given premisses, when once started, failed to receive a speedy and unanimous reply. Very few have been even the apparent exceptions; and those few have been due to the fact that it is only within the last half century that mathematicians have come to have a perfectly clear recognition of what is mathematical soil and what foreign to mathematics. Perhaps the nearest approximation to an exception was the dispute about the use of divergent series. Here neither party was in possession of sufficient pure mathematical reasons covering the whole ground; and such reasons as they had were not only of an extra-mathematical kind, but were used to support more or less vague positions. It appeared then, as we all know now, that divergent series are of the utmost utility.[8]

Struck by this circumstance, and making an inference, of which it is sufficient to say that it was not mathematical, many of the old mathematicians pushed the use of divergent series beyond all reason. This was a case of mathematicians disputing about the validity of a kind of inference that is not mathematical. No doubt, a sound logic (such as has not hitherto been developed) would have shown clearly that that non-mathematical inference was not a sound one. But this is, I believe, the only instance in which any large party in the mathematical world ever proposed to rely, in mathematics, upon unmathematical reasoning. My proposition is that true mathematical reasoning is so much more evident than it is possible to render any doctrine of logic proper—without just such reasoning—that an appeal in mathematics to logic could only embroil a situation. On the contrary, such difficulties as may arise concerning necessary reasoning have to be solved by the logician by reducing them to questions of mathematics. Upon those mathematical dicta, as we shall come clearly to see, the logician has ultimately to repose.

[8] It would not be fair, however, to suppose that every reader will know this. Of course, there are many series so extravagantly divergent that no use at all can be made of them. But even when a series is divergent from the very start, some use might commonly be made of it, if the same information could not otherwise be obtained more easily. The reason is—or rather, one reason is—that most series, even when divergent, approximate at last somewhat to geometrical series, at least, for a considerable succession of terms. The series $\log(1 + x) = x - \frac{1}{2}x^2 + \frac{1}{3}x^3 - \frac{1}{4}x^4 +$, etc., is one that would not be judiciously employed in order to find the natural logarithm of 3, which is 1.0986, its successive terms being $2 - 2 + \frac{8}{3} - 4 + \frac{32}{5} - \frac{32}{3} +$, etc. Still, employing the common device of substituting for the last two terms that are to be used, say M and N, the expression $M/(1 - N/M)$, the succession of the first six values is 0.667, 1.143, 1.067, 1.128, 1.067, which do show some approximation to the value. The mean of the last two, which any professional computer would use (supposing him to use this series, at all) would be 1.098, which is not very wrong. Of course, the computer would practically use the series $\log 3 = 1 + \frac{1}{12} + \frac{1}{80} + \frac{1}{448} +$, etc., of which the terms written give the correct value to four places, if they are properly used.

COMMENTARY ON
ERNST MACH

ERNST MACH (1838–1916) was an Austrian physicist, psychologist and philosopher whose opinions greatly stimulated Einstein and severely provoked Lenin. His philosophy of science, which affected these two so different men so differently, was the most important development in positivism after the founding work of Auguste Comte.

Mach held that the main object of science is to obviate the unnecessary expenditure of human thought. If man were immortal he would be foolish to devise time- and work-saving methods, since they would only add to the tedium of eternity. Man pays for the boon of mortality by inventing science. Its purpose is "to replace, or *save*, experiences, by the reproduction of facts in thought"; the duty of the scientist is to use the simplest, most direct means of drawing conclusions and to exclude all evidence not based upon observation.[1]

Mathematics inevitably occupies the central position in such a system. It is a model of frugality of expression; its methods, once proved sound, can be used over and over again to conserve effort; it affords innumerable short cuts, and it provides us with beautiful models for testing theories and guiding experiments. Mach shared Hume's conviction that it was improper to grant the relation of cause and effect anything more than psychological reality. It is something we require to explain phenomena, but it corresponds to no verifiable relationship. Here again mathematics provides a useful substitute: the concept of *function* permits phenomena to be accurately and succinctly described in terms of the interdependence of variables rather than by causal links.[2] Despite his positivist convictions and abhorrence of metaphysics, Mach was not so foolish as to argue that scientific hypotheses and tentative theoretical systems were unnecessary, or that natural laws could be "simply 'derived' from experience." This extreme Baconian view has mistakenly been imputed to him. Comte was willing to settle for anything, even "theological conceptions," to give impetus to scientific inquiry and "coherence" to its observations. It would be time enough to kick these conceptions out of the window *after* they had served their purpose. Mach acknowledged that imagination and "use-

---

[1] "According to Mach and his immediate followers, the fundamental laws of physics should be formulated so that they would contain only concepts which could be defined by direct observations or at least by a short chain of thoughts connected with direct observations." Phillip G. Frank, "Einstein, Mach and Logical Positivism," in *Albert Einstein: Philosopher-Scientist,* edited by Paul Arthur Schilpp, Evanston, 1949, p. 274.

[2] See Federigo Enriques, *The Historic Development of Logic,* New York, 1929, pp. 213–214, 220–222, 225–227.

ful images" were important to give research a start and to provide models, most of which would inevitably "succumb to the facts of an inexorable criticism . . ." (his words).[3] The principal danger to be avoided was that the conceptual system should ever gain ascendancy over actual research, or emancipate itself from the firm control of sensory experience. In such cases, as Einstein has said, scientific thinking must degenerate into "empty talk."[4]

Einstein has acknowledged that "the study of Mach and Hume has been directly and indirectly a great help in my work . . . Mach recognized the weak spots of classical mechanics and was not very far from requiring a general theory of relativity half a century ago [this was written in 1916]."[5] In particular, Mach's requirement "that every statement in physics has to state relations between observable quantities" prompted Einstein to reconsider the concept of simultaneity and to redefine it for the special theory of relativity. For the same reason he re-examined the concepts of mass, inertia and gravitation in connection with the formulation of his general theory.[6]

Mach's description of simple observations as "sensations" led to a common belief that he was a philosophical idealist. This, to be sure, is a small matter and in any case the interpretation was wrong. But it is not a small matter among Marxists, who recoil from the word "idealist" as from a poisonous insect. Lenin therefore took out after Mach in his book *Materialism and Empirio-Criticism*. Since this is one of the sacred writings, Mach's philosophy of science "has become a target of attack in every textbook and in every classroom in the Soviet Union where philosophy is being taught."[7] Russia expects every man to bear arms against the threat of idealistic subversion. Einstein himself was a questionable fellow because he associated with Mach and had openly praised his opinions.

Mach's major writings include *The Analysis of Sensations and the Relation of the Physical to the Psychical, Theory of Heat*, his excellent *Popular Scientific Lectures*, and *The Science of Mechanics*, a distinguished critical and historical work of considerable influence. *The Science of Mechanics*, which first appeared in 1883, is not a treatise upon the application of the principles of mechanics. Its aim, said Mach, "is to clear up ideas, expose the real significance of the matter, and get rid of metaphysical obscurities."[8] Mach was especially concerned with showing how the principles of mechanics evolved, and "how far they can be regarded

---

[3] Enriques, *op. cit.*, pp. 225–227; also Frank, *op. cit.*, p. 272.
[4] Albert Einstein in "Remarks on Bertrand Russell's Theory of Knowledge" in *The Philosophy of Bertrand Russell*, edited by Paul A. Schilpp, Evanston, 1944, p. 289. (Quoted by Frank, *op. cit.*)
[5] *Physikalische Zeitschrift*, XVII (1916), pp. 101 ff. (Quoted by Frank, *op. cit.*)
[6] Frank, *op. cit.*, p. 272.
[7] *Ibid.*
[8] Preface to the first edition, reprinted in the second English edition, Chicago, 1919.

as permanent acquisitions." This was in direct opposition to the widespread tendency of his time to regard the wonderfully exact mathematical
formulations of mechanics as somehow prescribing the only lawful course
of natural phenomena; thus observed departures from this logical system,
inconsistencies and other anomalies, were held to be due to experimental
errors rather than to shortcomings of the model itself. Mach suggested
that this was a topsy-turvy approach to mechanics; the model, though important as an aid to understanding and communication, must not be permitted to obscure the ultimate authority of the processes themselves. It is
the "positive and physical essence of mechanics" which must remain the
"chief and highest interest for a student of nature." *The Science of
Mechanics* was the first work in two hundred years to question the assumptions underlying Newton's formulation of fundamental dynamical
principles.[9] Mach's criticisms of the concepts of mass, inertia, absolute
motion, his insistence on the necessity of submitting scientific thought to
sharp epistemological scrutiny, his emphasis on the importance of the
"biologico-psychological" investigation of the development of science,
have had a profound effect on the evolution of modern science. It is from
this book that I have selected a section expressing Mach's views on the
economy of science and the uses of mathematics. It is a thoughtful and
penetrating essay. It oversimplifies certain issues, to my way of thinking,
but even at these points Mach does not distress me as he did Lenin.

[9] Sir William Dampier, *A History of Science*, Fourth Edition, New York, 1949, pp.
155–156. Newton defined mass as "the quantity of matter in a body as measured by the
product of its density and bulk." Mach pointed out "that Newton's definition of mass
and force leave us in a logical circle, for we only know matter in its effects on our
senses, and we can only define density as mass per unit volume. In summarizing the
history of the origins of dynamics, Mach shows that the dynamical work of Galileo,
Huygens and Newton really means the discovery of only one fundamental principle,
though, owing to the historical accidents inevitable in a completely new subject, it
was expressed in many seemingly independent laws or statements."

In scientific thought we adopt the simplest theory which will explain all the facts under consideration and enable us to predict new facts of the same kind. The catch in this criterion lies in the word "simplest." It is really an aesthetic canon such as we find implicit in our criticisms of poetry or painting. The layman finds such a law as

$$\frac{\partial x}{\partial t} = K \frac{\partial^2 x}{\partial y^2}$$

much less simple than "it oozes," of which it is the mathematical statement. The physicist reverses this judgment, and his statement is certainly the more fruitful of the two, so far as prediction is concerned. It is, however, a statement about something very unfamiliar to the plain man, namely, the rate of change of a rate of change. —J. B. S. HALDANE

3   The Economy of Science

By ERNST MACH

1. IT IS the object of science to replace, or *save*, experiences, by the reproduction and anticipation of facts in thought. Memory is handier than experience, and often answers the same purpose. This economical office of science, which fills its whole life, is apparent at first glance; and with its full recognition all mysticism in science disappears.

Science is communicated by instruction, in order that one man may profit by the experience of another and be spared the trouble of accumulating it for himself; and thus, to spare posterity, the experiences of whole generations are stored up in libraries.

Language, the instrument of this communication, is itself an economical contrivance. Experiences are analysed, or broken up, into simpler and more familiar experiences, and then symbolised at some sacrifice of precision. The symbols of speech are as yet restricted in their use within national boundaries, and doubtless will long remain so. But written language is gradually being metamorphosed into an ideal universal character. It is certainly no longer a mere transcript of speech. Numerals, algebraic signs, chemical symbols, musical notes, phonetic alphabets, may be regarded as parts already formed of this universal character of the future; they are, to some extent, decidedly conceptual, and of almost general international use. The analysis of colors, physical and physiological, is already far enough advanced to render an international system of color-signs perfectly practical. In Chinese writing, we have an actual example of a true ideographic language, pronounced diversely in different provinces, yet everywhere carrying the same meaning. Were the system and its signs only of a simpler character, the use of Chinese writing might become universal. The dropping of unmeaning and needless accidents of grammar, as English mostly drops them, would be quite requisite to the adoption of such a system. But universality would not be the sole merit of such a character; since to read it would be to understand it. Our children often read what they do not understand; but that which a Chinaman cannot understand, he is precluded from reading.

2. In the reproduction of facts in thought, we never reproduce the facts in full, but only that side of them which is important to us, moved to this directly or indirectly by a practical interest. Our reproductions are invariably abstractions. Here again is an economical tendency.

Nature is composed of sensations as its elements. Primitive man, however, first picks out certain compounds of these elements—those namely that are relatively permanent and of greater importance to him. The first and oldest words are names of "things." Even here, there is an abstractive process, an abstraction from the surroundings of the things, and from the continual small changes which these compound sensations undergo, which being practically unimportant are not noticed. No inalterable thing exists. The thing is an abstraction, the name a symbol, for a compound of elements from whose changes we abstract. The reason we assign a single word to a whole compound is that we need to suggest all the constituent sensations at once. When, later, we come to remark the changeableness, we cannot at the same time hold fast to the idea of the thing's permanence, unless we have recourse to the conception of a thing-in-itself, or other such like absurdity. Sensations are not signs of things; but, on the contrary, a thing is a thought-symbol for a compound sensation of relative fixedness. Properly speaking the world is not composed of "things" as its elements, but of colors, tones, pressures, spaces, times, in short what we ordinarily call individual sensations.

The whole operation is a mere affair of economy. In the reproduction of facts, we begin with the more durable and familiar compounds, and supplement these later with the unusual by way of corrections. Thus, we speak of a perforated cylinder, of a cube with beveled edges, expressions involving contradictions, unless we accept the view here taken. All judgments are such amplifications and corrections of ideas already admitted.

3. In speaking of cause and effect we arbitrarily give relief to those elements to whose connection we have to attend in the reproduction of a fact in the respect in which it is important to us. There is no cause nor effect in nature; nature has but an individual existence; nature simply *is*. Recurrences of like cases in which A is always connected with B, that is, like results under like circumstances, that is again, the essence of the connection of cause and effect, exist but in the abstraction which we perform for the purpose of mentally reproducing the facts. Let a fact become familiar, and we no longer require this putting into relief of its connecting

marks, our attention is no longer attracted to the new and surprising, and we cease to speak of cause and effect. Heat is said to be the cause of the tension of steam; but when the phenomenon becomes familiar we think of the steam at once with the tension proper to its temperature. Acid is said to be the cause of the reddening of tincture of litmus; but later we think of the reddening as a property of the acid.

Hume first propounded the question, How can a thing A act on another thing B? Hume, in fact, rejects causality and recognises only a wonted succession in time. Kant correctly remarked that a *necessary* connection between A and B could not be disclosed by simple observation. He assumes an innate idea or category of the mind, a *Verstandesbegriff*, under which the cases of experience are subsumed. Schopenhauer, who adopts substantially the same position, distinguishes four forms of the "principle of sufficient reason"—the logical, physical, and mathematical form, and the law of motivation. But these forms differ only as regards the matter to which they are applied, which may belong either to outward or inward experience.

The natural and common-sense explanation is apparently this. The ideas of cause and effect originally sprang from an endeavor to reproduce facts in thought. At first, the connection of A and B, of C and D, of E and F, and so forth, is regarded as familiar. But after a greater range of experience is acquired and a connection between M and N is observed, it often turns out that we recognise M as *made up of A, C, E,* and N of B, D, F, the connection of which was before a *familiar* fact and accordingly possesses with us a higher authority. This explains why a person of experience regards a new event with different eyes than the novice. The new experience is illuminated by the mass of old experience. As a fact, then, there really does exist in the mind an "idea" under which fresh experiences are subsumed; but that idea has itself been developed from experience. The notion of the *necessity* of the causal connection is probably created by our voluntary movements in the world and by the changes which these indirectly produce, as Hume supposed but Schopenhauer contested. Much of the authority of the ideas of cause and effect is due to the fact that they are developed *instinctively* and involuntarily, and that we are distinctly sensible of having personally contributed nothing to their formation. We may, indeed, say, that our sense of causality is not acquired by the individual, but has been perfected in the development of the race. Cause and effect, therefore, are things of thought, having an economical office. It cannot be said *why* they arise. For it is precisely by the abstraction of uniformities that we know the question "why."

4. In the details of science, its economical character is still more apparent. The so-called descriptive sciences must chiefly remain content with reconstructing individual facts. Where it is possible, the common

features of many facts are once for all placed in relief. But in sciences
that are more highly developed, rules for the reconstruction of great numbers of facts may be embodied in a *single* expression. Thus, instead of
noting individual cases of light-refraction, we can mentally reconstruct all
present and future cases, if we know that the incident ray, the refracted
ray, and the perpendicular lie in the same plane and that sin a/sin $\beta = n$.
Here, instead of the numberless cases of refraction in different combinations of matter and under all different angles of incidence, we have simply
to note the rule above stated and the values of n,—which is much easier.
The economical purpose is here unmistakable. In nature there is no *law*
of refraction, only different cases of refraction. The law of refraction is a
concise compendious rule, devised by us for the mental reconstruction of
a fact, and only for its reconstruction in part, that is, on its geometrical
side.

5. The sciences most highly developed economically are those whose
facts are reducible to a few numerable elements of like nature. Such is
the science of mechanics, in which we deal exclusively with spaces, times,
and masses. The whole previously established economy of mathematics
stands these sciences in stead. Mathematics may be defined as the economy
of counting. Numbers are arrangement-signs which, for the sake of per-
spicuity and economy, are themselves arranged in a simple system. Numerical operations, it is found, are independent of the kind of objects
operated on, and are consequently mastered once for all. When, for the
first time, I have occasion to add five objects to seven others, I count the
whole collection through, at once; but when I afterwards discover that I
can start counting from 5, I save myself part of the trouble; and still
later, remembering that 5 and 7 always count up to 12, I dispense with
the numeration entirely.

The object of all arithmetical operations is to *save* direct numeration,
by utilising the results of our old operations of counting. Our endeavor is,
having done a sum once, to preserve the answer for future use. The first
four rules of arithmetic well illustrate this view. Such, too, is the purpose
of algebra, which, substituting relations for values, symbolises and definitively fixes all numerical operations that follow the same rule. For example, we learn from the equation

$$\frac{x^2 - y^2}{x + y} = x - y,$$

that the more complicated numerical operation at the left may always be
replaced by the simpler one at the right, whatever numbers x and y stand
for. We thus save ourselves the labor of performing in future cases the
more complicated operation. Mathematics is the method of replacing in
the most comprehensive and *economical* manner possible, *new* numerical

operation by old ones done already with known results. It may happen in this procedure that the results of operations are employed which were originally performed centuries ago.

Often operations involving intense mental effort may be replaced by the action of semi-mechanical routine, with great saving of time and avoidance of fatigue. For example, the theory of determinants owes its origin to the remark, that it is not necessary to solve each time anew equations of the form

$$a_1 x + b_1 y + c_1 = 0$$
$$a_2 x + b_2 y + c_2 = 0,$$

from which result

$$x = -\frac{c_1 b_2 - c_2 b_1}{a_1 b_2 - a_2 b_1} = -\frac{P}{N}$$
$$y = -\frac{a_1 c_2 - a_2 c_1}{a_1 b_2 - a_2 b_1} = -\frac{Q}{N},$$

but that the solution may be effected by means of the coefficients, by writing down the coefficients according to a prescribed scheme and operating with them *mechanically*. Thus,

$$\begin{vmatrix} a_1 & b_1 \\ a_2 & b_2 \end{vmatrix} = a_1 b_2 - a_2 b_1 = N$$

and similarly

$$\begin{vmatrix} c_1 & b_1 \\ c_2 & b_2 \end{vmatrix} = P, \text{ and } \begin{vmatrix} a_1 & c_1 \\ a_2 & c_2 \end{vmatrix} = Q.$$

Even a *total* disburdening of the mind can be effected in mathematical operations. This happens where operations of counting hitherto performed are symbolised by mechanical operations with signs, and our brain energy, instead of being wasted on the repetition of old operations, is spared for more important tasks. The merchant pursues a like economy, when, instead of directly handling his bales of goods, he operates with bills of lading or assignments of them. The drudgery of computation may even be relegated to a machine. Several different types of calculating machines are actually in practical use. The earliest of these (of any complexity) was the difference-engine of Babbage, who was familiar with the ideas here presented.

A numerical result is not always reached by the *actual* solution of the problem; it may also be reached indirectly. It is easy to ascertain, for example, that a curve whose quadrature for the abscissa x has the value x^m, gives an increment $mx^{m-1}dx$ of the quadrature for the increment dx of the abscissa. But we then also know that $\int mx^{m-1}dx = x^m$; that is, we recognise the quantity x^m from the increment $mx^{m-1}dx$ as unmistakably as we recognise a fruit by its rind. Results of this kind, accidentally found by simple inversion, or by processes more or less analogous, are very extensively employed in mathematics.

That scientific work should be more useful the more it has been used, while mechanical work is expended in use, may seem strange to us. When a person who daily takes the same walk accidentally finds a shorter cut, and thereafter, remembering that it is shorter, always goes that way, he undoubtedly saves himself the difference of the work. But memory is really not work. It only places at our disposal energy within our present or future possession, which the circumstance of ignorance prevented us from availing ourselves of. This is precisely the case with the application of scientific ideas.

The mathematician who pursues his studies without clear views of this matter, must often have the uncomfortable feeling that his paper and pencil surpass him in intelligence. Mathematics, thus pursued as an object of instruction, is scarcely of more educational value than busying oneself with the Cabala. On the contrary, it induces a tendency toward mystery, which is pretty sure to bear its fruits.

6. The science of physics also furnishes examples of this economy of thought, altogether similar to those we have just examined. A brief reference here will suffice. The moment of inertia saves us the separate consideration of the individual particles of masses. By the force-function we dispense with the separate investigation of individual force-components. The simplicity of reasonings involving force-functions springs from the fact that a great amount of mental work had to be performed before the discovery of the properties of the force-functions was possible. Gauss's dioptrics dispenses us from the separate consideration of the single refracting surfaces of a dioptrical system and substitutes for it the principal and nodal points. But a careful consideration of the single surfaces had to precede the discovery of the principal and nodal points. Gauss's dioptrics simply *saves* us the necessity of often repeating this consideration.

We must admit, therefore, that there is no result of science which in point of principle could not have been arrived at wholly without methods. But, as a matter of fact, within the short span of a human life and with man's limited powers of memory, any stock of knowledge worthy of the name is unattainable except by the *greatest* mental economy. Science itself, therefore, may be regarded as a minimal problem, consisting of the completest possible presentment of facts with the *least possible expenditure of thought.*

7. The function of science, as we take it, is to replace experience. Thus, on the one hand, science must remain in the province of experience, but, on the other, must hasten beyond it, constantly expecting confirmation, constantly expecting the reverse. Where neither confirmation nor refutation is possible, science is not concerned. Science acts and only acts

in the domain of *uncompleted* experience. Exemplars of such branches of science are the theories of elasticity and of the conduction of heat, both of which ascribe to the smallest particles of matter only such properties as observation supplies in the study of the larger portions. The comparison of theory and experience may be farther and farther extended, as our means of observation increase in refinement.

Experience alone, without the ideas that are associated with it, would forever remain strange to us. Those ideas that hold good throughout the widest domains of research and that supplement the greatest amount of experience, are the *most scientific*. The principle of continuity, the use of which everywhere pervades modern inquiry, simply prescribes a mode of conception which conduces in the highest degree to the economy of thought.

8. If a long elastic rod be fastened in a vise, the rod may be made to execute slow vibrations. These are directly observable, can be seen, touched, and graphically recorded. If the rod be shortened, the vibrations will increase in rapidity and cannot be directly seen; the rod will present to the sight a blurred image. This is a new phenomenon. But the sensation of touch is still like that of the previous case; we can still make the rod record its movements; and if we mentally retain the *conception* of vibrations, we can still anticipate the results of experiments. On further shortening the rod the sensation of touch is altered; the rod begins to sound; again a new phenomenon is presented. But the phenomena do not all change at once; only this or that phenomenon changes; consequently the accompanying notion of vibration, which is not confined to any single one, is still serviceable, still economical. Even when the sound has reached so high a pitch and the vibrations have become so small that the previous means of observation are not of avail, we still *advantageously* imagine the sounding rod to perform vibrations, and can predict the vibrations of the dark lines in the spectrum of the polarised light of a rod of glass. If on the rod being further shortened *all* the phenomena suddenly passed into *new* phenomena, the conception of vibration would no longer be serviceable because it would no longer afford us a means of supplementing the new experiences by the previous ones.

When we mentally add to those actions of a human being which we can perceive, sensations and ideas like our own which we cannot perceive, the object of the idea we so form is economical. The idea makes experience intelligible to us; it supplements and supplants experience. This idea is not regarded as a great scientific discovery, only because its formation is so natural that every child conceives it. Now, this is exactly what we do when we imagine a moving body which has just disappeared behind a pillar, or a comet at the moment invisible, as continuing its motion and retaining its previously observed properties. We do this that we may not be surprised by its reappearance. We fill out the gaps in experience by the ideas that experience suggests.

9. Yet not all the prevalent scientific theories originated so naturally and artlessly. Thus, chemical, electrical, and optical phenomena are explained by atoms. But the mental artifice atom was not formed by the principle of continuity; on the contrary, it is a product especially devised for the purpose in view. Atoms cannot be perceived by the senses; like all substances, they are things of thought. Furthermore, the atoms are invested with properties that absolutely contradict the attributes hitherto observed in bodies. However well fitted atomic theories may be to reproduce certain groups of facts, the physical inquirer who has laid to heart Newton's rules will only admit those theories as *provisional* helps, and will strive to attain, in some more natural way, a satisfactory substitute.

The atomic theory plays a part in physics similar to that of certain auxiliary concepts in mathematics; it is a mathematical *model* for facilitating the mental reproduction of facts. Although we represent vibrations by the harmonic formula, the phenomena of cooling by exponentials, falls by squares of times, etc., no one will fancy that vibrations *in themselves* have anything to do with the circular functions, or the motion of falling bodies with squares. It has simply been observed that the relations between the quantities investigated were similar to certain relations obtaining between familiar mathematical functions, and these *more familiar* ideas are employed as an easy means of supplementing experience. Natural phenomena whose relations are not similar to those of functions with which we are familiar, are at present very difficult to reconstruct. But the progress of mathematics may facilitate the matter.

As mathematical helps of this kind, space of more than three dimensions may be used, as I have elsewhere shown. But it is not necessary to regard these, on this account, as anything more than mental artifices.[1]

---

[1] As the outcome of the labors of Lobatchévski, Bolyai, Gauss, and Riemann, the view has gradually obtained currency in the mathematical world, that that which we call *space* is a *particular, actual* case of a more *general*, conceivable case of multiple quantitative manifoldness. The space of sight and touch is a threefold manifoldness; it possesses three dimensions; and every point in it can be defined by three distinct and independent data. But it is possible to conceive of a quadruple or even multiple space-like manifoldness. And the character of the manifoldness may also be differently *conceived* from the manifoldness of actual space. We regard this discovery, which is chiefly due to the labors of Riemann, as a very important one. The properties of actual space are here directly exhibited as objects of *experience*, and the pseudo-theories of geometry that seek to excogitate these properties by metaphysical arguments are overthrown.

A thinking being is supposed to live in the surface of a sphere, with no other kind of space to institute comparisons with. His space will appear to him similarly constituted throughout. He might regard it as infinite, and could only be convinced of the contrary by experience. Starting from any two points of a great circle of the sphere and proceeding at right angles thereto on other great circles, he could hardly expect that the circles last mentioned would intersect. So, also, with respect to the space in which we live, only experience can decide whether it is finite, whether parallel lines intersect in it, or the like. The significance of this elucidation can scarcely be over-

This is the case, too, with *all* hypotheses formed for the explanation of new phenomena. Our conceptions of electricity fit in at once with the electrical phenomena, and take almost spontaneously the familiar course, the moment we note that things take place as if attracting and repelling fluids moved on the surface of the conductors. But these mental expedients have nothing whatever to do with the phenomenon *itself*.

---

rated. An enlightenment similar to that which Riemann inaugurated in science was produced in the mind of humanity at large, as regards the surface of the earth, by the discoveries of the first circumnavigators.

The theoretical investigation of the mathematical possibilities above referred to, has, primarily, nothing to do with the question whether things really exist which correspond to these possibilities; and we must not hold mathematicians responsible for the popular absurdities which their investigations have given rise to. The space of sight and touch is *three*-dimensional; that, no one ever yet doubted. If, now, it should be found that bodies vanish from this space, or new bodies get into it, the question might scientifically be discussed whether it would facilitate and promote our insight into things to conceive experiential space as part of a four-dimensional or multi-dimensional space. Yet in such a case, this fourth dimension would, none the less, remain a pure thing of thought, a mental fiction.

But this is not the way matters stand. The phenomena mentioned were not forth-coming until *after* the new views were published, and were then exhibited in the presence of certain persons at spiritualistic *séances*. The fourth dimension was a very opportune discovery for the spiritualists and for theologians who were in a quandary about the location of hell. The use the spiritualist makes of the fourth dimension is this. It is possible to move out of a finite straight line, without passing the extremities, through the second dimension; out of a finite closed surface through the third; and, analogously, out of a finite closed space, without passing through the enclosing bound-aries, through the fourth dimension. Even the tricks that prestidigitateurs, in the old days, harmlessly executed in three dimensions, are now invested with a new halo by the fourth. But the tricks of the spiritualists, the tying or untying of knots in endless strings, the removing of bodies from closed spaces, are all performed in cases where there is absolutely nothing at stake. All is purposeless jugglery. We have not yet found an *accoucheur* who has accomplished parturition through the fourth dimension. If we should, the question would at once become a serious one. Professor Simony's beautiful tricks in rope-tying, which, as the performance of a prestidigitateur, are very admirable, speak against, not for, the spiritualists.

Everyone is free to set up an opinion and to adduce proofs in support of it. Whether, though, a scientist shall find it worth his while to enter into serious investi-gations of opinions so advanced, is a question which his reason and instinct alone can decide. If these things, in the end, should turn out to be true, I shall not be ashamed of being the last to believe them. What I have seen of them was not calculated to make me less sceptical.

I myself regarded multi-dimensioned space as a mathematico-physical help even prior to the appearance of Riemann's memoir. But I trust that no one will employ what I have thought, said, and written on this subject as a basis for the fabrication of ghost stories. (Compare Mach, *Die Geschichte und die Wurzel des Satzes von der Erhaltung der Arbeit.*)

NORMAN ROBERT CAMPBELL

NORMAN ROBERT CAMPBELL was a British physicist and philosopher of science highly regarded by specialists in these subjects but not well known to the general public. Educated at Eton and Trinity College, Cambridge, Campbell had the advantage of being a pupil of Sir J. J. Thomson. Later, as a Cavendish Research Fellow, he worked at Leeds under Sir William Bragg. In 1919 he joined the staff of General Electric Co. Ltd. and there he continued the researches on ionization phenomena he had begun while a Fellow of Trinity. He also applied himself with conspicuous success to a variety of problems of photometry, statistics, the adjustment of observations and related matters. He died in 1949 at the age of sixty-nine.

Campbell's output of nine books and some eighty-nine research papers was of unfailing merit. Always clear and strikingly original, Campbell enjoyed the "ability to provoke reconsideration of fundamentals."

His most important work is *Physics, the Elements,* an examination of the logical and philosophical bases of this science. It exhibits Campbell's freshness of outlook and his mastery of a broad field, coupled with a sure instinct for essentials. It is a distinguished achievement. The complete original manuscript of the book was lost in the mail and Campbell was forced to rewrite it from scratch. He bore this misfortune, as he bore more tragic events in later years, with extraordinary courage. Ill health, the death of his wife and of his son while on active service, the destruction of his home and nearly all his possessions by a bomb: all these were visited upon Campbell within the last few years of his life.

The excerpts below have been taken from a popular little book, *What Is Science?*, published in 1921. Campbell states in the preface that it was written "with the hope of encouraging the study of science" among adult education groups to which, for many years, he gave freely of his time. The material deals, first, with one of Campbell's favorite topics, the important concept of measurement; second, with numerical laws and the use of mathematics in science. These first-rate introductions are examples of Campbell's skill in popularization.

I've measured it from side to side:
'Tis three feet long and two feet wide. —WILLIAM WORDSWORTH

4 Measurement

By NORMAN CAMPBELL

WHAT IS MEASUREMENT?

MEASUREMENT is one of the notions which modern science has taken over from common sense. Measurement does not appear as part of common sense until a comparatively high stage of civilization is reached; and even the common-sense conception has changed and developed enormously in historic times. When I say that measurement belongs to common sense, I only mean that it is something with which every civilized person to-day is entirely familiar. It may be defined, in general, as the assignment of numbers to represent properties. If we say that the time is 3 o'clock, that the price of coal is 56 shillings a ton, and that we have just bought 2 tons of it—in all such cases we are using numbers to convey important information about the "properties" of the day, of coal in general, of the coal in our cellar, or so on; and our statement depends somehow upon measurement.

The first point I want to notice is that it is only some properties and not all that can be thus represented by numbers. If I am buying a sack of potatoes I may ask what it weighs and what it costs; to those questions I shall expect a number in answer; it weighs 56 lbs. and costs 5 shillings. But I may also ask of what variety the potatoes are, and whether they are good cookers; to those questions I shall not expect a number in answer. The dealer may possibly call the variety "No. 11" in somebody's catalogue; but even if he does, I shall feel that such use of a number is not real measurement, and is not of the same kind as the use in connexion with weight or cost. What is the difference? Why are some properties measurable and others not? Those are the questions I want to discuss. And I will outline the answer immediately in order that the reader may see at what the subsequent discussion is aiming. The difference is this. Suppose I have two sacks of potatoes which are identical in weight, cost, variety, and cooking qualities; and that I pour the two sacks into one so that there is now only one sack of potatoes. This sack will differ from the two original sacks in weight and cost (the measurable properties), but will not differ from them in variety and cooking qualities (the properties that are not measurable). The measurable properties of a body are those which are changed by the combination of similar bodies; the non-measurable properties are those that are not changed. We shall see that this definition is rather too crude, but it will serve for the present.

NUMBERS

In order to see why this difference is so important we must inquire more closely into the meaning of "number." And at the outset we must note that confusion is apt to arise because that word is used to denote two perfectly different things. It sometimes means a mere name or word or symbol, and it sometimes means a property of an object. Thus, besides the properties which have been mentioned, the sack of potatoes has another definite property, namely the number of potatoes in it, and the number is as much a property of the object which we call a sack of potatoes as its weight or its cost. This property can be (and must be) "represented by a number" just as the weight can be; for instance, it might be represented by 200. But this "200" is not itself a property of the sack; it is a mere mark on the paper for which would be substituted, if I was speaking instead of writing, a spoken sound; it is a name or symbol for the property. When we say that measurement is the representation of properties by "numbers," we mean that it is the representation of properties, other than number, by the symbols which are always used to represent number. Moreover, there is a separate word for these symbols; they are called "numerals." We shall always use that word in future and confine "number" to the meaning of the property which is always represented by numerals.

These considerations are not mere quibbling over words; they bring out clearly an important point, namely, that the measurable properties of an object must resemble in some special way the property number, since they can be fitly represented by the same symbols; they must have some quality common with number. We must proceed to ask what this common quality is, and the best way to begin is to examine the property number rather more closely.

The number of a sack of potatoes, or, as it is more usually expressed, the number of potatoes contained in it, is ascertained by the process of counting. Counting is inseparably connected in our minds to-day with numerals, but the process can be, and at an earlier stage of civilization was, carried on without them. Without any use of numerals I can determine whether the number of one sack of potatoes is equal to that of another. For this purpose I take a potato from one sack, mark it in some way to distinguish it from the rest (e.g. by putting it into a box), and then perform a similar operation on a potato from the other sack. I then repeat this double operation continually until I have exhausted the potatoes from one sack. If the operation which exhausts the potatoes from one sack exhausts also the potatoes from the other, then I know the sacks had the same number of potatoes; if not, then the sack which is not exhausted had a larger number of potatoes than the other.

THE RULES FOR COUNTING

This process could be applied equally well if the objects counted against each other were not of the same nature. The potatoes in a sack can be counted, not only against another collection of potatoes, but also against the men in a regiment or against the days in the year. The "mark," which is used for distinguishing the objects in the process of counting, may have to be altered to suit the objects counted, but some other suitable mark could be found which would enable the process to be carried out. If, having never heard of counting before, we applied the process to all kinds of different objects, we should soon discover certain rules which would enable us to abbreviate and simplify the process considerably. These rules appear to us to-day so obvious as to be hardly worth stating, but as they are undoubtedly employed in modern methods of counting, we must notice them here. The first is that if two sets of objects, when counted against a third set, are found to have the same number as that third set, then, when counted against each other they will be found to have the same number. This rule enables us to determine whether two sets of objects have the same number without bringing them together; if I want to find out whether the number of potatoes in the sack I propose to buy is the same as that in a sack I have at home, I need not bring my sack to the shop; I can count the potatoes at the shop against some third collection, take this collection home, and count it against my potatoes. Accordingly the discovery of this first rule immediately suggests the use of portable collections which can be counted, first against one collection and then against another, in order to ascertain whether these two have the same number.

The value of this suggestion is increased greatly by the discovery of a second rule. It is that by starting with a single object and continually adding to it another single object, we can build up a series of collections of which one will have the same number as any other collection whatsoever. This rule helps us in two ways. First, since it states that it is possible to make a standard series of collections one of which will have the same number as any other collection, it suggests that it might be well to count collections, not against each other, but against a standard series of collections. If we could carry this standard series about with us, we could always ascertain whether any one collection had the same number as any other by observing whether the member of the standard series which had the same number as the first had also the same number as the second. Next, it shows us how to make such a standard series with the least possible cumbrousness. If we had to have a totally different collection for each member of the standard series, the whole series would be impossibly cumbrous; but our rule shows that the earlier members of the series (that is those with the smaller number) may be all parts of the later members. Suppose we have a collection of objects, each distinguishable from each other, and agree to take one of these objects as the first member of the series; this object together with some other as the next member; these objects with yet another as the next member; and so on. Then we shall obtain, according to our rule, a series, some member of which has the same number as any other collection we want to count, and yet the number of objects, in all the members of the standard series taken together, will not be greater than that of the largest collection we want to count.

And, of course, this is the process actually adopted. For the successive members of the standard series compounded in this way, primitive man chose, as portable, distinguishable objects, his fingers and toes. Civilized man invented numerals for the same purpose. Numerals are simply distinguishable objects out of which we build our standard series of collections by adding them in turn to previous members of the series. The first member of our standard series is 1, the next 1, 2, the next 1, 2, 3 and so on. We count other collections against these members of the standard series and so ascertain whether or no two collections so counted have the same number. By an ingenious convention we describe which member of the series has the same number as a collection counted against it by quoting simply the last numeral in that member; we describe the fact that the collection of the days of the week has the same number as the collection 1, 2, 3, 4, 5, 6, 7, by saying "that the number" of the days of the week is 7. But when we say that what we really mean, and what is really important, is that this collection has the same number as the collection of numerals (taken in the standard order) which ends in 7 and the same number as any other collection which also has the same number as the collection of numerals which ends in 7.[1]

The two rules that have been mentioned are necessary to explain what we mean by "the number" of a collection and how we ascertain that number. There is a third rule which is of great importance in the use of

---

[1] Numerals have also an immense advantage over fingers and toes as objects of which the standard series may be formed, in that the series can be extended indefinitely by a simple rule which automatically gives names to any new numerals that may be required. Even if we have never hitherto had reason to carry the series beyond (say) 131679 in order to count all the collections we have met with, when we do meet at last with a larger collection, we know at once that the objects we must add to our standard series are 131680, 131681, and so on. This is a triumph of conventional nomenclature, much more satisfactory than the old convention that when we have exhausted our fingers we must begin on our toes, but it is not essentially different.

numbers. We often want to know what is the number of a collection which is formed by combining two other collections of which the numbers are known, or, as it is usually called, adding the two collections. For instance we may ask what is the number of the collection made by adding a collection of 2 objects to a collection of 3 objects. We all know the answer, 5. It can be found by arguing thus: The first collection can be counted against the numerals 1, 2; the second against the numerals 1, 2, 3. But the numerals 1, 2, 3, 1, 2, a collection formed by adding the two first collections, can be counted against 1, 2, 3, 4, 5. Therefore the number of the combined collection is 5. However, a little examination will show that in reaching this conclusion we have made use of another rule, namely that if two collections A and *a*, have the same number, and two other collections B and *b*, have the same number, then the collection formed by adding A to B has the same number as that formed by adding *a* to *b*; in other words, equals added to equals produce equal sums. This is a third rule about numbers and counting; it is quite as important as the other two rules; all three are so obvious to us to-day that we never think about them, but they must have been definitely discovered at some time in the history of mankind, and without them all, our habitual use of numbers would be impossible.

WHAT PROPERTIES ARE MEASURABLE?

And now, after this discussion of number, we can return to the other measurable properties of objects which, like number, can be represented by numerals. We can now say more definitely what is the characteristic of these properties which makes them measurable. It is that there are rules true of these properties, closely analogous to the rules on which the use of number depends. If a property is to be measurable it must be such that (1) two objects which are the same in respect of that property as some third object are the same as each other; (2) by adding objects successively we must be able to make a standard series one member of which will be the same in respect of the property as any other object we want to measure; (3) equals added to equals produce equal sums. In order to make a property measurable we must find some method of judging equality and of adding objects, such that these rules are true.

Let me explain what is meant by using as an example the measurable property, weight.

Weight is measured by the balance. Two bodies are judged to have the same weight if, when they are placed in opposite pans, neither tends to sink; and two bodies are added in respect of weight when they are both placed on the same pan of the balance. With these definitions of equality and addition, it is found that the three rules are obeyed. (1) If the body A balances the body B, and B balances C, then A balances C. (2) By placing

a body in one pan and continually adding it to others, collections can be
built up which will balance any other body placed in the other pan.[2]
(3) If the body A balances the body B, and C balances D, then A and C
in the same pan will balance B and D in the other pan. To make the
matter yet clearer let us take another measurable property, length. Two
straight rods are judged equal in length, if they can be placed so that both
ends of one are contiguous to both ends of the other; they are added in
respect of length, when they are placed with one end of one contiguous
with one end of the other, while the two form a single straight rod. Here
again we find the three rules fulfilled. Bodies which are equal in length
to the same body are equal in length to each other. By adding successively
rods to each other, a rod can be built up which is equal to any other rod.
And equal rods added to equal rods produce equal rods. Length is therefore a measurable property.

It is because these rules are true that measurement of these properties is
useful and possible; it is these rules that make the measurable properties
so similar to numbers, that it is possible and useful to represent them by
numerals the primary purpose of which is to represent numbers. It is
because of them that it is possible to find one, and only one numeral,
which will fitly represent each property; and it is because of them, that
these numerals, when they are found, tell us something useful about the
properties. One such use arises in the combination of bodies possessing
the properties. We may want to know how the property varies when
bodies possessing it are added in the way characteristic of measurement.
When we have assigned numerals to represent the property we shall know
that the body with the property 2 added to that with the property 3 will
have the same property as that with the property 5, or as the combination
of the bodies with properties 4 and 1. This is not the place to examine
exactly how these conclusions are shown to be universally valid; but they
are valid only because the three rules are true.

THE LAWS OF MEASUREMENT

But what is the nature of these rules? They are laws established by
definite experiment. The word "rule" has been used hitherto, because it
is not quite certain whether they are truly laws in their application to
number; but they certainly are laws in their application to other measurable properties, such as weight or length. The fact that the rules are true
can be, and must be, determined by experiment in the same way as the
fact that any other laws are true. Perhaps it may have appeared to the
reader that the rules must be true; that it requires no experiment to determine that bodies which balance the same body will balance each other;

[2] See further, pp. 1804–1805.

and that it is inconceivable that this rule should not be true. But I think he will change his opinion, if it is pointed out that the rule is actually true only in certain conditions; for instance, it is only true if the balance is a good one, and has arms of equal length and pans of equal weight. If the arms were unequal, the rule would not be found to be true unless it were carefully prescribed in which pan the bodies were placed during the judgment of equality. Again, the rules would not be true of the property length, unless the rods were straight and were rigid. In implying that the balance is good, and the rods straight and rigid, we have implied definite laws which must be true if the properties are to be measurable, namely that it is possible to make a perfect balance, and that there are rods which are straight and rigid. These are experimental laws; they could not be known apart from definite experiment and observation of the external world; they are not self-evident.

Accordingly the process of discovering that a property is measurable in the way that has been described, and setting up a process for measuring it, is one that rests entirely upon experimental inquiry. It is a part, and a most important part, of experimental science. Whenever a new branch of physics is opened up (for, as has been said, physics is the science that deals with such processes of measurement), the first step is always to find some process for measuring the new properties that are investigated; and it is not until this problem has been solved, that any great progress can be made along the branch. Its solution demands the discovery of new laws. We can actually trace the development of new measurable properties in this way in the history of science. Before the dawn of definite history, laws had been discovered which made measurable some of the properties employed by modern science. History practically begins with the Greeks, but before their time the properties, weight, length, volume, and area had been found to be measurable; the establishment of the necessary laws had probably occurred in the great period of Babylonian and Egyptian civilization. The Greeks, largely in the person of Archimedes, found how to measure force by establishing the laws of the lever, and other mechanical systems. Again from the earliest era, there have been rough methods of measuring periods of time,[3] but a true method, really obeying the three rules, was not discovered till the seventeenth century; it arose out of Galileo's laws of the pendulum. Modern science has added greatly to the list of measurable properties; the science of electricity is based on the discovery, by Cavendish and Coulomb, of the law necessary to measure an electric charge; on the laws, discovered by Œrsted and Ampère, neces-

---

[3] By a period of time I mean the thing that is measured when we say that it took us 3 hours to do so-and-so. This is a different "time" from that which is measured when we say it is 3 o'clock. The difference is rather abstruse and cannot be discussed here; but it may be mentioned that the "measurement" involved in "3 o'clock" is more like that discussed later in the article.

sary to measure an electric current; and on the laws, discovered by Ohm and Kirchhoff, necessary to measure electrical resistance. And the discovery of similar laws has made possible the development of other branches of physics.

But, it may be asked, has there ever been a failure to discover the necessary laws? The answer is that there are certainly many properties which are not measurable in the sense that we have been discussing; there are more properties, definitely recognized by science, that are not so measurable than are so measurable. But, as will appear presently, the very nature of these properties makes it impossible that they should be measured in this way. For the only properties to which this kind of measurement seems conceivably applicable, are those which fulfil the condition stated provisionally on p. 1797; they must be such that the combination of objects possessing the property increases that property. For this is the fundamental significance of the property number; it is something that is increased by addition; any property which does not agree with number in this matter cannot be very closely related to number and cannot possibly be measured by the scheme that has been described. But it will be seen that fulfilment of this condition only makes rule (2) true; it is at least conceivable that a property might obey rule (2) and not rules (1) and (3). Does that every happen, or can we always find methods of addition and of judging equality such that, if rule (2) is true, the laws are such that rules (1) and (3) are also true? In the vast majority of cases we can find such methods and such laws; and it is a very remarkable fact that we can; it is only one more instance of the way in which nature kindly falls in with our ideas of what ought to be. But I think there is one case in which the necessary methods and laws have not yet been found and are not likely to be found. It is a very difficult matter concerning which even expert physicists might differ, and so no discussion of it can be entered on here. But it is mentioned in order to impress the reader with the fact that measurement does depend upon experimental laws; that it does depend upon the facts of the external world; and that it is not wholly within our power to determine whether we will or will not measure a certain property. That is the feature of measurement which it is really important to grasp for a proper understanding of science.

MULTIPLICATION

Before we pass to another kind of measurement reference must be made to a matter which space does not allow to be discussed completely. In stating the rules that were necessary in order that weight should be measurable (p. 1801), it was said that a collection having the same weight as any given body could be made by adding other bodies to that first selected. Now this statement is not strictly true; it is only true if the body first selected has a smaller weight than any other body it is desired to weigh; and even if this condition is fulfilled, it is not true if the bodies added successively to the collection are of the same weight as that first selected. Thus if my first body weighs 1 lb., I cannot by adding to it make a collection which weighs less than 1 lb., and by adding bodies which each weigh 1 lb., I cannot make a collection which has the same weight as a body weighing (say) 2½ lb.

These facts, to which there is no true analogy in connexion with number, force us to recognize "fractions." A considerable complication is thereby introduced, and the reader must accept my assurance that they can all be solved by simple developments of the process of measurement that has been sketched. But for a future purpose it is necessary to notice very briefly the processes of the multiplication and division of magnitudes on which the significance of fractions depends.

Suppose I have a collection of bodies, each of which has the same weight 3, the number of bodies in the collection being 4. I may ask what is the weight of the whole collection. The answer is given of course by multiplying 3 by 4, and we all know now that the result of that operation is 12. That fact, and all the other facts summed up in the multiplication table which we learn at school, can be proved from the rules on which weighing depend together with facts determined by counting numerals. But the point I want to make is that multiplication represents a definite experimental operation, namely the combination into a single collection, placed on one pan of the balance, of a set of bodies, all of the same weight, the number of those bodies being known. Division arises directly out of multiplication. In place of asking what will be the weight of a collection formed of a given number of bodies all of the same weight, we ask what must be the weight of each of a collection of bodies, having a given number, when the whole collection has a given weight. E.g. what must each body weigh in order that the whole collection of 4 bodies weighs 12? The answer is obtained by dividing 12 by 4. That answer is obtained, partly from the multiplication table, partly by inventing new numerals which we call fractions; but once again division corresponds to a definite experimental operation and has its primary significance because it corresponds to that operation. It is the conclusion that we shall use in the sequel. But it is worth while noting that the fractions which we obtain by this method of addition overcome the difficulty from which this paragraph started. If we make all possible fractions of our original weight (i.e. all possible bodies, such that some number of them formed into a single collection have the same weight as the original body), then, by adding together suitable collections of these fractions, we can make up a collection which will have the same weight as any body whatever that we desire to weigh. This result is an experimental fact which could not have been predicted without experimental inquiry. And the result is true, not only for the measurable property weight, but for all properties measurable by the process that is applicable to weight. Once more we see how much simpler and more conveniently things turn out than we have really any right to expect; measurement would have been a much more complex business if the law that has just been stated were not always true.

DERIVED MEASUREMENT

Measurement, it was said on p. 1797, is the assignment of numbers (or, as we say now, numerals) to represent properties. We have now considered one way in which this assignment is made, and have brought to light the laws which must be true if this way is to be possible. And it is the fundamental way. We are now going to consider some other ways in which numerals are assigned to represent properties; but it is important to insist at the outset, and to remember throughout, that these other ways are wholly dependent upon the fundamental process, which we have just been discussing, and must be so dependent if the numerals are to represent "real properties" and to tell us something scientifically significant about the bodies to which they are attached. This statement is confirmed by history; all properties measured in the definitely pre-scientific era were measured (or at least measurable) by the fundamental process; that is true of weight, length, volume, area and periods of time. The dependent measurement, which we are now about to consider, is a product of definitely and consciously scientific investigation, although the actual discovery may, in a few cases, be lost in the mists of the past.

The property which we shall take as an example of this dependent or, as it will be termed, derived measurement, is *density*. Every one has some idea of what density means and realizes, vaguely at least, why we say that iron is denser than wood or mercury than water; and most people probably know how density is measured, and what is meant when it is said that the density of iron is 8 times that of wood, and the density of mercury 13½ times of water. But they will feel also that there is something more scientific and less purely commonsense about the measurement of density than about the measurement of weight; as a matter of fact the discovery of the measurement of density certainly falls within the historic period and probably may be attributed to Archimedes (about 250 B.C.). And a little reflection will convince them that there is something essentially different in the two processes.

For what we mean when we say a body has a weight 2 is that a body of the same weight can be made by combining 2 bodies of the weight 1; that is the fundamental meaning of weight; it is what makes weight physically important and, as we have just seen, makes it measurable. But when we say that mercury has a density 13½ we do *not* mean that a body of the same density can be prepared by combining 13½ bodies of the density 1 (water). For, if we did mean that, the statement would not be true. However many pieces of water we take, all of the same density, we cannot produce a body with any different density. Combine water with water as we will, the resulting body has the density of water. And this, a little reflection will show, is part of the fundamental meaning of density; density is something that is characteristic of all pieces of water, large and small. The density of water, a "quality" of it, is something fundamentally independent of and in contrast with the weight of water, the "quantity" of it.

But the feature of density, from which it derives its importance, makes it totally impossible to measure density by the fundamental process discussed earlier in the chapter. How then do we measure it? Before we answer that question, it will be well to put another. As was insisted before, if measurement is really to mean anything, there must be some important resemblance between the property measured, on the one hand, and the numerals assigned to represent it, on the other. In fundamental measurement, this resemblance (or the most important part of it) arises from the fact that the property is susceptible to addition following the same rules as that of number, with which numerals are so closely associated. That resemblance fails here. What resemblance is left?

MEASUREMENT AND ORDER

There is left a resemblance in respect of "order." The numerals are characterized, in virtue of their use to represent numbers, by a definite order; they are conventionally arranged in a series in which the sequence is determined: "2" follows "1" and is before "3"; "3" follows "2" and is before "4" and so on. This characteristic order of numerals is applied usefully for many purposes in modern life; we "number" the pages of a book or the houses of a street, not in order to know the number of pages in the book or of houses in the street—nobody but the printer or the rate-surveyor cares about that—but in order to be able to find any given page or house easily. If we want p. 201 and the book opens casually at p. 153 we know in which direction to turn the pages.[4] Order then is characteristic of numerals; it is also characteristic of the properties represented by numerals in the manner we are considering now. This is our feature which makes the "measurement" significant. Thus, in our example, bodies have a natural order of density which is independent of actual measure-

---

[4] Numerals are also used to represent objects, such as soldiers or telephones, which have no natural order. They are used here because they provide an *inexhaustible* series of names, in virtue of the ingenious device by which new numerals can always be invented when the old ones have been used up.

ment. We might define the words "denser" or "less dense" as applied to liquids (and the definition could easily be extended to solids) by saying that the liquid A is denser than B, and B less dense than A, if a substance can be found which will float in A but not in B. And, if we made the attempt, we should find that by use of this definition we could place all liquids in a definite order, such that each member of the series was denser than the preceding and less dense than the following member. We might then assign to the first liquid the density 1, to the second 2, and so on; and we should then have assigned numerals in a way which would be physically significant and indicate definite physical facts. The fact that A was represented by 2 and B by 7 would mean that there was some solid body which would float in B, but not in A. We should have achieved something that might fairly be called measurement.

Here again it is important to notice that the possibility of such measurement depends upon definite laws; we could not have predicted beforehand that such an arrangement of liquids was possible unless we knew these laws. One law involved is this: If B is denser than A, and C denser than B, then C is denser than A. That sounds like a truism; but it is not. According to our definition it implies that the following statement is always true: If a body X floats in B and sinks in A, then if another body Y sinks in B it will also sink in A. That is a statement of facts; nothing but experiment could prove that it is true; it is a law. And if it were not true, we could not arrange liquids naturally in a definite order. For the test with X would prove that B was denser than A, while the test with Y (floating in A, but sinking in B) would prove that A was denser than B. Are we then to put A before or after B in the order of density? We should not know. The order would be indeterminate and, whether we assigned a higher or a lower numeral to A than to B, the assignment would represent no definite physical fact: it would be arbitrary.

In order to show that the difficulty might occur, and that it is an experimental law that it does not occur, an instance in which a similar difficulty has actually occurred may be quoted. An attempt has been made to define the "hardness" of a body by saying that A is harder than B if A will scratch B. Thus diamond will scratch glass, glass iron, iron lead, lead chalk, and chalk butter; so that the definition leads to the order of hardness: diamond, glass, iron, lead, chalk, butter. But if there is to be a definite order, it must be true in all cases that if A is harder than B and B than C, then A is harder than C; in other words, if A will scratch B and B C, then A will scratch C. But it is found experimentally that there are exceptions to this rule, when we try to include all bodies within it and not only such simple examples as have been quoted. Accordingly the definition does not lead to a definite order of hardness and does not permit the measurement of hardness.

There are other laws of the same kind that have to be true if the order is to be definite and the measurement significant; but they will not be given in detail. One of them the reader may discover for himself, if he will consider the property colour. Colour is not a property measurable in the way we are considering, and for this reason. If we take all reds (say) of a given shade, we can arrange them definitely in an order of lightness and darkness; but no colour other than red will fall in this order. On the other hand, we might possibly take all shades and arrange them in order of redness—pure red, orange, yellow, and so on; but in this order there would be no room for reds of different lightness. Colours cannot be arranged in a single order, and it is for this reason that colour is not measurable as is density.

NUMERICAL LAWS

But though arrangement in this manner in an order and the assignment of numerals in the order of the properties are to some extent measurement and represent something physically significant, there is still a large arbitrary element involved. If the properties A, B, C, D, are naturally arranged in that order, then in assigning numerals to represent the properties I must *not* assign to A 10, to B 3, to C 25, to D 18; for if I did so the order of the numerals would not be that of the properties. But I have an endless number of alternatives left; I might put A 1, B 2, C 3, D 4; or A 10, B 100, C 1,000, D 10,000; or A 3, B 9, C 27, D 81; and so on. In the true and fundamental measurement of the first part of the chapter there was no such latitude. When I had fixed the numeral to be assigned to one property, there was no choice at all of the numerals to be assigned to the others; they were all fixed. Can I remove this latitude here too and find a way of fixing definitely what numeral is to be assigned to represent each property?

In some cases, I can; and one of these cases is density. The procedure is this: I find that by combining the numerals representing other properties of the bodies, which can be measured definitely according to the fundamental process, I can obtain a numeral for each body, and that these numerals lie in the order of the property I want to measure. If I take these numerals as representing the property, then I still get numerals in the right order, but the numeral for each property is definitely fixed. An example will be clearer than this general statement. In the case of density, I find that if I measure the weight and the volume of a body (both measurable by the fundamental process and therefore definitely fixed), and I divide the weight by the volume, then the numerals thus obtained for different bodies lie in the order of their densities, as density was defined on p. 1808. Thus I find that 1 gallon of water weighs 10 lb., but 1 gallon of mercury

weighs 135 lb.; the weight divided by the volume for water is 10, for
mercury is 135; 135 is greater than 10; accordingly, if the method is
correct, mercury should be denser than water and any body which sinks
in mercury should sink in water. And that is actually found to be true. If
therefore I take as the measure of the density of a substance, its weight
divided by its volume, then I get a number which is definitely fixed,[5] and
the order of which represents the order of density. I have arrived at a
method of measurement which is as definitely fixed as the fundamental
process and yet conveys adequately the physically significant facts about
order.

The invention of this process of measurement for properties not suited
for fundamental measurement is a very notable achievement of deliberate
scientific investigation. The process was not invented by common sense;
it was certainly invented in the historic period, but it was not until the
middle of the eighteenth century that its use became widespread.[6] To-day
it is one of the most powerful weapons of scientific investigation; and it is
because so many of the properties of importance to other sciences are
measured in this way that physics, the science to which this process belongs, is so largely the basis of other sciences. But it may appear exceedingly obvious to the reader, and he may wonder why the invention was
delayed so long. He may say that the notion of density, in the sense that
a given volume of the denser substance weighs more than the same volume
of the less dense, is the fundamental notion; it is what we mean when we
speak of one substance being denser (or in popular language "heavier")
than another; and that all that has been discovered in this instance is
that the denser body, in this sense, is also denser in the sense of p. 1808.
This in itself would be a very noteworthy discovery, but the reader who
raises such an objection has overlooked a yet more noteworthy discovery
that is involved.

For we have observed that it is one of the most characteristic features
of density that it is the same for all bodies, large and small, made of the
same substance. It is this feature which makes it impossible to measure
it by the fundamental process. The new process will be satisfactory only
if it preserves this feature. If we are going to represent density by the
weight divided by the volume, the density of all bodies made of the same
substance will be the same, as it should be, only if for all of them the

[5] Except in so far as I may change the units in which I measure weights and volume. I should get a different number if I measured the volume in pints and the
weight in tons. But this latitude in the choice of units introduces a complication
which it will be better to leave out of account here. There is no reason why we should
not agree once and for all to use the same units; and if we did that the complication
would not arise.
[6] I think that until the eighteenth century only two properties were measured in
this way which were not measurable by the fundamental process, namely density and
constant acceleration.

weight divided by the density is the same, that is to say, in rather more technical language, if the weight is proportional to the density. In adopting the new process for measuring density and assigning numerals to represent it in a significant manner, we are, in fact, assuming that, for portions of the same substance, whether they are large or small, the weight is proportional to the volume. If we take a larger portion of the same substance and thereby double the weight, we must find, if the process of measurement is to be a success, that we also double the volume; and this law must be true for all substances to which the conception of density is applicable at all.

Of course every one knows that this relation is actually true; it is so familiar that we are apt to forget that it is an experimental truth that was discovered relatively late in the history of civilization, which easily might not be true. Perhaps it is difficult to-day to conceive that when we take "more" of a substance (meaning thereby a greater volume) the weight should not increase, but it is quite easy to conceive that the weight should not increase proportionally to the volume; and yet it is upon strict proportionality that the measurement of density actually depends. If the weight had not been proportional to the volume, it might still have been possible to measure density, so long as there was some fixed numerical relation between weight and volume. It is this idea of a fixed numerical relation, or, as we shall call it henceforward, a numerical law, that is the basis of the "derived" process of measurement that we are considering; and the process is of such importance to science because it is so intimately connected with such numerical laws. The recognition of such laws is the foundation of modern physics.

THE IMPORTANCE OF MEASUREMENT

For why is the process of measurement of such vital importance; why are we so concerned to assign numerals to represent properties. One reason doubtless is that such assignment enables us to distinguish easily and minutely between different but similar properties. It enables us to distinguish between the density of lead and iron far more simply and accurately than we could do by saying that lead is rather denser than iron, but not nearly so dense as gold—and so on. But for that purpose the "arbitrary" measurement of density, depending simply on the arrangements of the substances in their order (p. 1808), would serve equally well. The true answer to our question is seen by remembering that the terms between which laws express relationships are themselves based on laws and represent collections of other terms related by laws. When we measure a property, either by the fundamental process or by the derived process, the numeral which we assign to represent it is assigned as the result of experimental laws; the assignment implies laws. And therefore, in accordance with our principle, we should expect to find that other laws could be discovered relating the numerals so assigned to each other or to something else; while if we assigned numerals arbitrarily without reference to laws and implying no laws, then we should not find other laws involving these numerals. This expectation is abundantly fulfilled, and nowhere is there a clearer example of the fact that the terms involved in laws themselves imply laws. When we can measure a property truly, as we can volume (by the fundamental process) or density (by the derived process) then we are always able to find laws in which these properties are involved; we find, e.g., the law that volume is proportional to weight or that density determines, in a certain precise fashion, the sinking or floating of bodies. But when we cannot measure it truly, then we do not find a law. An example is provided by the property "hardness" (p. 1808); the difficulties met with in arranging bodies in order of hardness have been overcome; but we still do not know of any way of measuring, by the derived process, the property hardness; we know of no numerical law which leads to a numeral which always follows the order of hardness. And so, as we expect, we do not know any accurate and general laws relating hardness to other properties. It is because true measurement is essential to the discovery of laws that it is of such vital importance to science.

One final remark should be made before we pass on. In this chapter there has been much insistence on the distinction between fundamental measurement (such as is applicable to weight) and derived measurement (such as is applicable to density). And the distinction is supremely important, because it is the first kind of measurement which alone makes the second possible. But the reader who, when he studies some science in detail, tries, as he should, to discover which of the two processes is involved in the measurement of the various properties characteristic of that science, may occasionally find difficulty in answering the question. It should be pointed out, therefore, that it is quite possible for a property to be measurable by both processes. For all properties measurable by the fundamental process must have a definite order; for the physical property, number, to which they are so similar, has an order—the order of "more" or "less." This order of number is reflected in the order of the numerals used to represent number. But if it is to be measurable by the derived process, it must also be such that it is also a "constant" in a numerical law—a term that is just going to be explained in the next chapter.[7] There is nothing in the nature of fundamental measurement to show that a property to which it is applicable may not fulfil this condition also; and sometimes the condition is fulfilled, and then the property is measurable

[7] [See next selection. ED.]

either by the fundamental or the derived process. However, it must be remembered that the properties involved in the numerical law must be such that they are fundamentally measurable; for otherwise the law could not be established. The neglect of this condition is apt to lead to confusion; but with this bare hint the matter must be left.

*I do hate sums. There is no greater mistake than to call arithmetic an
exact science. There are permutations and aberrations discernible to minds
entirely noble like mine; subtle variations which ordinary accountants fail
to discover; hidden laws of number which it requires a mind like mine to
perceive. For instance, if you add a sum from the bottom up, and then
again from the top down, the result is always different.*
—MRS. LA TOUCHE (*19th century*)
(*Mathematical Gazette, vol. 12*)

5 Numerical Laws and the Use of Mathematics in Science

By NORMAN CAMPBELL

NUMERICAL LAWS

IN THE previous chapter [the preceding selection] we concluded that
density was a measurable property because there is a fixed numerical relation, asserted by a "numerical law," between the weight of a substance and
its volume. In this chapter we shall examine more closely the idea of a
numerical law, and discover how it leads to such exceedingly important
developments.

Let us first ask exactly what we do when we are trying to discover a
numerical law, such as that between weight and volume. We take various
portions of a substance, measure their weights and their volumes, and put
down the result in two parallel columns in our notebook. Thus I may find
these results:

TABLE I

WEIGHT	VOLUME	WEIGHT	VOLUME
1	7	4	28
2	14	10	70
3	21	29	203

I now try to find some fixed relation between the corresponding numbers
in the two columns; and I shall succeed in that attempt if I can find some
rule whereby, starting with the number in one column, I can arrive at the
corresponding number in the other. If I find such a rule—and if the rule
holds good for all the further measurements that I may make—then I
have discovered a numerical law.

In the example we have taken the rule is easy to find. I have only to
divide the numbers in the second column by 7 in order to arrive at those
in the first, or multiply those in the first by 7 in order to arrive at those in
the second. That is a definite rule which I can always apply whatever the

numbers are; it is a rule which might always be true, but need not always
be true; whether or no it is true is a matter for experiment to decide. So
much is obvious; but now I want to ask a further and important question.
How did we ever come to discover this rule; what suggested to us to try
division or multiplication by 7: and what is the precise significance of
division and multiplication in this connexion?

THE SOURCE OF NUMERICAL RELATIONS

The answer to the first part of this question is given by the discussion
on p. 1805.[1] Division and multiplication are operations of importance in
the counting of objects; in such counting the relation between 21, 7, 3
(the third of which results from the division of the first by the second)
corresponds to a definite relation between the things counted; it implies
that if I divide the 21 objects into 7 groups, each containing the same
number of objects, then the number of objects in each of the 7 groups is
3. By examining such relations through the experimental process of counting we arrive at the multiplication (or division) table. This table, when it
it completed, states a long series of relations between numerals, each of
which corresponds to an experimental fact; the numerals represent physical properties (numbers) and in any given relation (e.g. $7 \times 3 = 21$)
each numeral represents a different property. But when we have got the
multiplication table, a statement of relations between numerals, we can
regard it, and do usually regard it, *simply* as a statement of relations
between numerals; we can think about it without any regard to what those
numerals represented when we were drawing up the table. And if any
other numerals are presented to our notice, it is possible and legitimate to
ask whether these numerals, whatever they may represent, are in fact
related as are the numerals in the multiplication table. In particular, when
we are seeking a numerical relation between the columns of Table I, we
may inquire, and it is natural for us to inquire, whether by means of the
multiplication we can find a rule which will enable us to arrive at the
numeral in the second column starting from that in the first.

That explains why it is so natural to us to try division when we are
seeking a relation between numbers. But it does not answer the second
part of the question; for in the numerical law that we are considering, the
relation between the things represented by the numerals is *not* that which
we have just noticed between things counted. When we say that, by
dividing the volume by 7, we can arrive at the weight, we do *not* mean
that the weight *is* the volume of each of the things at which we arrive by
dividing the substance into 7 portions, each having the same volume. For
a weight can never *be* a volume, any more than a soldier can *be* a num-

---
[1] [See preceding selection. ED.]

ber; it can only be represented by the same numeral as a volume, as a soldier can be represented by a numeral which also represents a number.

The distinction is rather subtle, but if the reader is to understand what follows, he must grasp it. The relation which we have found between weight and volume is a pure numerical relation; it is suggested by the relation between actual things, namely collections which we count; but it is not that relation. The difference may be expressed again by means of the distinction between numbers and numerals. The relation between actual things counted is a relation between the numbers—which are physical properties—of those things; the relation between weight and volume is a relation between numerals, the numerals that are used to represent those properties. The physical relation in the second case is not between numbers at all, but between weight and volume which are properties quite different from numbers; it appears very similar to that between numbers only because we use numerals, originally invented to represent numbers, to represent other properties. The relation stated by a numerical law is a relation between numerals, and only between numerals, though the idea that there may be such a relation has been suggested to us by the study of the physical property, number.

If we understand this, we shall see what a very remarkable thing it is that there should be numerical laws at all, and shall see why the idea of such a law arose comparatively late in the history of science. For even when we know the relations between numbers, there is no reason to believe that there must be any relations of the same kind between the numerals which are used to represent, not only numbers, but also other properties. Until we actually tried, there was no reason to think that it must be possible to find at all numerical laws, stating numerical relations such as those of division and multiplication. The fact that there are such relations is a new fact, and ought to be surprising. As has been said so often, it does frequently turn out that suggestions made simply by our habits of mind are actually true; and it is because they are so often true that science is interesting. But every time they are true there is reason for wonder and astonishment.

And there is a further consequence yet more deserving of our attention at present. If we realize that the numerical relations in numerical laws, though suggested by relations between numbers, are not those relations, we shall be prepared to find also numerical relations which are not even suggested by relations between numbers, but only by relations between numerals. Let me take an example. Consider the pairs of numerals $(1, 1)$, $(2, 4)$, $(3, 9)$, $(4, 16)$ . . . Our present familiarity with numerals enables us to see at once what is the relation between each pair; it is that the second numeral of the pair is arrived at by multiplying the first numeral *by itself*; 1 is equal to 1×1, 4 to 2×2, 9 to 3×3; and so on.

But, if the reader will consider the matter, he will see that the multiplication of a number (the physical property of an object) by itself does not correspond to any simple relation between the things counted; by the mere examination of counted objects, we should never be led to consider such an operation at all. It is suggested to us only because we have drawn up our multiplication table and have reached the idea of multiplying one *numeral* by another, irrespective of what is represented by that numeral. We know what is the result of multiplying 3 × 3, when the two 3's represent different numbers and the multiplication corresponds to a physical operation on things counted; it occurs to us that the multiplication of 3 by *itself*, when the two 3's represent the same thing, although it does not correspond to a physical relation, may yet correspond to the numerical relation in a numerical law. And we find once more that this suggestion turns out to be true; there are numerical laws in which this numerical relation is found. Thus if we measure (1) the time during which a body starting from rest has been falling, (2) the distance through which it has fallen during that time, we should get in our notebook parallel columns like this:

TABLE II

TIME	DISTANCE		TIME	DISTANCE
1	.. 1		4	.. 16
2	.. 4		5	.. 25
3	.. 9		6	.. 36

The numerals in the second column are arrived at by multiplying those in the first by themselves; in technical language, the second column is the "square" of the first.

Another example. In place of dividing one column by some fixed number in order to get the other, we may use the multiplication table to divide some fixed number (e.g. 1) by that column. Then we should get the table

1	.. 1·00		3	.. 0·33
2	.. 0·50		4	.. 0·25
	5	..	0·20	

and so on. Here, again, is a pure numerical operation which does not correspond to any simple physical relation upon numbers; there is no collection simply related to another collection in such a way that the number of the first is equal to that obtained by dividing 1 by the number of the second. (Indeed, as we have seen that fractions have no application to number, and since this rule must lead to fractions, there cannot be such a relation.) And yet once more we find that this numerical relation does occur in a numerical law. If the first column represented the pressure on a given amount of gas, the second would represent the volume of that gas.

So far, all the relations we have considered were derived directly from the multiplication table. But an extension of the process that we are tracing leads to relations which cannot be derived directly and thus carries us further from the original suggestions indicated by mere counting. Let us return to Table II, and consider what would happen if we found for the numerals in the second column values intermediate between those given. Suppose we measured the distance first and found 2, 3, 5, 6, 7, 8, 10, 11, 12, 13, 14, 15 . . . ; what does the rule lead us to expect for the corresponding entries in the first column, the values of the time. The answer will be given if in the multiplication table we can find numerals which, when multiplied by themselves, give 2, 3, 5 . . . But a search will reveal that there are no such numerals. We can find numerals which, when multiplied by themselves give very nearly 2, 3, 5 . . . ; for instance, $1 \cdot 41$, $1 \cdot 73$, $2 \cdot 24$ give $1 \cdot 9881$, $2 \cdot 9929$, $5 \cdot 0166$, and we could find numerals which would come even closer to those desired. And that is really all we want, for our measurements are never perfectly accurate, and if we can get numerals which agree very nearly with our rule, that is all that we can expect. But the search for such numerals would be a very long and tedious business; it would involve our drawing up an enormously complicated multiplication table, including not only whole numbers but also fractions with many decimal places. And so the question arises if we cannot find some simpler rule for obtaining quickly the number which multiplied by itself will come as close as we please to 2, 3, 4 . . . Well, we can; the rule is given in every textbook of arithmetic; it need not be given here. The point which interests us is that, just as the simple multiplication of two numerals suggested a new process, namely the multiplication of a numeral by itself, so this new process suggests in its turn many other and more complicated processes. To each of these new processes corresponds a new rule for relating numerals and for arriving at one starting from another; and to each new rule may correspond a numerical law. We thus get many fresh forms of numerical law suggested, and some of them will be found to represent actual experiments.

This process for extending arithmetical operations beyond the simple division and multiplication from which we start; the consequent invention of new rules for relating numerals and deriving one from another; and the study of the rules, when they are invented—all this is a purely intellectual process. It does not depend on experiment at all; experiment enters only when we inquire whether there is an actual experimental law stating one of the invented numerical relations between measured properties. The process is, in fact, part of mathematics, not of experimental science; and one of the reasons why mathematics is useful to science is that it suggests possible new forms for numerical laws. Of course the examples that have been given are extremely elementary, and the actual mathematics of to-day has diverged very widely from such simple con-

siderations; but the invention of such rules leads, logically if not historically, to one of the great branches of modern mathematics, the Theory of
Functions. (When two numbers are related as in our tables, they are
technically said to be "functions" of each other.) It has been developed
by mathematicians to satisfy their own intellectual needs, their sense of
logical neatness and of form; but though great tracts of it have no bearing
whatever upon experimental science, it still remains remarkable how often
relations developed by the mathematician for his own purposes prove in
the end to have direct and immediate application to the experimental facts
of science.

NUMERICAL LAWS AND DERIVED MEASUREMENT

In this discussion there has been overlooked temporarily the feature
of numerical laws which, in the previous chapter, we decided gave rise
to their importance, namely, that they made possible systems of derived
measurement. In the first law, taken as an example (Table I), the rule by
which the numerals in the second column were derived from those in the
first involved a numeral 7, which was not a member of those columns,
but an additional number applicable equally to all members of the columns. This constant numeral, characteristic of the rule asserted by the
numerical law, represented a property of the system investigated and
permitted a derived measurement of that system. But in Table II, there is
no such constant numeral; the rule for obtaining the second from the
first column is simply that the numerals in the first column are to be
multiplied by themselves; no other numeral is involved. But this simplicity
is really misleading; we should not, except by a mere "fluke," ever get
such a table as Table II as a result of our measurements. The reason is
this. Suppose that, in obtaining Table II, we have measured the time in
seconds and the distance fallen in feet; and that we now propose to write
down the result of exactly the same measurements, measuring the time in
minutes and the distances in yards. Then the numerals in the first column,
representing exactly the same observations, would all be divided by 60
and those in the second would all be divided by 3; the observation which
was represented before by 60 in the first column would now be represented by 1; and the number in the second column represented before by
3 would now be represented by 1. If I now apply the rule to the two
columns I shall find it will not work; the second is *not* the first multiplied
by itself. But there will be a new rule, as the reader may see for himself;
it will be that the second column is the same as the first, when the first is
(1) multiplied by itself, and (2) the result multiplied by 1,200. And if
we measured the time and the distance in some other units (says hours
and miles), we should again have to amend our rule, but it would only
differ from the former rule in the substitution for 1,200 of some other

numeral. If we choose our units in yet a third way, we should get a third rule, and this time the constant numeral might be 1. We should have exactly Table II; but we should get that table exactly only because we had chosen our units of time and distance in a particular way.

These considerations are quite general. Whatever the numerical law, the rule involved in it will be changed by changing the unit in which we measure the properties represented by the two columns; but the change will only consist in the substitution of one constant numeral for another. If we chance to choose the units in some particular way, that constant numeral may turn out to be 1 and so will disappear from sight. But it will always be there. There must be associated with every numerical law, involving a rule for arriving at the numerals in one column from those in the other, some constant numeral which is applicable to all members of the column alike. And this constant may always, as in the case of density, be the measure of some property to which derived measurement is applicable. Every numerical law therefore—this is the conclusion to be enforced—may give rise to a system of derived measurement; and as a matter of fact all important numerical laws do actually so give rise.

CALCULATION

But though the establishment of system of derived measurement is one use of numerical laws, they have also another use, which is even more important. They permit *calculation*. This is an extremely important conception which deserves our close attention.

Calculation is the process of combining two or more numerical laws in such a way as to produce a third numerical law. The simplest form of it may be illustrated by the following example. We know the following two laws which, in rather different forms, have been quoted before: (1) the weight of a given volume of any substance is proportional to its density; (2) the density of a gas is proportional to the pressure upon it. From these two laws we can deduce the third law: the weight of a given volume of any gas is proportional to the pressure upon it. That conclusion seems to follow directly without any need for further experiments. Accordingly we appear to have arrived at a fresh numerical law without adducing any fresh experimental evidence. But is that possible? All our previous inquiry leads us to believe that laws, whether numerical or other, can only be proved by experimental inquiry and that the proof of a new law without new experimental evidence is impossible. How are we to reconcile the two conclusions? When we have answered that question we shall understand what is the importance of calculation for science.

Let us first note that it is possible, without violating the conclusions already reached, to deduce *something* from a numerical law by a process of mere thought without new experiment. For instance, from the law that the density of iron is 7, I can deduce that a portion of it which has a volume 1 will have a weight 7. But this deduction is merely stating in new terms what was asserted by the original law; when I said that the density of iron was 7, I meant (among other things) that a volume 1 had a weight 7; if I had not meant that I should never have asserted the law. The "deduction" is nothing but a translation of the law (or of part of it), into different language, and is of no greater scientific importance than a translation from (say) English into French. One kind of translation, like the other, may have useful results, but it is not the kind of useful result that is obtained from calculation. Pure deduction never achieves anything but this kind of translation; it never leads to anything new. But the calculation taken as an example does lead to something new. Neither when I asserted the first law, nor when I asserted the second did I mean what is asserted by the third; I might have asserted the first without knowing the second and the second without knowing the first (for I might have known what the density of a gas was under different conditions without knowing precisely how it is measured); and I might have asserted either of them, without knowing the third. The third law is not merely an expression in different words of something known before; it is a new addition to knowledge.

But we have added to knowledge only because we have introduced an assertion which was not contained in the two original statements. The deduction depends on the fact that if one thing (A) is proportional to another thing (B) and if B is proportional to a third thing (C), then A is proportional to C. This proposition was not contained in the original statements. But, the reader may reply, it *was* so contained, because it is involved in the very meaning of "proportional"; when we say that A is proportional to B, we mean to imply the fact which has just been stated. Now that is perfectly true if we are thinking of the mathematical meaning of "proportional," but it is not true if we are thinking of the physical meaning. The proposition which we have really used in making our deduction is this: If weight is proportional (in the mathematical sense) to density, when weight is varied by taking different substances, then it is also proportional to density when weight is varied by compressing more of the same substance into the same volume. That is a statement which experiment alone can prove, and it is because we have in fact assumed that experimental statement that we have been able to "deduce" a new piece of experimental knowledge. It is involved in the original statements only if, when it is said that density is proportional to pressure, it is implied that it has been ascertained by experiment that the law of density is true, and that there is a constant density of a gas, however compressed, given by dividing the weight by the volume.

The conclusion I want to draw is this. When we appear to arrive at new scientific knowledge by mere deduction from previous knowledge, we are always assuming some experimental fact which is not clearly involved in the original statements. What we usually assume is that some law is true in circumstances rather more general than those we have considered hitherto. Of course the assumption may be quite legitimate, for the great value of laws is that they are applicable to circumstances more general than those of the experiments on which they are based; but we can never be perfectly sure that it is legitimate until we try. Calculation, then, when it appears to add anything to our knowledge, is always slightly precarious; like theory, it suggests strongly that some law may be true, rather than proves definitely that some law must be true.

So far we have spoken of calculation as if it were merely deduction; we have not referred to the fact that calculation always involves a special type of deduction, namely mathematical deduction. For there are, of course, forms of deduction which are not mathematical. All argument is based, or should be based, upon the logical processes which are called deduction; and most of us are prepared to argue, however slight our mathematical attainments. I do not propose to discuss here generally what are the distinctive characteristics of mathematical argument; for an exposition of that matter the reader should turn to works in which mathematicians expound their own study.[2] I want only to consider why it is that this kind of deduction has such a special significance for science. And, stated briefly, the reason is this. The assumption, mentioned in the last paragraph, which is introduced in the process of deduction, is usually suggested by the form of the deduction and by the ideas naturally associated with it. (Thus, in the example we took, the assumption is suggested by the proposition quoted about proportionality which is the idea especially associated by the form of the deduction.) The assumptions thus suggested by mathematical deduction are almost invariably found to be actually true. It is this fact which gives to mathematical deduction its special significance for science.

THE NEWTONIAN ASSUMPTION

Again an example is necessary and we will take one which brings us close to the actual use of mathematics in science. Let us return to Table II which gives the relation between the time for which a body has fallen and the distance through which it has fallen. The falling body, like all moving bodies, has a "velocity." By the velocity of a body we mean the distance that it moves in a given time, and we measure the velocity by

---

[2] E.g. "An Introduction to Mathematics," by Prof. Whitehead, in the Home University Library.

dividing that distance by that time (as we measure density by dividing the weight by the volume). But this way of measuring velocity gives a definite result only when the velocity is constant, that is to say, when the distance travelled is proportional to the time and the distance travelled in any given time is always the same (compare what was said about density on p. 1809.[3] This condition is not fulfilled in our example; the distance fallen in the first second is 1, in the next 3, in the third 5, in the next 7—and so on. We usually express that fact by saying that the velocity increases as the body falls; but we ought really to ask ourselves whether there is such a thing as velocity in this case and whether, therefore, the statement can mean anything. For what is the velocity of the body at the end of the 3rd second—i.e. at the time called 3. We might say that it is to be found by taking the distance travelled in the second before 3 which is 5, or in the second after 3, which is 7, or in the second of which the instant "3" is the middle (from 2½ to 3½), which turns out to be 6. Or again we might say it is to be found by taking *half* the distance travelled in the two seconds of which "3" is the middle (from 2 to 4) which is again 6. We get different values for the velocity according to which of these alternatives we adopt. There are doubtless good reasons in this example for choosing the alternative 6, for two ways (and really many more than two ways, all of them plausible) lead to the same result. But if we took a more complicated relation between time and distance than that of Table II, we should find that these two ways gave different results, and that neither of them were obviously more plausible than any alternative. Do then we mean anything by velocity in such cases and, if so, what do we mean?

It is here that mathematics can help us. By simply thinking about the matter Newton, the greatest of mathematicians, devised a rule by which he suggested that velocity might be measured in all such cases.[4] It is a rule applicable to every kind of relation between time and distance that actually occurs; and it gives the "plausible" result whenever that relation is so simple that one rule is more plausible than another. Moreover it is a very pretty and ingenious rule; it is based on ideas which are themselves attractive and in every way it appeals to the æsthetic sense of the mathematician. It enables us, when we know the relation between time and distance, to measure uniquely and certainly the velocity at every instant, in however complicated a way the velocity may be changing. It is therefore strongly suggested that we take as the velocity the value obtained according to this rule.

But can there be any question whether we are right or wrong to take

---

[3] [See preceding selection. ED.]

[4] I purposely refrain from giving the rule, not because it is really hard to explain, but because I want to make clear that what is important is to have *some* rule, not any particular rule.

that value; can experiment show that we ought to take one value rather than another? Yes, it can; and in this way. When the velocity is constant and we can measure it without ambiguity, then we can establish laws between that velocity and certain properties of the moving body. Thus, if we allow a moving steel ball to impinge on a lead block, it will make a dent in it determined by its velocity; and when we have established by observations of this kind a relation between the velocity and the size of the dent, we can obviously use the size of the dent to measure the velocity. Suppose now our falling body is a steel ball, and we allow it to impinge on a lead block after falling through different distances; we shall find that its velocity, estimated by the size of the dent, agrees exactly with the velocity estimated by Newton's rule, and not with that estimated by any other rule (so long, of course, as the other rule does not give the same result as Newton's). That, I hope the reader will agree, is a very definite proof that Newton's rule is right.

On this account only Newton's rule would be very important, but it has a wider and much more important application. So far we have expressed the rule as giving the velocity at any instant when the relation between time and distance is known; but the problem might be reversed. We might know the velocity at any instant and want to find out how far the body has moved in any given time. If the velocity were the same at all instants, the problem would be easy; the distance would be the velocity multiplied by the time. But if it is not the same, the right answer is by no means easy to obtain; in fact the only way of obtaining it is by the use of Newton's rule. The form of that rule makes it easy to reverse it and, instead of obtaining the velocity from the distance, to obtain the distance from the velocity; but until that rule was given, the problem could not have been solved; it would have baffled the wisest philosophers of Greece. Now this particular problem is not of any very great importance, for it would be easier to measure by experiment the distance moved than to measure the velocity and calculate the distance. But there are closely analogous cases—one of which we shall notice immediately—in which the position will be reversed. Let us therefore ask what is the assumption which, in accordance with the conclusion reached on p. 1822, must be introduced, if the solution of the problem is to give new experimental knowledge.

We have seen that the problem could be solved easily if the velocity were constant; what we are asking, is how it is to be solved if the velocity does not remain constant. If we examined the rule by which the solution is obtained, we should find that it involves the assumption that the effect upon the distance travelled of a certain velocity at a given instant of time is the same as it would be if the body had at that instant the same *constant* velocity. We know how far the body would travel at that instant if the velocity were constant, and the assumption tells us that it will travel at that instant the same distance although the velocity is not constant. To obtain the whole distance travelled in any given time, we have to add up the distances travelled at the instants of which that time is made up; the reversed Newtonian rule gives a simple and direct method for adding up these distances, and thus solves the problem. It should be noted that the assumption is one that cannot possibly be proved by experiment; we are assuming that something would happen if things were not what they actually are; and experiment can only tell us about things as they are. Accordingly calculation of this kind must, in all strictness, always be confirmed by experiment before it is certain. But as a matter of fact, the assumption is one of which we are almost more certain than we are of any experiment. It is characteristic, not only of the particular example that we have been considering, but of the whole structure of modern mathematical physics which has arisen out of the work of Newton. We should never think it really necessary to-day to confirm by experiment the results of calculation based on that assumption; indeed if experiment and calculation did not agree, we should always maintain that the former and not the latter was wrong. But the assumption is there, and it is primarily suggested by the æsthetic sense of the mathematician, not dictated by the facts of the external world. Its certainty is yet one more striking instance of the conformity of the external world with our desires.

And now let us glance at an example in which such calculation becomes of real importance. Let us take a pendulum, consisting of a heavy bob at the end of a pivoted rod, draw it aside and then let it swing. We ask how it will swing, what positions the bob will occupy at various times after it is started. Our calculation proceeds from two known laws. (1) We know how the force on the pendulum varies with its position. That we can find out by actual experiments. We hang a weight by a string over a pulley, attach the other end of the string to the bob, and notice how far the bob is pulled aside by various weights hanging at the end of the string. We thus get a numerical law between the force and the angle which the rod of the pendulum makes with the vertical. (2) We know how a body will move under a constant force. It will move in accordance with Table II, the distance travelled being proportional to the "square" of the time during which the force acts. Now we introduce the Newtonian assumption. We know the force in each position; we know how it would move in that position if the force on it were constant; actually it is not constant, but we assume that the motion will be the same as it would be if, in that position, the force were constant. With that assumption, the general Newtonian rule (of which the application to velocity is only a special instance) enables us to sum up the effects of the motions in the different positions, and thus to arrive at the desired relation between the

time and the positions successively occupied by the pendulum. The whole
of the calculation which plays so large a part in modern science is nothing
but an elaboration of that simple example.

MATHEMATICAL THEORIES

We have now examined two of the applications of mathematics to
science. Both of them depend on the fact that relations which appeal to
the sense of the mathematician by their neatness and simplicity are found
to be important in the external world of experiment. The relations between
numerals which he suggests are found to occur in numerical laws, and the
assumptions which are suggested by his arguments are found to be true.
We have finally to notice a yet more striking example of the same fact,
and one which is much more difficult to explain to the layman.

This last application is in formulating theories. In an earlier chapter we
concluded that a theory, to be valuable, must have two features. It must
be such that laws can be predicted from it and such that it explains these
laws by introducing some analogy based on laws more familiar than those
to be explained. In recent developments of physics, theories have been
developed which conform to the first of these conditions but not to the
second. In place of the analogy with familiar laws, there appears the new
principle of mathematical simplicity. These theories explain the laws, as
do the older theories, by replacing less acceptable by more acceptable
ideas; but the greater acceptability of the ideas introduced by the theories
is not derived from an analogy with familiar laws, but simply from the
strong appeal they make to the mathematician's sense of form.

I do not feel confident that I can explain the matter further to those
who have not some knowledge of both physics and mathematics, but I
must try. The laws on the analogy with which theories of the older type
are based were often (in physics, usually) numerical laws, such laws for
example as that of the falling body. Now numerical laws, since they involve mathematical relations, are usually expressed, not in words, but in
the symbols in which, as every one knows, mathematicians express their
ideas and their arguments. I have been careful to avoid these symbols;
until this page there is hardly an "x" or a "y" in the book. And I have
done so because experience shows that they frighten people; they make
them think that something very difficult is involved. But really, of course,
symbols make things easier; it is conceivable that some super-human intellect might be able to study mathematics, and even to advance it, expressing all his thoughts in words. Actually, the wonderful symbolism
mathematics has invented make such efforts unnecessary; they make the
processes of reasoning quite easy to follow. They are actually inseparable
from mathematics; they make exceedingly difficult arguments easy to
follow by means of simple rules for juggling with these symbols—inter-

changing their order, replacing one by another, and so on. The consequence is that the expert mathematician has a sense about symbols, as symbols; he looks at a page covered with what, to anyone else, are unintelligible scrawls of ink, and he immediately realizes whether the argument expressed by them is such as is likely to satisfy his sense of form; whether the argument will be "neat" and the results "pretty." (I can't tell you what those terms mean, any more than I can tell you what I mean when I say that a picture is beautiful.)

Now sometimes, but not always, simple folk can understand what he means; let me try an example. Suppose you found a page with the following marks on it—never mind if they mean anything:

$$i = \frac{d\gamma}{dy} - \frac{d\beta}{dz} \qquad \frac{dX}{dt} = \frac{d\gamma}{dy} - \frac{d\beta}{dz}$$

$$j = \frac{d\alpha}{dz} - \frac{d\gamma}{dx} \qquad \frac{dY}{dt} = \frac{d\alpha}{dz} - \frac{d\gamma}{dx}$$

$$k = \frac{d\beta}{dx} - \frac{d\alpha}{dy} \qquad \frac{dZ}{dt} = \frac{d\beta}{dx} - \frac{d\alpha}{dy}$$

$$\frac{d\alpha}{dt} = \frac{dY}{dz} - \frac{dZ}{dy} \qquad \frac{d\alpha}{dt} = \frac{dY}{dz} - \frac{dZ}{dy}$$

$$\frac{d\beta}{dt} = \frac{dZ}{dx} - \frac{dX}{dz} \qquad \frac{d\beta}{dt} = \frac{dZ}{dx} - \frac{dX}{dz}$$

$$\frac{d\gamma}{dt} = \frac{dX}{dy} - \frac{dY}{dx} \qquad \frac{d\gamma}{dt} = \frac{dX}{dy} - \frac{dY}{dx}$$

I think you would see that the set of symbols on the right side are "prettier" in some sense than those on the left; they are more symmetrical. Well, the great physicist, James Clerk Maxwell, about 1870, thought so too; and by substituting the symbols on the right side for those on the left, he founded modern physics, and, among other practical results, made wireless telegraphy possible.

It sounds incredible; and I must try to explain a little more. The symbols on the left side represent two well-known electrical laws: Ampère's Law and Faraday's Law; or rather a theory suggested by an analogy with those laws. The symbols i, j, k represent in those laws an electric current. For these symbols Maxwell substituted $\frac{dX}{dt}\ \frac{dY}{dt}\ \frac{dZ}{dt}$; that substitution was roughly equivalent to saying that an electric current was related to the things represented by X, Y, Z, t (never mind what they are) in a way nobody had ever thought of before; it was equivalent to saying that so long as X, Y, Z, t were related in a certain way, there might be an electric current in circumstances in which nobody had believed that an electric current could flow. As a matter of fact, such a current would be one flowing in an absolutely empty space without any material conductor along which it might flow, and such a current was previously thought to be impossible. But Maxwell's feeling for symbolism suggested to him that there might be such a current, and when he worked out the consequences of supposing that there were such currents (not currents perceptible in the ordinary way, but theoretical currents, as molecules are theoretical hard particles), he arrived at the unexpected result that an alteration in an electric current in one place would be reproduced at another far distant from it by waves travelling from one to the other through absolutely empty space between. Hertz actually produced and detected such waves; and Marconi made them a commercial article.

That is the best attempt I can make at explaining the matter. It is one more illustration of the marvellous power of pure thought, aiming only at the satisfaction of intellectual desires, to control the external world. Since Maxwell's time, there have been many equally wonderful theories, the form of which is suggested by nothing but the mathematician's sense for symbols. The latest are those of Sommerfeld, based on the ideas of Niels Bohr, and of Einstein. Every one has heard of the latter, but the former (which concerns the constitution of the atom) is quite as marvellous. But of these I could not give, even if space allowed, even such an explanation as I have attempted for Maxwell's. And the reason is this: A theory by itself means nothing experimental; it is only when something is deduced from it that it is brought within the range of our material senses. Now in Maxwell's theory, the symbols, in the alteration of which the characteristic feature of the theory depends, are retained through the deduction and appear in the law which is compared with experiment. Accordingly it is possible to give some idea of what these symbols mean in terms of things experimentally observed. But in Sommerfeld's or Einstein's theory the symbols, which are necessarily involved in the assumption which differentiates their theories from others, disappear during the deduction; they leave a mark on the other symbols which remain and alter the relation between them; but the symbols on the relations of which the whole theory hangs, do not appear at all in any law deduced from the theory. It is quite impossible to give any idea of what they mean in terms of experiment.[5] Probably some of my readers will have read the very interesting and ingenious attempts to "explain Einstein" which have been published, and

---

[5] The same is true really of the exposition of the Newtonian assumption attempted on p. 1824. It is strictly impossible to state exactly what is the assumption discussed there without using symbols. The acute reader will have guessed already that on that page I felt myself skating on very thin ice.

will feel that they really have a grasp of the matter. Personally I doubt it; the only way to understand what Einstein did is to look at the symbols in which his theory must ultimately be expressed and to realize that it was reasons of symbolic form, and such reasons alone, which led him to arrange the symbols in the way he did and in no other.

But now I have waded into such deep water that it is time to retrace my steps and return to the safe shore of the affairs of practical life.

HERMANN WEYL

THERE are two directions of mathematical inquiry. It can either penetrate into the other sciences, making models, maps and bridges for
reasoning; or it can mind its own business, cultivate its own garden. Both
pursuits have been enormously fruitful. The success of mathematics as a
helper to science has been spectacular; the advances within its own
domain, though less widely appreciated, are no less spectacular. How is
this universality to be explained? Why has mathematics served so brilliantly in so many different undertakings—as a lamp, a tool, a language;
even in its curious, apelike preoccupation with itself? What, in other
words, is the mathematical way of thinking? It is with this question that
the next selection is concerned.

One cannot expect a simple answer, any more than one expects a simple
description of art or power or humor. To hear art defined as significant
form, or laughter as (in Bergson's words) "a slight revolt on the surface
of social life" is mildly exhilarating but not very enlightening. And so with
most definitions of mathematics. Felix Klein describes it as the science of
self-evident things; Benjamin Peirce, as the science which draws necessary
conclusions; Aristotle, as the study of "quantity"; Whitehead, as the development "of all types of formal, necessary and deductive reasoning";
Descartes, as the science of order and measure; Bacon, as the study which
makes men "subtile"; Bertrand Russell, as a subject identical with logic;
David Hilbert, as a meaningless formal game. None of these statements
gets a whole grip upon the subject, though one or two are important. It
is clear, I think, that any definition of mathematics, however elaborate
or epigrammatic, will fail to lay bare its fundamental structure and the
reasons for its universality. But a set of well-chosen examples of the main
mathematical ideas may well serve as a substitute. No one is better
qualified to choose such a set, and generalize about it, than Hermann
Weyl.

Weyl is an ornament to our time. The play of his powerful intelligence
has enriched mathematics, the natural sciences and philosophy; he has
invented new concepts and pressed forward to new insights. Weyl was
born in Elmshorn, Germany, in 1885 and educated at the universities of
Munich and Göttingen. At the latter institution he received his doctorate
and held his first teaching post in mathematics. From 1913 to 1930 he
was professor of mathematics at the Zürich *Technische Hochschule* (with
a year off, 1928–29, to serve as research professor in mathematical physics
at Princeton); from 1930–1933 he held a chair at Göttingen; in 1933 he

joined the Institute for Advanced Study, Princeton. He retired from the Institute in 1952.*

The researches and writings of this distinguished man span a wide field. He has worked in differential equations, function theory, group theory, topology, relativity, quantum mechanics, philosophy of mathematics; his books include the now-classic *Space-Time-Matter* (English translation, 1921),[1] *Mind and Nature* (1934), *The Classical Groups* (1939), *Algebraic Theory of Numbers* (1940), *Philosophy of Mathematics and Natural Science* (English edition, 1949). His most recent book, *Symmetry* (See p. 671), is a masterly integration of science and art. Among Weyl's distinctions here and abroad are membership in the National Academy of Sciences and one of the rare foreign fellowships of the Royal Society. Weyl is what Flaubert called a triple thinker—a thinker "to the nth degree." [2] His work is always fresh, always distinctive in approach. This is well demonstrated by the selection below, an address delivered in 1940 at the Bicentennial Conference of the University of Pennsylvania.

[1] Reprinted, New York, 1951.
[2] Edmund Wilson, *The Triple Thinkers* ("The Politics of Flaubert"), New York, 1948, p. 74.
* Note added in proof: Hermann Weyl died in Zurich, December 8, 1955. An obituary by Freeman J. Dyson, one of Weyl's former colleagues at the Institute for Advanced Study, appears in the March 10, 1956, issue of *Nature*. Dyson describes Weyl as the twentieth-century mathematician who made major contributions in the greatest number of different fields. "He alone could stand comparison with the last great universal mathematicians of the nineteenth century, Hilbert and Poincaré. So long as he was alive, he embodied a living contact between the main lines of advance in pure mathematics and in theoretical physics. Now he is dead, the contact is broken, and our hopes of comprehending the physical universe by a direct use of creative mathematical imagination are for the time being ended." Dyson offers a revealing remembrance: "Characteristic of Weyl was an aesthetic sense which dominated his thinking on all subjects. He once said to me, half-joking, 'My work always tried to unite the true with the beautiful; but when I had to choose one or the other, I usually chose the beautiful.' This remark sums up his personality perfectly. It shows his profound faith in an ultimate harmony of nature, in which laws should inevitably express themselves in a mathematically beautiful form. It shows also his recognition of human frailty, and his humour, which always stopped him short of being pompous."

Mathematicians are like Frenchmen: whatever you say to them they trans-late into their own language, and forthwith it is something entirely different.
—GOETHE

Although the whole of this life were said to be nothing but a dream and the physical world nothing but a phantasm, I should call this dream or phantasm real enough, if, using reason well, we were never deceived by it.
—LEIBNIZ

6 The Mathematical Way of Thinking

By HERMANN WEYL

BY the mathematical way of thinking I mean first that form of reasoning through which mathematics penetrates into the sciences of the external world—physics, chemistry, biology, economics, etc., and even into our everyday thoughts about human affairs, and secondly that form of reasoning which the mathematician, left to himself, applies in his own field. By the mental process of thinking we try to ascertain truth; it is our mind's effort to bring about its own enlightenment by evidence. Hence, just as truth itself and the experience of evidence, it is something fairly uniform and universal in character. Appealing to the light in our innermost self, it is neither reducible to a set of mechanically applicable rules, nor is it divided into watertight compartments like historic, philosophical, mathe-matical thinking, etc. We mathematicians are no Ku Klux Klan with a secret ritual of thinking. True, nearer the surface there are certain tech-niques and differences; for instance, the procedures of fact-finding in a courtroom and in a physical laboratory are conspicuously different. How-ever, you should not expect me to describe the mathematical way of thinking much more clearly than one can describe, say, the democratic way of life.

A movement for the reform of the teaching of mathematics, which some decades ago made quite a stir in Germany under the leadership of the great mathematician Felix Klein, adopted the slogan "functional think-ing." The important thing which the average educated man should have learned in his mathematics classes, so the reformers claimed, is thinking in terms of *variables and functions*. A function describes how one variable y depends on another x; or more generally, it maps one variety, the range of a variable element x, upon another (or the same) variety. This idea of function or mapping is certainly one of the most fundamental concepts, which accompanies mathematics at every step in theory and application.

Our federal income tax law defines the tax y to be paid in terms of the income x; it does so in a clumsy enough way by pasting several linear functions together, each valid in another interval or bracket of income. An archeologist who, five thousand years from now, shall unearth some of our income tax returns together with relics of engineering works and mathematical books, will probably date them a couple of centuries earlier, certainly before Galileo and Vieta. Vieta was instrumental in introducing a consistent algebraic symbolism; Galileo discovered the quadratic law of falling bodies, according to which the drop s of a body falling in a vacuum is a quadratic function of the time t elapsed since its release:

$$s = \tfrac{1}{2}gt^2, \tag{1}$$

g being a constant which has the same value for each body at a given place. By this formula Galileo converted a natural law inherent in the actual motion of bodies into an *a priori* constructed mathematical function, and that is what physics endeavors to accomplish for every phenomenon. The law is of much better design than our tax laws. It has been designed by Nature, who seems to lay her plans with a fine sense for mathematical simplicity and harmony. But then Nature is not, as our income and excess profits tax laws are, hemmed in by having to be comprehensible to our legislators and chambers of commerce.

Right from the beginning we encounter these characteristic features of the mathematical process: 1) variables, like t and s in the formula (1), whose possible values belong to a range, here the range of real numbers, which we can completely survey because it springs from our own free construction, 2) representation of these variables by symbols, and 3) functions or *a priori* constructed mappings of the range of one variable t upon the range of another s. Time is the independent variable *kat exochen.*

In studying a function one should let the independent variable run over its full range. A conjecture about the mutual interdependence of quantities in nature, even before it is checked by experience, may be probed in thought by examining whether it carries through over the whole range of the independent variables. Sometimes certain simple *limiting cases* at once reveal that the conjecture is untenable. Leibnitz taught us by his *principle of continuity* to consider rest not as contradictorily opposed to motion, but as a limiting case of motion. Arguing by continuity he was able *a priori* to refute the laws of impact proposed by Descartes. Ernst Mach gives this prescription: "After having reached an opinion for a special case, one gradually modifies the circumstances of this case as far as possible, and in so doing tries to stick to the original opinion as closely as one can. There is no procedure which leads more safely and with greater mental economy to the simplest interpretation of all natural events." Most of the variables with which we deal in the analysis of nature are continuous variables like time, but although the word seems to suggest it, the mathematical concept is not restricted to this case. The most important example of a discrete variable is given by the sequence of natural numbers or integers 1, 2, 3, . . . Thus the number of divisors of an arbitrary integer n is a function of n.

In Aristotle's logic one passes from the individual to the general by exhibiting certain abstract features in a given object and discarding the remainder, so that two objects fall under the same concept or belong to the same genus if they have those features in common. This descriptive classification, *e.g.*, the description of plants and animals in botany and zoology, is concerned with the actual existing objects. One might say that Aristotle thinks in terms of substance and accident, while the functional idea reigns over the formation of mathematical concepts. Take the notion of ellipse. Any ellipse in the x-y-plane is a set E of points (x, y) defined by a quadratic equation

$$ax^2 + 2bxy + cy^2 = 1$$

whose coefficients a, b, c satisfy the conditions

$$a > 0, \; c > 0, \; ac - b^2 > 0.$$

The set E depends on the coefficients a, b, c; we have a function $E(a, b, c)$ which gives rise to an individual ellipse by assigning definite values to the variable coefficients a, b, c. In passing from the individual ellipse to the general notion one does not discard any specific difference, one rather makes certain characteristics (here represented by the coefficients) variable over an *a priori* surveyable range (here described by the inequalities). The notion thus extends over all *possible*, rather than over all *actually existing*, specifications.[1]

From these preliminary remarks about functional thinking I now turn to a more systematic argument. Mathematics is notorious for the thin air of abstraction in which it moves. This bad reputation is only half deserved. Indeed, the first difficulty the man in the street encounters when he is taught to think mathematically is that he must learn to look things much more squarely in the face; his belief in words must be shattered; he must learn to think more concretely. Only then will he be able to carry out the second step, the step of abstraction where intuitive ideas are replaced by purely symbolic construction.

About a month ago I hiked around Longs Peak in the Rocky Mountain National Park with a boy of twelve, Pete. Looking up at Longs Peak he

---

[1] Compare about this contrast Ernst Cassirer, "Substanzbegriff und Funktionsbegriff," 1910, and my critical remark, "Philosophie der Mathematik und Naturwissenschaft," 1923, p. 111.

told me that they had corrected its elevation and that it is now 14,255 feet instead of 14,254 feet last year. I stopped a moment asking myself what this could mean to the boy, and should I try to enlighten him by some Socratic questioning. But I spared Pete the torture, and the comment then withheld, will now be served to you. Elevation is elevation above sea level. But there is no sea under Longs Peak. Well, in idea one continues the actual sea level under the solid continents. But how does one construct this ideal closed surface, the geoid, which coincides with the surface of the oceans over part of the globe? If the surface of the ocean were strictly spherical, the answer would be clear. However, nothing of this sort is the case. At this point dynamics comes to our rescue. Dynamically the sea level is a surface of constant potential $\phi = \phi_0$; more exactly ϕ denotes the gravitational potential of the earth, and hence the difference of ϕ at two points P, P' is the work one must put into a small body of mass 1 to transfer it from P to P'. Thus it is most reasonable to define the geoid by the dynamical equation $\phi = \phi_0$. If this constant value of ϕ fixes the elevation zero, it is only natural to define any fixed altitude by a corresponding constant value of ϕ, so that a peak P is called higher than P' if one gains energy by flying from P to P'. The geometric concept of altitude is replaced by the dynamic concept of potential or energy. Even for Pete, the mountain climber, this aspect is perhaps the most important: the higher the peak the greater—*ceteris paribus*—the mechanical effort in climbing it. By closer scrutiny one finds that in almost every respect the potential is the relevant factor. For instance the barometric measurement of altitude is based on the fact that in an atmosphere of given constant temperature the potential is proportional to the logarithm of the amospheric pressure, whatever the nature of the gravitational field. Thus atmospheric pressure, generally speaking, indicates potential and not altitude. Nobody who has learned that the earth is round and the vertical direction is not an intrinsic geometric property of space but the direction of gravity should be surprised that he is forced to discard the geometric idea of altitude in favor of the dynamic more concrete idea of potential. Of course there is a relationship to geometry: In a region of space so small that one can consider the force of gravity as constant throughout this region, we have a fixed vertical direction, and potential differences are proportional to differences of altitude measured in that direction. Altitude, height, is a word which has a clear meaning when I ask how high the ceiling of this room is above its floor. The meaning gradually loses precision when we apply it to the relative altitudes of mountains in a wider and wider region. It dangles in the air when we extend it to the whole globe, unless we support it by the dynamical concept of potential. Potential is more concrete than altitude because it is generated by and dependent on the mass distribution of the earth.

Words are dangerous tools. Created for our everyday life they may have their good meanings under familiar limited circumstances, but Pete and the man in the street are inclined to extend them to wider spheres without bothering about whether they then still have a sure foothold in reality. We are witnesses of the disastrous effects of this witchcraft of words in the political sphere where all words have a much vaguer meaning and human passion so often drowns the voice of reason. The scientist must thrust through the fog of abstract words to reach the concrete rock of reality. It seems to me that the science of economics has a particularly hard job, and will still have to spend much effort, to live up to this principle. It is, or should be, common to all sciences, but physicists and mathematicians have been forced to apply it to the most fundamental concepts where the dogmatic resistance is strongest, and thus it has become their second nature. For instance, the first step in explaining relativity theory must always consist in shattering the dogmatic belief in the temporal terms past, present, future. You can not apply mathematics as long as words still becloud reality.

I return to relativity as an illustration of this first important step preparatory to mathematical analysis, the step guided by the maxim, "Think concretely." As the root of the words *past, present, future,* referring to time, we find something much more tangible than time, namely, the causal structure of the universe. Events are localized in space and time; an event of small extension takes place at a space-time or world point, a here-now. After restricting ourselves to events on a plane E we can depict the events by a graphic timetable in a three-dimensional diagram with a horizontal E plane and a vertical t axis on which time t is plotted. A world point is represented by a point in this picture, the motion of a small body by a world line, the propagation of light with its velocity c radiating from a light signal at the world point O by a vertical straight circular cone with vertex at O (light cone). The *active future* of a given world point O, here-now, contains all those events which can still be influenced by what happens at O, while its *passive past* consists of all those world points from which any influence, any message, can reach O. I here-now can no longer change anything that lies outside the active future; all events of which I here-now can have knowledge by direct observation or any records thereof necessarily lie in the passive past. We interpret the words past and future in this causal sense where they express something very real and important, the causal structure of the world.

The new discovery at the basis of the theory of relativity is the fact that no effect may travel faster than light. Hence while we formerly believed that active future and passive past bordered on each other along the cross-section of *present*, the horizontal plane $t = $ const. going through O, Einstein taught us that the active future is bounded by the forward light

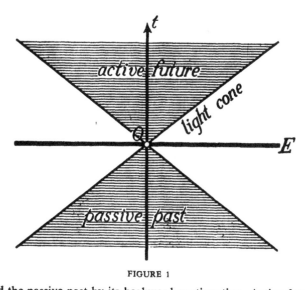

FIGURE 1

cone and the passive past by its backward continuation. Active future and passive past are separated by the part of the world lying between these cones, and with this part I am here-now not at all causally connected. The essential positive content of relativity theory is this new insight into the causal structure of the universe. By discussing the various interpretations of such a simple question as whether two men, say Bill on earth and Bob on Sirius, are contemporaries, as to whether it means that Bill can send a message to Bob, or Bob a message to Bill, or even that Bill can communicate with Bob by sending a message and receiving an answer, etc., I often succeed soon in accustoming my listener to thinking in terms of causal rather than his wonted temporal structure. But when I tell him that the causal structure is not a stratification by horizontal layers $t =$ const., but that active future and passive past are of cone-like shape with an interstice between, then some will discern dimly what I am driving at, but every honest listener will say: Now you draw a figure, you speak in pictures; how far does the simile go, and what is the naked truth to be conveyed by it? Our popular writers and news reporters, when they have to deal with physics, indulge in similes of all sorts; the trouble is that they leave the reader helpless in finding out how far these pungent analogies cover the real issue, and therefore more often lead him astray than enlighten him. In our case one has to admit that our diagram is no more than a picture, from which, however, the real thing emerges as soon as we replace the intuitive space in which our diagrams are drawn by its construction in terms of sheer symbols. Then the phrase that the world is a four-dimensional continuum changes from a figurative form of speech into a statement of what is literally true. At this second step the mathematician turns abstract, and here is the point where the layman's understanding most frequently breaks off: the intuitive picture must be exchanged for a symbolic construction. "By its geometric and later by its purely symbolic construction," says Andreas Speiser, "mathematics shook off the fetters of language, and one who knows the enormous work put into this process and its ever recurrent surprising successes can not help feeling that mathematics to-day is more efficient in its sphere of the intellectual world, than the modern languages in their deplorable state or even music are on their respective fronts." I shall spend most of my time to-day in an attempt to give you an idea of what this magic of symbolic construction is.

To that end I must begin with the simplest, and in a certain sense most profound, example: the natural numbers or *integers* by which we *count* objects. The symbols we use here are strokes put one after another. The objects may disperse, "melt, thaw and resolve themselves into a dew," but we keep this record of their number. What is more, we can by a constructive process decide for two numbers represented through such symbols which one is the larger, namely by checking one against the other, stroke by stroke. This process reveals differences not manifest in direct observation, which in most instances is incapable of distinguishing between even such low numbers as 21 and 22. We are so familiar with these miracles which the number symbols perform that we no longer wonder at them. But this is only the prelude to the mathematical step proper. We do not leave it to chance which numbers we shall meet by counting actual objects, but we generate the open sequence of *all possible* numbers which starts with 1 (or 0) and proceeds by adding to any number symbol n already reached one more stroke, whereby it changes into the following number n'. As I have often said before, being is thus projected onto the background of the possible, or more precisely onto a manifold of possibilities which unfolds by iteration and is open into infinity. Whatever number n we are given, we always deem it possible to pass to the next n'. "Number goes on." This intuition of the "ever one more," of the open countable infinity, is basic for all mathematics. It gives birth to the simplest example of what I termed above an *a priori* surveyable range of variability. According to this process by which the integers are created, functions of an argument ranging over all integers n are to be defined by so-called complete induction, and statements holding for all n are to be proved in the same fashion. The principle of this inference by complete induction is as follows. In order to show that every number n has a certain property V it is sufficient to make sure of two things:

1) 0 has this property;
2) If n is any number which has the property V, then the next number n' has the property V.

It is practically impossible, and would be useless, to write out in strokes the symbol of the number 10^{12}, which the Europeans call a billion and we in this country, a thousand billions. Nevertheless we talk about spending more than 10^{12} cents for our defense program, and the astronomers are still ahead of the financiers. In July the *New Yorker* carried this cartoon: man and wife reading the newspaper over their breakfast and she looking up in puzzled despair: "Andrew, how much *is* seven hundred billion dollars?" A profound and serious question, lady! I wish to point out that only by passing through the *infinite* can we attribute any significance to such figures. 12 is an abbreviation of

$$10^{12} = 10 \cdot 10 \cdot 10 \cdot 10 \cdot 10 \cdot 10 \cdot 10 \cdot 10 \cdot 10 \cdot 10 \cdot 10 \cdot 10$$

can not be understood without defining the function $10 \cdot n$ for *all* n, and this is done through the following definition by complete induction:

$$10 \cdot 0 = 0,$$
$$10 \cdot n' = (10 \cdot n)^{///////}.$$

The dashes constitute the explicit symbol for 10, and, as previously, each dash indicates transition to the next number. Indian, in particular Buddhist, literature indulges in the possibilities of fixing stupendous numbers by the decimal system of numeration which the Indians invented, *i.e.*, by a combination of sums, products and powers. I mention also Archimedes's treatise "On the counting of sand," and Professor Kasner's Googolplex in his recent popular book on "Mathematics and the Imagination."

Our conception of *space* is, in a fashion similar to that of natural numbers, depending on a constructive grip on all *possible* places. Let us consider a metallic disk in a plane E. Places on the disk can be marked *in concreto* by scratching little crosses on the plate. But relatively to two axes of coordinates and a standard length scratched into the plate we can also put ideal marks in the plane outside the disk by giving the numerical values of their two coordinates. Each coordinate varies over the *a priori* constructed range of real numbers. In this way astronomy uses our solid earth as a base for plumbing the sidereal spaces. What a marvelous feat of imagination when the Greeks first constructed the shadows which earth and moon, illumined by the sun, cast in empty space and thus explained the eclipses of sun and moon! In analyzing a continuum, like space, we shall here proceed in a somewhat more general manner than by measurement of coordinates and adopt the *topological* viewpoint, so that two continua arising one from the other by continuous deformation are the same to us. Thus the following exposition is at the same time a brief introduction to an important branch of mathematics, topology.

The symbols for the localization of points on the one-dimensional continuum of a straight line are the *real numbers*. I prefer to consider a *closed* one-dimensional continuum, the circle. The most fundamental statement about a continuum is that it may be divided into parts. We catch all the points of a continuum by spanning a net of division over it, which we refine by repetition of a definite process of subdivision *ad infinitum*. Let S be any division of the circle into a number of arcs, say l arcs. From S we derive a new division S' by the process of *normal subdivision*, which consists in breaking each arc into two. The number of arcs in S' will then be $2l$. Running around the circle in a definite sense (orientation) we may distinguish the two pieces, in the order in which we meet them, by the marks 0 and 1; more explicitly, if the arc is denoted by a symbol a then these two pieces are designated as $a0$ and $a1$. We start with the division S_0 of the circle into two arcs $+$ and $-$; either is topologically a cell, *i.e.*, equivalent to a segment. We then iterate the process of normal subdivision and thus obtain S_0', S_0'', . . . , seeing to it that the refinement of the division ultimately pulverizes the whole circle. If we had not renounced the use of metric properties we could decree that the normal subdivision takes place by cutting each arc into two *equal* halves. We introduce no such fixation; hence the actual performance of the process involves a wide measure of arbitrariness. However, the *combinatorial scheme* according to which the parts reached at any step border on each other, and according to which the division progresses, is unique and perfectly fixed. Mathematics cares for this symbolic scheme only. By our notation the parts occurring at the consecutive divisions are *catalogued* by symbols of this type

$$+ .011010001$$

with $+$ or $-$ before the dot and all following places occupied by either 0 or 1. We see that we arrive at the familiar symbols of binary (not

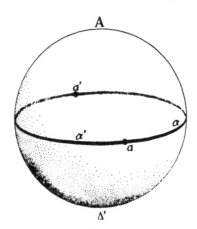

FIGURE 2

decimal) fractions. A point is caught by an infinite sequence of arcs of the consecutive divisions such that each arc arises from the preceding one by choosing one of the two pieces into which it breaks by the next normal subdivision, and the point is thus fixed by an infinite binary fraction.

Let us try to do something similar for two-dimensional continua, *e.g.*, for the surface of a sphere or a torus. The figures show how we may cast a very coarse net over either of them, the one consisting of two, the other of four meshes; the globe is divided into its upper and lower halves by the equator, the torus is welded together from four rectangular plates. The meshes are two-dimensional cells, or briefly, 2-cells which are topologically equivalent to a circular disk. The combinatorial description is facilitated by introducing also the vertices and edges of the division, which

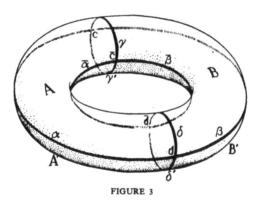

FIGURE 3

are 0- and 1-cells. We attach arbitrary symbols to them and state in symbols for each 2-cell which 1-cells bound it, and for each 1-cell by which 0-cells it is bounded. We then arrive at a *topological scheme* S_0. Here are our two examples:

Sphere. $A \rightarrow a, a'.\ A' \rightarrow a, a'.\ a \rightarrow a, a'.\ a' \rightarrow a, a'.$
        ($\rightarrow$ means: bound by)
Torus. $A \rightarrow a, \bar{a}, \gamma, \delta.\ A' \rightarrow a, \bar{a}, \gamma', \delta'.$
        $B \rightarrow \beta, \bar{\beta}, \gamma, \delta.\ B' \rightarrow \beta, \bar{\beta}, \gamma', \delta'.$
        $a \rightarrow c, d.\ \bar{a} \rightarrow \bar{c}, \bar{d}.\ \beta \rightarrow c, d.\ \bar{\beta} \rightarrow \bar{c}, \bar{d}.$
        $\gamma \rightarrow c, \bar{c}.\ \gamma' \rightarrow c, \bar{c}.\ \delta \rightarrow d, \bar{d}.\ \delta' \rightarrow d, \bar{d}.$

From this initial stage we proceed by iteration of a universal process of normal subdivision: On each 1-cell $a = ab$ we choose a point which serves as a new vertex a and divides the 1-cell into two segments aa and ab; in each 2-cell A we choose a point A and cut the cell into triangles by joining the newly created vertex A with the old and new vertices on its bounding 1-cells by lines within the 2-cell. Just as in elementary geometry we denote the triangles and their sides by means of their vertices. The figure shows a pentagon before and after subdivision; the triangle $A\beta c$ is bounded by the 1-cells βc, $A\beta$, Ac, the 1-cell Ac for instance by the vertices c and A. We arrive at the following general purely symbolic description of the process by which the subdivided scheme S' is derived from a given topological scheme S. Any symbol $e_2\, e_1\, e_0$ made up by the symbols of a 2-cell e_2, a 1-cell e_1 and a 0-cell e_0 in S such that e_2 is bounded by e_1 and e_1 bounded by e_0 represents a 2-cell e'_2 of S'. This 2-cell $e'_2 := e_2 e_1 e_0$ in S' is part of the 2-cell e_2 in S. The symbols of cells in S' which bound a given cell are derived from its symbol by dropping any one of its constituent letters. Through iteration of this symbolic process the initial scheme S_0 gives rise to a sequence of derived schemes S_0', S_0'', S_0''',$\cdots$. What we have done is nothing else than devise a systematic cataloguing of the parts created by consecutive subdivisions. A *point* of our continuum is caught by a sequence.

$$e\ e'\ e''\cdots \tag{2}$$

which starts with a 2-cell e of S_0 and in which the 2-cell $e^{(n)}$ of the scheme $S^{(n)}$ is followed by one of the 2-cells $e^{(n+1)}$ of $S^{(n+1)}$ into which $e^{(n)}$ breaks up by our subdivision. (To do full justice to the inseparability of parts in a continuum this description ought to be slightly altered. But for the present purposes our simplified description will do.) We are convinced that not only may each point be caught by such a sequence (Eudoxos), but that an arbitrarily constructed sequence of this sort always catches a point (Dedekind, Cantor). The fundamental concepts of *limit, convergence* and *continuity* follow in the wake of this construction.

We now come to the decisive step of mathematical abstraction: we forget about what the symbols stand for. The mathematician is concerned with the catalogue alone; he is like the man in the catalogue room who does not care what books or pieces of an intuitively given manifold the symbols of his catalogue denote. He need not be idle; there are many operations which he may carry out with these symbols, without ever having to look at the things they stand for. Thus, replacing the points by their symbols (2) he turns the given manifold into a *symbolic construct* which we shall call the *topological space* $\{S_0\}$ because it is based on the scheme S_0 alone.

The details are not important; what matters is that once the initial finite symbolic scheme S_0 is given we are carried along by an absolutely rigid symbolic construction which leads from S_0 to S_0', from S_0' to S_0'', etc. The idea of iteration, first encountered with the natural numbers, again plays a decisive role. The realization of the symbolic scheme for a given manifold, say a sphere or a torus, as a scheme of consecutive divisions involves a wide margin of arbitrariness restricted only by the requirement that the

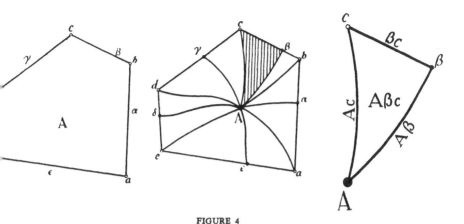

FIGURE 4

pattern of the meshes ultimately becomes infinitely fine everywhere. About this point and the closely affiliated requirement that each 2-cell has the topological structure of a circular disk, I must remain a bit vague. However, the mathematician is not concerned with applying the scheme or catalogue to a given manifold, but only with the scheme itself, which contains no haziness whatsoever. And we shall presently see that even the physicist need not care greatly about that application. It was merely for heuristic purposes that we had to go the way from manifold through division to pure symbolism.

In the same purely symbolic way we can evidently construct not only 1- and 2- but also 3, 4, 5, . . . -dimensional manifolds. An n-dimensional scheme S_0 consists of symbols distinguished as 0, 1, 2, . . . , n-cells and associates with each i-cell e_i ($i = 1, 2, \cdots, n$) certain $(i-1)$-cells of which one says that they bound e_i. It is clear how the process of normal subdivision carries over. *A certain such 4-dimensional scheme can be used for the localization of events*, of all possible here-nows; physical quantities which vary in space and time are functions of a variable point ranging over the corresponding symbolically constructed 4-dimensional topological space. In this sense the world *is* a 4-dimensional continuum. The causal structure, of which we talked before, will have to be constructed within the medium of this 4-dimensional world, *i.e.*, out of the symbolic material constituting our topological space. Incidentally the topological viewpoint has been adopted on purpose, because only thus our frame becomes wide enough to embrace both special and general relativity theory. The special theory envisages the causal structure as something geometrical, rigid, given once for all, while in the general theory it becomes flexible and dependent on matter in the same way as, for instance, the electromagnetic field.

In our analysis of nature we reduce the phenomena to simple elements each of which varies over a certain range of possibilities which we can survey *a priori* because we construct these possibilities *a priori* in a purely combinatorial fashion from some purely symbolic material. The manifold of space-time points is one, perhaps the most basic one, of these constructive elements of nature. We dissolve light into plane polarized monochromatic light beams with few variable characteristics like wave length which varies over the symbolically constructed continuum of real numbers. Because of this *a priori* construction we speak of a *quantitative* analysis of nature; I believe the word quantitative, if one can give it a meaning at all, ought to be interpreted in this wide sense. The power of science, as witnessed by the development of modern technology, rests upon the combination of *a priori* symbolic construction with systematic experience in the form of planned and reproducible reactions and their measurements. As material for the *a priori* construction, Galileo and Newton used certain features of reality like space and time which they considered as objective, in opposition to the subjective sense qualities, which they discarded. Hence the important role which geometric figures played in their physics. You probably know Galileo's words in the *Saggiatore* where he says that no one can read the great book of nature "unless he has mastered the code in which it is composed, that is, the mathematical figures and the necessary relations between them." Later we have learned that none of these features of our immediate observation, not even space and time, have a right to survive in a pretended truly objective world, and thus have gradually and ultimately come to adopt a purely symbolic combinatorial construction.

While a set of objects determines its number unambiguously, we have observed that a scheme of division S_0 with its consecutive derivatives $S_0', S_0'', \cdots$ can be established on a given manifold in many ways involving a wide margin of arbitrariness. But the question whether two schemes,

$$S_0, S_0', S_0'', \cdots \text{ and } T_0, T_0', T_0'', \cdots$$

are fit to describe the same manifold is decidable in a purely mathematical way: it is necessary and sufficient that the two topological spaces $\{S_0\}$ and $\{T_0\}$ can be mapped one upon the other by a continuous one-to-one transformation—a condition which ultimately boils down to a certain relationship called isomorphism between the two schemes S_0 and T_0. (Incidentally the problem of establishing the criterion of isomorphism for two finite schemes in finite combinatorial form is one of the outstanding unsolved mathematical problems.) The connection between a given continuum and its symbolic scheme inevitably carries with it this notion of *isomorphism*; without it and without our understanding that isomorphic schemes are to be considered as not intrinsically different, no more than congruent figures in geometry, the mathematical concept of a topological space would be incomplete. Moreover it will be necessary to formulate precisely the conditions which every topological scheme is required to satisfy. For instance, one such condition demands that each 1-cell be bounded by exactly *two* 0-cells.

I can now say a little more clearly why the physicist is almost as disinterested as the mathematician in the particular way how a certain combinatorial scheme of consecutive divisions is applied to the continuum of here-nows which we called the world. Of course, somehow our theoretical constructions must be put in contact with the observable facts. The historic development of our theories proceeds by heuristic arguments over a long and devious road and in many steps from experience to construction. But systematic exposition should go the other way: first develop the theoretical scheme without attempting to define individually by appropriate measurements the symbols occurring in it as space-time coordinates, electromagnetic field strengths, etc., then describe, as it were in one breath, the contact of the whole system with observable facts. The simplest example I can find is the observed angle between two stars. The symbolic construct in the medium of the 4-dimensional world from which theory determines and predicts the value of this angle includes: (1) the world-lines of the two stars, (2) the causal structure of the universe, (3) the world position of the observer and the direction of his world line at the moment of observation. But a continuous deformation, a one-to-one continuous transformation of this whole picture, does not affect the value of the angle. *Isomorphic pictures lead to the same results concerning observable facts.* This is, in its most general form, the *principle of relativity*. The arbitrariness involved in our ascent from the given manifold to the construct is expressed by this principle for the opposite descending procedure, which the systematic opposition should follow.

So far we have endeavored to describe how a mathematical construct is distilled from the given raw material of reality. Let us now look upon these products of distillation with the eye of a pure mathematician. One of them is the sequence of natural numbers and the other the general notion of a topological space $\{S_0\}$ into which a topological scheme S_0 develops by consecutive derivations $S_0, S_0', S_0'', \cdots$. In both cases *iteration* is the most decisive feature. Hence all our reasoning must be based on evidence concerning that completely transparent process which generates the natural numbers, rather than on any principles of formal logic like syllogism, etc. The business of the constructive mathematician is *not* to draw logical conclusions. Indeed his arguments and propositions are merely an accompaniment of his actions, his carrying out of constructions. For instance, we run over the sequence of integers 0, 1, 2, . . . by saying alternatingly even, odd, even, odd, etc., and in view of the possibility of this inductive construction which we can extend as far as we ever wish, we formulate the general arithmetical proposition: "Every integer is even or odd." Besides the idea of iteration (or the sequence of integers) we make constant use of mappings or of the functional idea. For instance, just now we have defined a function $\pi(n)$, called parity, with n ranging over all integers and π capable of the two values 0 (even) and 1 (odd), by this induction:

$$\pi(0) = 0;$$
$$\pi(n') = 1 \text{ if } \pi(n) = 0, \qquad \pi(n') = 0 \text{ if } \pi(n) = 1.$$

Structures such as the topological schemes are to be studied in the light of the idea of *isomorphism*. For instance, when it comes to introducing operators τ which carry any topological scheme S into a topological scheme $\tau(S)$ one should pay attention only to such operators or functions τ for which isomorphism of S and R entails isomorphism for $\tau(S)$ and $\tau(R)$.

Up to now I have emphasized the constructive character of mathematics. In our actual mathematics there vies with it the non-constructive *axiomatic method*. Euclid's axioms of geometry are the classical prototype. Archimedes employs the method with great acumen and so do later Galileo and Huyghens in erecting the science of mechanics. One defines all concepts in terms of a few undefined basic concepts and deduces all propositions from a number of basic propositions, the axioms, concerning the basic concepts. In earlier times authors were inclined to claim *a priori* evidence for their axioms; however this is an epistemological aspect which does not interest the mathematician. Deduction takes place according to the principles of formal logic, in particular it follows the syllogistic scheme. Such a treatment *more geometrico* was for a long time considered the ideal of every science. Spinoza tried to apply it to ethics. For the mathematician the meaning of the words representing the basic concepts is irrelevant; any interpretation of them which fits, *i.e.*, under which the axioms become true, will be good, and all the propositions of the discipline will hold for such an interpretation because they are all logical consequences of the axioms. Thus n-dimensional Euclidean geometry permits another interpretation where points are distributions of electric current in a given circuit consisting of n branches which connect at certain branch points. For instance, the problem of determining that distribution which results from given electromotoric forces inserted in the various branches of the net corresponds to the geometric construction of orthogonal projection of a point upon a linear subspace. From this standpoint mathematics treats of relations in a hypothetical-deductive manner without binding itself to any particular material interpretation. It is not concerned with the *truth* of axioms, but only with their *consistency*; indeed inconsistency would *a priori* preclude the possibility of our ever coming across a fitting interpretation. "Mathematics is the science which draws necessary conclusions," says B. Peirce in 1870, a definition which was in vogue for decades after. To me it seems that it renders very scanty information about the real nature of mathematics, and you are at present watching my struggle to give a fuller characterization. Past writers on the philosophy of mathematics have so persistently discussed the axiomatic methods that I don't think it necessary for me to dwell on it at any greater length, although my exposition thereby becomes somewhat lopsided.

However I should like to point out that since the axiomatic attitude has ceased to be the pet subject of the methodologists its influence has spread from the roots to all branches of the mathematical tree. We have seen before that topology is to be based on a full enumeration of the axioms which a *topological scheme* has to satisfy. One of the simplest and most basic axiomatic concepts which penetrates all fields of mathematics is that of *group*. Algebra with its *"fields," "rings,"* etc., is to-day from bottom to top permeated by the axiomatic spirit. Our portrait of mathematics would look a lot less hazy, if time permitted me to explain these mighty words which I have just uttered, group, field and ring. I shall not try it, as little as I have stated the axioms characteristic for a topological scheme. But such notions and their kin have brought it about that modern mathematical research often is a dexterous blending of the constructive and the axiomatic procedures. Perhaps one should be content to note their mutual interlocking. But temptation is great to adopt one of these two views as the genuine primordial way of mathematical thinking, to which the other merely plays a subservient role, and it is possible indeed to carry this standpoint through consistently whether one decides in favor of construction or axiom.

Let us consider the first alternative. Mathematics then consists primarily of construction. The occurring sets of axioms merely *fix the range of variables entering into the construction*. I shall explain this statement a little further by our examples of causal structure and topology. According to the special theory of relativity the causal structure is once for all fixed and can therefore be explicitly constructed. Nay, it is reasonable to construct it together with the topological medium itself, as for instance a circle together with its metric structure is obtained by carrying out the normal subdivision by cutting each arc into two *equal* halves. In the general theory of relativity, however, the causal structure is something flexible; it has only to satisfy certain axioms derived from experience which allow a considerable measure of free play. But the theory goes on by establishing laws of nature which connect the flexible causal structure with other flexible physical entities, distribution of masses, electromagnetic field, etc., and these laws in which the flexible things figure as variables are in their turn *constructed* by the theory in an explicit *a priori* way. Relativistic cosmology asks for the topological structure of the universe as a whole, whether it is open or closed, etc. Of course the topological structure can not be flexible as the causal structure is, but one must have a free outlook on all topological possibilities before one can decide by the testimony of experience which of them is realized by our actual world. To that end one turns to topology. There the topological scheme is bound only by certain axioms; but the topologist derives numerical characters from, or establishes universal connections between, arbitrary topological schemes, and again this is done by explicit construction into which the arbitrary schemes enter as variables. Wherever axioms occur, they ultimately serve to describe the range of variables in explicitly constructed functional relations.

So much about the first alternative. We turn to the opposite view, which subordinates construction to axioms and deduction, and holds that mathematics consists of systems of axioms freely agreed upon, and their necessary conclusions. In a completely axiomatized mathematics construction can come in only secondarily as construction of examples, thus forming the bridge between pure theory and its applications. Sometimes there is only *one* example because the axioms, at least up to arbitrary isomorphisms, determine their object uniquely; then the demand for translating the axiomatic set-up into an explicit construction becomes especially imperative. Much more significant is the remark that an axiomatic system, although it refrains from constructing the mathematical *objects*, constructs the mathematical *propositions* by combined and iterated application of logical rules. Indeed, drawing conclusions from given premises proceeds by certain logical rules which since Aristotle's day one has tried to enumerate completely. Thus on the level of propositions, the axiomatic method is undiluted constructivism. David Hilbert has in our day pursued the axiomatic method to its bitter end where all mathematical propositions, including the axioms, are turned into formulas and the game of deduction proceeds from the axioms by rules which take no account of the meaning of the formulas. The mathematical game is played in silence, without words, like a game of chess. Only the rules have to be explained and communicated in words, and of course any arguing about the possibilities of the game, for instance about its consistency, goes on in the medium of words and appeals to evidence.

If carried so far, the issue between explicit construction and implicit definition by axioms ties up with the last foundations of mathematics. Evidence based on construction refuses to support the principles of Aristotelian logic when these are applied to existential and general propositions in infinite fields like the sequence of integers or a continuum of points. And if the logic of the infinite is taken into account, it seems impossible to axiomatize adequately even the most primitive process, the transition $n \rightarrow n'$ from an integer n to its follower n'. As K. Gödel has shown, there will always be constructively evident arithmetical propositions which can not be deduced from the axioms however you formulate them, while at the same time the axioms, riding roughshod over the subtleties of the constructive infinite, go far beyond what is justifiable by evidence. We are not surprised that a concrete chunk of nature, taken in its isolated phenomenal existence, challenges our analysis by its inexhaustibility and incompleteness; it is for the sake of completeness, as we have seen, that physics projects what is given onto the background of the possible. However, it is surprising that a construct created by mind itself, the sequence of integers, the simplest and most diaphanous thing for the constructive mind, assumes a similar aspect of obscurity and deficiency when viewed from the axiomatic angle. But such is the fact; which casts an uncertain light upon the relationship of evidence and mathematics. In spite, or because, of our deepened critical insight we are to-day less sure than at any previous time of the ultimate foundations on which mathematics rests.

My purpose in this address has not been to show how the inventive mathematical intellect works in its manifold manifestations, in calculus, geometry, algebra, physics, etc., although that would have made a much more attractive picture. Rather, I have attempted to make visible the sources from which all these manifestations spring. I know that in an hour's time I can have succeeded only to a slight degree. While in other fields brief allusions are met by ready understanding, this is unfortunately seldom the case with mathematical ideas. But I should have completely failed if you had not realized at least this much, that mathematics, in spite of its age, is not doomed to progressive sclerosis by its growing complexity, but is still intensely alive, drawing nourishment from its deep roots in mind and nature.

PART XIII

Mathematics and Logic

Symbolic Logic, GEORGE BOOLE and a Horrible Dream

THE subject matter of symbolic logic is logic. This fact is worth stress-
ing because of the widespread but mistaken notion that the content of
symbolic logic differs from the content of ordinary logic, the discipline
systematized by Aristotle, elaborated by medieval thinkers and taught for
centuries in the higher schools. Both symbolic and traditional logic are
concerned with the general principles of reasoning; both make use of
symbols. But while traditional logic uses the familiar phonetic symbols of
language known as "words," symbolic logic employs a specially devised
set of marks ("ideographs") which symbolize *directly* the thing talked
about. It is because of this special and precise notation, and the accompanying body of rules for operating with or transforming these ideographic symbols, that symbolic logic is often called mathematical logic.
But the word "mathematical," let me repeat, describes the formal nature
of the presentation, not its substance. One of the essential characteristics
of symbolic or mathematical logic is generality; its principles do not in
any sense belong exclusively to the branch of rational procedure dealing
with numbers and quantity.[1]

Though the distinguishing mark of symbolic logic is thus its formal machinery, these symbols and symbolic procedures confer real and important
advantages. Improvements in notation have had a profound effect on the
progress of mathematics; the invention of a single symbol, the zero,
changed the history of the subject. Similarly, the notation of symbolic
logic has made possible powerful new methods of analysis and has greatly
deepened and widened the study of traditional logic. One clear gain from
the use of ideographic symbols is that meanings are sharpened. Each
separate character may be made to represent a relatively simple and distinct notion. Once this convention is adopted "the ambiguity with which
ordinary language is infected is then at a minimum." [2] A second advantage
is psychological: the symbol "enables us to concentrate upon what is
essential in a given context." We are not distracted by irrelevancies, such
as what the symbols stand for, but can concern ourselves with the abstract
relations between expressions, with the crucial matter of performing cor-

[1] See Clarence Irving Lewis, *A Survey of Symbolic Logic*, Berkeley (California),
1918, pp. 1–5.
[2] Morris R. Cohen and Ernest Nagel, *An Introduction to Logic and Scientific
Method*, New York, 1934, pp. 119–120. My discussion of the advantages of symbolism
are in large part drawn from this book.

rectly the various formal operations of the system. It is Logic, rather than Justice that needs the blindfold.

The simplicity and conciseness of purely symbolic expressions are also of considerable value. As an example of the sheer economy of symbolism compare the usual lengthy verbal definitions of mathematical induction [see selection by G. Polya, pp. 1980–1992] with the following tidy statement:

$$f (o) : n.f (n) \supset f (n + 1) : . \supset . (n) f (n).$$

But this conciseness of phrasing offers far more important benefits than eliminating irrelevancies and saving labor and paper. Symmetries, relations and resemblances which in ordinary language are hidden or veiled are made to stand out boldly in a symbolic idiom.[3] Expressions that look alike are apt in substance to be alike; for the skeleton, the essential inner linkage of meaning, is exhibited by an array of well-made ideographs. An efficient symbolism not only exposes errors previously unnoticed, but suggests new implications and conclusions, new and fruitful lines of thought. "The discovery of negative and imaginary numbers, Maxwell's introduction of the dielectric displacement and the subsequent discovery of ether waves, are directly due to the suggestiveness of symbols. For this reason, it has been said that 'in calculation the pen sometimes seems to be more intelligent than the user.' It is the ability of a properly constructed symbolism to function as a *calculus* which makes evident their importance in a striking manner." [4]

The next group of selections should give a good idea of how symbolic systems work and what their inventors intended in creating them. I have tried several different approaches because I know how difficult it is to comprehend abstractions in a strange idiom.

The first excerpt is from a little book by George Boole (1815–1864), an English mathematician and logician who taught for many years at Queens College, Cork. In his pamphlet, and in a later, better known work (*An Investigation of the Laws of Thought on Which Are Founded the Mathematical Theories of Logic and Probabilities*, 1854), Boole laid the foundations of the modern study of symbolic logic. Bertrand Russell once said "Pure mathematics was discovered by Boole, in a work which he

---

[3] The value of symbols as aids to perception "has long been recognized in mathematics. Thus, to take an elementary illustration, the difference in form between $4x^2 = 5x - 1$ and $4x^3 = 5x^2 - 1$, and the identity in form of $x + y = 1$ and $4x = 3y$, can be perceived at a glance. In the first pair of equations one is quadratic, the other cubic; both equations of the second pair are linear. It would be well-nigh humanly impossible to carry out a long series of inferences if such equations were stated in words. Thus the verbal statement of what is represented by Maxwell's equations would easily fill several pages and the essential relations between the various factors involved would thus be hidden." Cohen and Nagel, *op. cit.*, p. 120.

[4] Cohen and Nagel, *op. cit.*, p. 120.

called *The Laws of Thought."* [5] This is an epigram and an exaggeration—as you will see from the second selection—but it is certainly true that Boole's discoveries were of the first rank.

Boole was largely self-taught and showed in all his work an attractive independence of mind. This independence extended also to his religious views and social habits. Among the poorer people of the neighborhood he was regarded as an "innocent who should not be cheated. Among the higher classes he was admired as something of a saint, but rather odd." Whenever he met a person in a train or a shop whose conversation interested him he invited him to his home "to look through the telescope and talk science." [6] Although his wife was sometimes embarrassed by the motley of visitors thus recruited, she consoled herself with the thought that because of Boole's reputation for good will her household "would be safe in any Irish revolution." Boole resembled Augustus De Morgan (see pp. 2366–2368) in some of his interests and personal traits. He was less of a controversialist, but, like De Morgan, when he had decided to do a thing on principle, he was not easily dissuaded. His behavior during his last illness was typical. "Having caught a cold which turned to pneumonia he insisted on calling in a doctor who had recently been dismissed from a professorship of medicine at Queens College for some irregularity of conduct. This he did in order to show friendship and to give pleasure to a man who was in distress." [7] He died December 8, 1864. "One of his last wishes was that his children should not be allowed to fall into the hands of those who were commonly thought religious." [8]

The second selection affords a brief historical sketch of the subject from Leibniz through the work of Frege, Peano, Whitehead and Russell. The excerpt is taken from a standard text, *Symbolic Logic*,[9] by Clarence Irving Lewis and Cooper Harold Langford. Mr. Lewis, noted for his researches and writings in symbolic logic has since 1930 been professor of philosophy at Harvard. Mr. Langford, a specialist in logical theory and symbolic logic, has taught at Harvard and the University of Washington, served (1936–1940) as editor of the *Journal of Symbolic Logic*, and since 1933 has been professor of philosophy at the University of Michigan.

The third selection was written especially for this anthology by Ernest Nagel. In a remarkably ingenious way he shows the relationship between symbolic logic and language. He just translates into symbolism a famous passage from *Alice in Wonderland*, and a delightfully absurd anecdote from a book by Robert Graves and Alan Hodge—*The Reader over Your*

---

[5] The biographical details and associated quotes are from William Kneale, "Boole and the Revival of Logic," *Mind*, Vol. 57, April 1948, pp. 149–175. Another biographical essay is in E. T. Bell, *Men of Mathematics*, New York, 1937.

[6] Kneale, *op. cit.*, p. 156.

[7] Kneale, *op. cit.*, pp. 157–158.

[8] Kneale, *op. cit.*, p. 158.

[9] New York, 1932.

Shoulder.[10] The transformation of words and sentences into symbols illustrates vividly the confusions and ambiguities of ordinary language, how they are uncovered by logical analysis, and eliminated by the methods of symbolic logic. In the third section of his essay, Nagel then reverses the procedure and translates a celebrated theorem of Whitehead and Russell's *Principia Mathematica* from the original symbolism into plain language. This will give the reader a taste of one of the great books of our time. Perhaps I should turn aside for a moment and recount a story about the *Principia* told by G. H. Hardy: "I can remember Bertrand Russell telling me of a horrible dream. He was in the top floor of the University Library, about A.D. 2100. A library assistant was going round the shelves carrying an enormous bucket, taking down book after book, glancing at them, restoring them to the shelves or dumping them into the bucket. At last he came to three large volumes which Russell could recognize as the last surviving copy of *Principia Mathematica*. He took down one of the volumes, turned over a few pages, seemed puzzled for a moment by the curious symbolism, closed the volume, balanced. it in his hand and hesitated. . . ." [11]

The final selection of the group is a discussion of the use of variables in mathematical logic, and of the sentential calculus. The material consists of the first two chapers of the best all-round primer in the field: *Introduction to Logic and to the Methodology of Deductive Sciences*. The author, Alfred Tarski, is a distinguished Polish logician and mathematician, once professor at Zeromski's Lycée in Warsaw and since 1947 professor of mathematics at the University of California. Tarski was a member of the creative circle of eminent Polish logicians, several of whom were murdered during the last war by the Germans.

[10] New York, 1943.
[11] G. H. Hardy, *A Mathematician's Apology*, Cambridge, 1941, p. 23.

1   Mathematical Analysis of Logic

By GEORGE BOOLE

THEY who are acquainted with the present state of the theory of Symbolical Algebra, are aware, that the validity of the processes of analysis does not depend upon the interpretation of the symbols which are employed, but solely upon the laws of their combination. Every system of interpretation which does not affect the truth of the relations supposed, is equally admissible, and it is thus that the same process may, under one scheme of interpretation, represent the solution of a question on the properties of numbers, under another, that of a geometrical problem, and under a third, that of a problem of dynamics or optics. This principle is indeed of fundamental importance; and it may with safety be affirmed, that the recent advances of pure analysis have been much assisted by the influence which it has exerted in directing the current of investigation.

But the full recognition of the consequences of this important doctrine has been, in some measure, retarded by accidental circumstances. It has happened in every known form of analysis, that the elements to be determined have been conceived as measurable by comparison with some fixed standard. The predominant idea has been that of magnitude, or more strictly, of numerical ratio. The expression of magnitude, or of operations upon magnitude, has been the express object for which the symbols of Analysis have been invented, and for which their laws have been investigated. Thus the abstractions of the modern Analysis, not less than the ostensive diagrams of the ancient Geometry, have encouraged the notion, that Mathematics are essentially, as well as actually, the Science of Magnitude.

The consideration of that view which has already been stated, as embodying the true principle of the Algebra of Symbols, would, however, lead us to infer that this conclusion is by no means necessary. If every existing interpretation is shewn to involve the idea of magnitude, it is only by induction that we can assert that no other interpretation is possible. And it may be doubted whether our experience is sufficient to render such an induction legitimate. The history of pure Analysis is, it may be said, too recent to permit us to set limits to the extent of its applications. Should we grant to the inference a high degree of probability, we might still, and with reason, maintain the sufficiency of the definition to which the principle already stated would lead us. We might justly assign it as the definitive character of a true Calculus, that it is a method resting upon the employment of Symbols, whose laws of combination are known and general, and whose results admit of a consistent interpretation. That to the existing forms of Analysis a quantitative interpretation is assigned, is the result of the circumstances by which those forms were determined, and is not to be construed into a universal condition of Analysis. It is upon the foundation of this general principle, that I purpose to establish the Calculus of Logic, and that I claim for it a place among the acknowledged forms of Mathematical Analysis, regardless that in its object and in its instruments it must at present stand alone.

That which renders Logic possible, is the existence in our minds of general notions,—our ability to conceive of a class, and to designate its individual members by a common name. The theory of Logic is thus intimately connected with that of Language. A successful attempt to express logical propositions by symbols, the laws of whose combinations should be founded upon the laws of the mental processes which they represent, would, so far, be a step toward a philosophical language. But this is a view which we need not here follow into detail. Assuming the notion of a class, we are able, from any conceivable collection of objects, to separate by a mental act, those which belong to the given class, and to contemplate them apart from the rest. Such, or a similar act of election, we may conceive to be repeated. The group of individuals left under consideration may be still further limited, by mentally selecting those among them which belong to some other recognised class, as well as to the one before contemplated. And this process may be repeated with other elements of distinction, until we arrive at an individual possessing all the distinctive characters which we have taken into account, and a member, at the same time, of every class which we have enumerated. It is in fact a method similar to this which we employ whenever, in common language, we accumulate descriptive epithets for the sake of more precise definition.

Now the several mental operations which in the above case we have supposed to be performed, are subject to peculiar laws. It is possible to assign relations among them, whether as respects the repetition of a given operation or the succession of different ones, or some other particular, which are never violated. It is, for example, true that the result of two successive acts is unaffected by the order in which they are performed; and there are at least two other laws which will be pointed out in the proper place. These will perhaps to some appear so obvious as to be ranked among necessary truths, and so little important as to be undeserving of special notice. And probably they are noticed for the first time in this Essay. Yet it may with confidence be asserted, that if they were other than they are, the entire mechanism of reasoning, nay the very laws and constitution of the human intellect, would be vitally changed. A Logic might indeed exist, but it would no longer be the Logic we possess.

Such are the elementary laws upon the existence of which, and upon their capability of exact symbolical expression, the method of the following Essay is founded; and it is presumed that the object which it seeks to attain will be thought to have been very fully accomplished. Every logical proposition, whether categorical or hypothetical, will be found to be capable of exact and rigorous expression, and not only will the laws of conversion and of syllogism be thence deductible, but the resolution of the most complex systems of propositions, the separation of any proposed element, and the expression of its value in terms of the remaining elements, with every subsidiary relation involved. Every process will represent deduction, every mathematical consequence will express a logical inference. The generality of the method will even permit us to express arbitrary operations of the intellect, and thus lead to the demonstration of general theorems in logic analogous, in no slight degree, to the general theorems of ordinary mathematics. No inconsiderable part of the pleasure which we derive from the application of analysis to the interpretation of external nature, arises from the conceptions which it enables us to form of the universality of the dominion of law. The general formulæ to which we are conducted seem to give to that element a visible presence, and the multitude of particular cases to which they apply, demonstrate the extent of its sway. Even the symmetry of their analytical expression may in no fanciful sense be deemed indicative of its harmony and its consistency. Now I do not presume to say to what extent the same sources of pleasure are opened in the following Essay. The measure of that extent may be left to the estimate of those who shall think the subject worthy of their study. But I may venture to assert that such occasions of intellectual gratification are not here wanting. The laws we have to examine are the laws of one of the most important of our mental faculties. The mathematics we have to construct are the mathematics of the human intellect. Nor are the form and character of the method, apart from all regard to its interpretation, undeserving of notice. There is even a remarkable exemplification, in its general theorems, of that species of excellence which consists in freedom from exception. And this is observed where, in the corresponding cases of the received mathematics, such a character is by no means apparent. The few who think that there is that in analysis which renders it deserving of attention for its own sake, may find it worth while to study it under a form in which every equation can be solved and every solution interpreted. Nor will it lessen the interest of this study to reflect that every peculiarity which they will notice in the form of the Calculus represents a corresponding feature in the constitution of their own minds. . . .

Logic, like whiskey, loses its beneficial effect when taken in too large quantities. —LORD DUNSANY

God has not been so sparing of men to make them barely two-legged creatures, and left it to Aristotle to make them rational. —JOHN LOCKE

2 History of Symbolic Logic

By CLARENCE IRVING LEWIS
and COOPER HAROLD LANGFORD

THE study with which we are here concerned has not yet acquired any single and well-understood name. It is called 'mathematical logic' as often as 'symbolic logic,' and the designations 'exact logic,' 'formal logic,' and 'logistic' are also used. None of these is completely satisfactory; all of them attempt to convey a certain difference of this subject from the logic which comes down to us from Aristotle and was given its traditional form by the medieval scholastics. This difference, however, is not one of intent: so far as it exists, it is accidental or is due to the relative incompleteness and inexactness of the Aristotelian logic. The use of symbols has characterized logic from the beginning—for example, in the use of letters to represent the terms of the syllogism. Symbolic logic merely extends this use in ways which are required by, or conducive to, clarity and precision. Thus the subject matter of symbolic logic is merely *logic*—the principles which govern the validity of inference.

It is true, however, that the extended use of symbolic procedures opens up so much which is both new and important that symbolic logic becomes an immensely deeper and wider study than the logic of tradition. New implications of accepted logical principles come to light; subtle ambiguities and errors which have previously passed unnoticed are now detected and removed; new generalizations can be made which would be impossible of clear statement without a compact and precise symbolism: the subject of logic becomes broader in its scope, and enters into new relationships with other exact sciences, such as mathematics.

That such changes should come about as a result of a more auspicious mode of representation, will not be surprising if we consider a little the parallel case of mathematics. Arithmetic, at first, lacked any more appropriate medium than that of ordinary language. Ancient Greek mathematicians had no symbol for zero, and used letters of the alphabet for other numbers. As a result, it was impossible to state any general rule for division—to give only one example. Operations which any fourth-grade child can accomplish in the modern notation, taxed the finest mathematical minds of the Age of Pericles. Had it not been for the adoption of the new and more versatile ideographic symbols, many branches of mathematics could never have been developed, because no human mind could grasp the essence of their operations in terms of the phonograms of ordinary language.

Thus while the subject-matter which symbolic logic studies is merely that of logic in any form, it does not follow that there will be no important results of it which are not possible in the older terms. Quite the contrary. As a result of improved methods of notation, we find ourselves in a lively period of new discoveries: an old subject, which has been comparatively stagnant for centuries, has taken on new life. We stand to-day, with respect to logic, where the age of Leibnitz and Newton stood with respect to what can be accomplished in terms of number; or where Riemann and Lobatchevsky stood with respect to geometry. A wealth of new facts dawn upon us, the significance of which we are only beginning to explore. The manner in which the forms and principles characterizing inference admit of extension and generalization, and the connection between such general principles and the more special procedures of other exact sciences—these are matters concerning which the last four decades have produced more light than any preceding four centuries since Aristotle. It is probable that the near future will see further results of equal or greater importance.

As might be expected, it is difficult to mark any historical beginning of the distinctively symbolic method in logic. It is of the essence of the subject that its laws hold for all terms—or all propositions, or all syllogisms, or all conditional arguments, and so on. Thus it is natural that, in the expression of its laws, some indifferent terms or 'variables' should be used to designate those elements of any piece of reasoning which are irrelevant to the validity of it. Even the traditional logic, recognizing that only the *form* of the syllogism (its mood and figure) need be considered in determining its validity, most frequently used letters—A, B, C, or S, M, and P—for the terms of the propositions composing the syllogistic argument. Frequently, also, it was recognized that, in a hypothetical argument, validity did not depend upon the particularity or content of the statements connected by the 'if-then' relation. This recognition was exhibited by throwing the argument into such form as,

> If A, then B.
> But A is true.
> Hence B is true.

Here symbols are used for propositions, which is a step beyond the use of them for substantive terms.

However, the advances which the symbolic procedure makes upon traditional methods hardly begin to appear until symbols are used not only for terms, or for propositions, but for the relations between them. Here and there, through the history of logic in the middle ages, one finds such symbols for the logical relations, used as a kind of shorthand. But it is not until the time of Leibnitz that this use of symbolism begins to be studied with a view to the correction and extension of traditional logic.

Leibnitz may be said to be the first serious student of symbolic logic, though he recognized the *Ars Magna* of Raymond Lull and certain other studies as preceding him in the field. Leibnitz projected an extended and Utopian scheme for the reform of all science by the use of two instruments, a universal scientific language (*characteristica universalis*) and a calculus of reasoning (*calculus ratiocinator*) for the manipulation of it. The universal language was to achieve two ends; first, by being common to all workers in the sciences, it was to break down the barriers of alien speech, achieving community of thought and accelerating the circulation of new scientific ideas. Second, and more important, it was to facilitate the process of logical analysis and synthesis by the substitution of compact and appropriate ideograms for the phonograms of ordinary language.

In part, this second point of Leibnitz's scheme was thoroughly sound: the superiority of ideograms to phonograms in mathematics is an example; and science since Leibnitz's time has made continuously increasing use of ideograms devised for its particular purposes. But in part his program was based upon more dubious conceptions. He believed that all scientific concepts were capable of analysis into a relatively few, so that with a small number of original or undefined ideas, all the concepts that figure in science could be defined. The manner of such definition would, then, exhibit the component simple ideas used in framing a complex notion, somewhat as the algebraic symbolization of a product may exhibit its factors, or the formula of a chemical substance will exhibit its component elements and their relation. Finally, he conceived that the reasoning of science could then be carried out simply by that analysis and synthesis of concepts to which such ideographic symbolisms would supply the clue.

To criticize and evaluate these conceptions of Leibnitz would be a complicated business which is impossible here. Briefly, we may note that the kind of development which he here envisages for science in general coincides to a remarkable extent with the logistic analysis now achieved in mathematics and beginning to be made in other exact sciences. On the other hand, the notion that such development is the major business of science reflects that exaggeration of the rôle of deduction, as against induction, which is characteristic of Leibnitz's rationalistic point of view. And the conception that the primitive or simple concepts which would be disclosed by a just analysis of scientific ideas are uniquely determinable is not supported by developments since his time. Instead, we find that a plurality of possible analyses, and of primitive concepts, is the rule.

Leibnitz himself is more important for his prophetic insight, and for his stimulation of interest in the possibilities of logistic, than for any positive contribution. Though the projects above referred to were formulated before he was twenty years old,[1] and though he made innumerable studies looking toward the furtherance of them, the results are, without exception, fragmentary. The reformation of all science was, as he well understood, an affair which no man could accomplish single-handed. Throughout his life he besought the coöperation of learned societies to this end, but of course without success.

The *calculus ratiocinator* was a more restricted project. If this had been accomplished, it would have coincided with what we now know as symbolic logic. That is, it would have been an organon of reasoning in general, developed in ideographic symbols and enabling the logical operations to be performed according to precise rules. Here Leibnitz achieved some degree of success; and it would be interesting to examine some of his results if space permitted. He did not, however, lay a foundation which could be later built upon, because he was never able to free himself from certain traditional prepossessions or to resolve the difficulties which arose from them.[2]

Other scholars on the Continent were stimulated by these studies of Leibnitz to attempt a calculus of logic. The best of such attempts are those of Lambert and Holland.[3] But all of them are inferior to Leibnitz's own work.

The foundations from which symbolic logic has had continuous development were laid in Great Britain between 1825 and 1850. The signally important contribution was that of the mathematician George Boole; but this was preceded by, and in part resulted from, a renewed interest in logic the main instigators of which were Sir William Hamilton and Augustus De Morgan. Hamilton's "quantification of the predicate" is familiar to most students of logic. "All *A* is *B*" may mean "All *A* is *all B*" or "All *A* is *some B*." The other traditional forms of propositions may be similarily dealt with so as to render the 'quantity' of the predicate term unambiguous. The idea is a simple one, and really of little importance for exact logic: no use has been made of it in recent studies. It was not even

[1] They constitute the subject of his first published work, the intent of which is set forth on the title page as follows: "Dissertatio de Arte Combinatoria; In qua ex Arithmeticæ fundamentis Complicationum ac Transpositionum Doctrina novis præceptis exstruitur, & usus ambarum per universum scientiarum orbem ostenditur; nova etiam Artis Meditandi, seu Logicæ Inventionis femina sparguntur." This book was published in Leipzig, in 1656.

[2] For example, the conception that every universal proposition implies the corresponding particular; and the conception that the relations of terms in extension are always inversely parallel to their relations in intension.

[3] See Bibliography at the end of the selection.

new: Leibnitz, Holland, and others had previously quantified the predicate. But often the historical importance of an idea depends less upon its intrinsic merit than upon the stimulus exercised upon other minds; and this is a case in point. The sole significance of quantification of the predicate for exact logic is that it suggests a manner in which propositions can be treated as equations of terms; and the mere representation of propositions as equations keeps the thought of an analogy between logic and mathematics before the mind. This single fact, together with Hamilton's confident assumption that now at last logic was entering upon a new period of development, seems to have been a considerable factor in the renewal of logical studies in Great Britain.

Augustus De Morgan brought to the subject of logic a more versatile mind than Hamilton's, and one trained in mathematics. He, too, quantified the predicate, and he gives an elaborate table of the thirty-two different forms of propositions which then arise, with rules of transformation and statement of equivalents. His contributions are, in fact, literally too numerous to mention: one can only select the most important for brief notice. One of the most prophetic is his observation that the traditional restriction to propositions in which the terms are related by some form of the verb 'to be,' is artificial and is not dictated by any logical consideration. He wrote, "The copula performs certain functions; it is competent to those functions . . . because it has certain properties which validate its use. . . . Every transitive and convertible relation is as fit to validate the syllogism as the copula '*is*,' and by the same proof in every case. Some forms are valid when the relation is only transitive and not convertible; as in 'give.' Thus if $X — Y$ represent X and Y connected by a transitive copula, *Camestres* in the second figure is valid, as in

$$\text{'Every } Z — Y, \text{ No } X — Y, \text{ therefore No } X — Z.\text{'} \text{"} \, [4]$$

He also investigated many other non-traditional modes of inference, suggested new classifications, and indicated novel principles which govern these. In so doing, he provided the first clear demonstration that logical validity is not confined to the traditional modes and traditional principles.

De Morgan also made numerous and extended studies of the hitherto neglected logic of relations and relative terms. These investigations dropped from sight in the next twenty-five years—that is, until they were renewed by Charles S. Peirce. But as we now know, this study of relations in general is that extension of logic which is most important for the analysis of mathematics. De Morgan was first in this field, and he correctly judged the significance of it. In the concluding paper of one series of studies, he remarks:

[4] *Transactions of the Cambridge Philosophical Society*, X, 177.

"And here the general idea of relation emerges, and for the first time in the history of knowledge, the notion of relation and *relation* of *relation* are symbolized. And here again is seen the scale of graduation of forms, the manner in which what is difference of form at one step of the ascent is difference of *matter* at the next. But the relation of algebra to the higher developments of logic is a subject of far too great extent to be treated here. It will hereafter be acknowledged that, though the geometer did not think it necessary to throw his ever-recurring *principium et exemplum* into imitation of *Omnis homo est animal, Sortes est homo,* etc., yet the algebraist was living in the higher atmosphere of syllogism, the unceasing composition of relation, before it was admitted that such an atmosphere existed." [5]

To-day we can observe how accurately this trenchant prophecy has been fulfilled.

George Boole was really the second founder of symbolic logic. His algebra, first presented in 1847, is the basis of the whole development since. Intrinsically it may be of no greater importance than the work of De Morgan; and Leibnitz's studies exhibit at least an equal grasp of the logical problems to be met. But it is here for the first time that a complete and workable calculus is achieved, and that operations of the mathematical type are systematically and successfully applied to logic.

Three fundamental ideas govern the structure of Boole's system: (1) the conception of the operation of "election" and of "elective symbols"; (2) the "laws of thought" expressible as rules of operation upon these symbols; and (3) the observation that these rules of operation are the same which would hold in an algebra of the numbers 0 and 1.

The elective symbol x represents the result of electing all the x's in the universe: that is, x, y, z, etc., are symbols for classes, conceived as resulting from an operation of selection.

This operation of electing can be treated as analogous to algebraic multiplication. If we first select (from the world of existing things) the x's, and then from the result of that select the y's, the result of these two operations in succession, represented by $x \times y$, or $x\ y$, will be the class of things which are both x's and y's.

It is obvious that the order of operations of election does not affect the result: whether we first select the x's and then from the result of that select the y's; or first select the y's and from the result of that select the x's; in either case we have the class 'both x and y.' That is, 'both x and y' is the same as 'both y and x':

$$x\ y = y\ x.$$

The law that if $x = y$, then $z\ x = z\ y$, will also hold: if x and y are identical classes, or consist of the same members, then 'both z and x' will be the same class as 'both z and y.'

[5] *Loc. cit.,* p. 358.

Repetition of the same operation of election does not alter the result: selecting the x's and then from the result of that selecting again all the x's merely gives the class of x's. Thus

$$x\,x = x, \qquad \text{or} \qquad x^2 = x.$$

This is the one fundamental law peculiar to this algebra, which distinguishes it from the usual numerical algebra.

Boole uses the symbol $+$ as the sign of the operation of aggregation, expressible in language by "Either . . . or . . . ." The class of things which are either x's or y's (but not both) is represented by $x + y$. This operation is commutative:

$$x + y = y + x.$$

The class of things which are either x or y coincides with the class which are either y or x.

The operation of election, or 'multiplication,' is associative with respect to aggregation, or 'addition':

$$z(x + y) = z\,x + z\,y.$$

That is, if we select from the class z those things which are either x or y, the result is the same as if we select what is 'either both z and x or both z and y.'

The operation of 'exception' is represented by the sign of subtraction: if x is 'men' and y is 'Asiatics,' $x - y$ will be 'all men except Asiatics' or 'all men not Asiatics.'

Multiplication is associative with respect to subtraction: that is,

$$z(x - y) = z\,x - z\,y.$$

'White (men not Asiatics)' are the class 'white men except white Asiatics.'

Boole accepts the general algebraic principle that

$$-y + x = x - y,$$

regarding it as a convention that these two are to be equivalent in meaning.

The number 1 is taken to represent 'the universe' or 'everything,' and 0 to represent 'nothing,' or the class which has no members. These interpretations accord with the behavior of 0 and 1 in the algebra:

$$1 \cdot x = x.$$

Selecting the x's from the universe gives the class x.

$$0 \cdot x = 0.$$

Selecting the x's from the class 'nothing' gives 'nothing.'

The negative of any class x is expressible as $1 - x$; the 'not-x's' are 'everything except the x's.' Thus what is 'x but not y' is $x(1 - y)$. And the law holds that

$$x(1 - y) = x \cdot 1 - x\,y = x - x\,y.$$

What is 'x but not y' coincides with the class 'x excepting what is both x and y.'

The aggregate or sum of any class and its negative is 'everything':

$$x + (1 - x) = x + 1 - x = 1.$$

Everything is either x or not-x. The product of a class and its negative is 'nothing':

$$x(1 - x) = x - x^2 = x - x = 0.$$

Nothing is both x and not-x. This equation holds because of the previous law, $x^2 = x$.

An important consequence of the law, $x + (1 - x) = 1$, is that, for any class z,

$$z = z \cdot 1 = z[x + (1 - x)] = z\,x + z(1 - x).$$

That is, the class z coincides with 'either both z and x or both z and not-x.' This law allows any class-symbol x to be introduced into any expression which does not originally contain it—a procedure of great importance in manipulating this algebra, and one which accords with a fundamental logical fact.

It will be evident from the foregoing that at least a considerable number of the operations of ordinary algebra will be valid in this system. The results will be logically interpretable and logically valid. Leibnitz and his Continental successors had all of them tried to make use of the mathematical operations of addition and subtraction; and most of them had tried to use multiplication and division. Always they ran into insuperable difficulties—though they did not always recognize that fact. Boole's better success rests upon four features of his procedure: (1) he thinks of logical relations in extension exclusively, paying no regard to relations of intension; (2) he restricts 'sums,' $x + y$, to the adjunction of classes *having no members in common*; (3) in the law $x\,x = x$, he finds, at one and the same time, the distinctive principle of his algebra and the representation of a fundamental law of logic; (4) in hitting on the analogy of 1 to 'everything' and 0 to 'nothing,' he is able to express such basic logical principles as the laws of Contradiction and of the Excluded Middle Term in terms of the algebra.

The second of these features was later eliminated from his system—to its advantage—but for the time being it contributed to his success by avoiding certain difficulties which otherwise arise with respect to addition. These difficulties had been the principal reason for the failure of previous attempts at a mathematical calculus of logic. In good part, Boole's better success is due to superior ingenuity rather than greater logical acumen. This ingenuity is particularly evident in his meeting of certain difficulties which have, perhaps, already occurred to the reader:

(1) What will be the meaning of an expression such as $1 + x$? Recognizing that terms of a sum must represent classes having no members in common, and that no class x can be related in that manner to the class 'everything' except the class 0, we might say that $1 + x$ has no meaning unless $x = 0$. But in the manipulation of equations, expressions of this form can occur where $x = 0$ is false.

(2) What about $x + x$? Since the terms of this sum violate the requirement that they have no common members, such an expression will not be logically interpretable. But expressions of this form are bound to occur in the algebra. How are they to be dealt with?

(3) If the operations of ordinary algebra are to be used, then division will appear, as the inverse of multiplication. But if $x\ y = z$, and hence $x = z/y$, can the 'fraction' z/y be interpreted; and is the operation logically valid?

Boole surmounts these difficulties, and many others which are similar, by clever devices which depend upon recognizing that his system is completely interpretable as an algebra in which the "elective symbols" are restricted to the numbers 0 and 1. The distinctive law of the system is $x\ x = x$. This holds of 0 and 1, and of no other numbers. Suppose, then, we forget about the logical interpretation of the system, and treat it simply as a numerical algebra in which the variables, or literal symbols, are restricted to 0 or 1. All the ordinary laws and operations of algebra may then be allowed. As Boole says,

"But as the formal processes of reasoning depend only upon the laws of the symbols, and not upon the nature of their interpretation, we are permitted to treat the above symbols, x, y, z, as if they were quantitative symbols of the kind described. We may in fact lay aside the logical interpretation of the symbols in the given equation; convert them into quantitative symbols, susceptible only of the values 0 and 1; perform upon them as such all the requisite processes of solution; and finally restore them to their logical interpretation. . . .

"Now the above system of processes would conduct us to no intelligible result, unless the final equations resulting therefrom were in a form which should render their interpretation, after restoring to the symbols their logical significance, possible. There exists, however, a general method of reducing equations to such a form." [6]

With the devices which make up this general method we need not trouble ourselves. It is, as a matter of fact, always possible to secure a

---

[6] *Laws of Thought*, pp. 69–70.

finally interpretable result, even though the intermediate stages of manipulation involve expressions which are not interpretable; and such results are always in accord with the logical significance of the original data, representing a valid inference. It is not, of course, satisfactory that a logical calculus should involve expressions and operations which are not logically interpretable; and all such features of Boole's Algebra were eventually eliminated by his successors. But even in its original form, it is an entirely workable calculus.

The main steps of the transition from Boole's own system to the current form of Boolean Algebra may be briefly summarized.

W. S. Jevons did not make use of the algebraic methods, but formulated a procedure for solving logical problems by manipulations of the "logical alphabet," which represents what we shall later come to know as 'the expansion of 1'; that is, all possible combinations of the terms involved in any problem. Since this method of Jevons's, though entirely workable and valid, is less simple than the algebraic, and is no longer in use, our chief interest is in two differences of Jevons from Boole which were adopted later as modifications of Boole's algebra.

Jevons interprets $a + b$ to mean 'either a or b or both.' Logically it is merely a matter of convention whether we choose to symbolize by $a + b$ Boole's meaning of 'either-or' as mutually exclusive, or this non-exclusive meaning of Jevons. But adoption of the wider, non-exclusive meaning has three important consequences which are advantageous: it removes any difficulty about expressions of the form $a + a$; it leads to the law $a + a = a$, and it introduces the possibility of a very interesting symmetry between the relation of logical product, $a\, b$, and the logical sum, $a + b$. As we have seen, Boole's interpretation of the relation $+$ has the consequence that $a + a$ is logically uninterpretable. When such expressions occur in his system, they are treated by ordinary algebraic laws, which means that $a + a = 2a$; and such numerical coefficients must be got rid of by devices. If the other meaning be chosen, $a + a$ has meaning, and $a + a = a$ states an obvious principle. With this principle in the system, numerical coefficients do not occur. Thus the two laws, $a\, a = a$ and $a + a = a$, together result in the elimination of all notion of number or quantity from the algebra. Also, when $a + b$ means 'either a or b or both,' the negative of $a\, b$, 'both a and b,' is 'either not-a or not-b,' $(1 - a) + (1 - b)$, or as it would be written in the current notation, $-a + -b$. This law had already been stated by De Morgan, and, as we shall see, is a very important principle of the revised Boolean algebra.

John Venn and Charles S. Peirce both followed the practice of Jevons in this matter; and no important contributor since that time has returned to Boole's narrower meaning of $a + b$, or failed to recognize the law $a + a = a$.

Also, Boole's system was modified by eliminating the operations of subtraction and division. Peirce included these in certain of his early papers, but distinguished a relation of "logical subtraction" and "logical division" from "arithmetical subtraction" and "arithmetical division" (within a Boolean system). Later he discarded these operations altogether, and they have not been made use of since. As Peirce points out, a/b is uninterpretable unless the class b is contained in the class a. Even when this is the case, a/b is an ambiguous function of a and b, which can have any value between the 'limits' ab and $a + (1 - b)$, or $a + -b$. When $a + b$ is given the wider meaning 'either a or b or both,' $a - b$, defined as the value of x such that $a = b + x$, is also an ambiguous function. Nothing is lost in the discarding of these two operations: anything expressible by means of them can likewise be expressed—and better expressed—in terms of the relations of logical product and logical sum alone. The only remaining use of either is in the function 'negative of.' The negative of a, which Boole expressed by $1 - a$, is now written $-a$; and it does not behave, in the algebra, like a minus quantity.

Peirce added a new relation, the inclusion of one class in another; "All a is b," now symbolized by $a \subset b$. This is not a mathematical alteration of the algebra, since the same relation is expressible in Boole's terms as $a(1 - b) = 0$. But it is an obvious advantage that this simplest and most frequent of logical relations should be simply represented.

These changes, then, mark the development from Boole's own system to the algebra of logic as we now know it: (1) substitution of the meaning 'either a or b or both' for 'either a or b but not both' as the relation symbolized by $a + b$; (2) addition of the law, $a + a = a$, eliminating numerical coefficients; (3) as a result of the two preceding, the systematic connection of sums and products, according to De Morgan's Theorem; (4) elimination of the operations of subtraction and division; (5) addition of Peirce's relation, $a \subset b$. The changes (1), (2), and (4) together result in the disappearance from the algebra of all expressions and operations which are not logically interpretable. Other and less fundamental changes, consisting mostly of further developments which do not affect the mathematical character of the system, had also been introduced, principally by Peirce and Schröder.

The next stage in the development of the subject is the bringing together of symbolic logic and the methodology of rigorous deduction, as exhibited in pure mathematics. This turns principally upon development of the logic of propositions and propositional functions, and the logic of relations. Two lines of research, more or less independent, lead up to this. Starting from beginnings made by Boole himself, Peirce and Schröder developed the logic of propositions, propositional functions, and relations, up to the point where these approximate to a calculus which is adequate to the actual deductive procedures exemplified by mathematics. Peano and his collaborators, in the *Formulaire de mathématiques,* began at the other end. They took mathematics in its ordinary deductive form and rigorously analyzed the processes of proof which they found in such deductions. By symbolizing the logical relations upon which proof depends, they gave mathematics the 'logistic' form, in which the principles of logic actually become the vehicles of the demonstration. These general logical principles, set forth in the earlier sections of the *Formulaire,* coincide, to a degree, with the logical formulæ resulting from the work of Peirce and Schröder; but it was found necessary to make certain important additions. (Frege also had developed arithmetic in the logistic form, at an earlier date, and had made most penetrating analyses of the logic of mathematics. But Frege's work passed unnoticed until 1901, when attention was called to it by Mr. Bertrand Russell.)

Boole had indicated a second interpretation of his algebra. In this second interpretation, any term a is taken to represent the times when (or cases in which) the proposition a is true. The negative $1 - a$ or $-a$, will then represent the times, or cases, when the proposition a is false; $a \times b$, or $a\,b$, will represent the cases in which a and b are both true; and $a + b$, the cases in which one of the two, a and b, is true. The entire algebra will then hold for the logical relations of propositions. This will be true whether we take the system in its original form, or as modified by Schröder; with the sole fundamental difference that, in the former case, $a + b$ must be interpreted as the times when a or b *but not* both are true, whereas in the latter case it represents the times when a or b *or* both are true.

However, we should recognize to-day that a statement which is some-times true and sometimes false is not a proposition but a 'propositional function.' Thus the interpretation of the algebra made by Boole is in reality for propositional functions rather than propositions; or more accurately, Boole makes both applications, not distinguishing them, but providing a discussion which strictly applies only to functions. Peirce and Schröder distinguished these two. The entire algebra applies to *both.* But it is a distinguishing feature of a proposition, as against a propositional function, that if it is ever true, it is always true, and if it is ever false, then it is always false. In other words, a proposition is either definitely true or definitely false; while a propositional function may be sometimes true and sometimes false. Since in the algebra, $a = 1$ will mean "The class of cases in which a is true is *all* cases" and $a = 0$ will mean "a is true in *no* case," $a = 1$ represents "a is true," and $a = 0$ represents "a is false." Thus when the distinction is made between propositions and propositional functions, the whole of the algebra holds of both, but there is an *additional* law holding of propositions, but not of propositional functions: "If $a \neq 0$, then $a = 1$."

Peirce and Schröder made the distinction in question. They developed the calculus of propositions, incorporating this additional principle. And they developed the calculus of propositional functions, incorporating the ideas "*a* is sometimes true," "*a* is always true," "*a* is sometimes false," and "*a* is always false." The result of symbolizing these conceptions, and developing the laws which hold of them, is a considerable expansion of the calculus of propositional functions beyond what can be symbolized in terms of the Boolean algebra. What is important for understanding the development of symbolic logic is to remember that substantially such calculuses were at hand in the period when Peano and his collaborators began their work.

The aim of the *Formulaire de mathématiques* is stated in the opening sentence of the Preface: "Le Formulaire de Mathématiques a pour but de publier les propositions connues sur plusieurs sujets des sciences mathématiques. Ces propositions sont exprimées en formules par les notations de la Logique mathématique, expliquées dans l'Introduction au Formulaire."

The authors had before them, on the one hand, mathematics in the ordinary but non-logistic deductive form. It was by this time (1895) generally recognized as the ideal of mathematical form that each branch of the subject should be derived from a small number of assumptions, by a rigorous deduction. It was also recognized that pure mathematics is abstract; that is, that its development is independent of the nature of any concrete empirical things to which it may apply. If, for example, Euclidean geometry is true of our space, and Riemannian geometry is false of space, this fact of material truth or falsity is irrelevant to the mathematical development of Euclid's system or of Riemann's. As it became customary to say, the only truth with which pure mathematics is concerned is the truth that certain postulates imply certain theorems. The emphasis which this conception throws upon the *logic* of mathematics, is obvious: the logic of it is the *truth* of it; no other kind of truth is claimed for pure mathematics.

On the other hand, Peano and his collaborators had before them the developed logic of Peirce and Schröder, which had now become a sufficiently flexible and extended system to be capable (almost) of representing all those relations which hold amongst the assumed entities of mathematical systems, by virtue of which postulates deductively give rise to theorems. It becomes, therefore, an obvious step to use the notations of symbolic logic for representing those relations, and those steps of reasoning, which, in the usual deductive but non-logistic form, are expressed in ordinary language.

Let us take a sample of mathematics as exhibited in the *Formulaire*. For the development of arithmetic, the following undefined ideas are assumed:

"No signifies 'number,' and is the common name of 0, 1, 2, etc.

"0 signifies 'zero.'

"+ signifies 'plus.' If a is a number, $a+$ signifies 'the number succeeding a.'"

In terms of these ideas, the following postulates are assumed:

"1·0   No ϵ Cls

"1·1   0 ϵ No

"1·2 $a \epsilon$ No . $\supset$. $a+ \epsilon$ No

"1·3 $s \epsilon$ Cls . 0 ϵs . $\supset_a$. $a+ \epsilon s$: $\supset$. No ϵs

"1·4 $a, b \epsilon$ No . $a+ = b+$: $\supset$. $a = b$

"1·5 $a \epsilon$ No . $\supset$. $a+ - = 0$"

Let us translate these into English, thus giving the meaning of the symbolism:

1·0   No is a class, or 'number' is a common name.

1·1   0 is a number.

1·2   If a is a number, then the successor of a is a number; that is, every number has a next successor.

1·3   If s is a class, and 0 is a member of s; and if, for all values of a, "a is an s" implies "The successor of a is an s," then 'number' belongs to the class s. In other words, every class s which is such that 0 belongs to it, and that if a belongs to it, the successor of a belongs to it, is such that every number belongs to it.

1·4   If a and b are numbers, and the successor of a is identical with the successor of b, then a is identical with b. That is, no two distinct numbers have the same successor.

1·5   If a is a number, then the successor of a is not identical with 0.

The ideas of 'zero,' 0, and 'successor of a,' $a+$, being assumed, the other numbers are defined in the obvious way: $1 = 0+$, $2 = 1+$, etc.

Further ideas which are necessary to arithmetic are defined as they are introduced. Sometimes such definitions take a form in which they are distinguished from postulates only by convention. For example, the sum of two numbers, represented by the relation $+$ (which must be distinguished from the idea 'successor of a,' $a+$) is defined by the two assumptions:

$$a \epsilon \text{No} . \supset . a + 0 = a$$
$$a, b \epsilon \text{No} . \supset . a + (b+) = (a + b)+$$

These are read: (1) If a is a number, then $a + 0 = a$. (2) If a and b are numbers, then a plus the successor of b, is the successor of (a plus b).

The product of two numbers is defined by:

$$a, b \epsilon \text{No} . \supset . a \times 0 = 0$$
$$a \times (b + 1) = (a \times b) + a$$

It will be fairly obvious how, from these assumptions, the laws governing the sums and products of numbers in general may be deduced.

With the detail of this first development of mathematics in the logistic form we cannot deal here. But several important facts which stand out may be noted. First, in translating the logical relations and operations actually exhibited in mathematical deductions from ordinary language to precise symbols, the authors of the *Formulaire* were obliged to note logical relations and to make distinctions which had previously passed unnoticed. One example of this occurs in the above: they had to distinguish the relation of a member of a class to the class itself (symbolized by ϵ) from the relation of a subclass to the class including it—the relation of *a* to *b* when all *a* is *b*. Another example is the fact that they had very frequently to make use of the idea of a singular subject, '*the* so-and-so,' distinguishing statements about such a subject from those about '*some* so-and-so' or about '*every* so-and-so.' The greater precision of the logistic method revealed the necessity for such distinctions, and for such additions to logical principles, through the fact that, in the absence of them, fallacious mathematical consequences would follow from perfectly good assumptions, or valid mathematical consequences, which should be capable of being proved, could not actually be deduced. Thus, starting with the aim to take mathematics as already developed, and to express the logical relations and operations of proof in an exact and terse symbolism, the authors found that they must first expand logic beyond any previously developed form. The head and front of mathematical logic is found in the calculus of propositional functions, as developed by Peirce and Schröder; but even this logic is not completely adequate to the task in hand.

A second result which stands out in the *Formulaire* is one which was already implicit in the recognition of pure mathematics as abstract and independent of any concrete subject-matter to which it may apply. One of two things must be true of mathematics: either (1) it will be impossible to express the logic of it completely in terms of principles which are universally true—laws of logic which are true of every subject-matter, and not simply of deductions in geometry or in arithmetic—or (2) mathematics must consist of purely analytic statements. If the essential truth asserted by mathematics is that certain assumptions imply certain theorems, and if every such implication in mathematics is an instance of some universally valid principle of deduction (principle of logic), then there can be no step in a proof in mathematics which depends, for example, upon the particular character of our space, or the empirical properties of countable collections. That is to say, Kant's famous dictum that mathematical judgments are *synthetic a priori* truths, must be false. Such mathematical truths will be *a priori*, since they can be certified by logic alone. But they will be *a priori* precisely because they are *not synthetic,* but are merely instances of general logical principles.

As has been said, this is implicit in the previously recognized ideal of mathematics as at once deductive and abstract; for example, in the recognition that non-Euclidean geometry has the same mathematical validity as Euclid. The importance of the logistic development, in this connection, is that the possibility of achieving the logistic form depends upon the elimination from mathematics of any deductive step which is not simply an instance of some universal logical principle; and that symbolic representation is a very reliable check upon this purely analytic character of the demonstration. Thus the actual achievement of the logistic form is a concrete demonstration that it *is a priori* and that it is *not,* in any least detail, *synthetic.* Since this result is of immense importance for our understanding of mathematics, it constitutes one significant consequence of modern logistic.

This bearing of their work is one with which the authors of the *Formulaire* are not explicitly concerned. It had, however, been noted by Frege as a central point of his logistic development of arithmetic, and it was again recognized by the authors of *Principia Mathematica* as an essential bearing of their work.

The *Principia* may, in good part, be taken as identical in its aim with the *Formulaire.* So far as this is the case, the difference between the two works is much as might be expected between a first attempt at the realization of certain aims and a later work having the advantage of the earlier results. Peano had taken mathematics as he found it and translated it completely into precise symbols, and in so doing achieved a degree of explicit analysis and explicit formulation of principles which was not attained in the non-logistic form. The *Principia* goes further in the same direction: the analysis of mathematics is here more extended and more meticulous; the logical connections are more firmly made; the demonstrations achieve rigor in a higher degree.

However, there is also a shift of emphasis. The purpose is no longer that of a compendious presentation of mathematics in the neatest form; it is, rather, the demonstration of the nature of mathematics and its relation to logic. On this point, moreover, the authors of the *Principia* found it possible to go far beyond Peano in the derivation of mathematical truth from logical truth. In the *Formulaire,* as we have seen, the development of arithmetic requires, in addition to the general principles of logic, (1) the undefined ideas, 'number,' 'zero,' and 'the successor of (any given number)'; (2) five postulates in terms of these; (3) definitions of arithmetical relations or operations, such as $+$ and $\times$, in a fashion which is hardly distinguishable from additional postulates. By contrast, in the *Principia,* the development of arithmetic is achieved in a fashion such that: (1) *All* ideas of arithmetic are defined; the only undefined ideas in the whole work being those of logic itself. 'Number,' 'zero,' 'the successor of,' the relations + and ×, and all other ideas of arithmetic are defined in terms of the *logical* ideas assumed, such as 'proposition,' 'negation,' and 'either-or.' This achievement is so extraordinary that one hardly credits it without examining the actual development in the *Principia* itself. (2) *Postulates of arithmetic are eliminated.* There are some exceptions to this, but exceptions which affect only a certain class of theorems concerning transfinite numbers. These exceptions are made explicit by stating the assumption made (there are two such) as an hypothesis to the theorem in question. With these somewhat esoteric exceptions, the propositions of mathematics are proved to be merely *logical consequences of the truths of logic itself.* When mathematical ideas have been defined—defined in terms of logical ideas—the postulates previously thought necessary for mathematics, such as Peano's postulates for arithmetic, which we have given, can all be *deduced.*[7]

Logic itself, in a form sufficient for all further demonstrations, is developed in the early sections. And it is developed in the same deductive fashion, from a small number of undefined ideas and a few postulates in terms of these. Thus it is proved that these primitive ideas and postulates for logic are the only assumptions required for the whole of mathematics.[8]

It is difficult or impossible to convey briefly the significance of this achievement. It will be increasingly appreciated, as time goes on, that the publication of *Principia Mathematica* is a landmark in the history of mathematics and of philosophy, if not of human thought in general. Speculation and plausible theories of the nature and basis of mathematical knowledge have been put forward time out of mind, and have been confronted by opposed conceptions, equally plausible. Here, in a manner peculiarly final, the nature and basis of mathematical truth is definitely determined, and demonstrated *in extenso.* It follows logically from the truths of logic alone, and has whatever characters belong to logic itself.

Although contributions to logic and to mathematics in the logistic form have been more numerous in the period since *Principia Mathematica* than in any other equal period of time, it is hardly possible as yet to have perspective upon these. Three outstanding items may be mentioned. First, by the work of Sheffer and Nicod, considerably greater economy of assumption for any such development as that of the *Principia* has been achieved: two undefined ideas and five symbolic postulates of logic (as in

---

[7] The steps by which Peano's five postulates are rendered unnecessary is clearly and interestingly explained in the first three chapters of Russell's *Introduction to Mathematical Philosophy*. This book is written without symbols, and is intelligible to any reader having an elementary knowledge of mathematics.

[8] Various fundamental ideas of geometry are developed in the later portion of Volume III of the *Principia*; but Volume IV, intended to complete that subject, has not appeared.

the first edition of the *Principia*) have been replaced by one undefined idea and a single postulate. Second, the nature of logical truth itself has become more definitely understood, largely through the discussions of Wittgenstein. It is 'tautological'—such that any law of logic is equivalent to some statement which exhausts the possibilities; whatever is affirmed in logic is a truth to which no alternative is conceivable. From the relation of mathematics to logic, it follows that mathematical truth is similarly tautological.[9] Third, we are just beginning to see that logical truth has a variety and a multiform character hardly to be suspected from the restricted conceptions in terms of which, up till now, it has always been discussed. The work of Lukasiewicz, Tarski, and their school demonstrates that an unlimited variety of systems share that same tautological and undeniable character which we recognize in the logic of our familiar deductions. The logic which we readily recognize as such has been restricted to those forms which we have found useful in application—somewhat as geometry was restricted up to the time of Riemann and Lobatchevsky. The field of logical truth, we begin to understand, is as much wider than its previously recognized forms as modern geometry is wider than the system of Euclid.

BIBLIOGRAPHY

Leibnitz, G. W., *Philosophische Schriften* (hrsg. v. C. I. Gerhardt, Berlin, 1887), Bd. VII; see esp. fragments XX and XXI.

Couturat, L., *La logique de Leibniz, d'après des documents inédits* (Paris, 1901).

Lambert, J. H., *Deutscher Gelehrter Briefwechsel*, 4 vols. (hrsg. v. Bernouilli, Berlin, 1781–1784).

——, *Logische und philosophische Abhandlungen*, 2 vols. (hrsg. v. Bernouilli, Berlin, 1872–1887).

Hamilton, Sir W., *Lectures on Logic* (Edinburgh, 1860).

De Morgan, A., *Formal Logic; or, the Calculus of Inference, Necessary and Probable* (London, 1847).

——, "On the Syllogism," etc., five papers, *Transactions of the Cambridge Philosophical Society*, Vols. VIII, IX, X (1846–1863).

——, *Syllabus of a Proposed System of Logic* (London, 1860).

Boole, G., *The Mathematical Analysis of Logic* (Cambridge, 1847).

——, *An Investigation of the Laws of Thought* (London, 1854; reprinted, Chicago, 1916).

Jevons, W. S., *Pure Logic, or the Logic of Quality Apart from Quantity* (London, 1864).

Venn, J., "Boole's Logical System," *Mind*, I (1876), 479–491.

——, *Symbolic Logic*, 2d ed. (London, 1894).

Peirce, C. S., "Description of a Notation for the Logic of Relatives," *Memoirs of the American Academy of Arts and Sciences*, IX (1870), 317–378.

——, "On the Algebra of Logic," *American Journal of Mathematics*, III (1880), 15–57.

[9] With the further detail of Wittgenstein's conceptions the present authors would not completely agree.

——, "On the Logic of Relatives," in *Studies in Logic by Members of Johns Hopkins University* (Boston, 1883).

——, "On the Algebra of Logic; a Contribution to the Philosophy of Notation," *American Journal of Mathematics*, VII (1885), 180–202.

Frege, G., *Begriffschrift, eine der arithmetischen nachgebildete Formelsprache des reinen Denkens* (Halle, 1879).

——, *Die Grundlagen der Arithmetik* (Breslau, 1884).

——, *Grundgesetze der Arithmetik*, 2 vols. (Jena, 1893–1903).

Schröder, E., *Vorlesungen über die Algebra der Logik*, 3 vols. (Leipzig, 1890–1905).

——, *Abriss der Algebra der Logik*, in parts (hrsg. E. Müller, Leipzig, 1909).

Peano, G., *Formulaire de mathématiques*, 5 vols. (Turin, 1895–1908).

Russell, B. A. W., *Principles of Mathematics*, Vol. I (Cambridge, 1903).

Whitehead, A. N., and Russell, B. A. W., *Principia Mathematica*, 3 vols. (Cambridge, 1910–1913; 2d ed., 1925–1927).

Sheffer, H. M., "A Set of Five Independent Postulates for Boolean Algebras," etc., *Transactions of the American Mathematical Society*, XIV (1913), 481–488.

Nicod, J., "A Reduction in the Number of the Primitive Propositions of Logic," *Proceedings of the Cambridge Philosophical Society*, XIX (1916), 32–42.

Wittgenstein, L., *Tractatus Logico-philosophicus* (New York, 1922).

Lukasiewicz, J., "Philosophische Bemerkungen zu mehrwertigen Systemen des Aussagenkalküls," *Comptes Rendus des séances de la Société des Sciences et des Lettres de Varsovie*, XXIII (1930), Classe III, 51–77.

——, and Tarski, A., "Untersuchungen über den Aussagenkalkül," *ibid.*, 1–21.

Carnap, R., *Der logische Aufbau der Welt* (Berlin, 1928).

> *Good, too, Logic, of course; in itself, but not in fine weather.*
> —ARTHUR HUGH CLOUGH *(1819-1861)*
> *(The Bothie of Tober-na-vuolich)*

3   Symbolic Notation, Haddocks' Eyes and the Dog-Walking Ordinance

By ERNEST NAGEL

I. EXPLANATION OF NOTATION

THE notation of modern formal (or mathematical) logic is an instrument with the help of which the meanings of ordinary statements—whether in common discourse or in the special sciences—may be analyzed, clarified, and more precisely formulated. That notation enables us to state what we wish to state more explicitly than common speech ordinarily permits, and to express complex ideas so as to avoid ambiguities which the structure of every-day language is not equipped to circumvent with ease. There is, however, a price that must be paid for these advantages: unfamiliar distinctions must be learned; a new notation must be mastered; and the reformulations of many apparently simple statements in terms of this notation usually become enormously complex and lengthy. Like the microscope, modern logical notation has great resolving power, and it makes prominent many features in our discourse that are only of negligible interest in the normal business of life; and were we to employ it on all occasions, our vision could easily be rendered myopic. Mathematical logic has undoubtedly contributed to the development of various calculi and of modes of analysis that are of great mathematical significance, and that have a direct bearing in some instances on technological problems. Nevertheless, the methods and the notation of modern formal logic have thus far revealed themselves as indispensable instruments primarily in investigations concerning the foundations of pure logic and pure mathematics—in researches where subtle distinctions must be made and codified, and where maximum explicitness in statement and utmost rigor in demonstration are at a premium.

It is nonetheless amusing, and perhaps even illuminating, to apply current logical distinctions and notation to the analysis of some examples of every-day discourse. This is attempted in the sequel, though certainly not with the hope of thereby revealing the full power or the possible values of modern logic. In preparation for this analysis, there is first introduced what is perhaps a minimum apparatus of distinctions and notation that is essential for making explicit in a convenient manner the meanings of many ordinary statements.

1. STATEMENT FORMS

A brief inspection of the following statements (that is, of bits of discourse concerning which it is significant to raise questions of truth and falsity)

> The Empire State Building is tall
> The Eiffel Tower is tall
> The Woolworth Building is tall

reveals that each of them is an instance of the expression:

> x is tall.

This latter expression is clearly not a statement; but it contains a variable 'x' (other letters may be used as variables) such that if individual names are substituted for the variable, the resulting expression will be a statement.

Again, the statements:

> Paris is older than New York
> The sun is older than Altair
> The Appalachian Range is older than the Rocky Mountains

are easily seen to be instances of the form:

> x is older than y

which contains the two variables 'x' and 'y'.

Expressions like 'x is tall' and 'x is older than y,' which contain one or more variables, such that when individual names are substituted for the variables the resulting expressions are statements, are called "statement forms" (or "statement functions" or "propositional functions").

2. STATEMENTAL CONNECTIVES

From a given set of statements other statements may be formed with the help of certain logical particles called "statemental connectives." Five such connectives and their symbolic surrogates will be mentioned.

(i) *Negation.* Given the statement 'Mary is at home,' its negation or contradictory is formed with the help of the particle 'not': 'Mary is not at home.' In mathematical logic the particle 'not' is replaced by the sign '~' (the tilde), and the negation of a statement is formed by prefixing this sign to the statement. Thus, the negation of 'Mary is at home' is written as: '∼(Mary is at home).' More generally, if 'S' is a statement, its negation is '∼S.'

(ii) *Conjunction.* Two statements (for example, 'Mary is at home' and 'John is happy') may be combined to form a conjunctive statement with the help of the particle 'and,' thus: 'Mary is at home and John is happy.' In current logic the particle is replaced by the dot '.', which is placed between the statements when forming their conjunction; thus:

(Mary is at home) . (John is happy)

The parentheses are introduced to avoid possible confusion, but they are usually omitted when confusion is not likely to arise. This observation applies to parentheses used with other statemental connectives as well.

(iii) *Alternation.* Two statements may be combined with the help of the particles 'either—or,' to form an alternative statement; for example: 'Either Mary is at home or John is happy.' In mathematical logic the expression 'either—or' is replaced by the wedge 'v' (from the initial letter of the Latin *vel*), and is placed between the statements when forming their alternation; thus:

(Mary is at home) v (John is happy)

(iv) *Conditional.* Any two statements may be combined with the help of the particles 'if—then' to form a conditional statement; for example: 'If Mary is at home, then John is happy.' Current logical notation replaces the phrase 'if—then' by the horseshoe '⊃', which is placed between the two statements forming the conditional; thus:

(Mary is at home) ⊃ (John is happy)

(v) *Biconditional.* Two statements may be combined with the help of the phrase 'if and only if' to form a biconditional statement; for example; 'Mary is at home if and only if John is happy.' In mathematical logic, the phrase 'if and only if' is replaced by the triple bar '≡' when constructing a biconditional; for example:

(Mary is at home) ≡ (John is happy)

3. QUANTIFIERS—UNIVERSAL AND EXISTENTIAL

The notation thus far introduced is not sufficient to represent the structure of general statements, such as 'All men are mortal.' However, this example of a general statement may also be written as:

For every x, if x is a man, then x is mortal

where the expression following the phrase 'For every x' is a statement form. Phrases like 'For every x' occur frequently, and it is convenient to replace them by a special combination of signs, thus: '(x).' This combination of signs is called a "universal quantifier," and is formed by flanking a variable (which may be other than the variable 'x') by parentheses. Accordingly, the above general statement when transcribed into logical notation takes the form:

$$(x)(x \text{ is a man} \supset x \text{ is mortal})$$

Again, the statement 'Some men are mortal' may also be written as:

For some x, x is a man and x is mortal

or

There is at least one x such that x is a man and x is mortal.

Phrases like 'There is at least one x' (or more briefly, 'There is an x') also occur frequently, and are replaced in modern logic by the combination of signs: '$(\exists x)$.' This combination of signs is called an "existential quantifier," and is formed by enclosing in parentheses the inverted capital letter 'E' followed by a variable. Accordingly, the statement 'Some men are mortal' when transcribed into logical notation takes the form:

$$(\exists x)(x \text{ is a man. } x \text{ is mortal})$$

A single quantifier does not always suffice in transcribing statements into logical notation. Thus, the statement 'Everybody loves somebody' in effect says that for any individual x whatever, there is some individual y such that x loves y. Consequently, in logical notation the statement takes the form:

$$(x)(\exists y)(x \text{ loves } y)$$

On the other hand, the statement 'Someone is loved by everybody' in effect says that there is at least one individual y such that for any x whatever, y is loved by x. Hence in logical dress this statement appears as:

$$(\exists y)(x)(x \text{ loves } y)$$

It will be evident, therefore, that the order in which quantifiers occur is in general not irrelevant.

4. DEFINITE DESCRIPTIONS

Consider the statement 'The discoverer of argon was Lord Rayleigh.' It contains the singular name 'The discoverer of argon' which, though it is not a proper name, describes a unique individual. Such names have the form 'The so-and-so,' and are called "definite descriptions"; and statements containing them are often construed as asserting (or assuming) the existence of a unique individual (or "object") that satisfies a stipulated condition. For example, the above statement is commonly interpreted to mean that there is one and only one individual who discovered argon, this individual being identical with Lord Rayleigh.

It is sometimes convenient and even necessary to analyze the structure of definite descriptions, and current writers in the main follow the lead of Bertrand Russell, who was the first modern logician to do so. Russell was motivated in his analysis by certain puzzles that arise when definite descriptions are assumed to be simply a species of proper names. Consider the statement 'The proof of Fermat's Last Theorem does not exist.' Russell argued in effect that if the definite description 'The proof of Fermat's Last Theorem' is taken to have the logical status of a proper name, the expression refers to something assumed to be actual or "existing." But, in that case, the statement just mentioned will be self-contradictory, since its subject-term assumes the existence of a unique proof, although the statement denies that there exists such a proof. Russell therefore proposed a way of treating definite descriptions so that, in contradistinction to proper names, their significant use does not commit one to assuming the actual existence of anything which they ostensibly describe. His proposal is essentially a general rule for eliminating definite descriptions by translating statements containing them into logically equivalent statements in which definite descriptions no longer occur.

Russell's analysis can be rendered briefly as follows.

In the first place, it is readily seen that the definite descriptions:

> The discoverer of argon
> The discoverer of oxygen
> The discoverer of the neutron

are all instances of

> The discoverer of x.

This last expression is obviously not a definite description, since it describes no unique individual, and such expressions are frequently called "descriptive functions" (or "descriptive forms").

In the second place, it is possible to construct definite descriptions out of statement forms with the help of a special "descriptional operator." For example, we prefix the descriptional operator '$(\imath x)$,' which is short for 'the one and only one x such that,' to the statement-form 'x discovered argon,' so as to obtain

$$(\imath x)\,(x \text{ discovered argon}).$$

This expression is to be read as: 'The one and only one x such that x discovered argon,' or more briefly as 'The discoverer of argon.' The descriptional operator is itself constructed by enclosing in parentheses the inverted Greek letter iota followed by a variable.

And finally, descriptional operators and definite descriptions cannot be defined explicitly in terms of notions that have already been introduced in this exposition. However, if the notion of strict identity is taken for granted and is represented by the customary sign '=', it is possible to eliminate definite descriptions from any statement in which they occur, in favor of locutions that have already been presented. For example, the statement 'The discoverer of argon was Lord Rayleigh,' which can also be rendered with the help of the descriptional operator as:

$$(\imath x)\,(x \text{ discovered argon}) \text{ was Lord Rayleigh,}$$

can be transformed into the statement that no longer contains a definite description:

$$(\exists x)\,(y)\,((y \text{ discovered argon} \equiv x \text{ discovered}$$
$$\text{argon}) \,.\, (x = \text{Lord Rayleigh}))$$

Read literally, this statement says: There is an individual x such that for any individual y, y discovered argon if and only if. x discovered argon, and x is identical with Lord Rayleigh. When translated into ordinary English, it says that there is just one individual who discovered argon, that individual being Lord Rayleigh.

5. THE USE AND MENTION OF SIGNS

One does not ordinarily confuse a name (or other linguistic expression) with what the name designates. Thus, probably no one will have difficulty in recognizing that in the following two statements:

George Eliot was a woman

George Eliot was a pseudonym

the subject-term of the first refers to a human being, while the subject-term of the second refers to a linguistic expression. (In terms of a locution that has gained wide currency, the expression 'George Eliot' in the first statement is being *used* but is not being *mentioned*, for it is a certain human being who is being mentioned; on the other hand, in the second statement the expression 'George Eliot' is being *mentioned* but is not being *used*, for it is not being used to designate the human being whom the expression normally names.) Nevertheless, there are many contexts in which this distinction is not so readily seen, so that serious confusions sometimes arise in consequence.

A mildly amusing example of the confusions that may arise, is contained in the following anecdote. A teacher of English composition, rebuking his students for their use of cheap language, remarked that there were two words in particular he wished them to avoid in the future. "These words are 'lousy' and 'awful,' " he added. The brief silence that followed this chiding was interrupted by one student's question: "But what *are* the words, Professor?" The point of the story is of course that while the instructor was *mentioning* but not using the words, the student understood him to be *using* rather than mentioning them. A more serious confusion is illustrated by the following fallacious argument. Since $\frac{2}{3} = \frac{14}{21}$ and 7 is greater than $\frac{14}{21}$, it follows (substituting equals for equals) that 7 is greater than $\frac{2}{3}$. Similarly, so it is argued, since $\frac{2}{3} = \frac{14}{21}$ and 7 is a divisor of the numerator of $\frac{14}{21}$, it follows (substituting equals for equals) that 7 is a divisor of the numerator of $\frac{2}{3}$. The mistake here arises from confounding numbers with their names, and in particular from identifying ratios with fractions (which are commonly used as names or designations of ratios). Thus, in the equation '$\frac{2}{3} = \frac{14}{21}$' we are using but not mentioning the fractional expression '$\frac{14}{21}$'; but in the statement '7 is a divisor of the numerator of $\frac{14}{21}$' we are mentioning but not using the fraction '$\frac{14}{21}$,' though we *are* using the phrase 'the numerator of $\frac{14}{21}$' to refer to a certain *numeral* (namely, '14') which in turn is the name of a certain *number* (namely, 14). Although $\frac{2}{3} = \frac{14}{21}$, it is clearly not the case that the *fraction* '$\frac{2}{3}$' is identical with the *fraction* '$\frac{14}{21}$'; and the error in the argument consists in confusing these two statements. Much more serious errors arising from the neglect of the distinction under discussion, but too involved for brief mention, occur in the literature on the foundations of logic and mathematics, as well as in the writings of philosophers.

In order to prevent such confusions, it is useful to formulate a general principle underlying the distinction between *use* and *mention*; and it is also convenient to introduce a notational device which can prevent us from violating the principle.

In a correctly formed English sentence about anything whatsoever, the things talked about never appear in the sentence, but must instead be represented by their names or other designatory expressions. Thus, the individual who was a writer of novels, who lived in England, and so on, is obviously not a constituent of the statement 'George Eliot was a woman,' though her pen-name is such a constituent. This general rule must be observed whether we are talking about non-linguistic things and events, or whether we are constructing statements about linguistic expressions. Accordingly, since in the statement 'George Eliot was a pseudonym' we are asserting something about a linguistic expression, the statement should contain as a constituent not *that expression* but a *name* for the expression. Now there is a widely used current device for manufacturing names for written and printed expressions: it consists in placing an expression within single quotation marks, and using the complex made up out of the expression and its enclosing quotation marks as the name for the expression itself. In agreement with this convention, the statement last mentioned must be formulated as:

'George Eliot' was a pseudonym.

If this convention as to the use of single quotation marks is adopted, it is evident that of the following three statements only the third is correct:

George Eliot contains quotation marks
'George Eliot' contains quotation marks
' 'George Eliot' ' contains quotation marks.

The first of these is obviously false, if not nonsensical. The second is false, because the subject-term refers to the name of a British authoress, and this name contains no quotation marks. But the third statement is true, since its subject-term refers to the name of an expression, and this name does contain quotation marks.

The following locutions and notation, which in part embody the distinctions just explained, will be useful in the immediate sequel. We shall employ the descriptive function 'the name of x' as an abbreviation for '$(\imath y)(y$ names $x)$', and shall construe the sentential form 'y names x' to mean that any expression which is substituted for 'y' is the *conventional name* for the "object" designated by the expression which is substituted for 'x.' Thus, it will be correct to say that:

'Napoleon' names the victor of Austerlitz

or alternatively:

The name of the victor of Austerlitz = 'Napoleon';

but it will be incorrect to say that:

Napoleon names the victor of Austerlitz

or:

The name of the victor of Austerlitz = Napoleon.

Moreover, we shall employ the descriptive function 'the call-name of x' as an abbreviation for '$(\imath y)(y$ call-names $x)$', and shall understand the statement form 'y call-names x' to mean that any expression which is substituted for 'y' is *not* the conventional name for, but is what one *calls* the "object," designated by the expression substituted for 'x.' Thus, it will be correct to say that:

The call-name of George Washington = 'The father of his country,'
but not correct to say either that:

The call-name of George Washington = The father of his country
or that:

The call-name of George Washington = 'George Washington.'

FROM "THROUGH THE LOOKING GLASS" BY LEWIS CARROLL

"You are sad," the Knight said in an anxious tone: "let me sing you a song to comfort you."

"Is it very long?" Alice asked, for she had heard a good deal of poetry that day.

"It's long," said the Knight, "but it's very, *very* beautiful. Everybody that hears me sing it—either it brings the *tears* into their eyes, or else—"

"Or else what?" said Alice, for the Knight had made a sudden pause.

"Or else it doesn't, you know. The name of the song is called '*Haddocks' Eyes.*' "

"Oh, that's the name of the song, is it?" Alice said, trying to feel interested.

"No, you don't understand," the Knight said, looking a little vexed. "That's what the name is *called*. The name really is '*The Aged Aged Man.*' "

"Then I ought to have said 'That's what the *song* is called'?" Alice corrected herself.

"No, you oughtn't: that's quite another thing! The *song* is called '*Ways and Means*': but that's only what it's *called*, you know!"

"Well, what *is* the song, then?" said Alice, who was by this time completely bewildered.

"I was coming to that," the Knight said. "The song really is '*A-sitting On A Gate*': and the tune's my own invention."

So saying, he stopped his horse and let the reins fall on its neck: then, slowly beating time with one hand, and with a faint smile lighting up his gentle foolish face, as if he enjoyed the music of his song, he began.

II. HADDOCKS' EYES

The selection above and that on p. 1890 illustrate in an amusing way two types of difficulty that are frequently encountered in every-day speech, though doubtless rarely in such exaggerated form. It is generally possible to overcome these difficulties without serious trouble, but ordinary language possesses no explicit, systematic technique for doing so. Modern logical theory does have such a technique, and it is instructive to see how it can serve for resolving the difficulties. It must be emphasized, however, that the power and significance of current logical techniques are not fully revealed in these quite elementary examples of their application. The objectives for whose realization every-day speech is an admirable instrument can normally be achieved without employing the elaborate and cumbersome machinery of modern formal logic. No one in his right mind would recommend the systematic use of the interferometer for measuring lumber required in making a table. On the other hand, it would be equally absurd to maintain that since the physicist's techniques of high precision measurement are pointless in the carpenter's shop, those techniques are completely without a *raison d'être*.

The selection from Lewis Carroll illustrates the type of misunderstanding that may arise from the failure to distinguish between fragments of discourse (such as names) and what linguistic expressions are about or designate. Perhaps the grossest example of this type of error is the textbook fallacy: Tigers eat meat, Meat is a word, therefore Tigers eat words. Such misunderstandings can be eliminated by strict adherence to the distinction between the use and mention of symbols, and to the notational convention of employing quotation marks for manufacturing names for linguistic expressions, both of which have been explained above.

The selection from Robert Graves makes evident the lack in every-day speech of a standard procedure for expressing unambiguously the necessary or sufficient conditions for occurrences, and for stating explicitly the time-dependence of events. For example, it is not clear in the statement "The price of corn will be high and the rainfall will increase if the prevailing winds continue" whether what is being asserted is that a consequence of the prevailing winds will be *both* higher corn prices and increased precipitation, or only the latter. Again, it would usually be assumed that in the statement "He took off his clothes and went to bed," the action reported in the first clause preceded the one reported in the second. But the order in which clauses are written does not always suffice to fix the time order of events, as is evident from the statement "His house was destroyed and he disappeared." Such ambiguities can be obviated through the use of devices that are standard in modern logic: precise rules of punctuation (i.e., rules for grouping expressions with the help of parentheses or some other notation), rigid prescriptions for employing statemental connectives, and the introduction of variables and quantifying operators.

In the interchange between Alice and the White Knight, the latter offers to sing her a song whose tune, as is eventually revealed, is claimed by him to be his own invention. Since the statement form: 'x is a song whose tune is claimed by the Knight to be his own invention' has a length too unwieldy for comfort, let us abbreviate it into the shorter form:

$$\text{'}x \text{ is a KS.'}$$

Apparently there is only one such song, so that the definite description

$$\text{'}(\imath x)(x \text{ is a KS})\text{'}$$

uniquely identifies it. It will be convenient occasionally to abbreviate this mode of writing the definite description into the quite brief phrase: 'The Song.'

1. Under Alice's questioning, the Knight admits that The Song is long, but that it is nevertheless very, very beautiful. Let us see what this admission looks like when reformulated in logical notation.

In the first place, we obtain:

(The Song is long) . (The Song is very, very beautiful)
If, next, we replace the abbreviated form of the definite description by the
initial formulation, we get:
(($ıx$)(x is a KS) is long).(($ıx$)(x is a KS) is very, very beautiful)
But suppose we decide to eliminate the definite descriptions altogether.
We then obtain:

$$(\exists x)(y)((y \text{ is a KS} \equiv x \text{ is a KS}) . (x \text{ is long}) . (x \text{ is very,}$$
$$\text{very beautiful}))$$

2. However, the Knight has another confession to make: The Song
either brings tears to the eyes of anyone who hears him sing it, or it does
not. How does this statement look in logical dress? At the first try we get:

(z)(t)(z hears the Knight sing The Song at time $t \supset$ ((The Song brings
tears to the eyes of z at time t) v $\sim$ (The Song brings tears to the
eyes of z at time t))).

But if we replace the abbreviation of the definite description by the un-
adulterated original, the result is:

(z)(t)(z hears the Knight sing ($ıx$)(x is a KS) at time $t \supset$ ((($ıx$)(x
is a KS) brings tears to the eyes of z at time t) v $\sim$ (($ıx$)(x is a KS)
brings tears to the eyes of z at time t))).

And finally, if we prefer to eliminate the definite description, we are left
with the following rather formidable statement containing four quantifiers:

($\exists x$)(y)(z)(t)((y is a KS $\equiv x$ is a KS) . (z hears the Knight sing x
at time $t \supset$ ((x brings tears to the eyes of z at time t) v $\sim$ (x brings
tears to the eyes of z at time t))).

3. According to the Knight, the name of The Song is called 'Haddocks'
Eyes.' And if we recall the locution and notation explained at the close of
Section I, this can be stated as follows:

The call-name of the name of The Song = 'Haddocks' Eyes'

Just for the exercise, let us transcribe this into logical notation, replacing the abbreviated forms of the various definite descriptions contained in this statement by their originals. We then get:

($ıu$)(u call-names ($ız$)(z names ($ıx$)(x is a KS))) =
'Haddocks' Eyes'

And if we insist upon eliminating the definite descriptions, we obtain:

($\exists u$)(w)($\exists z$)(r)($\exists x$)(y)((w call-names $z \equiv u$ call-names z) . (r names
$x \equiv z$ names x) . (y is a KS $\equiv x$ is a KS) . (u = 'Haddocks' Eyes'))

In any event, Alice is obviously wrong when she suggests that

The name of The Song = 'Haddocks' Eyes'

4. What then is the name of The Song? The Knight says that it is 'The
Aged Aged Man.' Accordingly,

The name of The Song = 'The Aged Aged Man'

Alice is therefore clearly in error once more when she thinks that

The call-name of The Song = 'The Aged Aged Man'

5. For as the Knight points out, The Song is called 'Ways and Means.' That is:

The call-name of The Song = 'Ways and Means'

6. Finally, the Knight resolves Alice's bewilderment as to what The Song really is by declaring: "The Song really *is* 'A-sitting On A Gate'." But here the author or the printer who recorded this statement has slipped badly, for the Knight is made to say:

The Song = 'A-sitting On A Gate'

which is quite absurd. For the "object" designated by the left-hand expression in this identity could not possibly be the "object" designated by the right-hand expression, since the former is the song whose tune is claimed by the Knight to be his own invention, while the latter is a certain linguistic expression whose name is exhibited on the right-hand side of the identity. What the Knight obviously meant to say is that:

The Song = A-sitting On A Gate.

7. We can now bring together some of the information supplied by the Knight in the following identities:

The call-name of the name of The Song = 'Haddocks' Eyes'
The name of The Song   = 'The Aged Aged Man'
The call-name of The Song   = 'Ways and Means'
The Song   = A-sitting On A Gate

8. Lewis Carroll's text, with a partial transcription of it into current logical notation can now be set out as follows:

"You are sad," the Knight said in an anxious tone: "let me sing you a song to comfort you."

"Is it very long?" Alice asked, for she had heard a good deal of poetry that day.

"It's long," said the Knight, "but it's very, *very* beautiful. Everybody that hears me sing it—either it brings the *tears* into their eyes, or else—"

The Knight:
$(\exists x)(y)((y$ is a KS $\equiv x$ is a KS$).(x$ is long$).(x$ is very, very beautiful$))$.

"Or else what?" said Alice, for the Knight had made a sudden pause.

"Or else it doesn't, you know. The name of the song is called 'Haddocks' Eyes.'"

The Knight:
$(\exists x)(y)(z)(t)((y$ is a KS $\equiv x$ is a KS$).(z$ hears the Knight sing x at time $t \supset ((x$ brings tears to the eyes of z at time $t)$ v $\sim (x$ brings tears to the eyes of z at time $t)))$.

The Knight:
The call-name of the name of The Song = 'Haddocks' Eyes.'

"Oh, that's the name of the song, is it?" Alice said, trying to feel interested.

"No, you don't understand," the Knight said, looking a little vexed. "That's what the name is *called.* The name really is *'The Aged Aged Man.'* "

"Then I ought to have said 'That's what the *song* is called'?" Alice corrected herself.

"No, you oughtn't: that's quite another thing! The *song* is called 'Ways and Means'; but that's only what it's *called*, you know!"

"Well, what *is* the song, then?" said Alice, who was by this time completely bewildered.

"I was coming to that," the Knight said. "The song really is *'A-sitting On A Gate'*: and the tune's my own invention."

So saying, he stopped his horse and let the reins fall on its neck: then, slowly beating time with one hand, and with a faint smile lighting up his gentle foolish face, as if he enjoyed the music of his song, he began.

Alice:
The name of The Song = 'Haddocks' Eyes.'

The Knight:
The name of The Song = 'The Aged Aged Man.'

Alice:
The call-name of The Song = 'The Aged Aged Man.'

The Knight:
The call-name of The Song = 'Ways and Means.'

The Knight (as reported):
The Song = 'A-sitting On A Gate.'

The Knight (what he doubtless intended to say):
The Song = A-sitting On A Gate.

FROM "THE READER OVER YOUR SHOULDER," BY ROBERT GRAVES
AND ALAN HODGE

From the Minutes of a Borough Council Meeting:

Councillor Trafford took exception to the proposed notice at the entrance of South Park: 'No dogs must be brought to this Park except on a lead.' He pointed out that this order would not prevent an owner from releasing his pets, or pet, from a lead when once safely inside the Park.

The Chairman (Colonel Vine): What alternative wording would you propose, Councillor?

Councillor Trafford: 'Dogs are not allowed in this Park without leads.'

Councillor Hogg: Mr. Chairman, I object. The order should be addressed to the owners, not to the dogs.

Councillor Trafford: That is a nice point. Very well then: 'Owners of dogs are not allowed in this Park unless they keep them on leads.'

Councillor Hogg: Mr. Chairman, I object. Strictly speaking, this would prevent me as a dog-owner from leaving my dog in the back-garden at home and walking with Mrs. Hogg across the Park.

Councillor Trafford: Mr. Chairman, I suggest that our legalistic friend be asked to redraft the notice himself.

Councillor Hogg: Mr. Chairman, since Councillor Trafford finds it so difficult to improve on my original wording, I accept. 'Nobody without his dog on a lead is allowed in this Park.'

Councillor Trafford: Mr. Chairman, I object. Strictly speaking, this notice would prevent me, as a citizen, who owns no dog, from walking in the Park without first acquiring one.

Councillor Hogg (with some warmth): Very simply, then: 'Dogs must be led in this Park.'

Councillor Trafford: Mr. Chairman, I object: this reads as if it were a general injunction to the Borough to lead their dogs into the Park.

Councillor Hogg interposed a remark for which he was called to order; upon his withdrawing it, it was directed to be expunged from the Minutes.

The Chairman: Councillor Trafford, Councillor Hogg has had three tries; you have had only two . . .

Councillor Trafford: 'All dogs must be kept on leads in this Park.'

The Chairman: I see Councillor Hogg rising quite rightly to raise another objection. May I anticipate him with another amendment: 'All dogs in this Park must be kept on the lead.'

This draft was put to the vote and carried unanimously, with two abstentions.

III. THE DOG WALKING ORDINANCE

1. The proposed ordinance in its initial form, which apparently is the fruit of Councillor Hogg's efforts, reads: 'No dogs must be brought to this Park except on a lead.' That is:

$(x)(t)((x$ is a dog. x is brought to this Park at time $t) \supset (x$ is required to be on a lead at time $t))$

But Councillor Trafford is right in noting that this rule is quite compatible with the lead being removed when once the dog is in the Park.

2. Councillor Trafford's first try at an amended version: 'Dogs are not allowed in this Park without leads.'

If Councillor Hogg is correct that this formulation is addressed to dogs rather than their owners, its transcription into logical notation is:

$(t)((\text{You are a dog. You are in this Park at time } t) \supset (\text{You are required to be on a lead at time } t))$

3. Councillor Trafford's second try: 'Owners of dogs are not allowed in this Park unless they keep them on leads'; that is:

$(x)(y)(t)((x$ is a dog. y is owner of x. y is in this Park at time $t) \supset y$ is required to keep x on a lead at time $t))$

However, this rule is consistent with an owner of a dog being in the Park without the dog, the owner being nevertheless required to keep his absent dog on a lead.

4. Councillor Hogg's second try: 'Nobody without his dog on a lead is allowed in this Park.' But as Councillor Trafford construes this formulation, it says:

$(x)(t)(\exists y)(x$ is in this Park at time $t \supset (y$ is a dog. x is owner of y. x is required to keep y on a lead at time $t))$

Hence on this reading of the proposed ordinance, a necessary condition for being in the Park at any time is the possession of a dog who is on a lead at that time.

5. Councillor Hogg's third try: 'Dogs must be led in this Park.' On Councillor Trafford's interpretation, however, this says:

$(x)(t)((x$ is a dog. x is led at time $t)$ ⊃ $(x$ is required to be led
in this Park at time $t))$

Hence a necessary condition for a dog being led is that the animal be
led in the Park.

6. Councillor Trafford's third try: 'All dogs must be kept on leads in
this Park.' This can be construed quite naturally as asserting:

$(x)(t)(x$ is a dog ⊃ x is required to be kept on a lead in this
Park at time $t)$

Accordingly, Councillor Hogg's unvoiced objection must be that this
formulation makes it a necessary condition for a dog that the animal be
kept at all times on a lead in the Park.

7. The Chairman's proposal: 'All dogs in this Park must be kept on
the lead.' The adopted form of the ordinance thus says:

$(x)(t)((x$ is a dog. x is in this Park at time $t)$ ⊃ x is required
to be kept on a lead at time $t))$

8. Comparing these various versions of the ordinance, we note that the
final one is the only one (with the exception of the second, which is unsatisfactory for stylistic reasons) which states as the sufficient condition
for a dog being on a lead while in the Park at any time, the presence of
the dog in the Park at that time. Thus, the first version neglects this condition entirely, since it specifies as sufficient condition for a dog being on
a lead at a given time merely that the animal be brought to the Park at
that time. The third version stipulates as the sufficient condition the presence in the Park of the owner, but not of the dog; and the fourth version
does likewise. The fifth and sixth versions also fail to make the presence
of a dog in the Park a condition for its being on a lead.

9. The text of the discussion in the Borough Council Meeting, with the
translation into logical notation of the proposed ordinances, can be exhibited as follows:

Councillor Trafford took exception to the proposed notice at the
entrance of South Park: 'No dogs
must be brought to this Park except
on a lead.' He pointed out that this
order would not prevent an owner
from releasing his pets, or pet, from
a lead when once safely inside the
Park.

Councillor Hogg:
$(x)(t)((x$ is a dog. x is brought
into this Park at time $t)$ ⊃ $(x$ is required to be on a lead at time $t))$

The Chairman (Colonel Vine):
What alternative wording would
you propose, Councillor?

Councillor Trafford: 'Dogs are
not allowed in this Park without
leads.'

Councillor Trafford:
$(t)((You are a dog. You are in
this Park at time $t)$ ⊃ (You are re-

Councillor Hogg: Mr. Chairman, I object. The order should be addressed to the owners, not to the dogs.

Councillor Trafford: That is a nice point. Very well then: 'Owners of dogs are not allowed in this Park unless they keep them on leads '

Councillor Hogg: Mr. Chairman, I object. Strictly speaking, this would prevent me as a dog-owner from leaving my dog in the back-garden at home and walking with Mrs. Hogg across the Park.

Councillor Trafford: Mr. Chairman, I suggest that our legalistic friend be asked to redraft the notice himself.

Councillor Hogg: Mr. Chairman, since Councillor Trafford finds it so difficult to improve on my original wording, I accept. 'Nobody without his dog on a lead is allowed in this Park.'

Councillor Trafford: Mr. Chairman, I object. Strictly speaking, this notice would prevent me, as a citizen who owns no dog, from walking in the Park without first acquiring one.

Councillor Hogg (with some warmth): Very simply, then: 'Dogs must be led in this Park.'

Councillor Trafford: Mr. Chairman, I object: this reads as if it were a general injunction to the Borough to lead their dogs into the Park.

Councillor Hogg interposed a remark for which he was called to order; upon his withdrawing it, it was directed to be expunged from the Minutes.

The Chairman: Councillor Trafford, Councillor Hogg has had three tries; you have had only two . . .

Councillor Trafford: 'All dogs must be kept on leads in this Park.'

The Chairman: I see Councillor Hogg rising quite rightly to raise another objection. May I anticipate him with another amendment: 'All dogs in this Park must be kept on the lead.'

This draft was put to the vote and carried unanimously, with two abstentions.

quired to be on a lead at time t))

Councillor Trafford:
$(x)(y)(t)((x$ is a dog. y is owner of x. y is in this Park at time $t) \supset y$ is required to keep x on a lead at time t))

Councillor Hogg:
$(x)(t)(\exists y)(x$ is in this Park at time $t \supset (y$ is a dog. x is owner of y. x is required to keep y on a lead at time t))

Councillor Hogg:
$(x)(t)((x$ is a dog. x is led at time $t) \supset (x$ is required to be led in this Park at time t))

Councillor Trafford:
$(x)(t)(x$ is a dog $\supset x$ is required to be kept on a lead in this Park at time t)

The Chairman:
$(x)(t)((x$ is a dog. x is in this Park at time $t) \supset x$ is required to be kept on the lead at time t))

IV. ONE PLUS ONE EQUALS TWO

Principia Mathematica by Alfred North Whitehead and Bertrand Russell is published in three quarto volumes, the first of which appeared in 1910. It is the culmination of the task successfully initiated by George Boole in 1847 to enlarge the scope of traditional Aristotelian formal logic, and to introduce into this discipline the methods and something like the notation and algorithm of modern algebra. Although since the publication of *Principia* revolutionary changes have taken place in the study of formal logic, the work is one of the great classics in the subject. For Whitehead and Russell not only generalized, extended and systematized traditional logic almost beyond recognition; they also carried through to a climactic conclusion the 19th century of "arithmetizing" mathematical analysis, by showing that arithmetic itself can be regarded as simply an extension (or a branch) of formal logic. They were able to define all of the ideas occurring in arithmetic (for example, such basic ones as those designated by "cardinal number," "the immediate successor of," and "zero," which are the fundamental notions in Peano's [1] axiomatization of arithmetic) exclusively in terms of a handful of purely logical notions (such as those designated by "not," "either-or," "for every x"). Furthermore, they also succeeded in demonstrating all propositions of arithmetic as the logical consequences of a small number of purely logical axioms (such as those represented by "If p, then p or q," and "If any x has the property P, then there is an x such that x has the property P"). Accordingly, since 19th century mathematicians had already shown in all essentials that mathematical analysis is "reducible" to arithmetic, it was relatively easy for Whitehead and Russell to establish in painstaking detail and with hitherto almost unrivalled rigor, the further thesis that all mathematical analysis is reducible to formal logic.

In carrying through this program of "logicizing" all of mathematics—Volume IV of *Principia*, which had been planned to work out this thesis for geometry, was never completed, apparently because of altered circumstances in the lives of the authors—Whitehead and Russell were compelled to introduce an enormous number of new distinctions and definitions, to construct demonstrations for preliminary theorems whose number easily runs into the hundreds, and to resolve serious difficulties and puzzles that threatened in various ways the success of the enterprise. The first volume of *Principia* contains well-nigh 700 pages, the second more than this number, and the third almost 500 pages. The proposition that $1 + 1 = 2$ appears as Theorem *110.643, and its proof is not given until page 83

---

[1] Giuseppe Peano (1858–1932) was an Italian mathematician noted particularly for his researches in vector algebra and formal logic. The symbolism of his ideographic language was widely adopted by mathematical logicians because of its simplicity.

of Volume II; Theorem *54.43, which is an important prior lemma, is not demonstrated until page 362 of Volume I. These theorems with their proofs are here reproduced; and appended to each a translation is given into more familiar, though hardly colloquial, English prose.

However, these theorems appear relatively late in *Principia*; and the structure of proofs is not revealed so clearly to the general reader in the demonstrations of these theorems as it is in the case of earlier theorems. Three very early theorems with their demonstrations are therefore also reproduced, accompanied by a prose translation and brief commentary. To make these demonstrations entirely intelligible, a brief account is given of some of the "primitive" (undefined) ideas and "primitive" (undemonstrated) propositions that constitute the logical starting point of Whitehead and Russell's analysis.

The strictly mathematico-logical discussion of *Principia* begins with what its authors call "The Theory of Deduction." Whitehead and Russell first list their primitive ideas. These include the notions of *negation* (symbolized by the tilde: "$\sim$") and *alternation* (symbolized by the wedge "v"); the notion of *elementary propositional functions*[2] (represented by the letters "p," "q," "r," and "s"); and the notion of *assertion* (denoted by the sign "$\vdash$", so that "$\vdash.p$" may be read as "It is true that p").

The notion of *implication* as a relation between propositions is now defined as follows:

$$*1.01 \quad p \supset q . = . \sim p \text{ v } q \text{ Df.}$$

(In words: "If p, then q" is the defined equivalent of "either not p or q." "If the prevailing winds continue then the rainfall will increase" is accordingly short for "either the prevailing winds do not continue or the rainfall will increase.")

Next follow the list of primitive propositions. Some of them, Whitehead and Russell recognize, cannot be expressed by means of the symbolism of their system—these primitive propositions are in fact *rules of deduction*, which specify under what conditions a statement is derivable from other statements. The list of rules of deduction given in the *Principia* is not complete, since the authors use rules (in particular, rules of substitution) in their demonstration of theorems that are not mentioned in their list. One of the important rules that is mentioned is:

*1.1   Anything implied by a true elementary proposition is true. This principle expresses what is traditionally known as the rule of *modus*

---

[2] A propositional function, as defined by Russell, is an "expression containing one or more undetermined constituents such that, when values are assigned to these constituents, the expression becomes a proposition." For example, "All x's are y's" is a propositional function, therefore neither true nor false nor a proposition. But "All men are mortal," a statement got by assigning values to x and y is a proposition, and, as it happens, true.

ponens (currently also called "the rule of detachment"), which permits the derivation of a statement S_2 from two statements of the form S_1 and $S_1 \supset S_2$.

Five primitive propositions are then stated which are expressed entirely by means of the symbolism of the system. Whitehead and Russell adopt a convention of using dots as well as parentheses for punctuation marks. The rules for this convention are a trifle involved; but for present purposes it suffices to note that a larger number of dots has a stronger binding or grouping force than a smaller number. The five "symbolical" primitive propositions are as follows:

*1.2 $\vdash : p \vee p . \supset . p$ (This is called the "principle of tautology" and is referred to by the abbreviated title "Taut.")

*1.3 $\vdash : q . \supset . p \vee q$ (This is called the "principle of addition," and is referred to by the abbreviation "Add.")

*1.4 $\vdash : p \vee q . \supset . q \vee p$ (This is called the "principle of permutation," and is referred to as "Perm.")

*1.5 $\vdash : p \vee (q \vee r) . \supset . q \vee (p \vee r)$ (This is called the "associative principle" and is referred to as "Assoc.")

*1.6 $\vdash : . q \supset r . \supset : p \vee q . \supset . p \vee r$ (This is called the "principle of summation" and is referred to as "Sum.")

It is of some general interest to note that primitive proposition *1.5 is now known to be unnecessary, since it can be derived from the remaining "symbolical" primitives. But it can also be shown that the remaining four of these primitives are independent of one another: no one of them is derivable from the other three. However, when *Principia* was first published, methods for showing the independence of logical primitives were not available.

We can now turn to the theorems and their demonstration.

*2.05 $\vdash : . q \supset r . \supset : p \supset q . \supset . p \supset r$

$$Dem. \left[Sum \frac{\sim p}{p} \right] \vdash : . q \supset r . \supset : p \vee q . \supset . p \vee r \qquad (1)$$

$$[(1) . (*1.01)] \vdash : . q \supset r . \supset : p \supset q . \supset . p \supset r$$

This theorem is one form of the principle of the syllogism. It asserts the following as true: If q implies r, then if p implies q then p implies r.

The demonstration is straightforward. The first line asserts what is obtained by *substitution* from the primitive statement Sum, when '$\sim p$' is substituted for 'p' uniformly (that is, for every occurrence of this letter) in that primitive.

The final line, which is the theorem to be established, is obtained from line (1) with the help of the definition *1.01 (which permits the replacement of a defining expression by the defined expression, and vice versa).

$$*2.07 \quad \vdash : p . \supset . p \vee p \left[*1.3 \; \frac{p}{q} \right]$$

Here we put nothing beyond "*1.3 $\frac{p}{q}$," because the proposition to be proved is what *1.3 becomes when p is written in place of q.

This theorem asserts the following as true: If p, then p or p. No comment beyond that supplied by Whitehead and Russell is needed.

$$*2.08 \quad \vdash . p \supset p$$

Dem.

$$\left[*2.05 \; \frac{p \vee p, p}{q, \quad r} \right] \quad \vdash : : p \vee p . \supset . p : \supset : . p . \supset . p \vee p : \supset . p \supset p \quad (1)$$

$$[\text{Taut}] \qquad \vdash : p \vee p . \supset . p \qquad (2)$$

$$[(1) . (2) . *1.11] \quad \vdash : . p . \supset . p \vee p : \supset . p \supset p \qquad (3)$$

$$[*2.07] \qquad \vdash : p . \supset . p \vee p \qquad (4)$$

$$[(3) . (4) *1.11] \quad \vdash . p \supset p$$

This theorem simply asserts that: If p then p. It is one form of what is sometimes called "the principle of identity."

The demonstration begins with making a substitution in theorem *2.05. The substitution consists in putting '$p \vee p$' for every occurrence of 'q' in that theorem, and putting 'p' for every occurrence of 'r.' Line (1) is the result of that substitution.

The second line is simply the assertion of the primitive proposition Taut.

The third line of the demonstration is obtained from the first two lines with the help of primitive proposition *1.11. This primitive is a slight generalization of the rule of the *modus ponens*. It will be noted that the second line of the proof is identical with that part of the expression in the first line which is to the left of the *main* implication sign—i.e., to the left of the horse-shoe sign which is flanked by the largest number of dots. In accordance with the *modus ponens* it is therefore possible to assert what is to the right of the main implication line—and this is line (3).

The fourth line of the demonstration is simply theorem *2.07.

The final line of the demonstration is obtained from lines (3) and (4) with the help of the primitive proposition *1.11—in exactly the same manner as line (3) is obtained from lines (1) and (2).

$$*54.43 \quad \vdash : . \alpha , \beta \, \epsilon \, 1 . \supset : \alpha \cap \beta = \Lambda . \equiv . \alpha \cup \beta \, \epsilon \, 2$$

Dem.

⊢ . *54.26 . ⊃ ⊢ : . α = ι'x . β = ι'y . ⊃ : α ∪ β ε 2 . ≡ . x ≠ y .

[*51.231]   ≡ . ι'x∩⫴ι'y = Λ

[*13.12]   ≡ . α∩⫴β = Λ (1)

⊢ . (1) . *11.11.35 . ⊃

⊢ : . (Ǝx, y) . α = ι'x . β = ι'y . ⊃ : α ∪ β ε 2 . ≡ . α∩⫴β = Λ (2)

⊢ . (2) . *11.54 . *52.1 . ⊃ ⊢ . Prop

From this proposition it will follow, when arithmetical addition has been defined, that $1 + 1 = 2$.

To understand what is being said here, some new notation must first be explained. The Greek letter "ε" signifies the relation of *class-membership,* so that in general "A ε B" is to be read as "A is a member of (the class) B," and in particular "Frege ε Men" is to be read as "Frege is a member of the class of men." If A and B are classes, "A ∩ B" represents their *logical product,* i.e., A∩⫴B is the class whose members belong to *both* A and B; thus, Male ∩⫴Parent is the class of those individuals who are both males and parents, i.e., the class of fathers. If A and B are classes, "A∪⫴B" represents their *logical sum,* i.e., A∪⫴B is the class whose members belong *either to A or to B or to both*; thus, Male ∪⫴ Parent is the class of those individuals who are either male or parents or both. "Λ" signifies the *null-class,* that is, the class which contains *no* members; thus, since the equation "$x^2 + 1 = 0$" has no real root, the phrase "the real roots of the equation '$x^2 + 1 = 0$'" signifies the null-class. "ι'x" is the name of a *unit-class* whose sole member is x; thus, ι 'Plato is the unit-class whose only member is the individual Plato. "1" symbolizes the class of all unit-classes; and "2" symbolizes the class of all classes each of which is a *couple,* that is, each of which has just two elements as members.

Theorem *54.43 accordingly asserts the following: If the classes α and β are members of the class of unit-classes, then the logical product of α and β is the null-class if, and only if, the logical-sum of α and β is a member of the class of couples.

The demonstration, translated into English, says the following.

Since *54.26 is a theorem and can be asserted, it follows (and hence can also be asserted) that if α is identical with the unit-class whose only member is x and β is identical with the unit-class whose sole member is y, then the logical-sum of α and β is a member of the class of couples if, and only if, x and y are not identical.

But by *51.231, x is not identical with y if, and only if, the logical-product of the unit-classes whose sole members are x and y respectively is the null-class. Accordingly, under the hypothesis, the logical-sum of α and β is a member of the class of couples if, and only if, the logical-product of the unit-classes whose sole members are x and y respectively is the null-class.

And by *13.12, the logical-product of the unit-classes whose sole members are x and y respectively is the null-class, if and only if the logical-product of α and β is the null-class. Hence under the hypothesis, the logical-sum of α and β is a member of the class of couples, if and only if the logical product of α and β is the null-class.   (1)

But since (1) and *11.11 and *11.35 are theorems and can be asserted, it follows (and hence can also be asserted) that

If there is an x and a y, such that α is the unit-class whose sole member is x and β is the unit-class whose sole member is y, then the logical-sum of α and β is a member of the class of couples if, and only if, the logical-product of α and β is the null-class.   (2)

But since (2) and *11.54 and *52.1 are theorems and can be asserted, *54.43 (which is required to be proved) can be asserted.

*110.643. $\vdash . 1 +_c 1 = 2$

Dem.

$$\vdash . *110.632 . *101.21.28 . \supset$$
$$\vdash . 1 +_c 1 - \hat{\xi} \{ (\exists y) . y \in \xi . \xi - \iota\, ' y \in 1 \}$$
$$[*54.3] = 2 . \supset \vdash . \text{Prop}$$

Some further new notation must now be explained. The expression "$+_c$" signifies arithmetical (or cardinal) addition. A letter used as a variable with a "cap" over it (e.g., $\hat{x}$) serves to specify a *class* whose members must satisfy the condition indicated by the expression following the capped-letter. For example, $\hat{x} \{ (x \in \text{Prime Number}) . (x > 25) . (x < 75) \}$ is the class of prime numbers greater than 25 and less than 75. If A is a class, $-$A is the negative (or the negation) of A, and is the class whose members are all elements which are not members of A. Thus, $-\iota\,'$ Plato is the class of all individuals who are not members of the unit-class whose sole member is Plato. If A and B are classes, B $-$ A is the logical product of B and the negation of A; thus, Men $-$ Parents is the class of all those individuals who are men but are not parents.

Theorem *110.643 accordingly asserts the following: The arithmetical sum of the cardinal number 1 and the cardinal number 1 is the cardinal number 2.

The demonstration, translated into English, says the following.

Since *110.632 and *101.21 and *101.28 are theorems and can be asserted, it follows (and can hence be asserted) that

The arithmetical sum of the cardinals 1 and 1 is identical with the class of classes ξ such that for some y or other, y is a member of ξ, and the class which is the logical-product of ξ and the negation of the unit-class whose sole member is y, is a member of the class of unit-classes.

But since *54.3 has already been demonstrated, it follows that this class of classes ξ is identical with the class of couples (that is, with the cardinal number 2), so that the proposition which requires demonstration follows.

Blake wrote: "I have heard many People say, 'Give me the Ideas. It is no matter what Words you put them into.'" To this he replies, "Ideas cannot be Given but in their minutely Appropriate Words."
—WILLIAM BLAKE (Quoted by Agnes Arber, "The Mind and the Eye")

4   Symbolic Logic

By ALFRED TARSKI

ON THE USE OF VARIABLES

CONSTANTS AND VARIABLES

EVERY scientific theory is a system of sentences which are accepted as true and which may be called LAWS or ASSERTED STATEMENTS or, for short, simply STATEMENTS. In mathematics, these statements follow one another in a definite order according to certain principles, and they are, as a rule, accompanied by considerations intended to establish their validity. Considerations of this kind are referred to as PROOFS, and the statements established by them are called THEOREMS.

Among the terms and symbols occurring in mathematical theorems and proofs we distinguish CONSTANTS and VARIABLES.

In arithmetic, for instance, we encounter such constants as *"number,"* *"zero"* ("0"), *"one"* ("1"), *"sum"* ("+"), and many others.[1] Each of these terms has a well-determined meaning which remains unchanged throughout the course of the considerations.

As variables we employ, as a rule, single letters, e.g., in arithmetic the small letters of the English alphabet: *"a," "b," "c," · · · , "x," "y," "z."* As opposed to the constants, the variables do not possess any meaning by themselves. Thus, the question:

does zero have such and such a property?

e.g.:

is zero an integer?

can be answered in the affirmative or in the negative; the answer may be

---

[1] By "arithmetic" we shall here understand that part of mathematics which is concerned with the investigation of the general properties of numbers, relations between numbers and operations on numbers. In place of the word "arithmetic" the term "algebra" is frequently used, particularly in high-school mathematics. We have given preference to the term "arithmetic" because, in higher mathematics, the term "algebra" is reserved for the much more special theory of algebraic equations. (In recent years the term "algebra" has obtained a wider meaning, which is, however, still different from that of "arithmetic.")—The term "number" will here always be used with that meaning which is normally attached to the term "real number" in mathematics; that is to say, it will cover integers and fractions, rational and irrational, positive and negative numbers, but not imaginary or complex numbers.

true or false, but at any rate it is meaningful. A question concerning x, on the other hand, for example the question:

is x an integer?

cannot be answered meaningfully.

In some textbooks of elementary mathematics, particularly the less recent ones, one does occasionally come across formulations which convey the impression that it is possible to attribute an independent meaning to variables. Thus it is said that the symbols *"x," "y,"* $\cdots$ also denote certain numbers or quantities, not "constant numbers" however (which are denoted by constants like "0," "1," $\cdots$), but the so-called "variable numbers" or rather "variable quantities." Statements of this kind have their source in a gross misunderstanding. The "variable number" x could not possibly have any specified property, for instance, it could be neither positive nor negative nor equal to zero; or rather, the properties of such a number would change from case to case, that is to say, the number would sometimes be positive, sometimes negative, and sometimes equal to zero. But entities of such a kind we do not find in our world at all; their existence would contradict the fundamental laws of thought. The classification of the symbols into constants and variables, therefore, does not have any analogue in the form of a similar classification of the numbers.

EXPRESSIONS CONTAINING VARIABLES—SENTENTIAL AND DESIGNATORY FUNCTIONS

In view of the fact that variables do not have a meaning by themselves, such phrases as:

x is an integer

are not sentences, although they have the grammatical form of sentences; they do not express a definite assertion and can be neither confirmed nor refuted. From the expression:

x is an integer

we only obtain a sentence when we replace *"x"* in it by a constant denoting a definite number; thus, for instance, if *"x"* is replaced by the symbol "1," the result is a true sentence, whereas a false sentence arises on replacing *"x"* by "½." An expression of this kind, which contains variables and, on replacement of these variables by constants, becomes a sentence, is called a SENTENTIAL FUNCTION. But mathematicians, by the way, are not very fond of this expression, because they use the term "function" with a different meaning. More often the word "CONDITION" is employed in this sense; and sentential functions and sentences which are composed entirely of mathematical symbols (and not of words of everyday language), such as:

$$x + y = 5,$$

are usually referred to by mathematicians as FORMULAS. In place of "sentential function" we shall sometimes simply say "sentence"—but only in cases where there is no danger of any misunderstanding.

The role of the variables in a sentential function has sometimes been compared very adequately with that of the blanks left in a questionnaire; just as the questionnaire acquires a definite content only after the blanks have been filled in, a sentential function becomes a sentence only after constants have been inserted in place of the variables. The result of the replacement of the variables in a sentential function by constants—equal constants taking the place of equal variables—may lead to a true sentence; in that case, the things denoted by those constants are said to SATISFY the given sentential function. For example, the numbers 1, 2 and 2½ satisfy the sentential function:

$$x < 3,$$

but the numbers 3, 4 and 4½ do not.

Besides the sentential functions there are some further expressions containing variables that merit our attention, namely, the so-called DESIGNATORY OR DESCRIPTIVE FUNCTIONS. They are expressions which, on replacement of the variables by constants, turn into designations ("descriptions") of things. For example, the expression:

$$2x + 1$$

is a designatory function, because we obtain the designation of a certain number (e.g., the number 5), if in it we replace the variable "x" by an arbitrary numerical constant, that is, by a constant denoting a number (e.g., "2").

Among the designatory functions occurring in arithmetic, we have, in particular, all the so-called algebraic expressions which are composed of variables, numerical constants and symbols of the four fundamental arithmetical operations, such as:

$$x - y, \qquad \frac{x + 1}{y + 2}, \qquad 2(x + y - z).$$

Algebraic equations, on the other hand, that is to say, formulas consisting of two algebraic expressions connected by the symbol "=", are sentential functions. As far as equations are concerned, a special terminology has become customary in mathematics; thus, the variables occurring in an equation are referred to as the unknowns, and the numbers satisfying the equation are called the roots of the equation. E.g., in the equation:

$$x^2 + 6 = 5x$$

the variable "x" is the unknown, while the numbers 2 and 3 are roots of the equation.

Of the variables "x," "y," $\cdots$ employed in arithmetic it is said that they STAND FOR DESIGNATIONS OF NUMBERS or that numbers are VALUES of these variables. Thereby approximately the following is meant: a sentential function containing the symbols "x," "y," $\cdots$ becomes a sentence, if these symbols are replaced by such constants as designate numbers (and not by expressions designating operations on numbers, relations between numbers or even things outside the field of arithmetic like geometrical configurations, animals, plants, etc.). Likewise, the variables occurring in geometry stand for designations of points and geometrical figures. The designatory functions which we meet in arithmetic can also be said to stand for designations of numbers. Sometimes it is simply said that the symbols "x," "y," $\cdots$ themselves, as well as the designatory functions made up out of them, denote numbers or are designations of numbers, but this is then a merely abbreviative terminology.

FORMATION OF SENTENCES BY MEANS OF VARIABLES—UNIVERSAL
AND EXISTENTIAL SENTENCES

Apart from the replacement of variables by constants there is still another way in which sentences can be obtained from sentential functions. Let us consider the formula:

$$x + y = y + x.$$

It is a sentential function containing the two variables "x" and "y" that is satisfied by any arbitrary pair of numbers; if we put any numerical constants in place of "x" and "y," we always obtain a true formula. We express this fact briefy in the following manner:

for any numbers x and y,   x + y = y + x.

The expression just obtained is already a genuine sentence and, moreover, a true sentence; we recognize in it one of the fundamental laws of arithmetic, the so-called commutative law of addition. The most important theorems of mathematics are formulated similarly, namely, all so-called UNIVERSAL SENTENCES, or SENTENCES OF A UNIVERSAL CHARACTER, which assert that arbitrary things of a certain category (e.g., in the case of arithmetic, arbitrary numbers) have such and such a property. It has to be noticed that in the formulation of universal sentences the phrase *"for any things* (or *numbers*) *x*, *y*, ⋯" is often omitted and has to be inserted mentally; thus, for instance, the commutative law of addition may simply be given in the following form:

$$x + y = y + x.$$

This has become a well accepted usage, to which we shall generally adhere in the course of our further considerations.

Let us now consider the sentential function:

$$x > y + 1.$$

This formula fails to be satisfied by every pair of numbers; if, for instance, "3" is put in place of "*x*" and "4" in place of "*y*," the false sentence:

$$3 > 4 + 1$$

is obtained. Therefore, if one says:

for any numbers x and y,   x > y + 1,

one does undoubtedly state a meaningful, though obviously false, sentence. There are, on the other hand, pairs of numbers which satisfy the sentential function under consideration; if, for example, "*x*" and "*y*" are replaced by "4" and "2," respectively, the result is the true formula:

$$4 > 2 + 1.$$

This situation is expressed briefly by the following phrase:

for some numbers x and y,   x > y + 1,

or, using a more frequently employed form:

there are numbers x and y such that   x > y + 1.

The expressions just given are true sentences; they are examples of EX-ISTENTIAL SENTENCES, or SENTENCES OF AN EXISTENTIAL CHARACTER, stating the existence of things (e.g., numbers) with a certain property.

With the help of the methods just described we can obtain sentences from any given sentential function; but it depends on the content of the sentential function whether we arrive at a true or a false sentence. The following example may serve as a further illustration. The formula:

$$.x = x + 1$$

is satisfied by no number; hence, no matter whether the words *"for any number x"* or *"there is a number x such that"* are prefixed, the resulting sentence will be false.

In contradistinction to sentences of a universal or existential character we may denote sentences not containing any variables, such as:

$$3 + 2 = 2 + 3,$$

as SINGULAR SENTENCES. This classification is not at all exhaustive, since there are many sentences which cannot be counted among any of the three categories mentioned. An example is represented by the following statement:

for any numbers x and y there is a number z such that
$$x = y + z.$$

Sentences of this type are sometimes called CONDITIONALLY EXISTENTIAL SENTENCES (as opposed to the existential sentences considered before, which may also be called ABSOLUTELY EXISTENTIAL SENTENCES); they state the existence of numbers having a certain property, but on condition that certain other numbers exist.

UNIVERSAL AND EXISTENTIAL QUANTIFIERS; FREE AND BOUND VARIABLES

Phrases like:

for any x, y, · · ·

and

there are x, y, · · · such that

are called QUANTIFIERS; the former is said to be a UNIVERSAL, the latter an EXISTENTIAL QUANTIFIER. Quantifiers are also known as OPERATORS; there are, however, expressions counted likewise among operators, which are different from quantifiers. In the preceding section we tried to explain the meaning of both quantifiers. In order to emphasize their significance it may be pointed out that, only due to the explicit or implicit employment of operators, an expression containing variables can occur as a sentence, that is, as the statement of a well-determined assertion. Without the help of operators, the usage of variables in the formulation of mathematical theorems would be excluded.

In everyday language it is not customary (though quite possible) to use variables, and quantifiers are also, for this reason, not in use. There are, however, certain words in general usage which exhibit a very close connection with quantifiers, namely, such words as *"every," "all," "a certain," "some."* The connection becomes obvious when we observe that expressions like:

all men are mortal

or

some men are wise

have about the same meaning as the following sentences, formulated with the help of quantifiers:

for any x, if x is a man, then x is mortal

and

there is an x, such that x is both a man and wise,

respectively.

For the sake of brevity, the quantifiers are sometimes replaced by symbolic expressions. We can, for instance, agree to write in place of:

for any things (or *numbers*) *x, y, ⋯*

and

there exist things (or *numbers*) *x, y, ⋯ such that*

the following symbolic expressions:

$$\mathop{\mathsf{A}}_{x,y,\ldots} \quad \text{and} \quad \mathop{\mathsf{E}}_{x,y,\ldots}$$

respectively (with the understanding that the sentential functions following the quantifiers are put in parentheses). According to this agreement, the statement which was given at the end of the preceding section as an example of a conditionally existential sentence, for instance, assumes the following form:

(I) $$\mathop{\mathsf{A}}_{x,y} [\mathop{\mathsf{E}}_{z}(x = y + z)]$$

A sentential function in which the variables "x," "y," "z," ⋯ occur automatically becomes a sentence as soon as one prefixes to it one or several operators containing all those variables. If, however, some of the variables do not occur in the operators, the expression in question remains a sentential function, without becoming a sentence. For example, the formula:

$$x = y + z$$

changes into a sentence if preceded by one of the phrases:

for any numbers x, y and z;
there are numbers x, y and z such that;
for any numbers x and y, there is a number z such that;

and so on. But if we merely prefix the quantifier:

there is a number z such that or $\mathop{\mathsf{E}}_{z}$

we do not yet arrive at a sentence; the expression obtained, namely:

(II) $$\mathop{\mathsf{E}}_{z}(x = y + z)$$

is, however, undoubtedly a sentential function, for it immediately becomes a sentence when we substitute some constants in the place of "*x*" and "*y*" and leave "*z*" unaltered, or else, when we prefix another suitable quantifier, e.g.:

$$\text{for any numbers } x \text{ and } y \quad \text{or} \quad \bigwedge_{x,y}$$

It is seen from this that, among the variables which may occur in a sentential function, two different kinds can be distinguished. The occurrence of variables of the first kind—they will be called FREE or REAL VARIABLES—is the decisive factor in determining that the expression under consideration is a sentential function and not a sentence; in order to effect the change from a sentential function to a sentence it is necessary to replace these variables by constants or else to put operators in front of the sentential function that contain those free variables. The remaining, so-called BOUND or APPARENT VARIABLES, however, are not to be changed in such a transformation. In the above sentential function (II), for instance, "*x*" and "*y*" are free variables, and the symbol "*z*" occurs twice as a bound variable; on the other hand, the expression (I) is a sentence, and thus contains bound variables only.

THE IMPORTANCE OF VARIABLES IN MATHEMATICS

Variables play a leading role in the formulation of mathematical theorems. From what has been said it does not follow, however, that it would be impossible in principle to formulate the latter without the use of variables. But in practice it would scarcely be feasible to do without them, since even comparatively simple sentences would assume a complicated and obscure form. As an illustration let us consider the following theorem of arithmetic:

$$\text{for any numbers } x \text{ and } y, \quad x^3 - y^3 = (x - y) \cdot (x^2 + xy + y^2).$$

Without the use of variables, this theorem would look as follows:

> *the difference of the third powers of any two numbers is equal to the product of the difference of these numbers and a sum of three terms, the first of which is the square of the first number, the second the product of the two numbers, and the third the square of the second number.*

An even more essential significance, from the standpoint of the economy of thought, attaches to variables as far as mathematical proofs are concerned. This fact will be readily confirmed by the reader if he attempts to eliminate the variables in any of the proofs which he will meet in the course of our further considerations. And it should be pointed out that these proofs are much simpler than the average considerations to be found in the various fields of higher mathematics; attempts at carrying the latter through without the help of variables would meet with very considerable difficulties. It may be added that it is to the introduction of variables that we are indebted for the development of so fertile a method for the solution of mathematical problems as the method of equations. Without exaggeration it can be said that the invention of variables constitutes a turning point in the history of mathematics; with these symbols man acquired a tool that prepared the way for the tremendous development of the mathematical sciences and for the solidification of its logical foundations.[2]

ON THE SENTENTIAL CALCULUS

LOGICAL CONSTANTS; THE OLD LOGIC AND THE NEW LOGIC

The constants with which we have to deal in every scientific theory may be divided into two large groups. The first group consists of terms which are specific for a given theory. In the case of arithmetic, for instance, they are terms denoting either individual numbers or whole classes of numbers, relations between numbers, operations on numbers, etc.; the constants which we used in Section 1 as examples belong here among others. On the other hand, there are terms of a much more general character occurring in most of the statements of arithmetic, terms which are met constantly both in considerations of everyday life and in every possible field of science, and which represent an indispensable means for conveying human thoughts and for carrying out inferences in any field whatsoever; such words as *"not," "and," "or," "is," "every," "some"* and many others belong here. There is a special discipline, namely LOGIC, considered the basis for all the other sciences, whose concern it is to establish the precise meaning of such terms and to lay down the most general laws in which these terms are involved.

Logic developed into an independent science long ago, earlier even than arithmetic and geometry. And yet it has only been recently—after a long period of almost complete stagnation—that this discipline has begun an intensive development, in the course of which it has undergone a complete transformation with the effect of assuming a character similar to that of

---

[2] Variables were already used in ancient times by Greek mathematicians and logicians,—though only in special circumstances and in rare cases. At the beginning of the 17th century, mainly under the influence of the work of the French mathematician F. Vieta (1540–1603), people began to work systematically with variables and to employ them consistently in mathematical considerations. Only at the end of the 19th century, however, due to the introduction of the notion of a quantifier, was the role of variables in scientific language and especially in the formulation of mathematical theorems fully recognized; this was largely the merit of the outstanding American logician and philosopher Ch. S. Peirce (1839–1914).

the mathematical disciplines; in this new form it is known as MATHE-
MATICAL or DEDUCTIVE or SYMBOLIC LOGIC, and sometimes it is also called
LOGISTIC. The new logic surpasses the old in many respects—not only
because of the solidity of its foundations and the perfection of the methods
employed in its development, but mainly on account of the wealth of concepts and theorems that have been established. Fundamentally, the old
traditional logic forms only a fragment of the new, a fragment moreover
which, from the point of view of the requirements of other sciences, and
of mathematics in particular, is entirely insignificant. . . .

SENTENTIAL CALCULUS; NEGATION OF A SENTENCE, CONJUNCTION
AND DISJUNCTION OF SENTENCES

Among the terms of a logical character there is a small distinguished
group, consisting of such words as *"not," "and," "or," "if* · · ·, *then* · · ·."
All these words are well-known to us from everyday language, and serve
to build up compound sentences from simpler ones. In grammar, they are
counted among the so-called sentential conjunctions. If only for this
reason, the presence of these terms does not represent a specific property
of any particular science. To establish the meaning and usage of these
terms is the task of the most elementary and fundamental part of logic,
which is called SENTENTIAL CALCULUS, or sometimes PROPOSITIONAL CAL-
CULUS or (less happily) THEORY OF DEDUCTION.[3]

We will now discuss the meaning of the most important terms of sentential calculus.

With the help of the word *"not"* one forms the NEGATION of any sentence; two sentences, of which the first is a negation of the second, are
called CONTRADICTORY SENTENCES. In sentential calculus, the word *"not"*
is put in front of the whole sentence, whereas in everyday language it is
customary to place it with the verb; or should it be desirable to have it at
the beginning of the sentence, it must be replaced by the phrase *"it is not
the case that."* Thus, for example, the negation of the sentence:

$$1 \text{ } is \text{ } a \text{ } positive \text{ } integer$$

reads as follows:

$$1 \text{ } is \text{ } not \text{ } a \text{ } positive \text{ } integer,$$

or else:

$$it \text{ } is \text{ } not \text{ } the \text{ } case \text{ } that \text{ } 1 \text{ } is \text{ } a \text{ } positive \text{ } integer.$$

Whenever we utter the negation of a sentence, we intend to express the

---

[3] The historically first system of sentential calculus is contained in the work
Begriffsschrift (Halle 1879) of the German logician G. Frege (1848–1925) who, without doubt, was the greatest logician of the 19th century. The eminent contemporary
Polish logician and historian of logic, J. Lukasiewicz, succeeded in giving sentential
calculus a particularly simple and precise form and caused extensive investigations
concerning this calculus.

idea that the sentence is false; if the sentence is actually false, its negation is true, while otherwise its negation is false.

The joining of two sentences (or more) by the word *"and"* results in
their so-called CONJUNCTION or LOGICAL PRODUCT; the sentences joined
in this manner are called the MEMBERS OF THE CONJUNCTION or the
FACTORS OF THE LOGICAL PRODUCT. If, for instance, the sentences:

$$2 \text{ is a positive integer}$$

and

$$2 < 3$$

are joined in this way, we obtain the conjunction:

$$2 \text{ is a positive integer and } 2 < 3.$$

The stating of the conjunction of two sentences is tantamount to stating
that both sentences of which the conjunction is formed are true. If this is
actually the case, then the conjunction is true, but if at least one of its
members is false, then the whole conjunction is false.

By joining sentences by means of the word *"or"* one obtains the DIS-
JUNCTION of those sentences, which is also called the LOGICAL SUM; the
sentences forming the disjunction are called the MEMBERS OF THE DIS-
JUNCTION or the SUMMANDS OF THE LOGICAL SUM. The word *"or,"* in
everyday language, possesses at least two different meanings. Taken in the
so-called NON-EXCLUSIVE meaning, the disjunction of two sentences merely
expresses that at least one of these sentences is true, without saying anything as to whether or not both sentences may be true; taken in another
meaning, the so-called EXCLUSIVE one, the disjunction of two sentences
asserts that one of the sentences is true but that the other is false. Suppose
we see the following notice put up in a bookstore:

Customers who are teachers or college students are entitled
to a special reduction.

Here the word *"or"* is undoubtedly used in the first sense, since it is not
intended to refuse the reduction to a teacher who is at the same time a
college student. If, on the other hand, a child has asked to be taken on a
hike in the morning and to a theater in the afternoon, and we reply:

no, we are going on a hike or we are going to the theater,

then our usage of the word *"or"* is obviously of the second kind since
we intend to comply with only one of the two requests. In logic and
mathematics, the word *"or"* is always used in the first, non-exclusive
meaning; the disjunction of two sentences is considered true if both or at
least one of its members are true, and otherwise false. Thus, for instance,
it may be asserted:

every number is positive or less than 3,

although it is known that there are numbers which are both positive and less than 3. In order to avoid misunderstandings, it would be expedient, in everyday as well as in scientific language, to use the word *"or"* by itself only in the first meaning, and to replace it by the compound expression *"either* $\cdots$ *or* $\cdots$*"* whenever the second meaning is intended.

Even if we confine ourselves to those cases in which the word *"or"* occurs in its first meaning, we find quite noticeable differences between the usages of it in everyday language and in logic. In common language, two sentences are joined by the word *"or"* only when they are in some way connected in form and content. (The same applies, though perhaps to a lesser degree, to the usage of the word *"and."*) The nature of this connection is not quite clear, and a detailed analysis and description of it would meet with considerable difficulties. At any rate, anybody unfamiliar with the language of contemporary logic would presumably be little inclined to consider such a phrase as:

$$2 \cdot 2 = 5 \quad or \quad New \; York \; is \; a \; large \; city$$

as a meaningful expression, and even less so to accept it as a true sentence. Moreover, the usage of the word *"or"* in everyday English is influenced by certain factors of a psychological character. Usually we affirm a disjunction of two sentences only if we believe that one of them is true but wonder which one. If, for example, we look upon a lawn in normal light, it will not enter our mind to say that the lawn is green or blue, since we are able to affirm something simpler and, at the same time, stronger, namely that the lawn is green. Sometimes even, we take the utterance of a disjunction as an admission by the speaker that he does not know which of the members of the disjunction is true. And if we later arrive at the conviction that he knew at the time that one—and, specifically, which—of the members was false, we are inclined to look upon the whole disjunction as a false sentence, even should the other member be undoubtedly true. Let us imagine, for instance, that a friend of ours, upon being asked when he is leaving town, answers that he is going to do so today, tomorrow or the day after. Should we then later ascertain that, at that time, he had already decided to leave the same day, we shall probably get the impression that we were deliberately misled and that he told us a lie.

The creators of contemporary logic, when introducing the word *"or"* into their considerations, desired, perhaps unconsciously, to simplify its meaning and to render the latter clearer and independent of all psychological factors, especially of the presence or absence of knowledge. Consequently, they extended the usage of the word *"or,"* and decided to consider the disjunction of any two sentences as a meaningful whole, even should no connection between their contents or forms exist; and they also decided to make the truth of a disjunction—like that of a negation or conjunction—dependent only and exclusively upon the truth of its members. Therefore, a man using the word *"or"* in the meaning of contemporary logic will consider the expression given above:

$$2 \cdot 2 = 5 \quad or \quad New \; York \; is \; a \; large \; city$$

as a meaningful and even a true sentence, since its second part is surely true. Similarly, if we assume that our friend, who was asked about the date of his departure, used the word *"or"* in its strict logical meaning, we shall be compelled to consider his answer as true, independent of our opinion as to his intentions.

IMPLICATION OR CONDITIONAL SENTENCE; IMPLICATION
IN MATERIAL MEANING

If we combine two sentences by the words *"if* $\cdots$, *then* $\cdots$,*"* we obtain a compound sentence which is denoted as an IMPLICATION or a CONDITIONAL SENTENCE. The subordinate clause to which the word *"if"* is prefixed is called ANTECEDENT, and the principal clause introduced by the word *"then"* is called CONSEQUENT. By asserting an implication one asserts that it does not occur that the antecedent is true and the consequent is false. An implication is thus true in any one of the following three cases: (i) both antecedent and consequent are true, (ii) the antecedent is false and the consequent is true, (iii) both antecedent and consequent are false; and only in the fourth possible case, when the antecedent is true and the consequent is false, the whole implication is false. It follows that, whoever accepts an implication as true, and at the same time accepts its antecedent as true, cannot but accept its consequent; and whoever accepts an implication as true and rejects its consequent as false, must also reject its antecedent.

As in the case of disjunction, considerable differences between the usages of implication in logic and everyday language manifest themselves. Again, in ordinary language, we tend to join two sentences by the words *"if* $\cdots$, *then* $\cdots$*"* only when there is some connection between their forms and contents. This connection is hard to characterize in a general way, and only sometimes is its nature relatively clear. We often associate with this connection the conviction that the consequent follows necessarily from the antecedent, that is to say, that if we assume the antecedent to be true we are compelled to assume the consequent, too, to be true (and that possibly we can even deduce the consequent from the antecedent on the basis of some general laws which we might not always be able to quote explicitly). Here again, an additional psychological factor manifests itself;

usually we formulate and assert an implication only if we have no exact knowledge as to whether or not the antecedent and consequent are true. Otherwise the use of an implication seems unnatural and its sense and truth may raise some doubt.

The following example may serve as an illustration. Let us consider the law of physics:

every metal is malleable,

and let us put it in the form of an implication containing variables:

if x is a metal, then x is malleable.

If we believe in the truth of this universal law, we believe also in the truth of all its particular cases, that is, of all implications obtainable by replacing "*x*" by names of arbitrary materials such as iron, clay or wood. And, indeed, it turns out that all sentences obtained in this way satisfy the conditions given above for a true implication; it never happens that the antecedent is true while the consequent is false. We notice, further, that in any of these implications there exists a close connection between the antecedent and the consequent, which finds its formal expression in the coincidence of their subjects. We are also convinced that, assuming the antecedent of any of these implications, for instance, "*iron is a metal,*" as true, we can deduce from it its consequent "*iron is malleable,*" for we can refer to the general law that every metal is malleable.

Nevertheless, some of the sentences discussed just now seem artificial and doubtful from the point of view of common language. No doubt is raised by the universal implication given above, or by any of its particular cases obtained by replacing "*x*" by the name of a material of which we do not know whether it is a metal or whether it is malleable. But if we replace "*x*" by "*iron,*" we are confronted with a case in which the antecedent and consequent are undoubtedly true; and we shall then prefer to use, instead of an implication, an expression such as:

since iron is a metal, it is malleable.

Similarly, if for "*x*" we substitute "*clay,*" we obtain an implication with a false antecedent and a true consequent, and we shall be inclined to replace it by the expression:

although clay is not a metal, it is malleable.

And finally, the replacement of "*x*" by "*wood*" results in an implication with a false antecedent and a false consequent; if, in this case, we want to retain the form of an implication, we should have to alter the grammatical form of the verbs:

if wood were a metal, then it would be malleable.

The logicians, with due regard for the needs of scientific languages, adopted the same procedure with respect to the phrase *"if* ···, *then* ···*"* as they had done in the case of the word *"or."* They decided to simplify and clarify the meaning of this phrase, and to free it from psychological factors. For this purpose they extended the usage of this phrase, considering an implication as a meaningful sentence even if no connection whatsoever exists between its two members, and they made the truth or falsity of an implication dependent exclusively upon the truth or falsity of the antecedent and consequent. To characterize this situation briefly, we say that contemporary logic uses IMPLICATIONS IN MATERIAL MEANING, or simply, MATERIAL IMPLICATIONS; this is opposed to the usage of IMPLICATION IN FORMAL MEANING or FORMAL IMPLICATION, in which case the presence of a certain formal connection between antecedent and consequent is an indispensable condition of the meaningfulness and truth of the implication. The concept of formal implication is not, perhaps, quite clear, but, at any rate, it is narrower than that of material implication; every meaningful and true formal implication is at the same time a meaningful and true material implication, but not vice versa.

In order to illustrate the foregoing remarks, let us consider the following four sentences:

$$\text{if} \quad 2 \cdot 2 = 4, \quad \text{then} \quad \textit{New York is a large city;}$$
$$\text{if} \quad 2 \cdot 2 = 5, \quad \text{then} \quad \textit{New York is a large city;}$$
$$\text{if} \quad 2 \cdot 2 = 4, \quad \text{then} \quad \textit{New York is a small city;}$$
$$\text{if} \quad 2 \cdot 2 = 5, \quad \text{then} \quad \textit{New York is a small city.}$$

In everyday language, these sentences would hardly be considered as meaningful, and even less as true. From the point of view of mathematical logic, on the other hand, they are all meaningful, the third sentence being false, while the remaining three are true. Thereby it is, of course, not asserted that sentences like these are particularly relevant from any viewpoint whatever, or that we apply them as premisses in our arguments.

It would be a mistake to think that the difference between everyday language and the language of logic, which has been brought to light here, is of an absolute character, and that the rules, outlined above, of the usage of the words *"if* ···, *then* ···*"* in common language admit of no exceptions. Actually, the usage of these words fluctuates more or less, and if we look around, we can find cases in which this usage does not comply with our rules. Let us imagine that a friend of ours is confronted with a very difficult problem and that we do not believe that he will ever solve it. We can then express our disbelief in a jocular form by saying:

if you solve this problem, I shall eat my hat.

The tendency of this utterance is quite clear. We affirm here an implication whose consequent is undoubtedly false; therefore, since we affirm the truth of the whole implication, we thereby, at the same time, affirm the falsity of the antecedent; that is to say, we express our conviction that our friend will fail to solve the problem in which he is interested. But it is also quite clear that the antecedent and the consequent of our implication are in no way connected, so that we have a typical case of a material and not of a formal implication.

The divergency in the usage of the phrase *"if · · ·, then · · ·"* in ordinary language and mathematical logic has been at the root of lengthy and even passionate discussions,—in which, by the way, professional logicians took only a minor part.[4] (Curiously enough, considerably less attention was paid to the analogous divergency in the case of the word *"or."*) It has been objected that the logicians, on account of their employment of the material implication, arrived at paradoxes and even plain nonsense. This has resulted in an outcry for a reform of logic to the effect of bringing about a far-reaching rapprochement between logic and ordinary language regarding the use of implication.

It would be hard to grant that these criticisms are well-founded. There is no phrase in ordinary language that has a precisely determined sense. It would scarcely be possible to find two people who would use every word with exactly the same meaning, and even in the language of a single person the meaning of the same word varies from one period of his life to another. Moreover, the meaning of words of everyday language is usually very complicated; it depends not only on the external form of the word but also on the circumstances in which it is uttered and sometimes even on subjective psychological factors. If a scientist wants to introduce a concept from everyday life into a science and to establish general laws concerning this concept, he must always make its content clearer, more precise and simpler, and free it from inessential attributes; it does not matter here whether he is a logician concerned with the phrase *"if · · ·, then · · ·"* or, for instance, a physicist establishing the exact meaning of the word *"metal."* In whatever way the scientist realizes his task, the usage of the term as it is established by him will deviate more or less from the practice of everyday language. If, however, he states explicitly in what meaning he decides to use the term, and if he consistently acts in

---

[4] It is interesting to notice that the beginning of this discussion dates back to antiquity. It was the Greek philosopher Philo of Megara (in the 4th century B.C.) who presumably was the first in the history of logic to propagate the usage of material implication; this was in opposition to the views of his master, Diodorus Cronus, who proposed to use implication in a narrower sense, rather related to what is called here the formal meaning. Somewhat later (in the 3d century B.C.)—and probably under the influence of Philo—various possible conceptions of implication were discussed by the Greek philosophers and logicians of the Stoic School (in whose writings the first beginnings of sentential calculus are to be found).

conformity to this decision, nobody will be in a position to object that his procedure leads to nonsensical results.

Nevertheless, in connection with the discussions that have taken place, some logicians have undertaken attempts to reform the theory of implication. They do not, generally, deny material implication a place in logic, but they are anxious to find also a place for another concept of implication, for instance, of such a kind that the possibility of deducing the consequent from the antecedent constitutes a necessary condition for the truth of an implication; they even desire, so it seems, to place the new concept in the foreground. These attempts are of a relatively recent date, and it is too early to pass a final judgment as to their value.[5] But it appears today almost certain that the theory of material implication will surpass all other theories in simplicity, and, in any case, it must not be forgotten that logic, founded upon this simple concept, turned out to be a satisfactory basis for the most complicated and subtle mathematical reasonings.

THE USE OF IMPLICATION IN MATHEMATICS

The phrase "*if* $\cdots$, *then* $\cdots$" belongs to those expressions of logic which are used most frequently in other sciences and, especially, in mathematics. Mathematical theorems, particularly those of a universal character, tend to have the form of implications; the antecedent is called in mathematics the HYPOTHESIS, and the consequent is called the CONCLUSION.

As a simple example of a theorem of arithmetic, having the form of an implication, we may quote the following sentence:

if x is a positive number, then 2x is a positive number

in which "*x is a positive number*" is the hypothesis, while "*2x is a positive number*" is the conclusion.

Apart from this, so to speak, classical form of mathematical theorems, there are, occasionally, different formulations, in which hypothesis and conclusion are connected in some other way than by the phrase "*if* $\cdots$, *then* $\cdots$*.*" The theorem just mentioned, for instance, can be paraphrased in any of the following forms:

from: x is a positive number, it follows: 2x is a positive number;

the hypothesis: x is a positive number, implies (or has as a consequence) the conclusion: 2x is a positive number;

the condition: x is a positive number, is sufficient for 2x to be a positive number;

for 2x to be a positive number it is sufficient that x be a positive number;

---

[5] The first attempt of this kind was made by the contemporary American philosopher and logician C. I. Lewis.

> *the condition: 2x is a positive number, is necessary for x to be a positive number;*
>
> *for x to be a positive number it is necessary that 2x be a positive number.*

Therefore, instead of asserting a conditional sentence, one might usually just as well say that the hypothesis IMPLIES the conclusion or HAS it AS A CONSEQUENCE, or that it is a SUFFICIENT CONDITION for the conclusion; or one can express it by saying that the conclusion FOLLOWS from the hypothesis, or that it is a NECESSARY CONDITION for the latter. A logician may raise various objections against some of the formulations given above, but they are in general use in mathematics.

EQUIVALENCE OF SENTENCES

We shall consider one more expression from the field of sentential calculus. It is one which is comparatively rarely met in everyday language, namely, the phrase *"if, and only if."* If any two sentences are joined up by this phrase, the result is a compound sentence called an EQUIVALENCE. The two sentences connected in this way are referred to as the LEFT and RIGHT SIDE OF THE EQUIVALENCE. By asserting the equivalence of two sentences, it is intended to exclude the possibility that one is true and the other false; an equivalence, therefore, is true if its left and right sides are either both true or both false, and otherwise the equivalence is false.

The sense of an equivalence can also be characterized in still another way. If, in a conditional sentence, we interchange antecedent and consequent, we obtain a new sentence which, in its relation to the original sentence, is called the CONVERSE SENTENCE (or the CONVERSE OF THE GIVEN SENTENCE). Let us take, for instance, as the original sentence the implication:

(I)   *if x is a positive number, then 2x is a positive number;*

the converse of this sentence will then be:

(II)   *if 2x is a positive number, then x is a positive number.*

As is shown by this example, it occurs that the converse of a true sentence is true. In order to see, on the other hand, that this is not a general rule, it is sufficient to replace "$2x$" by "x^2" in (I) and (II); the sentence (I) will remain true, while the sentence (II) becomes false. If, now, it happens that two conditional sentences, of which one is the converse of the other, are both true, then the fact of their simultaneous truth can also be expressed by joining the antecedent and consequent of any one of the two sentences by the words *"if, and only if."* Thus, the above two implications —the original sentence (I) and the converse sentence (II)—may be replaced by a single sentence:

x is a positive number if, and only if, 2x is a positive number

(in which the two sides of the equivalence may yet be interchanged).

There are, incidentally, still a few more possible formulations which may serve to express the same idea, e.g.:

> *from: x is a positive number, it follows: 2x is a positive number, and conversely;*
>
> *the conditions that x is a positive number and that 2x is a positive number are equivalent with each other;*
>
> *the condition that x is a positive number is both necessary and sufficient for 2x to be a positive number;*
>
> *for x to be a positive number it is necessary and sufficient that 2x be a positive number.*

Instead of joining two sentences by the phrase *"if, and only if,"* it is therefore, in general, also possible to say that the RELATION OF CONSEQUENCE holds between these two sentences IN BOTH DIRECTIONS, or that the two sentences are EQUIVALENT, or, finally, that each of the two sentences represents a NECESSARY AND SUFFICIENT CONDITION for the other.

THE FORMULATION OF DEFINITIONS AND ITS RULES

The phrase *"if, and only if"* is very frequently used in laying down DEFINITIONS, that is, conventions stipulating what meaning is to be attributed to an expression which thus far has not occurred in a certain discipline, and which may not be immediately comprehensible. Imagine, for instance, that in arithmetic the symbol "$\leq$" has not as yet been employed but that one wants to introduce it now into the considerations (looking upon it, as usual, as an abbreviation of the expression *"is less than or equal to"*). For this purpose it is necessary to define this symbol first, that is, to explain exactly its meaning in terms of expressions which are already known and whose meanings are beyond doubt. To achieve this, we lay down the following definition,—assuming that "$>$" belongs to the symbols already known:

> *we say that x $\leq$ y if, and only if, it is not the case that x $>$ y.*

The definition just formulated states the equivalence of the two sentential functions:

$$x \leq y$$

and

$$\text{it is not the case that} \quad x > y;$$

it may be said, therefore, that it permits the transformation of the formula "$x \leq y$" into an equivalent expression which no longer contains the symbol "$\leqq$" but is formulated entirely in terms already comprehensible to us. The same holds for any formula obtained from "$x \leqq y$" by replacing "x" and "y" by arbitrary symbols or expressions designating numbers. The formula:

$$3 + 2 \leqq 5,$$

for instance, is equivalent with the sentence:

it is not the case that $3 + 2 > 5$;

since the latter is a true assertion, so is the former. Similarly, the formula:

$$4 \leqq 2 + 1$$

is equivalent with the sentence:

it is not the case that $4 > 2 + 1$,

both being false assertions. This remark applies also to more complicated sentences and sentential functions; by transforming, for instance, the sentence:

if $x \leqq y$ *and* $y \leqq z$,   *then* $x \leqq z$,

we obtain:

if it is not the case that $x > y$ *and if it is not the case that*
$y > z$,   *then it is not the case that* $x > z$.

In short, by virtue of the definition given above, we are in a position to transform any simple or compound sentence containing the symbol "$\leqq$" into an equivalent one no longer containing it; in other words, so to speak, to translate it into a language in which the symbol "$\leqq$" does not occur. And it is this very fact which constitutes the role which definitions play within the mathematical disciplines.

If a definition is to fulfil its proper task well, certain precautionary measures have to be observed in its formulation. To this effect special rules are laid down, the so-called RULES OF DEFINITION, which specify how definitions should be constructed correctly. Since we shall not here go into an exact formulation of these rules, it may merely be remarked that, on their basis, every definition may assume the form of an equivalence; the first member of that equivalence, the DEFINIENDUM, should be a short, grammatically simple sentential function containing the constant to be defined; the second member, the DEFINIENS, may be a sentential function of an arbitrary structure, containing, however, only constants whose meaning either is immediately obvious or has been explained previously. In particular, the constant to be defined, or any expression previously defined with its help, must not occur in the definiens; otherwise the definition is

incorrect, it contains an error known as a VICIOUS CIRCLE IN THE DEFINI-
TION (just as one speaks of a VICIOUS CIRCLE IN THE PROOF, if the argument meant to establish a certain theorem is based upon that theorem
itself, or upon some other theorem previously proved with its help). In
order to emphasize the conventional character of a definition and to
distinguish it from other statements which have the form of an equivalence, it is expedient to prefix it by words such as *"we say that."* It is
easy to verify that the above definition of the symbol "$\leq$" satisfies all
these conditions; it has the definiendum:

$$x \leq y,$$

whereas the definiens reads:

it is not the case that $x > y$.

It is worth noticing that mathematicians, in laying down definitions,
prefer the words *"if"* or *"in case that"* to the phrase *"if, and only if."* If,
for example, they had to formulate the definition of the symbol "$\leq$,"
they would, presumably, give it the following form:

we say that $x \leq y$,   if it is not the case that $x > y$.

It looks as if such a definition merely states that the definiendum follows
from the definiens, without emphasizing that the relation of consequence
also holds in the opposite direction, and thus fails to express the equivalence of definiendum and definiens. But what we actually have here is a
tacit convention to the effect that *"if"* or *"in case that,"* if used to join
definiendum and definiens, are to mean the same as the phrase *"if, and
only if"* ordinarily does. It may be added that the form of an equivalence
is not the only form in which definitions may be laid down.

LAWS OF SENTENTIAL CALCULUS

After having come to the end of our discussion of the most important
expressions of sentential calculus, we shall now try to clarify the character
of the laws of this calculus.

Let us consider the following sentence:

*if 1 is a positive number and $1 < 2$,   then 1 is a positive
number.*

This sentence is obviously true, it contains exclusively constants belonging
to the field of logic and arithmetic, and yet the idea of listing this sentence
as a special theorem in a textbook of mathematics would not occur to
anybody. If one reflects why this is so, one comes to the conclusion that
this sentence is completely uninteresting from the standpoint of arithmetic;

it fails to enrich in any way our knowledge about numbers, its truth does not at all depend upon the content of the arithmetical terms occurring in it, but merely upon the sense of the words *"and," "if," "then."* In order to make sure that this is so, let us replace in the sentence under consideration the components:

> 1 *is a positive number*

and

$$1 < 2$$

by any other sentences from an arbitrary field; the result is a series of sentences, each of which, like the original sentence, is true; for example:

> *if the given figure is a rhombus and if the same figure is a rectangle, then the given figure is a rhombus;*
> *if today is Sunday and the sun is shining, then today is Sunday.*

In order to express this fact in a more general form, we shall introduce the variables *"p"* and *"q,"* stipulating that these symbols are not designations of numbers or any other things, but that they stand for whole sentences; variables of this kind are denoted as SENTENTIAL VARIABLES. Further, we shall replace in the sentence under consideration the phrase:

> 1 *is a positive number*

by *"p"* and the formula:

$$1 < 2$$

by *"q"*; in this manner we arrive at the sentential function:

> *if   p and q,   then p.*

This sentential function has the property that only true sentences are obtained if arbitrary sentences are substituted for *"p"* and *"q."* This observation may be given the form of a universal statement:

> *for any p and q,   if   p and q,   then p.*

We have here obtained a first example of a law of sentential calculus, which will be referred to as the LAW OF SIMPLIFICATION for logical multiplication. The sentence considered above was merely a special instance of this universal law—just as, for instance, the formula:

$$2 \cdot 3 = 3 \cdot 2$$

is merely a special instance of the universal arithmetical theorem:

> *for arbitrary numbers x and y, $x \cdot y = y \cdot x$.*

In a similar way, other laws of sentential calculus can be obtained. We give here a few examples of such laws; in their formulation we omit the universal quantifier *"for any p, q, . . ."*—in accordance with the usage mentioned earlier, which becomes almost a rule throughout sentential calculus.

> *If p, then p.*
> *If p, then   q or p.*
> *If   p implies q   and   q implies p,   then   p if, and only if, q.*
> *If   p implies q   and   q implies r,   then   p implies r.*

The first of these four statements is known as the LAW OF IDENTITY, the second as the LAW OF SIMPLIFICATION for logical addition, and the fourth as the LAW OF THE HYPOTHETICAL SYLLOGISM.

Just as the arithmetical theorems of a universal character state something about the properties of arbitrary numbers, the laws of sentential calculus assert something, so one may say, about the properties of arbitrary sentences. The fact that in these laws only such variables occur as stand for quite arbitrary sentences is characteristic of sentential calculus and decisive for its great generality and the scope of its applicability.

SYMBOLISM OF SENTENTIAL CALCULUS; TRUTH FUNCTIONS
AND TRUTH TABLES

There exists a certain simple and general method, called METHOD OF TRUTH TABLES or MATRICES, which enables us, in any particular case, to recognize whether a given sentence from the domain of the sentential calculus is true, and whether, therefore, it can be counted among the laws of this calculus.[6]

In describing this method it is convenient to apply a special symbolism. We shall replace the expressions:

not;   and;   or;   if · · ·, then · · ·;   if, and only if

by the symbols.

$\sim$; $\wedge$; $\vee$; $\rightarrow$; $\longleftrightarrow$

respectively. The first of these symbols is to be placed in front of the expression whose negation one wants to obtain; the remaining symbols are always placed between two expressions ("$\rightarrow$" stands therefore in the place of the word *"then,"* while the word *"if"* is simply omitted). From one or two simpler expressions we are, in this way, led to a more complicated expression; and if we want to use the latter for the construction of further still more complicated expressions, we enclose it in parentheses.

With the help of variables, parentheses and the constant symbols listed above (and sometimes also additional constants of a similar character), we are able to write down all sentences and sentential functions belonging

[6] This method originates with Charles S. Peirce.

to the domain of sentential calculus. Apart from the individual sentential variables the simplest sentential functions are the expressions:

$$\sim p, \quad p \wedge q, \quad p \vee q, \quad p \rightarrow q, \quad p \leftrightarrow q$$

(and other similar expressions which differ from these merely in the shape of the variables used). As an example of a compound sentential function let us consider the expression:

$$(p \vee q) \rightarrow (p \wedge r),$$

which we read, translating symbols into common language:

if p or q, then p and r.

A still more complicated expression is the law of the hypothetical syllogism given above, which now assumes the form:

$$[(p \rightarrow q) \wedge (q \rightarrow r)] \rightarrow (p \rightarrow r).$$

We can easily make sure that every sentential function occurring in our calculus is a so-called TRUTH FUNCTION. This means to say that the truth or falsehood of any sentence obtained from that function by substituting whole sentences for variables depends exclusively upon the truth or falsehood of the sentences which have been substituted. As for the simplest sentential functions "$\sim p$," "$p \wedge q$," and so on, this follows immediately from the remarks made in earlier discussions concerning the meaning attributed in logic to the words *"not," "and,"* and so on. But the same applies, likewise, to compound functions. Let us consider, for instance, the function "$(p \vee q) \rightarrow (p \wedge r)$." A sentence obtained from it by substitution is an implication, and, therefore, its truth depends on the truth of its antecedent and consequent only; the truth of the antecedent, which is a disjunction obtained from "$p \vee q$," depends only on the truth of the sentences substituted for "p" and "q," and similarly the truth of the consequent depends only on the truth of the sentences substituted for "p" and "r." Thus, finally, the truth of the whole sentence obtained from the sentential function under consideration depends exclusively on the truth of the sentences substituted for "p," "q," and "r."

In order to see quite exactly how the truth or falsity of a sentence obtained by substitution from a given sentential function depends upon the truth or falsity of the sentences substituted for variables, we construct what is called the TRUTH TABLE or MATRIX for this function. We shall begin by giving such a table for the function "$\sim p$":

p	$\sim p$
T	F
F	T

And here is the joint truth table for the other elementary functions "$p \wedge q$," "$p \vee q$," and so on:

p	q	$p \wedge q$	$p \vee q$	$p \to q$	$p \leftrightarrow q$
T	T	T	T	T	T
F	T	F	T	T	F
T	F	F	T	F	F
F	F	F	F	T	T

The meaning of these tables becomes at once comprehensible if we take the letters "T" and "F" to be abbreviations of "true sentence" and "false sentence," respectively. In the second table, for instance, we find, in the second line below the headings "p," "q" and "$p \to q$," the letters "F," "T" and "T," respectively. We gather from that that a sentence obtained from the implication "$p \to q$" is true if we substitute any false sentence for "p" and any true sentence for "q"; this, obviously, is entirely consistent with the remarks made on pp. 1913–1917.—The variables "p" and "q" occurring in the tables can, of course, be replaced by any other variables.

With the help of the two above tables, called FUNDAMENTAL TRUTH TABLES, we can construct DERIVATIVE TRUTH TABLES for any compound sentential function. The table for the function "$(p \vee q) \to (p \wedge r)$," for instance, looks as follows:

p	q	r	$p \vee q$	$p \wedge r$	$(p \vee q) \to (p \wedge r)$
T	T	T	T	T	T
F	T	T	T	F	F
T	F	T	T	T	T
F	F	T	F	F	T
T	T	F	T	F	F
F	T	F	T	F	F
T	F	F	T	F	F
F	F	F	F	F	T

In order to explain the construction of this table, let us concentrate, say, on its fifth horizontal line (below the headings). We substitute true sentences for "p" and "q" and a false sentence for "r." According to the second fundamental table, we then obtain from "$p \vee q$" a true sentence and from "$p \wedge r$" a false sentence. From the whole function "$(p \vee q) \to (p \wedge r)$" we obtain then an implication with a true antecedent and a false consequent; hence, again with the help of the second fundamental table (in which we think of "p" and "q" being for the moment replaced by "$p \vee q$" and "$p \wedge r$"), we conclude that this implication is a false sentence.

The horizontal lines of a table that consist of symbols "T" and "F" are called ROWS of the table, and the vertical lines are called COLUMNS. Each row or, rather, that part of each row which is on the left of the vertical bar represents a certain substitution of true or false sentences for the variables. When constructing the matrix of a given function, we take care to exhaust all possible ways in which a combination of symbols "T" and "F" could be correlated to the variables; and, of course, we never write in a table two rows which do not differ either in the number or in the order of the symbols "T" and "F." It can then be seen very easily that the number of rows in a table depends in a simple way on the number of different variables occurring in the function; if a function contains 1, 2, 3, $\cdots$ variables of different shape, its matrix consists of $2^1 = 2$, $2^2 = 4$, $2^3 = 8$, $\cdots$ rows. As for the number of columns, it is equal to the number of partial sentential functions of different form contained in the given function (where the whole function is also counted among its partial functions).

We are now in a position to say how it may be decided whether or not a sentence of sentential calculus is true. As we know, in sentential calculus, there is no external difference between sentences and sentential functions; the only difference consisting in the fact that the expressions considered to be sentences are always completed mentally by the universal quantifier. In order to recognize whether the given sentence is true, we treat it, for the time being, as a sentential function, and construct the truth table for it. If, in the last column of this table, no symbol "F" occurs, then every sentence obtainable from the function in question by substitution will be true, and therefore our original universal sentence (obtained from the sentential function by mentally prefixing the universal quantifier) is also true. If, however, the last column contains at least one symbol "F," our sentence is false.

Thus, for instance, we have seen that in the matrix constructed for the function "$(p \lor q) \to (p \land r)$" the symbol "F" occurs four times in the last column. If, therefore, we considered this expression as a sentence (that is, if we prefixed to it the words *"for any p, q and r"*), we would have a false sentence. On the other hand, it can be easily verified with the help of the method of truth tables that all the laws of sentential calculus stated above, that is, the laws of simplification, identity, and so on, are true sentences. The table for the law of simplification:

$$(p \land q) \to p,$$

for instance, is as follows:

p	q	$p \land q$	$(p \land q) \to p$
T	T	T	T
F	T	F	T
T	F	F	T
F	F	F	T

We give here a number of other important laws of sentential calculus whose truth can be ascertained in a similar way:

$\sim[p \wedge (\sim p)]$,
$(p \wedge p) \longleftrightarrow p$,
$(p \wedge q) \longleftrightarrow (q \wedge p)$,
$[p \wedge (q \wedge r)] \longleftrightarrow [(p \wedge q) \wedge r]$,

$p \vee (\sim p)$,
$(p \vee p) \longleftrightarrow p$,
$(p \vee q) \longleftrightarrow (q \vee p)$,
$[p \vee (q \vee r)] \longleftrightarrow [(p \vee q) \vee r]$.

The two laws in the first line are called the LAW OF CONTRADICTION and the LAW OF EXCLUDED MIDDLE; we next have the two LAWS OF TAUTOLOGY (for logical multiplication and addition); we then have the two COMMUTATIVE LAWS, and finally the two ASSOCIATIVE LAWS. It can easily be seen how obscure the meaning of these last two laws becomes if we try to express them in ordinary language. This exhibits very clearly the value of logical symbolism as a precise instrument for expressing more complicated thoughts.

It occurs that the method of matrices leads us to accept sentences as true whose truth seemed to be far from obvious before the application of this method. Here are some examples of sentences of this kind:

$$p \rightarrow (q \rightarrow p),$$
$$(\sim p) \rightarrow (p \rightarrow q),$$
$$(p \rightarrow q) \vee (q \rightarrow p).$$

That these sentences are not immediately obvious is due mainly to the fact that they are a manifestation of the specific usage of implication characteristic of modern logic, namely, the usage of implication in material meaning.

These sentences assume an especially paradoxical character if, when reading them in words of common language, the implications are replaced by phrases containing "*implies*" or "*follows*," that is, if we give them, for instance, the following form:

> *if p is true, then p follows from any q* (in other words: *a true sentence follows from every sentence*);
> *if p is false, then p implies any q* (in other words: *a false sentence implies every sentence*);
> *for any p and q, either p implies q or q implies p* (in other words: *at least one of any two sentences implies the other*).

In this formulation, these statements have frequently been the cause of misunderstandings and superfluous discussions [arising from the failure to distinguish between the meaning of words as used in common language and mathematical logic].

APPLICATION OF LAWS OF SENTENTIAL CALCULUS IN INFERENCE

Almost all reasonings in any scientific domain are based explicitly or implicitly upon laws of sentential calculus; we shall try to explain by means of an example in what way this happens.

Given a sentence having the form of an implication, we can, apart from its converse of which we had already spoken (p. 1918), form two further sentences: the INVERSE SENTENCE (or the INVERSE OF THE GIVEN SENTENCE) and the CONTRAPOSITIVE SENTENCE. The inverse sentence is obtained by replacing both the antecedent and the consequent of the given sentence by their negations. The contrapositive is the result of interchanging the antecedent and the consequent in the inverse sentence; the contrapositive sentence is, therefore, the converse of the inverse sentence and also the inverse of the converse sentence. The converse, the inverse and the contrapositive sentences, together with the original sentence, are referred to as CONJUGATE SENTENCES. As an illustration we may consider the following conditional sentence:

(I) *if x is a positive number, then 2x is a positive number,*

and form its three conjugate sentences:

if 2x is a positive number, then x is a positive number;
if x is not a positive number, then 2x is not a positive number;
if 2x is not a positive number, then x is not a positive number.

In this particular example, all the conjugate sentences obtained from a true sentence turn out to be likewise true. But this is not at all so in general; in order to see that it is quite possible that not only the converse sentence but also the inverse sentence may be false, although the original sentence is true, it is sufficient to replace "$2x$" by "x^2" in the above sentences.

Thus it is seen that from the validity of an implication nothing definite can be inferred about the validity of the converse or the inverse sentence. The situation is quite different in the case of the fourth conjugate sentence; whenever an implication is true, the same applies to the corresponding contrapositive sentence. This fact may be confirmed by numerous examples, and it finds its expression in a general law of sentential calculus, namely the so-called LAW OF TRANSPOSITION or OF CONTRAPOSITION.

In order to be able to formulate this law with precision, we observe that every implication may be given the schematic form:

if p, then q;

the converse, the inverse and the contrapositive sentences will then assume the forms:

if q, then p;   *if not p, then not q;*   *if not q, then not p.*

The law of contraposition, according to which any conditional sentence implies the corresponding contrapositive sentence, may hence be formulated as follows:

$$if: \quad if \ p, \ then \ q, \quad then: \quad if \ not \ q, \ then \ not \ p.$$

In order to avoid the accumulation of the words "*if*" it is expedient to make a slight change in the formulation:

(II) *from:* if p, then q, *it follows that:* if not q, then not p.

We now want to show how, with the help of this law, we can, from a statement having the form of an implication—for instance, from statement (I)—derive its contrapositive statement.

(II) applies to arbitrary sentences "*p*" and "*q*," and hence remains valid if for "*p*" and "*q*" the expressions:

$$x \ is \ a \ positive \ number$$

and

$$2x \ is \ a \ positive \ number$$

are substituted. Changing, for stylistic reasons, the position of the word "*not*," we obtain:

(III) *from:* if x is a positive number, then 2x is a positive number, it *follows that:* if 2x is not a positive number, then x is not a positive number.

Now compare (I) and (III): (III) has the form of an implication, (I) being its hypothesis. Since the whole implication as well as its hypothesis have been acknowledged as true, the conclusion of the implication must likewise be acknowledged as true; but that is just the contrapositive statement in question:

(IV) *if 2x is not a positive number, then x is not a positive number.*

Anyone knowing the law of contraposition can, in this way, recognize the contrapositive sentence as true, provided he has previously proved the original sentence. Further, as one can easily verify, the inverse sentence is contrapositive with respect to the converse of the original sentence (that is to say, the inverse sentence can be obtained from the converse sentence by replacing antecedent and consequent by their negations and then interchanging them); for this reason, if the converse of the given sentence has been proved, the inverse sentence may likewise be considered valid. If, therefore, one has succeeded in proving two sentences—the original and its converse—a special proof for the two remaining conjugate sentences is superfluous.

It may be mentioned that several variants of the law of contraposition are known; one of them is the converse of (II):

from: if not q, then not p, *it follows that:* if p, then q.

This law makes it possible to derive the original sentence from the con-trapositive, and the inverse from the converse sentence.

We shall now consider in a little more detail the mechanism itself of the proof by means of which the sentence (IV) had been demonstrated in the preceding section. Besides the rules of definition, of which we have already spoken, we have other rules of a somewhat similar character, namely, the RULES OF INFERENCE or RULES OF PROOF. These rules, which must not be mistaken for logical laws, amount to directions as to how sentences already known as true may be transformed so as to yield new true sentences. In the proof carried out above, two rules of demonstration have been made use of: the RULE OF SUBSTITUTION and the RULE OF DE-TACHMENT (also known as the MODUS PONENS RULE).

The content of the rule of substitution is as follows. If a sentence of a universal character, that has already been accepted as true, contains sen-tential variables, and if these variables are replaced by other sentential variables or by sentential functions or by sentences—always substituting equal expressions for equal variables throughout—, then the sentence ob-tained in this way may also be recognized as true. It was by applying this very rule that we obtained the sentence (III) from sentence (II). It should be emphasized that the rule of substitution may also be applied to other kinds of variables, for example, to the variables "x," "y," $\cdots$ desig-nating numbers: in place of these variables, any symbols or expressions denoting numbers may be substituted.

The rule of detachment states that, if two sentences are accepted as true, of which one has the form of an implication while the other is the antecedent of this implication, then that sentence may also be recognized as true which forms the consequent of the implication. (We "detach" thus, so to speak, the antecedent from the whole implication.) By means of this rule, the sentence (IV) had been derived from the sentences (III) and (I).

It can be seen from this that, in the proof of the sentence (IV) as carried out above, each step consisted in applying a rule of inference to sentences which were previously accepted or recognized as true. A proof of this kind will be called COMPLETE. A little more precisely a complete proof may also be characterized as follows. It consists in the construction of a chain of sentences with these properties: the initial members are sentences which were already previously accepted as true; every subse-quent member is obtainable from preceding ones by applying a rule of inference; and finally the last member is the sentence to be proved.

It should be observed what an extremely elementary form—from the psychological point of view—all mathematical reasonings assume, due to the knowledge and application of the laws of logic and the rules of inference; complicated mental processes are entirely reducible to such simple activities as the attentive observation of statements previously accepted as true, the perception of structural, purely external, connections among these statements, and the execution of mechanical transformations as prescribed by the rules of inference. It is obvious that, in view of such a procedure, the possibility of committing mistakes in a proof is reduced to a minimum.

.

PART XIV

The Unreasonableness of Mathematics

The Importance of Being Absurd

A PARADOX is defined as a statement either seemingly or essentially absurd. But how is one to know that a statement really is absurd? A widely accepted test is public opinion. The majority thinks that what the majority thinks is true, and the opposite absurd. On the other hand there is a sophisticated minority that looks on the majority position as absurd by definition. Neither view is a satisfactory criterion for scientific statements. A reasonable course is to match statements against experience. It often happens when a proposition is tested experimentally that what seemed absurd turns out to be true. Thus the paradox vanishes with increase of understanding. Heavy objects fall no faster than light ones, a solid table consists of electrical waves, man is descended from apes, heat is motion, the earth spins on its axis, malaria is caused by mosquitoes: all these are hair-raisingly improbable statements—paradoxes—proven to be fact. The process of validating a paradox, getting it accepted as true, is a difficult process not infrequently attended by much folly and cruelty. Many of those who advanced new views were scorned; some were tortured and burned. Even in our day it is prudent to be sensible and wrong. Mathematics is commonly believed to be above this sort of vulgar and dangerous commotion. To be sure, Gauss said that he refrained from publishing his conclusions about non-Euclidean geometry because he feared the outcries of the Boeotians; Newton had a morbid dread of controversy and was disinclined (because "philosophy is such an impertinently litigious lady") to announce his theories; and other mathematicians have withheld their results to avoid ridicule. It is improbable, however, that anyone has suffered thumb screws for attacking the parallel postulate or reforming algebra.

Mathematics has escaped embroilment partly because so much of the subject is esoteric. It is hard to tell offhand whether any of its propositions are ridiculous; moreover, mathematics is seen as a harmless discipline, offering no threat—as even physics and astronomy have done—to established religious and political beliefs. But there are other reasons why mathematics does not fan controversy. The most important of these is that its statements are generally thought to be beyond dispute. Mathematical propositions, it is frequently asserted, are either so self-evident that it is frivolous to doubt them; or, where less self-evident, e.g., $89 \times 43 = 3827$, $e^{i\pi} + 1 = 0$, can be proved by logical reasoning. Yet a closer scrutiny of this assertion discloses that the matter is not quite so simple. Just as nothing bearing on the essential absurdity of a statement is settled by calling it absurd, so nothing is settled as to a statement's validity by calling it self-evident. (See pp. 1614–1618 [introduction to selections on the nature of mathematics].) Furthermore, even if it is true that mathematical propositions are untouchable when deduced by correct logical reasoning, what of the validity of the axioms and postulates from which the propositions have been deduced? Obviously if the axioms are shaky, the edifice resting on them will be even shakier. And finally, what of logical reasoning itself? Who is to vouch for it? The question in another form is very old: *Quis ipsos custodiet custodies?*

The fact is mathematics has its paradoxes, and they have stirred tempests. Some pose logical dilemmas; others are merely bizarre and incredible, resembling the statement that the moon is made of green cheese. Zeno's little fables about motion (I use the word "little" admiringly, not depreciatingly; the fables have caused a good deal of trouble and are still causing it) are perhaps the most famous paradoxes; others, though not so well known, have distressed mathematicians and philosophers, shaken preconceived notions, and led to far-reaching reforms in both logic and mathematics. Paradoxes are skeletons at the feast, but as will appear from the selections, they are useful skeletons. They come to chasten us in our convictions as to what is evident, impossible, certain, absurd. It is true, of course, that an all-knowing intellect would encounter no paradoxes in logic or mathematics.[1] He could not be led astray by either seeming or essential absurdities. But he would be deprived, as we are not, of one of the most entertaining and exciting aspects of the history of thought.

The two selections which follow present a representative sample of paradoxes. The chapter from *Mathematics and the Imagination* deals with absurd geometric propositions, fantastications of point-set theory, arithmetic fallacies and the logical paradoxes. The second selection is a lecture by the mathematician and positivist, Hans Hahn (see pp. 1591–1592). He was an active participant in the affairs of the Vienna Circle, which played a vigorous part in the modern re-examination of the foundations of logic and natural science, in uncovering philosophical confusions, sharpening the methods of linguistic analysis, and furthering the mathematical reforms alluded to above. Hahn's lecture discusses the cherished faculty of intuition, its role in mathematics, the nest of paradoxes it got us into, and how successful we have been in crawling out.

[1] For an interesting discussion of this point see A. J. Ayer, *Language, Truth and Logic*, London, 1948, pp. 85–86; also, Hans Hahn, "Logik, Mathematik, und Naturerkennen," *Einheits Wissenschaft*, Heft II, p. 18.

I don't quite hear what you say, but I beg to differ entirely with you.
—Augustus De Morgan
    (Report on an old Cambridge don, a little deaf but not a wit less contentious)

I am not for imposing any sense on your words: you are at liberty to explain them as you please. Only, I beseech you, make me understand something by them. —Bishop Berkeley

1   Paradox Lost and Paradox Regained

By EDWARD KASNER
and JAMES R. NEWMAN

How quaint the ways of paradox—
At common sense she gaily mocks.
        —W. S. Gilbert

PERHAPS the greatest paradox of all is that there are paradoxes in mathematics. We are not surprised to discover inconsistencies in the experimental sciences, which periodically undergo such revolutionary changes that although only a short time ago we believed ourselves descended from the gods, we now visit the zoo with the same friendly interest with which we call on distant relatives. Similarly, the fundamental and age-old distinction between matter and energy is vanishing, while relativity physics is shattering our basic concepts of time and space. Indeed, the testament of science is so continuously in a flux that the heresy of yesterday is the gospel of today and the fundamentalism of tomorrow. Paraphrasing Hamlet—what was once a paradox is one no longer, but may again become one. Yet, because mathematics builds on the old but does not discard it, because it is the most conservative of the sciences, because its theorems are deduced from postulates by the methods of logic, in spite of its having undergone revolutionary changes we do not suspect it of being a discipline capable of engendering paradoxes.

Nevertheless, there are three distinct types of paradoxes which do arise in mathematics. There are contradictory and absurd propositions, which arise from fallacious reasoning. There are theorems which seem strange and incredible, but which, because they are logically unassailable, must be accepted even though they transcend intuition and imagination. The third and most important class consists of those logical paradoxes which arise in connection with the theory of aggregates, and which have resulted in a re-examination of the foundations of mathematics. These logical paradoxes have created confusion and consternation among logicians and mathematicians and have raised problems concerning the nature of mathematics and logic which have not yet found a satisfactory solution.

PARADOXES—STRANGE BUT TRUE

This section will be devoted to apparently contradictory and absurd propositions which are nevertheless true.[1] The paradoxes of Zeno were explained by means of infinite series and the transfinite mathematics of Cantor. There are yet others involving motion, but unlike Zeno's puzzles, they do not consist of logical demonstrations that motion is impossible.

FIGURE 1

However, they graphically illustrate how false our ideas about motion may be; how easily, for example, one may be deceived by the path of a moving object.

In Figure 1, there are two identical coins. If we roll the coin at the left along half the circumference of the other, following the path indicated by the arrow, we might suspect that its final position, when it reaches the extreme right, should be with the head inverted and not in an upright position. That is to say, after we revolved the coin through a semicircle (half of its circumference), the head on the face of the coin, having started from an upright position, should now be upside down. If, however, we perform the experiment, we shall see that the final position will be as illustrated in Figure 1, just as though the coin had been revolved once completely about its own circumference.

The following enigma is similar. The circle in Figure 2 has made one complete revolution in rolling from A to B. The distance AB is therefore equal in length to the circumference of the circle. The smaller circle inside the larger one has also made one complete revolution in traversing

[1] Strictly speaking, mathematical propositions are neither true nor false; they are merely implied by the axioms and postulates which we assume. If we accept these premises and employ legitimate logical arguments, we obtain legitimate propositions. The postulates are not characterized by being true or false; we simply agree to abide by them. But we have used the word *true* without any of its philosophical implications to refer unambiguously to propositions logically deduced from commonly accepted axioms.

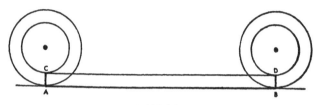

FIGURE 2

the distance *CD*. Since the distance *CD* is equal to the distance *AB* and each distance is apparently equal to the circumference of the circle which has been unrolled upon it, we are confronted with the evident absurdity that the circumference of the small circle is equal to the circumference of the large circle.

In order to explain these paradoxes, and several others of a similar nature, we must turn our attention for a moment to a famous curve— the *cycloid*. (See Figure 3.) The cycloid is the path traced by a fixed point on the circumference of a wheel as it rolls without slipping upon a fixed straight line.

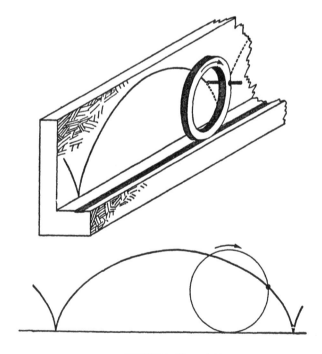

FIGURE 3—The cycloid.

In Figure 4, as the wheel rolls on the line *MN*, the points *A* and *B* describe a cycloid. After the wheel has made half a revolution, the point A_1 is at A_3, and B_1 is at B_3. At this juncture, there is nothing to indicate that the point *A* and the point *B* have not traveled throughout at the same speed, since it is evident that they have covered the same distance. But, if we examine the intermediary points A_2 and B_2 which show the respective positions of *A* and *B* after a *quarter*-turn of the wheel, it is clear that in the same time *A* has traveled a greater distance than *B*. This difference is compensated for in the second quarter turn in which *B*, traveling from B_2 to B_3, covers the same distance that *A* covered moving from A_1 to A_2; it is obvious that the distance along the curve from B_2 to B_3 is equal in length to the distance from A_1 to A_2. Hence in one-half revolution, both *A* and *B* have traversed exactly the same distance.

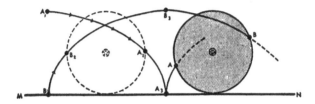

FIGURE 4—When the rolling wheel is in the dotted position it has completed one-quarter of a turn, and *A* has traveled from A_1 to A_2, but *B* only from B_1 to B_2. The shaded circle indicates the wheel has completed three-quarters of a revolution.

This strange behavior of the cycloid explains the fact that when a wheel is in motion, the part furthest from the ground, at any instant, actually moves along horizontally faster than the part in contact with the ground. It can be shown that as the point of a wheel in contact with the road starts moving up, it travels more and more quickly, reaching its maximum horizontal speed when its position is furthest from the ground.

Another interesting property of the cycloid was discovered by Galileo. It was pointed out earlier that the area of a circle could only be expressed with the aid of π, the transcendental number.[2] Since the numerical value of π can only be approximated (although as closely as we please, by taking as many terms of the infinite series as we wish), the area of a circle can also only be expressed as an approximation. Remarkably enough, however, with the aid of a cycloid, we may *construct* an area *exactly* equal to the area of a given circle. Based upon the fact that the length of a cycloid, from cusp to cusp, is equal to four times the length of the diameter of the generating circle, it may be shown that the area bound by the portion of the cycloid between the two cusps and the straight line joining the

[2] [In an earlier chapter on π, *i* and *e*. ED.]

cusps, is equal to three times the area of the circle. From which it follows that the enclosed space (shaded in Figure 5) on *either* side of the circle in the center is *exactly* equal to the area of the circle itself.

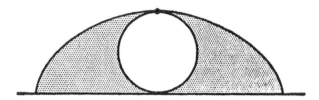

FIGURE 5—When the rolling circle is in the indicated position, the shaded areas on each side are exactly equal to the area of the circle.

The paradox resulting from the pseudo-proof that the circumference of a small circle is equal to that of a larger circle can be explained with the aid of another member of the cycloid family—the prolate cycloid (Figure 6).

An inner point of a wheel which rolls on a straight line describes the prolate cycloid. Thus, a point on the circumference of a smaller circle concentric with a larger one will generate this curve. The small circle in Figure 2 makes only one complete revolution in moving from *C* to *D* and a point on the circumference of this circle will describe a prolate cycloid. However, by comparing the prolate cycloid with the cycloid, we observe

FIGURE 6—The prolate cycloid is generated by the point *P* on the smaller circle as the larger circle rolls along the line *MN*.

that the small circle would not cover the distance *CD* merely by making one revolution as the large circle does. Part of the distance is covered by the circle while it is unrolling, but simultaneously, it is being carried forward by the large circle as this moves from *A* to *B*. This may be seen even more clearly if we regard the *center* of the large circle in Figure 2. The center of a circle, being a mathematical point and having no dimensions, does not revolve at all, but is carried the entire distance from *A* to *B* by the wheel.

With regard to the problems arising from a wheel rolling on a straight line, we have discussed the trajectory (path) of a point on the circumference of the wheel and found this path to be a cycloid, and we have considered the curve traced by a point on the inside of the wheel and discovered the prolate cycloid. In addition, it is interesting to mention the path traced by a point *outside* of the circumference of a wheel, such as the outermost point of the flanged wheels used on railway trains. Such a point is not actually in contact with the rail upon which the wheel is revolving. The curve generated is called a curtate cycloid (Figure 7) and explains the curious paradox that, at any instant of time, a railroad train never moves entirely in the direction in which the engine is pulling. There are always parts of the train which are traveling in the opposite direction!

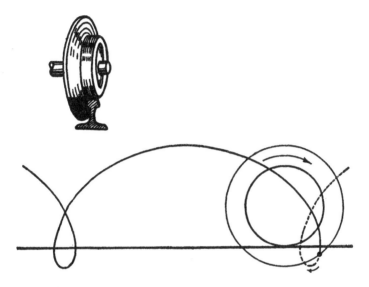

FIGURE 7—The Curtate Cycloid. A point on the flange of a moving railway wheel generates this curve. The part of a railway train which moves *backward* when the train moves forward is the shaded portion of the wheel.

Among the innovations in mathematics of the last quarter-century, none overshadows in importance the development of the theory of point sets and the theory of functions of a real variable. Based entirely upon the new methods of mathematical analysis, a greater rigor and generality in geometry was achieved than could have been imagined had science been developed entirely by intuitive means. It was found that all conventional geometrical ideas could be redefined with increased accuracy by drawing upon the theory of aggregates and the powerful new tools of analysis. In rubber-sheet geometry, curves are defined in such a way as to eliminate every naïve appeal to intuition and experience. A simple closed curve is defined as a set of points possessing the property that it divides the plane into exactly two regions: an *inside* and an *outside*, where *inside* and *out-*

side are precisely formulated by analytical methods without reference to our customary notions of space. By just such means, figures far more complex than had ever before been studied were developed and investigated. Indeed, although analytical geometry is limited to contours which can be described by algebraic equations whose variables are the co-ordinates of the points of the configuration, the new analysis made possible the study of forms which *cannot* be described by *any* algebraic equation.[8]

Extended studies were also undertaken of certain classes of points—like the points in space—and the notion of dimensionality was freshly re-examined. In connection with this study, one of the great accomplishments of recent years has been to assign to each configuration a number: 0, 1, 2, or 3, to denote its dimensionality. The established belief had been that this was a simple and obvious matter which did not require mathematical analysis and could be solved intuitively. Thus, a point would be said to have zero dimension, a line or a curve—one dimension, a plane or a surface—two dimensions, and a solid—three dimensions. It must be conceded that the problem of determining whether an object has 0, 1, 2, or 3 dimensions does not look very formidable. However, one remarkable paradox which was uncovered is sufficient in itself to show that this is not the case at all and that our intuitive ideas about dimensionality, as well as area, are not only lacking in precision, but are often wholly misleading.

The paradox appeared in trying to ascertain whether a number (called a measure) could be uniquely assigned to every figure in the plane so that the following three conditions would be satisfied: (1) the word "congruent" being used in the same sense in which it was learned in elementary geometry,[4] two congruent figures were to have the same measure; (2) if a figure were divided into two parts, the sum of the measures assigned to each of the two parts was to be exactly equal to the measure assigned to the original figure; (3) as a model for determining the method of assigning a measure to each figure in the plane, it was agreed that the measure 1 should be assigned to the square whose side has a length of one unit.

What is this concept of measure? From the foregoing, it would seem to follow that the *measure* to be assigned to each figure in the plane is nothing more than the *area* of that figure. In other words, the problem is to ascertain whether the *area* of *every figure* in the plane, regardless of its complexity, can be uniquely determined. It need hardly be pointed out that this was intended as a general and theoretical exercise and not as the vast and obviously impossible undertaking of actually measuring every

---

[3] [See selection by Hans Hahn, *Crisis in Intuition*, p. 1956.]

[4] Two point sets (configurations) are called congruent if, to every pair of points *P*, *Q* of one set, there uniquely corresponds a pair of points *P'*, *Q'* of the other set, such that the distance between *P'* and *Q'* equals the distance between *P* and *Q*.

conceivable figure. The problem was to be considered solved if a theoretical proof were given that *every figure* could be assigned a unique measure. But it should be noted that the principal aim was to keep this investigation free from the traditional concepts of classical geometry—the notion of area understood in the old way was taboo, and the customary methods of determining it specifically excluded; the approach was to be *analytic* (by means of point sets), rather than geometric. Adhering to just such restrictions, it was proved that no matter how complex a figure is, no matter how many times the boundary crosses and recrosses itself, a unique measure can be assigned to it.

Then came the débacle. For the amazing fact was uncovered that the same problem, when extended to surfaces, was not only unsolvable but led to the most stunning paradoxes. Indeed, the very same methods which had been so fruitful in the investigations in the plane, when applied to the surface of a sphere proved inadequate to determine a unique measure.

Does this really mean that the area of the surface of a sphere cannot be uniquely determined? Does not the conventional formula $4\pi r^2$ give correctly the area of the surface of a sphere? Unfortunately, we cannot undertake to answer these questions in detail, for to do so would carry us far afield and require much technical knowledge. We admit that the area of a spherical surface as determined by the old classical methods is $4\pi r^2$. But the old methods were lacking in generality; they were found to be inadequate to determine the area of complex figures; furthermore, we already gave warning that the naïve concept of area was deliberately to be omitted from the measuring attempt. While the advance in function theory and the new methods of analysis did overcome some of these difficulties, they also introduced new problems closely connected with the infinite, and as mathematicians have long realized, the presence of that concept is by no means an unmixed blessing. Though it has enabled mathematics to make great strides forward, these have always been in the shadow of uncertainty. One may continue to employ such formulae as $4\pi r^2$, for the very good reason that they work; but if one wishes to keep pace with the bold and restless mathematical spirit, one is faced with the comfortless alternatives of abandoning logic to preserve the classical concepts, or of accepting the paradoxical results of the new analysis and casting horse sense to the winds.

The conditions for assigning a measure to a surface are similar to the conditions for assigning a measure to figures in the plane: (1) The same measure shall be assigned to congruent surfaces; (2) The sum of the measures assigned to each of two component parts of a surface shall be equal to the measure assigned to the original surface; (3) If S denotes the entire surface of a sphere of radius r, the measure assigned to S shall be $4\pi r^2$.

The German mathematician Hausdorff showed that this problem is insoluble, that a measure *cannot* uniquely be assigned to the portions of the surface of a sphere so that the above conditions will be satisfied. He showed that if the surface of a sphere were divided into three separate and distinct parts: A, B, C, so that A is congruent to B and B is congruent to C, a strange paradox arises which is strongly reminiscent of, and indeed, related to some of the paradoxes of transfinite arithmetic. For Hausdorff proved that not only is A congruent to C (as might be expected), *but also* that A is congruent to $B + C$. What are the implications of this startling result?

If a measure is assigned to A, the same measure must be assigned to B and to C, because A is congruent to B, B is congruent to C and A is congruent to C. But, on the other hand, since A is congruent to $B + C$, the measure assigned to A would also have to be equal to the sum of the measures assigned to B and C. Obviously, such a relationship could only hold if the measures assigned to A, B, and C were all equal to 0. But that is impossible by condition (3), according to which the sum of the measures assigned to the parts of the surface of a sphere must be equal to $4\pi r^2$. How then is it possible to assign a measure?

From a slightly different viewpoint, we see that if A, B, and C are congruent to each other and together make up the surface of the entire sphere, the measure of any one of them must be the measure of one-third of the surface of the entire sphere. But if A is not only congruent to B and C, but also to $B + C$ (as Hausdorff has shown), the measure assigned to A and the measure assigned to $B + C$ must each be equal to half the surface of the sphere. Thus, whichever way we look at it, assigning measures to portions of the surface of a sphere involves us in a hopeless contradiction.

Two distinguished Polish mathematicians, Banach and Tarski, extended the implications of Hausdorff's paradoxical theorem to three-dimensional space, with results so astounding and unbelievable that their like may be found nowhere else in the whole of mathematics. And the conclusions, though rigorous and unimpeachable, are almost as incredible for the mathematician as for the layman.

Imagine two bodies in three-dimensional space: one very large, like the sun; the other very small, like a pea. Denote the sun by S and the pea by S'. Remember now that we are referring not to the surfaces of these two spherical objects, but to the *entire solid spheres of both the sun and the pea*. The theorem of Banach and Tarski holds that the following operation can theoretically be carried out:

Divide the sun S into a great many small parts. Each part is to be separate and distinct and the totality of the parts is to be finite in number. They may then be designated by $s_1, s_2, s_3, \ldots s_n$, and together these small parts will make up the entire sphere S. Similarly, S'—the pea—is to be divided into an equal number of mutually exclusive parts—s'_1, s'_2, s'_3, . . . s'_n, which together will make up the pea. Then the proposition goes on to say that if the sun and the pea have been cut up in a suitable manner, so that the little portion s_1 of the sun is congruent to the little portion s'_1 of the pea, s_2 congruent to s'_2, s_3 congruent to s'_3, up to s_n congruent to s'_n, this process will exhaust not only all the little portions of the pea, *but all the tiny portions of the sun as well.*

In other words, the sun and the pea may both be divided into a finite number of disjoint parts *so that every single part of one is congruent to a unique part of the other, and so that after each small portion of the pea has been matched with a small portion of the sun, no portion of the sun will be left over.*[5] To express this giant bombshell in terms of a small firecracker: *There is a way of dividing a sphere as large as the sun into separate parts, so that no two parts will have any points in common, and yet without compressing or distorting any part, the whole sun may at one time be fitted snugly into one's vest pocket.* Furthermore the pea may have its component parts so rearranged that without expansion or distortion, no two parts having any points in common, *they will fill the entire universe solidly, no vacant space remaining either in the interior of the pea, or in the universe.*

Surely no fairy tale, no fantasy of the Arabian nights, no fevered dream can match this theorem of hard, mathematical logic. Although the theorems of Hausdorff, Banach, and Tarski cannot, at the present time, be put to any practical use, not even by those who hope to learn how to pack their overflowing belongings into a week-end grip, they stand as a magnificent challenge to imagination and as a tribute to mathematical conception.[6]

*   *   *   *   *

As distinguished from the paradoxes just considered, there are those which are more properly referred to as mathematical fallacies. They arise in both arithmetic and geometry and are to be found sometimes, although not often, even in the higher branches of mathematics as, for instance, in the calculus or in infinite series. Most mathematical fallacies are too trivial to deserve attention; nevertheless, the subject is entitled to some consideration because, apart from its amusing aspect, it shows how a

[5] We recognize this, of course, to be a simple one-to-one correspondence between the elements of one set which make up the sun, and the elements of another set which make up the pea. The paradox lies in the fact that each element is matched with one which is completely congruent to it (at the risk of repeating, congruent means identical in size and shape) and that there are enough elements in the set making up the pea to match exactly the elements which make up the sun.

[6] In the version given of the theorems of Hausdorff, Banach, and Tarski, we have made liberal use of the lucid explanation given by Karl Menger in his lecture: "Is the Squaring of the Circle Solvable?" in *Alte Probleme—Neue Lösungen*, Vienna: Deuticke, 1934.

chain of mathematical reasoning may be entirely vitiated by one fallacious step.

ARITHMETIC FALLACIES

I. A proof that 1 is equal to 2 is familiar to most of us. Such a proof may be extended to show that *any* two numbers or expressions are equal. The error common to all such frauds lies in dividing by zero, an operation strictly forbidden. For the fundamental rules of arithmetic demand that every arithmetic process (addition, subtraction, multiplication, division, evolution, involution) should yield a unique result. Obviously, this requirement is essential, for the operations of arithmetic would have little value, or meaning, if the results were ambiguous. If $1 + 1$ were equal to 2 *or* 3; if 4×7 were equal to 28 *or* 82; if $7 \div 2$ were equal to 3 or 3½, mathematics would be the Mad Hatter of the sciences. Like fortunetelling or phrenology, it would be a suitable subject to exploit at a boardwalk concession at Coney Island.

Since the results of the operation of division are to be unique, division by 0 *must* be excluded, for the result of this operation is anything that you may desire. In general, division is so defined that if *a, b,* and *c* are three numbers, $a \div b = c$, *only when* $c \times b = a$. From this definition, what is the result of $5 \div 0$? It cannot be any number from zero to infinity, for no number when multiplied by 0 will be equal to 5. Thus $5 \div 0$ is meaningless. And even $5 \div 0 = 5 \div 0$ is a meaningless expression.

Of course, fallacies resulting from division by 0 are rarely presented in so simple a form that they may be detected at a glance. The following example illustrates how paradoxes arise whenever we divide by an expression, the value of which is 0:

Assume $A + B = C$, and assume $A = 3$ and $B = 2$.

Multiply both sides of the equation $A + B = C$ by $(A + B)$.

We obtain $A^2 + 2AB + B^2 = C(A + B)$.

Rearranging the terms, we have
$$A^2 + AB - AC = -AB - B^2 + BC.$$

Factoring out $(A + B - C)$, we have
$$A(A + B - C) = -B(+A + B - C).$$

Dividing both sides by $(A + B - C)$, that is, dividing by zero, we get $A = -B$, or $A + B = 0$, which is evidently absurd.

II. In extracting square roots, it is necessary to remember the algebraic rule that the square root of a positive number is equal to *both* a negative and a positive number. Thus, the square root of 4 is -2 as well as $+2$ (which may be written $\sqrt{4} = \pm 2$), and the square root of 100 is equal to $+10$ and -10 (or, $\sqrt{100} = \pm 10$). Failure to observe this rule may generate the following contradiction: [7]

(a) $(n + 1)^2 = n^2 + 2n + 1$
(b) $(n + 1)^2 - (2n + 1) = n^2$
(c) Subtracting $n(2n + 1)$ from both sides and factoring, we have
(d) $(n + 1)^2 - (n + 1)(2n + 1) = n^2 - n(2n + 1)$
(e) Adding $\frac{1}{4}(2n + 1)^2$ to both sides of (d) yields
$$(n + 1)^2 - (n + 1)(2n + 1) + \frac{1}{4}(2n + 1)^2 =$$
$$n^2 - n(2n + 1) + \frac{1}{4}(2n + 1)^2.$$

This may be written:

(f) $[(n + 1) - \frac{1}{2}(2n + 1)]^2 = [n - \frac{1}{2}(2n + 1)]^2.$

Taking square roots of both sides,

(g) $n + 1 - \frac{1}{2}(2n + 1) = n - \frac{1}{2}(2n + 1)$

and, therefore,

(h) $n = n + 1.$

III. The following arithmetic fallacy the reader may disentangle for himself: [8]

(1) $\sqrt{a} \times \sqrt{b} = \sqrt{a \times b}$ ....true
(2) $\sqrt{-1} \times \sqrt{-1} = \sqrt{(-1) \times (-1)}$ ....true
(3) Therefore, $(\sqrt{-1})^2 = \sqrt{1}$; i.e., $-1 = 1$ ....?

IV. A paradox which cannot be solved by the use of elementary mathematics is the following: Assume that $\log(-1) = x$. Then, by the law of logs,

$$\log(-1)^2 = 2 \times \log(-1) = 2x.$$

But, on the other hand, $\log(-1)^2 = \log(1)$, which is equal to 0. Therefore, $2 \cdot x = 0$. Therefore, $\log(-1) = 0$, which is obviously not the case. The explanation lies in the fact that the function that represents the log of a negative, or complex, number is not *single-valued*, but is *many-valued*. That is to say, if we were to make the usual functional table for the logarithm of negative and complex numbers, there would be an *infinitude* of values corresponding to each number.[9]

V. The infinite in mathematics is always unruly unless it is properly treated. Instances of this were found in the development of the theory of

[7] Lietzmann, *Lustiges und Merkwürdiges von Zahlen und Formen*, Breslau: Ferd. Hirt, 1930.
[8] Ball, W. W. R., *Mathematical Recreations and Essays*, Eleventh Edition, N. Y., 1939.
[9] Weismann, *Einführung in das mathematische Denken*, Vienna, 1937.

aggregates and further examples will be seen in the logical paradoxes. One instance is appropriate here.

Just as transfinite arithmetic has its own laws differing from those of finite arithmetic, special rules are required for operating with *infinite series.* Ignorance of these rules, or failure to observe them brings about inconsistencies. For instance, consider the series equivalent to the natural logarithm of 2:

$$\text{Log } 2 = 1 - \tfrac{1}{2} + \tfrac{1}{3} - \tfrac{1}{4} + \tfrac{1}{5} - \tfrac{1}{6} \ldots$$

If we rearrange these terms as we would be prompted to do in finite arithmetic, we obtain:

$$\text{Log } 2 = (1 + \tfrac{1}{3} + \tfrac{1}{5} + \tfrac{1}{7} \ldots) - (\tfrac{1}{2} + \tfrac{1}{4} + \tfrac{1}{6} + \tfrac{1}{8} \ldots)$$

Thus,

$$\begin{aligned}
\text{Log } 2 &= \{(1 + \tfrac{1}{3} + \tfrac{1}{5} + \tfrac{1}{7} \ldots) + (\tfrac{1}{2} + \tfrac{1}{4} + \tfrac{1}{6} + \tfrac{1}{8})\} \\
&\quad - 2(\tfrac{1}{2} + \tfrac{1}{4} + \tfrac{1}{6} + \tfrac{1}{8} \ldots) \\
&= \{1 + \tfrac{1}{2} + \tfrac{1}{3} + \tfrac{1}{4} + \tfrac{1}{5} + \ldots\} \\
&\quad - \{1 + \tfrac{1}{2} + \tfrac{1}{3} + \tfrac{1}{4} + \tfrac{1}{5} + \ldots\} \\
&= 0
\end{aligned}$$

Therefore, log 2 = 0.

On the other hand,

$$\log 2 = 1 - \tfrac{1}{2} + \tfrac{1}{3} - \tfrac{1}{4} + \tfrac{1}{5} - \tfrac{1}{6} \ldots = 0.69315,$$

an answer that can be obtained from any logarithmic table.

Rearranging the terms in a slightly different way:

$$\begin{aligned}
\log 2 &= 1 + \tfrac{1}{3} - \tfrac{1}{2} + \tfrac{1}{5} + \tfrac{1}{7} - \tfrac{1}{4} + \tfrac{1}{9} + \tfrac{1}{11} - \tfrac{1}{6} \ldots \\
&= \tfrac{3}{2} \times 0.69315 \text{ or, in other words,} \\
\log 2 &= \tfrac{3}{2} \times \log 2.
\end{aligned}$$

A famous series which had troubled Leibniz is the beguilingly simple: $+ 1 - 1 + 1 - 1 + 1 - 1 + 1 \ldots$ By pairing the terms differently, a variety of results is obtained; for example:

$$(1 - 1) + (1 - 1) + (1 - 1) + \ldots = 0,$$

but

$$1 - (1 - 1) + (1 - 1) \ldots = 1.$$

GEOMETRIC FALLACIES

Optical illusions [10] concerning geometric figures account for many deceptions. We confine our attention to fallacies which do not arise from physiological limitations, but from errors in mathematical argument. A well-known geometric "proof" is that every triangle is isosceles. It assumes

[10] For additional illustrative material see Plates I–V on pp. 1954–1955. [ED.]

that the line bisecting an angle of the triangle and the line which is the perpendicular bisector of the side opposite this angle intersect at a point inside the triangle.

The following is a similarly fallacious proof, namely, that a right angle is equal to an angle greater than a right angle.[11]

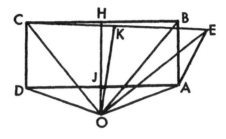

FIGURE 8

In Figure 8, *ABCD* is a rectangle. If *H* is the midpoint of *CB*, through *H* draw a line at right angles to *CB*, which will bisect *DA* at *J* and be perpendicular to it. From *A* draw the line *AE* outside of the rectangle and equal to *AB* and *DC*. Connect *C* and *E*, and let *K* be the midpoint of this line. Through *K* construct a perpendicular to *CE*. *CB* and *CE* not being parallel, the lines through *H* and *K* will meet at a point *O*. Join *OA*, *OE*, *OB*, *OD* and *OC*. It will be made clear that the triangle *ODC* and *OAE* are equal in all respects. Since *KO* is the perpendicular bisector of *CE* and thus any point on *KO* is equidistant from *C* and *E*, *OC* is equal to *OE*. Similarly, since *HO* is the perpendicular bisector of *CB* and *DA*, *OD* equals *OA*. As *AE* was constructed to equal *DC*, the three sides of the triangle *ODC* are equal respectively to the three sides of the triangle *OAE*. Hence, the two triangles are equal, and therefore, the angle *ODC* is equal to the angle *OAE*. But angle *ODA* is equal to angle *OAD*, because side *AO* is equal to side *OD* in the triangle *OAD* and the base angles of the isosceles triangle are equal. Therefore, the angle *JDC*, which is equal to the difference of *ODC* and *ODJ*, equals *JAE*, which is the difference between *OAE* and *OAJ*. But the angle *JDC* is a right angle, whereas the angle *JAE* is greater than a right angle, and hence the result is contradictory. Can you find the flaw? Hint: Try drawing the figure exactly.

LOGICAL PARADOXES

Like folk tales and legends, the logical paradoxes had their forerunners in ancient times. Having occupied themselves with philosophy and with the foundations of logic, the Greeks formulated some of the logical conundrums which, in recent times, have returned to plague mathematicians and

[11] Ball, *op. cit.*

philosophers. The Sophists made a specialty of posers to bewilder and confuse their opponents in debate, but most of them rested on sloppy thinking and dialectical tricks. Aristotle demolished them when he laid down the foundations of classical logic—a science which has outworn and outlasted all the philosophical systems of antiquity, and which, for the most part, is perfectly valid today.

But there were troublesome riddles that stubbornly resisted unraveling.[12] Most of them are caused by what is known as "the vicious circle fallacy," which is "due to neglecting the fundamental principle that what involves the whole of a given totality cannot itself be a member of the totality." [13] Simple instances of this are those pontifical phrases, familiar to everyone, which seem to have a great deal of meaning, but actually have none, such as "never say never," or "every rule has exceptions," or, "every generality is false." We shall consider a few of the more advanced logical paradoxes involving the same basic fallacy, and then discuss their importance from the mathematician's point of view.

(A) Poaching on the hunting preserves of a powerful prince was punishable by death, but the prince further decreed that anyone caught poaching was to be given the privilege of deciding whether he should be hanged or beheaded. The culprit was permitted to make a statement—if it were false, he was to be hanged; if it were true, he was to be beheaded. One logical rogue availed himself of this dubious prerogative—to be hanged if he didn't and to be beheaded if he did—by stating: "I shall be hanged." Here was a dilemma not anticipated. For, as the poacher put it, "If you now hang me, you break the laws made by the prince, for my statement is true, and I ought to be beheaded; but if you behead me, you are also breaking the laws, for then what I said was false and I should, therefore, be hanged." As in Frank Stockton's story of the lady and the tiger, the ending is up to you. However, the poacher probably fared no worse at the hands of the executioner than he would have at the hands of a philosopher, for until this century philosophers had little time to waste on such childish riddles—especially those they could not solve.

(B) The village barber shaves everyone in the village who does not shave himself. But this principle soon involves him in a dialectical plight analogous to that of the executioner. Shall he shave himself? If he does, then he is shaving someone who shaves himself and breaks his own rule. If he does not, besides remaining unshaven, he also breaks his rule by failing to shave a person in the village who does not shave himself.

(C) Consider the fact that every integer may be expressed in the English language without the use of symbols. Thus, (a) 1400 may be written

---

[12] For instance, the riddle of the Epimenides concerning the Cretan who says that all Cretans are liars.
[13] Ramsay, Frank Plumpton. Articles on "Mathematics," and "Logic," *Encyclopaedia Britannica*, 13th edition.

as one thousand, four hundred, or (b) 1769823 as one million, seven hundred and sixty-nine thousand, eight hundred and twenty-three. It is evident that certain numbers require more syllables than others; in general, the larger the integer, the more syllables needed to express it. Thus, (a) requires 6 syllables, and (b) 21. Now, it may be established that certain numbers will require 19 syllables or less, while others will require more than 19 syllables. Furthermore, it is not difficult to show that among those integers requiring exactly 19 syllables to be expressed in the English language, there must be a smallest one. Now, "it is easy to see" [14] that "*The least integer not nameable in fewer than nineteen syllables*" is a phrase which must denote the specific number, 111777. But the italicized expression above is itself an unambiguous means of denoting the smallest integer expressible in nineteen syllables in the English language. Yet, the italicized statement has only eighteen syllables! Thus, we have a contradiction, for the least integer expressible in nineteen syllables can be expressed in eighteen syllables.

(D) The simplest form of the logical paradox which arises from the indiscriminate use of the word *all* may be seen in Figure 9.

What is to be said about the statement numbered 3? 1 and 2 are false, but 3 is both a wolf dressed like a sheep and a sheep dressed like a wolf. It is neither the one thing nor the other: It is neither false nor true.

An elaboration appears in the famous paradox of Russell about the

FIGURE 9

class of all classes not members of themselves. The thread of the argument is somewhat elusive and will repay careful attention:

(E) Using the word class in the customary sense, we can say that there are classes made up of tables, books, peoples, numbers, functions, ideas, etc. The class, for instance, of all the Presidents of the United

---

[14] This expression may, perhaps, be taken in the sense in which Laplace employed it. When he wrote his monumental *Mécanique Céleste*, he made abundant use of the expression, "It is easy to see" often prefixing it to a mathematical formula which he had arrived at only after months of labor. The result was that scientists who read his work almost invariably recognized the expression as a danger signal that there was very rough going ahead.

States has for its members every person, living or dead, who was ever President of the United States. Everything in the world other than a person who was or is a President of the United States, including the *concept* of the class itself is *not* a member of this class. This then, is an example of a class which is *not* a member of itself. Likewise, the class of all members of the Gestapo, or German secret police, which contains some, but not all, of the scoundrels in Germany; or the class of all geometric figures in a plane bounded by straight lines; or the class of all integers from one to four thousand inclusive, have for members, the things described, but the classes are not members of themselves.

Now, if we consider a class as a *concept*, then the *class* of *all concepts* in the world is itself a concept, and thus is a class which is a member of itself. Again, the class of all ideas brought to the attention of the reader in this book is a class which contains itself as a member, since in mentioning this class, it is an idea which we bring to the attention of the reader. Bearing this distinction in mind, we may divide all classes into two types. Those which are members of themselves and those which are *not* members of themselves. Indeed, we may form a class which is composed of *all those classes which* are *not* members of themselves (note the dangerous use of the word "all"). The question is presented: Is this class (composed of those classes which are not members of themselves) a member of itself, or not? Either an affirmative or a negative answer involves us in a hopeless contradiction. If the class in question *is* a member of itself, it ought not be by definition, for it should contain only those classes which are not members of themselves. But if it is *not* a member of itself, it ought to be a member of itself, for the same reason.

It cannot be too strongly emphasized that the logical paradoxes are not idle or foolish tricks. They were not included in this volume to make the reader laugh, unless it be at the limitations of logic. The paradoxes are like the fables of La Fontaine which were dressed up to look like innocent stories about fox and grapes, pebbles and frogs. For just as all ethical and moral concepts were skillfully woven into their fabric, so all of logic and mathematics, of philosophy and speculative thought, is interwoven with the fate of these little jokes.

Modern mathematics, in attempting to avoid the paradoxes of the theory of aggregates, was squarely faced with the alternatives of adopting annihilating skepticism in regard to all mathematical reasoning, or of reconsidering and reconstructing the foundations of mathematics as well as logic. It should be clear that if paradoxes can arise from apparently legitimate reasoning about the theory of aggregates, they may arise *anywhere* in mathematics. Thus, even if mathematics could be reduced to logic, as Frege and Russell had hoped, what purpose would be served if logic itself were insecure? In proposing their "Theory of Types" Whitehead and

Russell, in the *Principia Mathematica,* succeeded in avoiding the contradictions by a formal device. Propositions which were *grammatically* correct but contradictory, were branded as meaningless. Furthermore, a principle was formulated which specifically states what form a proposition must take to be meaningful; but this solved only half the difficulty, for although the contradictions could be recognized, the arguments leading to the contradictions could not be invalidated without affecting certain accepted portions of mathematics. To overcome this difficulty, Whitehead and Russell postulated the *axiom of reducibility* which, however, is too technical to be considered here. But the fact remains that the axiom is not acceptable to the great majority of mathematicians and that the logical paradoxes, having divided mathematicians into factions unalterably opposed to each other, have still to be disposed of.[15]

*   *   *   *   *

It has been emphasized throughout that the mathematician strives always to put his theorems in the most general form. In this respect, the aims of the mathematician and the logician are identical—to formulate propositions and theorems of the form: if *A* is true, *B* is true, where *A* and *B* embrace much more than merely cabbages and kings. But if this is a high aim, it is also dangerous, in the same way that the concept of the infinite is dangerous. When the mathematician says that such and such a proposition is true of one thing, it may be interesting, and it is surely safe. But when he tries to extend his proposition to *everything*, though it is much more interesting, it is also much more dangerous. In the transition from *one* to *all*, from the specific to the general, mathematics has made its greatest progress, and suffered its most serious setbacks, of which the logical paradoxes constitute the most important part. For, if mathematics is to advance securely and confidently it must first set its affairs in order at home.

[15] As was pointed out in discussing the googol,* there are the followers of Russell who are satisfied with the theory of types and the axiom of reducibility; there are the Intuitionists, led by Brouwer and Weyl, who reject the axiom and whose skepticism about the infinite in mathematics has carried them to the point where they would reject large portions of modern mathematics as meaningless, because they are interwoven with the infinite; and there are the Formalists, led by Hilbert, who, while opposed to the beliefs of the Intuitionists, differ considerably from Russell and the Logistic school. It is Hilbert who considers mathematics a meaningless game, comparable to chess, and he has created a subject of metamathematics which has for its program the discussion of this meaningless game and its axioms.

* [For the meaning of "googol" see selection by Kasner and Newman, "New Names for Old," p. 2007.]

THE FOLLOWING OPTICAL ILLUSIONS, WHILE NOT PROPERLY PART OF A
BOOK ON MATHEMATICS, MAY BE OF SOME INTEREST—
AT LEAST TO THE IMAGINATION

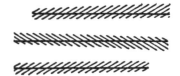

PLATE I—Are the three horizontal lines parallel?

PLATE II—The white square is of course larger than the black. Or is it smaller?

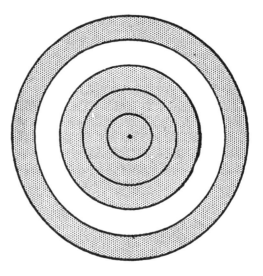

PLATE III—The two shaded regions have the same area.

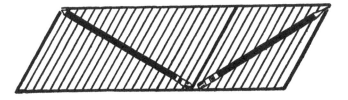

PLATE IV—Which of the two pencils is longer? Measure them and find out.

PLATE V—What do you see? Now look again.

Truth can never be told so as to be understood, and not be believ'd.
    —WILLIAM BLAKE

Earthly minds, like mud walls, resist the strongest batteries; and though,
perhaps, sometimes the force of a clear argument may make some impres-
sion, yet they nevertheless stand firm, keep out the enemy, truth, that would
captivate or disturb them.   —JOHN LOCKE

How often have I said to you that when you have eliminated the impossible,
whatever remains, however improbable, *must be the truth.*
    —SIR ARTHUR CONAN DOYLE (*The Sign of Four*)

2   The Crisis in Intuition

By HANS HAHN

OF ALL the leading philosophers Immanuel Kant was undoubtedly the
one who assigned the greatest importance to the part played by intuition
in what we call knowledge. He observed that two opposite factors are
basic to our knowledge: a passive factor of simple receptivity and an
active factor of spontaneity. In his "Critique of Pure Reason," at the beginning of the section entitled "Transcendental theory of elements; Part
two: Transcendental logic," we find the following: "Our knowledge comes
from two basic sources in the mind, of which the first is the faculty of
receiving sensations (receptivity to impressions), the second the ability
to recognize an object by these perceptions (spontaneity in forming concepts). Through the first an object is *given* to us, through the second this
object is *thought* in relation to these perceptions, as a simple determination of the mind. Thus, intuition and concepts constitute the elements of
all our knowledge . . ." That is to say, we conduct ourselves passively
when through intuition we receive impressions, and actively when we deal
with them in our thought. Further, according to Kant, we must distinguish
between two ingredients of intuition. One of these, the empirical, *a posteriori* part, arises from experience and forms the *content* of intuition,
such as colors, sounds, smells, and sensations of touch (hardness, softness,
roughness, etc.). The other is a pure, *a priori* part, independent of all
experience; it constitutes the *form* of intuition. We possess two such pure
intuitional forms: *space*, the intuitional form of our external sense by
means of which we "picture things as outside ourselves"; and *time*, the
intuitional form of our inner sense "by means of which the mind observes
itself or its inner state."

In Kant's system, as I have said, this pure intuition plays an extremely
important role. He believed that mathematics is founded on pure intuition,
not on thought. Geometry, as it has been taught since ancient times, deals

with the properties of the space that is fully and exactly presented to us by pure intuition; arithmetic (the study of the real numbers) rests on our pure and fully exact intuition of time. The intuitional forms of space and time constitute the *a priori* frame into which we fit all physical happenings that experience presents to us. Every physical event has its precise and exactly determined place in space and time.

However plausible these ideas may at first seem, and however well they corresponded to the state of science in Kant's day, their foundations have been shaken by the course that science has taken since then.

The physical side of the question has already been treated in the first two lectures, so I can here confine myself to mentioning it briefly. Kant's ideas about the place of space and time in physics correspond with Newtonian physics, which was supreme in Kant's day and which remained so down to very recent times. This conception received its first serious jolt from Einstein's theory of relativity. According to Kant, space and time have nothing to do with each other, for they stem from quite different sources. Space is the intuitional form of our outer sense, time of our inner sense. We have an absolutely stationary space and an absolute time that flows independent of it. The theory of relativity holds, on the contrary, that there is no absolute space and no absolute time; it is only a combination of space and time—the "universe"—that has absolute physical meaning.

A much worse blow was struck at Kant's conception of space and time as *a priori* intuitional forms by the most recent developments of physics. We have already noted that, according to Kant's conception, every physical event has its precisely fixed location in space and time. But there has always been a certain difficulty about this. We know physical events only through experience; but all experience is inexact and every observation involves observational errors. Thus the earlier conception embodies the inconsistency that, while every physical event has its exact place in space and time, we can never precisely determine those places. Let us take, for example, a circular piece of chalk; once a unit of length is chosen, the distance between any two points on this piece of chalk is measured by an exact real number. Imagine that the distance has been determined between every pair of points on this piece of chalk and call the greatest of these distances its "diameter." Assuming that the chalk occupies an exact fixed portion of the space given to us by precise intuition, it would be reasonable to ask, "Is the diameter of the chalk disc expressed by a rational or an irrational number?" But the question could never be answered, for the difference between rational and irrational is much too fine ever to be determined by observation. Thus what might be called the classical conception raises questions that are fundamentally unanswerable; which is to say that the conception is metaphysical.

For a long time this difficulty was not taken very seriously. It was answered somewhat as follows: "Even if every single observation is inexact and subject to observational errors, yet our methods of observation are becoming more and more accurate. Let us now imagine a specific physical quantity measured over and over again with more and more precise observational methods. The results thus obtained, though each one is inexact, will nevertheless approach without limit a definite limiting value, and this limit is the exact value of the physical quantity in question." This argument is scarcely satisfactory from a philosophical standpoint, and the recent advances of physics seem to prove that it is also untenable on purely physical grounds. For it now seems that on purely physical grounds the location of an event in space and time cannot be determined with unlimited precision. If, then, measurements cannot be pushed beyond certain limits of exactness we are left with this result: The doctrine of the exact location of physical events in space and time is metaphysical, and therefore meaningless. This most recent and revolutionary development of physics will necessarily come as a shock to most persons—including most physicists—grounded as they are in dogmatic and metaphysical theories; but for the thinker trained in empirical philosophy it contains nothing paradoxical. He will recognize it at once as something familiar and will welcome it as a major step forward along the road toward the "physicalization" of physics, toward cleansing physics of metaphysical elements.

After this very brief reference to the physical side of the question we turn to the field of mathematics where the opposition to Kant's doctrine of pure intuition manifested itself considerably earlier than in physics. From here on I shall deal exclusively with the subject "mathematics and intuition"; moreover, even within this sphere I shall pass over a whole group of questions, as important as they are difficult, which Menger will deal with in the final lecture of this series. I shall not discuss the vehement and successful opposition to Kant's thesis that arithmetic, the study of numbers, also rests on pure intuition—an opposition inextricably bound up with the name of Bertrand Russell, and which has set out to prove that, in complete contradiction to Kant's thesis, arithmetic belongs exclusively to the domains of the intellect and of logic. Thus I have narrowed my subject to "geometry and intuition," and I shall attempt to show how it came about that, even in the branch of mathematics which would seem to be its original domain, intuition gradually fell into disrepute and at last was completely banished.

One of the outstanding events in this development was the discovery that, in apparent contradiction to what had previously been accepted as intuitively certain, there are curves that possess no tangent at any point, or (and we shall see that this amounts to the same thing) that it is possible to imagine a point moving in such a manner that at no instant does it have a definite velocity. Mathematicians were tremendously impressed when their great Berlin colleague K. Weierstrass made this discovery known in the year 1861. But manuscripts preserved in the Vienna National Library show that the fact was recognized considerably earlier by the Austrian philosopher, theologian and mathematician, Bernhard Bolzano. Since some of the questions involved here directly affect the foundations of the differential calculus as developed by Newton and Leibniz, I shall first say a few words about the basic concepts of that discipline.

Newton started with the concept of velocity. Imagine a point moving along a straight line, as shown in Figure 1. At time *t* the moving point

FIGURE 1

will be, say, at *q*. What is to be understood by the expression "the velocity of the moving point at the instant *t*"? If we determine the position of the point at a second instant *t'* (at this second instant think of the point as being at *q'*), then we can ascertain the distance *qq'* that it has traversed in the time that has elapsed between the instants *t* and *t'*. We now divide the distance *qq'* that the point has traversed, by the time that has elapsed between the instants *t* and *t'*, and get the so-called "mean velocity" of the moving point between *t* and *t'*. This "mean" velocity is in no sense the velocity at time *t* itself. The mean velocity may, for instance, turn out to be very great, even though the velocity at instant *t* is quite small, if the point moved very rapidly during the greater part of the time interval in question. But if the second instant *t'* is chosen sufficiently close to the first instant *t*, then the mean velocity between *t* and *t'* will provide a good approximation to the velocity at time *t* itself, and this approximation will be closer, the closer *t'* is to *t*. Newton's reasoning about this matter ran somewhat as follows: Think of the instant *t'* chosen closer and closer to *t*; then the average velocity between *t* and *t'* will approach closer and closer to a certain definite value; it will—to use the language of mathematics—tend toward a definite limit, which limit is called the "velocity of the moving point at the instant *t*." In other words, the velocity at *t* is the limiting value approached by the average velocity between t and t', as t' approaches t without limit.

Leibniz started from the so-called tangent problem. Consider the curve shown in Figure 2; what is its slope (relative to the horizontal) at point p?

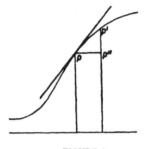

FIGURE 2

Choose a second point, p', on the curve and construct the "average slope" of the curve between p and p'. This is obtained by dividing the height (represented in Figure 2 by the line $p''p'$) gained in ascending the section of the curve from p to p' by the horizontal projection of the distance passed over (represented in Figure 2 by the line pp'', which indicates how far one moves in the horizontal direction by following along the section of the curve from p to p'). The average slope of the curve between p and p' is not, of course, identical with its slope at the point p itself (in Figure 2 the slope at p is obviously greater than the average slope between p and p'). However, it will give a good approximation to the slope at p, if only p' is chosen sufficiently close to p; and the approximation will be more accurate, the closer p' is to p. Now again as in the Newtonian example: If p' is permitted to approach p without limit, the average slope of the curve between p and p' will tend toward a definite limit, which limit is called the "slope of the curve at the point p." That is to say, the slope at p is the limiting value approached by the average slope between p and p' as p' approaches p without limit. One designates as the "tangent of the curve at the point p" the straight line passing through p which (throughout its entire length) has the same slope as the curve at p.

There is thus a striking resemblance between the procedure for obtaining the slope of a curve and the procedure for determining the velocity of a moving point. In fact, the problem of determining the velocity of a moving point at a given instant becomes identical with the problem of determining the slope of a curve at a given point if we employ a simple device, familiar from its use in [German] railroad timetables. [Hahn apparently refers to a graphic type of timetable with which Americans are not acquainted. ED.] Along a horizontal straight line (a "time axis") mark off time intervals so that every point on the line represents a definite

point of time; and on the straight line in Figure 1, along the path of the
moving point in question select an arbitrary point *o*. If at instant *t* the
moving point is at *q*, erect at right angles to the time axis at *t* the line
segment *oq*. (See Figure 2.) It can be seen that the point *p* thus obtained
will represent as in Figure 2 the position of the moving point at instant *t*.
If this process were carried out for every single instant of time, one would
obtain a smooth curve portraying the path of the moving point, namely its
"time-distance curve." From this curve one can derive all the particulars
of the motion of the point, just as one can work out a train schedule from
the graphic type of timetable referred to above. Now it is evident that
the average slope of the time-distance curve between *p* and *p'* is identical
with the average velocity of the moving point between *t* and *t'*, and thus
the slope of the time-distance curve at *p* is identical with the velocity of
the moving point at the instant *t*. This is the simple connection between
the velocity problem and the tangent problem; the two are the same in
principle. The fundamental problem of the differential calculus is this:
Let the path of a moving point be known; from these data its velocity at
any instant is to be calculated; or, let a curve be given—for each of its
points the slope is to be calculated (at every point the tangent is to be
found). We shall now examine the tangent problem, bearing in mind that
everything we say about this problem can, on the basis of the foregoing,
be carried over directly to the velocity problem.

We noted that if the point *p'* on the curve in question approaches the
point *p* without limit, the average slope between *p* and *p'* will approach
more and more closely a definite limiting value, which will represent the
slope of the curve at the point *p* itself. It may now be asked whether this
is true for every curve. The principle holds for the standard curves that
have been studied since early times: circles, ellipses, hyperbolas, parabolas,
cycloids, etc. But a relatively simple example will show that it is not true
of every curve. Take the curve shown in Figure 3; it is a wave curve, and
in the neighborhood of the point *p* it has infinitely many waves. The wave

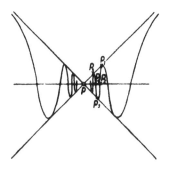

FIGURE 3

length as well as the amplitude of the separate waves decrease without limit as they approach p. Using the method described above we shall attempt to ascertain the slope of the curve at the point p. We take a second point p' on the curve and obtain the average slope between p and p'. If we take p_1 as the point p' (Figure 3) the average slope between p and p_1 turns out to be equal to 1. If p' is permitted to approach p along the curve, it can be seen that the average slope at once decreases; when p' reaches p_2 the average slope (between p and p_2) becomes 0. If p' moves farther along the curve toward p, the average slope between p and p' decreases further, becoming negative, and drops to -1 when p' reaches p_3. If p' moves still closer to p the average slope now begins to increase: it becomes 0 again when p' reaches p_4; then keeps increasing and attains the value 1 when p' reaches p_5. And if p' moves farther along the curve toward p the same cycle is repeated: As p' approaches p and traverses a complete wave of the wave curve, the average slope between p and p' drops from 1 to -1, only to rise again from -1 to 1. Observe that if p' approaches p without limit, it must travel through infinitely many waves, since the curve as we have defined it generates this pattern. That is, as p' approaches p without limit the average slope between p and p' keeps oscillating between the values 1 and -1. Thus, as regards this slope, there can be no question of its limit nor of a definite slope of the curve at the point p. In other words the curve we have been considering has no tangent at p.

This relatively simple intuitable illustration demonstrates that a curve does not have to have a tangent at every point. It used to be thought, however, that intuition forced us to acknowledge that such a deficiency could occur only at isolated and exceptional points of a curve, never at all points. It was believed that a curve must possess an exact slope, or tangent, if not at every point, at least at an overwhelming majority of them. The mathematician and physicist Ampère, whose contribution to the theory of electricity is well known, attempted to prove this conclusion. His proof was false, and it was therefore a great surprise when Weierstrass announced the existence of a curve that lacked a precise slope or tangent

FIGURE 4

at any point. Weierstrass invented the curve by an intricate and arduous calculation, which I shall not attempt to reproduce. But his result can today be achieved in a much simpler way, and this I shall attempt to explain, at least in outline.

We start with the simple figure shown in Figure 4, which consists of an ascending and a descending line. The ascending line we shall replace, as shown in Figure 5, by a broken line of six parts, which first rises to half

FIGURE 5

the height of the original line, then drops all the way down, then again rises to half height, continues on to full height, drops back again to half height, and finally rises once more to full height. Similarly we replace the descending line of Figure 4 by a broken line of six parts, which drops from full height to half height, rises again to full height, then again drops to half height and continues all the way down, rises once more to half height, and finally drops all the way down. From this figure composed of 12 line segments, we evolve by an analogous method the figure of 72 line segments, shown in Figure 6; that is by replacing every line segment of

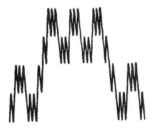

FIGURE 6

Figure 5 by a broken line of 6 parts. It is easy to see how this procedure can be repeated, and that it will lead to more and more complicated figures. There exists a rigorous proof (though I cannot give it here) that the succession of geometric objects constructed according to this rule approach without limit a definite curve possessing the desired property: namely, at no point will it have a precise slope, and hence at no point a tangent. The character of this curve entirely eludes intuition; indeed after a few repetitions of the segmenting process the evolving figure has grown so intricate that intuition can scarcely follow; and it forsakes us completely as regards the curve that is approached as a limit. The fact is that only thought, or logical analysis can pursue this strange object to its final form. Thus, had we relied on intuition in this instance, we should have remained in error, for intuition seems to force the conclusion that there cannot be curves lacking a tangent at any point.

This first example of the failure of intuition involves the fundamental concepts of differentiation; a second example can be derived from the fundamental concepts of integration. The basic problem of differentiation is: given the path of a moving point, to calculate its velocity, or given a curve, to calculate its slope; the basic problem of integration is the inverse: given the velocity of a moving point at every instant, to calculate its path, or given the slope of a curve at each of its points, to calculate the curve. This latter problem, however, has meaning only if the path of the moving point is in fact determined by its velocity, if the curve itself is actually determined by its slope. The question facing us can be more precisely phrased as follows: If two movable points whose track is a single straight line are set in motion in the same direction, at the same instant, from the same place on the line, and at every instant have the same velocity, must they remain together or can they become separated?—or: If two curves in a plane start from a common origin and continuously have the same slope, must they coincide in their entire course or can one of them rise above the other? The dictate of intuition is that the two moving points must always remain together, and that the two curves must coincide in their entire course; yet logical analysis shows that this is not necessarily so. The intuitive answer is true of course for ordinary curves and motions, but we can conceive of certain rather complicated motions and curves for which it is not true. I am sorry not to have the space to go into this in more detail; we must content ourselves with recognizing that in this instance also the apparent certainty of intuition proves to be deceiving.[1]

The foregoing examples of the failure of intuition involve the concepts of the calculus—a subject whose difficulty is acknowledged by the customary designation of "higher mathematics." Lest it be supposed, therefore, that intuition fails only in the more complex branches of mathematics, I propose to examine an occurrence of failure in the elementary branches.

At the very threshold of geometry lies the concept of the curve; everyone believes that he has an intuitively clear notion of what a curve is, and since ancient times it has been held that this idea could be expressed by the following definition: Curves are geometric figures generated by the

---

[1] [Hahn showed in a technical article (*Monatshefte f. Mathematik und Physik*, 16 (1905), p. 161) that the motions (or curves) in question assume infinite velocities (or slopes of infinite value). ED.]

motion of a point.[2] But, attend! In the year 1890 the Italian mathematician Giuseppe Peano (who is also renowned for his investigations in logic) proved that the geometric figures that can be generated by a moving point also include entire plane surfaces. For instance, it is possible to imagine a point moving in such a way that in a finite time it will pass through all the points of a square—and yet no one would consider the entire area of a square as simply a curve. With the aid of a few diagrams I shall attempt to give at least a general idea of how this space-filling motion is generated.

Divide a square into four small squares of equal size (as shown in

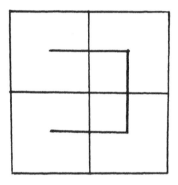

FIGURE 7

Figure 7) and join the center points of these squares by a continuous curve composed of line segments; now imagine a point moving in such a way that at uniform velocity it will traverse the curve in a finite time—say, in some particular unit of time. Next, divide (Figure 8) each of the four small squares of Figure 7 into four smaller squares of equal size and

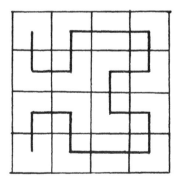

FIGURE 8

---

[2] [Hahn refers to continuous motion in which the point makes no "jumps" and the curve it generates therefore shows no "discontinuities" or gaps. ED.]

connect the center points of these 16 squares by a similar line, and again imagine the point moving so that in unit time it will traverse this second curve at uniform velocity. Repeat this procedure (Figure 9) by dividing

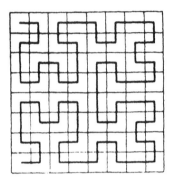

FIGURE 9

each of the 16 small squares of Figure 8 into four still smaller squares, connecting the center points of these 64 squares by a third curve, and imagining the point to move so that in unit time it will traverse this new system of lines at a uniform velocity. It is easy to see how this procedure is to be continued; Figure 10 shows one of the later stages, when the

FIGURE 10

original square has been divided into 4096 small squares. It is now possible to give a rigorous proof that the successive motions considered here approach without limit a definite course, or curve, that takes the moving point through all the points of the large square in unit time. This motion cannot possibly be grasped by intuition; it can only be understood by logical analysis.

While certain geometric objects which no one regards as curves (e.g., a square) can, contrary to intuition, be generated by the motion of a point, other objects that one would not hesitate to classify as curves cannot be so generated. Observe, for instance, the geometric shape shown in Figure 11. It is a wave curve which in the neighborhood of the line segment *ab* (as shown in the figure) consists of infinitely many waves,

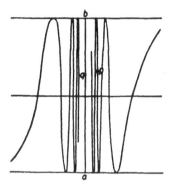

FIGURE 11

whose lengths decrease without limit but whose amplitudes (in contrast to the curve of Figure 3), do not decrease. It is not difficult to prove that this figure, in spite of its linear character cannot be generated by the motion of a point; for no motion of a point is conceivable that would carry it through all the points of this wave curve in a finite time.

Two important questions now suggest themselves. 1. Since the time-honored definition of a curve fails to cover the fundamental concept, what other more serviceable definition can be substituted for it? 2. Since the class of geometric objects that can be produced by the motion of a point does not coincide with the class of all curves, how shall the former class be defined? Today both questions are satisfactorily answered; I shall withhold for a moment the answer to the first question and speak briefly about the second. This was solved with the aid of a new geometric concept, "connectivity in the small" [*"Zusammenhang im Kleinem"*] or "local connectivity." [3] Consider certain figures that can be generated by the motion of a point, such as (Figure 12) a line, a circle, or a square. In each of

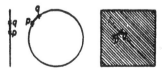

FIGURE 12

[3] [The German expression, not easy to translate, may also be rendered as "connection . . ." and "connectedness in the small," and the function describing the property is sometimes given as a "piecewise continuous function." ED.]

these figures we fix our attention on two points p and q that lie very close together. It is evident that in each case we can move from p to q along a path that does not leave the confines of the figure and remains throughout in close proximity to p and q. This property is called (in an appropriately precise formulation) "connectivity in the small." The structure shown in Figure 11 does not have this property. Take for example the neighboring points p and q; in order to move from p to q without leaving the curve it is necessary to traverse the infinitely many waves lying between them. This path does not remain in close proximity to p and q, since all the intervening waves have the same amplitude. Now it is important to realize that "connectivity in the small" is the basic characteristic of figures that can be generated by the motion of a point. A line, a circle, and a square can be generated by the motion of a point, because they are connected in the small; the construction in Figure 11 cannot be generated by the motion of a point, because it is not connected in the small.

We can convince ourselves by a second example of the undependability of intuition even as regards very elementary geometrical questions. Think of a map (Figure 13) showing the areas of three countries. On this map there will be boundary points at which two of the countries touch each other, but there may also be points at which all three countries come

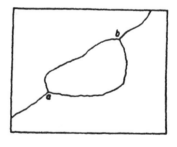

FIGURE 13

together—so-called "three-country corners," like the points a and b in Figure 13. Intuition seems to indicate that these three-country corners can occur only at isolated points, that at the great majority of boundary points on the map only two countries will be in contact. Yet the Dutch mathematician L. E. J. Brouwer showed in 1910 how a map can be divided into three countries in such a way that at every boundary point all three countries will touch one another.[4] Let us attempt briefly to describe how this is done.

We start with the map shown in Figure 14, on which there are three

---

[4] [A point is called a "boundary point" if in each of its neighborhoods lie points of various countries. Three countries meet in a boundary point if in each of its neighborhoods points are to be found of each of the three countries. ED.]

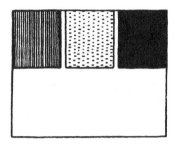

FIGURE 14

different countries, one hatched (A), one dotted (B), and one solid (C); the unmarked remainder is unoccupied land. Country A, seeking to bring this land into its sphere of influence, decides to push through a corridor (Figure 15) which approaches within one mile of every point of the

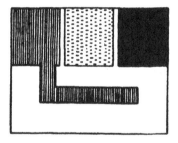

FIGURE 15

unoccupied territory; but—in order to avoid trouble—the corridor is not to impinge upon either of the two other countries. After this has been done country B decides that it must make a similar move, and proceeds to drive a corridor into the remaining unoccupied territory (Figure 16)— a corridor that comes within one-half mile of all the unoccupied points but does not touch either of the other two countries. Thereupon country

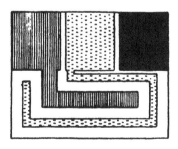

FIGURE 16

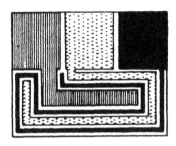

FIGURE 17

C, deciding that it cannot lag behind, also extends a corridor (Figure 17) into the territory as yet unoccupied, which comes to within a third of a mile of every point of this territory but does not touch the other countries. Country *A* now feels that it has been outwitted and proceeds to push a second corridor into the remaining unoccupied territory, which comes within a quarter of a mile of all points of this territory but does not touch the other two countries. The process continues: country *B* extends a corridor that comes within a fifth of a mile of every unoccupied point; country *C* extends one that comes within a sixth of a mile of every unoccupied point; country *A* starts over again, and so on and on. And since we are giving imagination free rein, let us assume further that country *A* required a year for the construction of its first corridor, country *B* the following half-year for its first corridor, country *C* the next quarter year for its first corridor; country *A* the next eighth of a year for its second, and so on; thus each succeeding extension is completed in half the time of its predecessor. It can easily be seen that after the passage of two years none of the originally unoccupied territory will remain unclaimed; moreover the entire map will then be divided among the three countries in such a fashion that at no point will only two of the countries touch each other, but instead all three countries will meet at every boundary point. Intuition cannot comprehend this pattern, although logical analysis requires us to accept it. Once more intuition has led us astray.

Because intuition turned out to be deceptive in so many instances, and because propositions that had been accounted true by intuition were repeatedly proved false by logic, mathematicians became more and more sceptical of the validity of intuition. They learned that it is unsafe to accept any mathematical proposition, much less to base any mathematical discipline on intuitive convictions. Thus a demand arose for the expulsion of intuition from mathematical reasoning, and for the complete formalization of mathematics. That is to say, every new mathematical concept was to be introduced through a purely logical definition; every mathematical proof was to be carried through by strictly logical means. The pioneers of this program (to mention only the most famous) were Augustin Cauchy (1789–1857), Bernhard Bolzano (1781–1848), Karl Weierstrass (1815–1897), Georg Cantor (1845–1918) and Richard Dedekind (1831–1916).

The task of completely formalizing mathematics, of reducing it entirely to logic, was arduous and difficult; it meant nothing less than a reform in root and branch. Propositions that had formerly been accepted as intuitively evident had to be painstakingly proved. To cite one example: the simple geometric proposition that "every closed polygon that does not cross itself divides the plane into two separate parts" requires a lengthy and highly complicated proof. This is true to an even greater degree of the analogous proposition of solid geometry: "every closed polyhedron that does not intersect itself divides space into two separate parts."

As the prototype of an *a priori* synthetic judgment based on pure intuition Kant cites the proposition that space is three-dimensional. But by present-day standards even this statement calls for searching logical analysis. First it is necessary to define purely logically what is meant by the "dimensionality" of a geometric figure, or "point set," and then it must be logically proved that the space of ordinary geometry—which is also the space of Newtonian physics—as embraced in this definition, is in fact three-dimensional. This proof was not achieved till recent times, in 1922, and then simultaneously by the Viennese mathematician K. Menger and the Russian mathematician P. Urysohn—the latter having since succumbed to a tragic accident at the height of his creative powers. I wish to give at least a sketchy explanation of how the dimensionality of a point set is defined.

A point set is called null-dimensional if for each of its points there exists an arbitrarily small neighborhood whose boundary contains no point of the set: for example, every set consisting of a finite number of points is null-dimensional (cf. Figure 18), but there are also a great many

FIGURE 18

very complicated null-dimensional point sets that consist of infinitely many points. A point set that is not null-dimensional is called one-dimensional if for each of its points there is an arbitrarily small neighborhood whose boundary has only a null-dimensional set in common with the point set. Every straight line, every figure composed of a finite number of straight lines, every circle, every ellipse, in short all geometrical constructs that we ordinarily designate as curves are one-dimensional in this

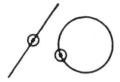

FIGURE 19

sense (cf. Figure 19); in fact, even the geometric object shown in Figure 11, which we saw could not be generated by the motion of a point, is one-dimensional. Similarly, a point set that is neither null-dimensional nor one-dimensional is called two-dimensional if for each of its points there is an arbitrarily small neighborhood whose boundary has at the most a one-dimensional set in common with the point set. Every plane, every polygonic or circular plane area, every spherical surface, in short every geometric construct ordinarily classified as a surface is two-dimensional in this sense. A point set that is neither null-dimensional, one-dimensional, nor two-dimensional is called three-dimensional if for each of its points there is an arbitrarily small neighborhood whose boundary has at most a two-dimensional set in common with the point set. It can be proved—not at all simply, however—that the space of ordinary geometry is a three-dimensional point set in the foregoing theory.

This theory provides what we have been seeking, a fully satisfactory definition of the concept of a curve. The essential characteristic of a curve turns out to be its one-dimensionality. But beyond that the theory also makes possible an unusually precise and subtle analysis of the structure of curves, about which I should like to comment briefly. A point on a curve is called an end point if there are arbitrarily small neighborhoods surrounding it, each of whose boundaries has only a single point in common with the curve (cf. points *a* and *b* in Figure 20). A point on the curve

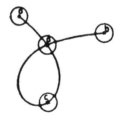

FIGURE 20

that is not an end point is called an ordinary point if it has arbitrarily small neighborhoods each of whose boundaries has exactly two points in common with the curve (cf. point *c* in Figure 20). A point on a curve is called a branch point if the boundary of any of its arbitrarily small neighborhoods has more than two points in common with the curve (cf. point *d* in Figure 20).

Intuition seems to indicate that the end points and branch points of a curve are exceptional cases; that they can occur only sporadically, that it is impossible for a curve to be made up of nothing but end points or of branch points. This intuitive conviction is specifically confirmed by logical analysis as far as end points are concerned; but as regards branch points it has been refuted. The Polish mathematician W. Sierpinski proved in 1915 that there are curves *all of whose points are branch points*. Let us attempt to visualize how this comes about.

Suppose that an equilateral triangle has been inscribed within another equilateral triangle (as shown in Figure 21) and the interior of the in-

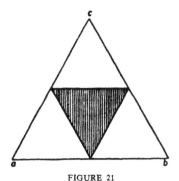

FIGURE 21

scribed triangle erased (hatched in Figure 21); there remain three equilateral triangles with their sides. In each of these three triangles (Figure 22) inscribe an equilateral triangle and again erase its interior; there are now left nine equilateral triangles together with their sides. In each of

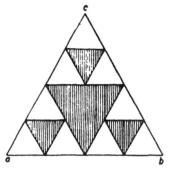

FIGURE 22

these nine triangles an equilateral triangle is to be inscribed and its interior erased so that 27 equilateral triangles are left. Imagine this process continued indefinitely. (Figure 23 shows the fifth step, where 243 tri-

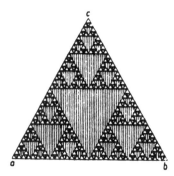

FIGURE 23

angles remain.) The points of the original equilateral triangle that survive the infinitely numerous erasures can be shown to form a curve, and specifically a curve all of whose points, with the exception of the vertex points *a, b, c,* of the original triangle, are branch points. From this it is easy to obtain a curve with all its points branch points; for instance, by distorting the entire figure so that the three vertices *a, b, c,* of the original triangle are brought together in a single point.

But enough of examples—let us now summarize what has been said. Again and again we have found that in geometric questions, and indeed even in simple and elementary geometric questions, intuition is a wholly unreliable guide. It is impossible to permit so unreliable an aid to serve as the starting point or basis of a mathematical discipline. The space of geometry is not a form of pure intuition, but a logical construct.

The way is then open for other non-contradictory logical constructs in the form of spaces differing from the space of ordinary geometry; spaces, for instance, in which the so-called Euclidean parallel postulate is replaced by a contrary postulate (non-Euclidean spaces), spaces whose dimensionality is greater than three, non-Archimedean spaces. I shall say a few words about the last named.

The possibility of measuring the length of a line segment by a real number, and the possibility that follows from this, namely, fixing the position of a point, as is done in analytic geometry, by assigning real numbers as its "coordinates," rests on the so-called "postulate of Archimedes." [5] This postulate reads as follows: given lengths, there is always a multiple of the first that is greater than the second. As logical constructs,

[5] [It should properly be called the postulate of Eudoxus. Eudoxus' dates are 408–355 B.C.; Archimedes: 287–212 B.C. ED.]

however, spaces can be devised in which the Archimedean postulate is replaced by its opposite, that is, in which there are lengths that are greater than any multiple of a given length.[6] Hence in these spaces infinitely large and infinitely small lengths can exist (as determined by any arbitrarily chosen unit of measure); while in the space of ordinary geometry there are no infinitely large and infinitely small lengths. In a "non-Archimedean" space, lengths can be measured, and a system of analytical geometry can be developed. Of course, the real numbers of ordinary arithmetic are of no help in this geometry, but one uses "non-Archimedean" number systems, which can be interpreted and applied in calculation as well as the real numbers of ordinary arithmetic.

But what are we to say to the often heard objection that the multi-dimensional, non-Euclidean, non-Archimedean geometries, though consistent as logical constructs, are useless in arranging our experience because they are non-intuitional? For projecting our experience, it is said, only the conventional three-dimensional, Euclidean, Archimedean geometry is usable, for it is the only one that satisfies intuition. My first comment on this score—and this is the point of my entire lecture—is that ordinary geometry itself by no means constitutes a supreme example of the intuitive process. The fact is that *every* geometry—three-dimensional as well as multi-dimensional, Euclidean as well as non-Euclidean, Archimedean as well as non-Archimedean—is a logical construct. Traditional physics is responsible for the fact that until recently the logical construction of three-dimensional, Euclidean, Archimedean space has been used exclusively for the ordering of our experience. For several centuries, almost up to the present day, it served this purpose admirably; thus we grew used to operating with it. This habituation to the use of ordinary geometry for the ordering of our experience explains why we regard this geometry as intuitive; and every departure from it unintuitive, contrary to intuition, intuitively impossible. But as we have seen, such "intuitional impossibilities," also occur in ordinary geometry. They appear as soon as we no longer restrict ourselves to the geometrical entities with which we have long been familiar, but instead reflect upon objects that we had not thought about before.

Modern physics now makes it appear appropriate to avail ourselves of the logical constructs of multi-dimensional and non-Euclidean geometries for the ordering of our experience. (Although we have as yet no indication that the inclusion of non-Archimedean geometry might prove useful, this possibility is by no means excluded.) But these advances in physics are so recent that we are not yet accustomed to the manipulation of these logical constructs; hence they still seem an affront to intuition.

[6] The first to investigate thoroughly the properties of non-Archimedean spaces was the Italian mathematician G. Veronese. [H.H.]

The theory that the earth is a sphere was also once an affront to intuition. This hypothesis was widely rejected on the grounds that the existence of the antipodes was contrary to intuition. However, we have got used to the idea and today it no longer occurs to anyone to pronounce it impossible because it conflicts with intuition.

Physical concepts are also logical constructs and here too we can see clearly how concepts whose application is familiar to us acquire an intuitive status which is denied to those whose application is unfamiliar. The concept "weight" is so much a part of common experience that almost everyone regards it as intuitive. The concept "moment of inertia," however, does not enter into most persons' activities and is therefore not regarded by them as intuitive. Yet among experimental physicists and engineers, who constantly work with it, "moment of inertia" possesses an intuitive status equal to that generally accorded to "weight." Similarly, the concept "potential difference" is intuitive for the electrical technician, but not for most men.

If the use of multi-dimensional and non-Euclidean geometries for the ordering of our experience continues to prove itself so that we become more and more accustomed to dealing with these logical constructs; if they penetrate into the curriculum of the schools; if we, so to speak, learn them at our mother's knee, as we now learn three-dimensional Euclidean geometry—then nobody will think of saying that these geometries are contrary to intuition. They will be considered as deserving of intuitive status as three-dimensional Euclidean geometry is today. For it is not true, as Kant urged, that intuition is a pure *a priori* means of knowledge, but rather that it is force of habit rooted in psychological inertia.

PART XV

How to Solve It

The Tears of Mathematics

THE most painful thing about mathematics is how far away you are from being able to use it after you have learned it. It differs in this respect from golf, say, or piano playing. In these activities you learn as you go, but in mathematics it is possible to acquire an impressive amount of information as to theorems and methods and yet be totally incapable of solving the simplest problems. Entrants for tripos, the difficult competitive examinations which used to be held at Cambridge, customarily enrolled with a mathematical coach whose job it was to drill them until they were blue in the face and, more important, could work out problems quickly and almost by instinct. It is debatable whether this grind enriched the spirit but it obviously improved one's arithmetic.

Besides learning how problems should be solved, can one learn how to solve problems? In 1945 Princeton University Press published a little book, *How to Solve It*, which promised that the art could be learned. The book jacket carried the blurb: "A System of thinking which can help you solve any problem." The claim is too large, but the book is both delightful and instructive. George Polya, its author and a professor of mathematics at Stanford University, describes his book as *heuristic*, a word which means "serving to discover." To solve a problem, Polya says, is to make a discovery: a great problem means a great discovery but "there is a grain of discovery in the solution of any problem. Your problem may be modest; but if it challenges your curiosity and brings into play your inventive faculties, and if you solve it by your own means, you may experience the tension and enjoy the triumph of discovery."

How to Solve It considers the general questions that problem solvers face and proposes general methods to deal with them. It also ferrets out the small, often unspoken, questions that gnaw at the mind and leave it a ruin of confusion before the problem has even been put on paper. The book is full of valuable hints and urbane, soothing suggestions. Problems can be "decomposed" into their elements and "recombined," the new arrangement often being easier to solve; good use can be made of analogy and related methods; there are tricks to setting up equations and to paring the inessentials; it is often profitable to work backwards. Polya does not scorn the value of having a bright idea but does not assume that you will get one. "If you cannot solve the proposed problem," he says cheerfully, "do not let this failure afflict you too much but try to find consolation with some easier success, try to solve first some related problem; then you may find courage to attack your original problem again." I have selected a few items from the book almost at random: *Mathematical Induction*—the best exposition I have ever seen; *Setting up Equations; The Traditional Mathematics Professor; Working Backwards.* I am sorry I cannot give more.

Polya is an American born in Budapest in 1887. He was educated abroad, has taught at Princeton, Brown and Stanford, and has done important research in function theory, probability, applications of mathematics and mathematical method. Judging from his book, he has thought deeply about improving the teaching of mathematics and he cares what is learned in his classes. His students are to be envied.

1 How to Solve It

By G. POLYA

INDUCTION AND MATHEMATICAL INDUCTION. Induction is the process of discovering general laws by the observation and combination of particular instances. It is used in all sciences, even in mathematics. Mathematical induction is used in mathematics alone to prove theorems of a certain kind. It is rather unfortunate that the names are connected because there is very little logical connection between the two processes. There is, however, some practical connection; we often use both methods together. We are going to illustrate both methods by the same example.

1. We may observe, by chance, that

$$1 + 8 + 27 + 64 = 100$$

and, recognizing the cubes and the square, we may give to the fact we observed the more interesting form:

$$1^3 + 2^3 + 3^3 + 4^3 = 10^2.$$

How does such a thing happen? Does it often happen that such a sum of successive cubes is a square?

In asking this we are like the naturalist who, impressed by a curious plant or a curious geological formation, conceives a general question. Our general question is concerned with the sum of successive cubes

$$1^3 + 2^3 + 3^3 + \ldots + n^3.$$

We were led to it by the "particular instance" $n = 4$.

What can we do for our question? What the naturalist would do; we can investigate other special cases. The special cases $n = 2$, 3 are still simpler, the case $n = 5$ is the next one. Let us add, for the sake of uniformity and completeness, the case $n = 1$. Arranging neatly all these cases, as a geologist would arrange his specimens of a certain ore, we obtain the following table:

$$
\begin{aligned}
1 &= 1 = 1^2 \\
1 + 8 &= 9 = 3^2 \\
1 + 8 + 27 &= 36 = 6^2 \\
1 + 8 + 27 + 64 &= 100 = 10^2 \\
1 + 8 + 27 + 64 + 125 &= 225 = 15^2.
\end{aligned}
$$

It is hard to believe that all these sums of consecutive cubes are squares by mere chance. In a similar case, the naturalist would have little doubt that the general law suggested by the special cases heretofore observed is correct; the general law is almost proved by *induction*. The mathematician expresses himself with more reserve although fundamentally, of course, he thinks in the same fashion. He would say that the following theorem is strongly suggested by induction:

The sum of the first n *cubes is a square.*

2. We have been led to conjecture a remarkable, somewhat mysterious law. Why should those sums of successive cubes be squares? But, apparently, they are squares.

What would the naturalist do in such a situation? He would go on examining his conjecture. In so doing, he may follow various lines of investigation. The naturalist may accumulate further experimental evidence; if we wish to do the same, we have to test the next cases, $n = 6, 7,$ .... The naturalist may also reexamine the facts whose observation has led him to his conjecture; he compares them carefully, he tries to disentangle some deeper regularity, some further analogy. Let us follow this line of investigation.

Let us reexamine the cases $n = 1, 2, 3, 4, 5$ which we arranged in our table. Why are all these sums squares? What can we say about these squares? Their bases are 1, 3, 6, 10, 15. What about these bases? Is there some deeper regularity, some further analogy? At any rate, they do not seem to increase too irregularly. How do they increase? The difference between two successive terms of this sequence is itself increasing,

$$3 - 1 = 2, \; 6 - 3 = 3, \; 10 - 6 = 4, \; 15 - 10 = 5.$$

Now these differences are conspicuously regular. We may see here a surprising analogy between the bases of those squares, we may see a remarkable regularity in the numbers 1, 3, 6, 10, 15:

$$1 = 1$$
$$3 = 1 + 2$$
$$6 = 1 + 2 + 3$$
$$10 = 1 + 2 + 3 + 4$$
$$15 = 1 + 2 + 3 + 4 + 5.$$

If this regularity is general (and the contrary is hard to believe) the theorem we suspected takes a more precise form:
It is, for n = 1, 2, 3, ...

$$1^3 + 2^3 + 3^3 + \ldots + n^3 = (1 + 2 + 3 + \ldots + n)^2.$$

3. The law we just stated was found by induction, and the manner in which it was found conveys to us an idea about induction which is neces-

sarily one-sided and imperfect but not distorted. Induction tries to find regularity and coherence behind the observations. Its most conspicuous instruments are generalization, specialization, analogy. Tentative generalization starts from an effort to understand the observed facts; it is based on analogy, and tested by further special cases.

We refrain from further remarks on the subject of induction about which there is wide disagreement among philosophers. But it should be added that many mathematical results were found by induction first and proved later. Mathematics presented with rigor is a systematic deductive science but mathematics in the making is an experimental inductive science.

4. In mathematics as in the physical sciences we may use observation and induction to discover general laws. But there is a difference. In the physical sciences, there is no higher authority than observation and induction but in mathematics there is such an authority: rigorous proof.

After having worked a while experimentally it may be good to change our point of view. Let us be strict. We have discovered an interesting result but the reasoning that led to it was merely plausible, experimental, provisional, heuristic; let us try to establish it definitively by a rigorous proof.

We have arrived now at a "problem to prove": to prove or to disprove the result stated before (see 2, p. 1981).

There is a minor simplification. We may know that

$$1 + 2 + 3 + \ldots + n = \frac{n(n + 1)}{2}.$$

At any rate, this is easy to verify. Take a rectangle with sides n and $n + 1$, and divide it in two halves by a zigzag line as in Figure 1a which shows the case $n = 4$. Each of the halves is "staircase-shaped" and its area

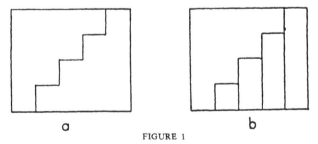

a   b

FIGURE 1

has the expression $1 + 2 + \ldots + n$; for $n = 4$ it is $1 + 2 + 3 + 4$, see Figure 1b. Now, the whole area of the rectangle is $n(n + 1)$ of which the staircase-shaped area is one half; this proves the formula.

We may transform the result which we found by induction into

$$1^3 + 2^3 + 3^3 + \ldots + n^3 = \left(\frac{n(n+1)}{2} \right)^2.$$

5. If we have no idea how to prove this result, we may at least test it. Let us test the first case we have not tested yet, the case $n = 6$. For this value, the formula yields

$$1 + 8 + 27 + 64 + 125 + 216 = \left(\frac{6 \times 7}{2} \right)^2$$

and, on computation, this turns out to be true, both sides being equal to 441.

We can test the formula more effectively. The formula is, very likely, generally true, true for all values of n. Does it remain true when we pass from any value n to the next value $n + 1$? Along with the formula as written above we should also have

$$1^3 + 2^3 + 3^3 + \ldots + n^3 + (n+1)^3 = \left(\frac{(n+1)(n+2)}{2} \right)^2.$$

Now, there is a simple check. Subtracting from this the formula written above, we obtain

$$(n+1)^3 = \left(\frac{(n+1)(n+2)}{2} \right)^2 - \left(\frac{n(n+1)}{2} \right)^2.$$

This is, however, easy to check. The right hand side may be written as

$$\left(\frac{n+1}{2} \right)^2 [(n+2)^2 - n^2] = \left(\frac{n+1}{2} \right)^2 [n^2 + 4n + 4 - n^2]$$

$$\frac{(n+1)^2}{4} (4n + 4) = (n+1)^2 (n+1) = (n+1)^3.$$

Our experimentally found formula passed a vital test.

Let us see clearly what this test means. We verified beyond doubt that

$$(n+1)^3 = \left(\frac{(n+1)(n+2)}{2} \right)^2 - \left(\frac{n(n+1)}{2} \right)^2.$$

We do not know yet whether

$$1^3 + 2^3 + 3^3 + \ldots + n^3 = \left(\frac{n(n+1)}{2} \right)^2.$$

is true. But *if* we knew that this *was* true we could infer, by adding the equation which we verified beyond doubt, that

$$1^3 + 2^3 + 3^3 + \ldots + n^3 + (n+1)^3 = \left(\frac{(n+1)(n+2)}{2} \right)^2$$

is *also* true which is the same assertion for the next integer $n + 1$. Now, we actually know that our conjecture is true for $n = 1, 2, 3, 4, 5, 6$. By virtue of what we have just said, the conjecture, being true for $n = 6$, must also be true for $n = 7$; being true for $n = 7$ it is true for $n = 8$; being true for $n = 8$ it is true for $n = 9$; and so on. It holds for all n, it is proved to be true generally.

6. The foregoing proof may serve as a pattern in many similar cases. What are the essential lines of this pattern?

The assertion we have to prove must be given in advance, in precise form.

The assertion must depend on an integer n.

The assertion must be sufficiently "explicit" so that we have some possibility of testing whether it remains true in the passage from n to the next integer $n + 1$.

If we succeed in testing this effectively, we may be able to use our experience, gained in the process of testing, to conclude that the assertion must be true for $n + 1$ provided it is true for n. When we are so far it is sufficient to know that the assertion is true for $n = 1$; hence it follows for $n = 2$; hence it follows for $n = 3$, and so on; passing from any integer to the next, we prove the assertion generally.

This process is so often used that it deserves a name. We could call it "proof from n to $n + 1$" or still simpler "passage to the next integer." Unfortunately, the accepted technical term is "mathematical induction." This name results from a random circumstance. The precise assertion that we have to prove may come from any source, and it is immaterial from the logical viewpoint what the source is. Now, in many cases, as in the case we discussed here in detail, the source is induction, the assertion is found experimentally, and so the proof appears as a mathematical complement to induction; this explains the name.

7. Here is another point, somewhat subtle, but important to anybody who desires to find proofs by himself. In the foregoing, we found two different assertions by observation and induction, one after the other, the first under 1, the second under 2; the second was more precise than the first. Dealing with the second assertion, we found a possibility of checking the passage from n to $n + 1$, and so we were able to find a proof by "mathematical induction." Dealing with the first assertion, and ignoring the precision added to it by the second one, we should scarcely have been able to find such a proof. In fact, the first assertion is less precise, less "explicit," less "tangible," less accessible to testing and checking than the second one. Passing from the first to the second, from the less precise to the more precise statement, was an important preparative for the final proof.

This circumstance has a paradoxical aspect. The second assertion is stronger; it implies immediately the first, whereas the somewhat "hazy" first assertion can hardly imply the more "clear-cut" second one. Thus, the stronger theorem is easier to master than the weaker one; this is the inventor's paradox.

SETTING UP EQUATIONS is like translation from one language into another. This comparison, used by Newton in his *Arithmetica Universalis,* may help to clarify the nature of certain difficulties often felt both by students and by teachers.

1. To set up equations means to express in mathematical symbols a condition that is stated in words; it is translation from ordinary language into the language of mathematical formulas. The difficulties which we may have in setting up equations are difficulties of translation.

In order to translate a sentence from English into French two things are necessary. First, we must understand thoroughly the English sentence. Second, we must be familiar with the forms of expression peculiar to the French language. The situation is very similar when we attempt to express in mathematical symbols a condition proposed in words. First, we must understand thoroughly the condition. Second, we must be familiar with the forms of mathematical expression.

An English sentence is relatively easy to translate into French if it can be translated word for word. But there are English idioms which cannot be translated into French word for word. If our sentence contains such idioms, the translation becomes difficult; we have to pay less attention to the separate words, and more attention to the whole meaning; before translating the sentence, we may have to rearrange it.

It is very much the same in setting up equations. In easy cases, the verbal statement splits almost automatically into successive parts, each of which can be immediately written down in mathematical symbols. In more difficult cases, the condition has parts which cannot be immediately translated into mathematical symbols. If this is so, we must pay less attention to the verbal statement, and concentrate more upon the meaning. Before we start writing formulas, we may have to rearrange the condition, and we should keep an eye on the resources of mathematical notation while doing so.

In all cases, easy or difficult, we have to understand the condition, to *separate the various parts of the condition,* and to ask: *Can you write them down?* In easy cases, we succeed without hesitation in dividing the condition into parts that can be written down in mathematical symbols; in difficult cases, the appropriate division of the condition is less obvious.

The foregoing explanation should be read again after the study of the following examples.

2. *Find two quantities whose sum is* 78 *and whose product is* 1296.

We divide the page by a vertical line. On one side, we write the verbal statement split into appropriate parts. On the other side, we write algebraic signs, opposite to the corresponding part of the verbal statement. The original is on the left, the translation into symbols on the right.

STATING THE PROBLEM

in English	*in algebraic language*
Find two quantities	$x, \quad y$
whose sum is 78 and	$x + y = 78$
whose product is 1296	$xy = 1296.$

In this case, the verbal statement splits almost automatically into successive parts, each of which can be immediately written down in mathematical symbols.

3. *Find the breadth and the height of a right prism with square base, being given the volume, 63 cu. in., and the area of the surface, 102 sq. in.*

What are the unknowns? The side of the base, say x, and the altitude of the prism, say y.

What are the data? The volume, 63, and the area, 102.

What is the condition? The prism whose base is a square with side x and whose altitude is y must have the volume 63 and the area 102.

Separate the various parts of the condition. There are two parts, one concerned with the volume, the other with the area.

We can scarcely hesitate in dividing the whole condition just in these two parts; but we cannot write down these parts "immediately." We must know how to calculate the volume and the various parts of the area. Yet, if we know that much geometry, we can easily restate both parts of the condition so that the translation into equations is feasible. We write on the left hand side of the page an essentially rearranged and expanded statement of the problem, ready for translation into algebraic language.

Of a right prism with square base	
find the side of the base	x
and the altitude.	y
First. The volume is given.	63
The area of the base which is a square	
with side x	x^2
and the altitude	y
determine the volume which is their	
product.	$x^2 y = 63$
Second. The area of the surface is given.	102
The surface consists of two squares with	
side x	$2x^2$
and of four rectangles, each with base x	
and altitude y,	$4xy$
whose sum is the area.	$2x^2 + 4xy = 102.$

4. *Being given the equation of a straight line and the coordinates of a point, find the point which is symmetrical to the given point with respect to the given straight line.*

This is a problem of plane analytic geometry.

What is the unknown? A point, with coordinates, say p, q.

What is given? The equation of a straight line, say $y = mx + n$, and a point with coordinates, say a, b.

What is the condition? The points (a, b) and (p, q) are symmetrical to each other with respect to the line $y = mx + n$.

We now reach the essential difficulty which is to divide the condition into parts each of which can be expressed in the language of analytic geometry. The nature of this difficulty must be well understood. A decomposition of the condition into parts may be logically unobjectionable and nevertheless useless. What we need here is a decomposition into parts which are fit for analytic expression. In order to find such a decomposition we must *go back to the definition* of symmetry, but keep an eye on the resources of analytic geometry. What is meant by symmetry with respect to a straight line? What geometric relations can we express simply in analytic geometry? We concentrate upon the first question, but we should not forget the second. Thus, eventually, we may find the decomposition which we are going to state.

The given point and the point required are so related that first, the line joining them is perpendicular to the given line, and second, they lie on opposite sides of the given line but are at equal distance from it.	(a, b) (p, q) $\dfrac{q - b}{p - a} = -\dfrac{1}{m}$ $\dfrac{b - ma - n}{\sqrt{1 + m^2}} = -\dfrac{q - mp - n}{\sqrt{1 + m^2}}.$

THE TRADITIONAL MATHEMATICS PROFESSOR of the popular legend is absentminded. He usually appears in public with a lost umbrella in each hand. He prefers to face the blackboard and to turn his back on the class. He writes a, he says b, he means c; but it should be d. Some of his sayings are handed down from generation to generation.

"In order to solve this differential equation you look at it till a solution occurs to you."

"This principle is so perfectly general that no particular application of it is possible."

"Geometry is the art of correct reasoning on incorrect figures."

"My method to overcome a difficulty is to go round it."

"What is the difference between method and device? A method is a device which you use twice."

After all, you can learn something from this traditional mathematics professor. Let us hope that the mathematics teacher from whom you cannot learn anything will not become traditional.

WORKING BACKWARDS. If we wish to understand human behavior we should compare it with animal behavior. Animals also "have problems" and "solve problems." Experimental psychology has made essential progress in the last decades in exploring the "problem-solving" activities of various animals. We cannot discuss here these investigations but we shall describe sketchily just one simple and instructive experiment and our description will serve as a sort of comment upon the method of analysis, or method of "working backwards." . . .[1]

1. Let us try to find an answer to the following tricky question: *How can you bring up from the river exactly six quarts of water when you have only two containers, a four quart pail and a nine quart pail, to measure with?*

Let us visualize clearly the given tools we have to work with, the two containers. (*What is given?*) We imagine two cylindrical containers having equal bases whose altitudes are as 9 to 4, see Figure 2. *If* along the lateral surface of each container there were a scale of equally spaced horizontal

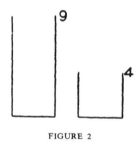

FIGURE 2

lines from which we could tell the height of the waterline, our problem would be easy. Yet there is no such scale and so we are still far from the solution.

We do not know yet how to measure exactly 6 quarts; but could we measure something else? (*If you cannot solve the proposed problem try to solve first some related problem. Could you derive something useful from the data?*) Let us do something, let us play around a little. We could fill the larger container to full capacity and empty so much as we can into the smaller container; then we could get 5 quarts. Could we also get 6 quarts? Here are again the two empty containers. We could also . . .

We are working now as most people do when confronted with this puzzle. We start with the two empty containers, we try this and that, we empty and fill, and when we do not succeed, we start again, trying something else. We are *working forwards*, from the given initial situation to

---

[1] [The Greek mathematician Pappus, Polya points out, gave an important description of the method. ED.]

the desired final situation, from the data to the unknown. We may succeed, after many trials, accidentally.

2. But exceptionally able people, or people who had the chance to learn in their mathematics classes something more than mere routine operations, do not spend too much time in such trials but turn round, and start working backwards.

What are we required to do? (*What is the unkown?*) Let us visualize the final situation we aim at as clearly as possible. Let us imagine that we have here, before us, the larger container with exactly 6 quarts in it and the smaller container empty as in Figure 3. (Let us *start from what is required* and *assume what is sought as already found*, says Pappus.)

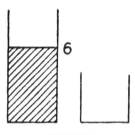

FIGURE 3

From what foregoing situation could we obtain the desired final situation shown in Figure 3? (Let us *inquire from what antecedent the desired result could be derived*, says Pappus.) We could, of course, fill the larger container to full capacity, that is, to 9 quarts. But then we should be able to pour out exactly three quarts. To do that . . . we must have just one quart in the smaller container! That's the idea. See Figure 4.

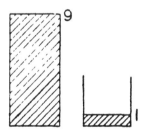

FIGURE 4

(The step that we have just completed is not easy at all. Few persons are able to take it without much foregoing hesitation. In fact, recognizing the significance of this step, we foresee an outline of the following solution.)

But how can we reach the situation that we have just found and illustrated by Figure 4? (Let us *inquire again what could be the antecedent of that antecedent.*) Since the amount of water in the river is, for our purpose, unlimited, the situation of Figure 4 amounts to the same as the next one in Figure 5

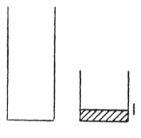

FIGURE 5

or the following in Figure 6.

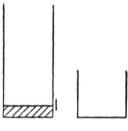

FIGURE 6

It is easy to recognize that if any one of the situations in Figures 4, 5, 6 is obtained, any other can be obtained just as well, but it is not so easy to hit upon Figure 6, unless we *have seen it before,* encountered it accidentally in one of our initial trials. Playing around with the two containers, we may have done something similar and remember now, in the right moment, that the situation of Figure 6 can arise as suggested by Figure 7: We fill the large container to full capacity, and pour from it four quarts

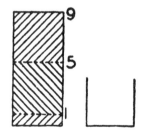

FIGURE 7

into the smaller container and then into the river, twice in succession. We *came eventually upon something already known* (these are Pappus's words) and following the method of analysis, *working backwards*, we have discovered the appropriate sequence of operations.

It is true, we have discovered the appropriate sequence in retrogressive order but all that is left to do is to *reverse the process* and *start from the point which we reached last of all in the analysis* (as Pappus says). First, we perform the operations suggested by Figure 7 and obtain Figure 6; then we pass to Figure 5, then to Figure 4, and finally to Figure 3. *Retracing our steps, we finally succeed in deriving what was required.*

3. Greek tradition attributed to Plato the discovery of the method of analysis. The tradition may not be quite reliable but, at any rate, if the method was not invented by Plato, some Greek scholar found it necessary to attribute its invention to a philosophical genius.

There is certainly something in the method that is not superficial. There is a certain psychological difficulty in turning around, in going away from the goal, in working backwards, in not following the direct path to the desired end. When we discover the sequence of appropriate operations, our mind has to proceed in an order which is exactly the reverse of the actual performance. There is some sort of psychological repugnance to this reverse order which may prevent a quite able student from understanding the method if it is not presented carefully.

Yet it does not take a genius to solve a concrete problem working backwards; anybody can do it with a little common sense. We concentrate upon the desired end, we visualize the final position in which we would like to be. From what foregoing position could we get there? It is natural to ask this question, and in so asking we work backwards. Quite primitive problems may lead naturally to working backwards.

Working backwards is a common-sense procedure within the reach of everybody and we can hardly doubt that it was practiced by mathematicians and nonmathematicians before Plato. What some Greek scholar may have regarded as an achievement worthy of the genius of Plato is to state the procedure in general terms and to stamp it as an operation typically useful in solving mathematical and nonmathematical problems.

4. And now, we turn to the psychological experiment—if the transition from Plato to dogs, hens, and chimpanzees is not too abrupt. A fence forms three sides of a rectangle but leaves open the fourth side as shown in Figure 8. We place a dog on one side of the fence, at the point *D*, and some food on the other side, at the point *F*. The problem is fairly easy for the dog. He may first strike a posture as if to spring directly at the food but then he quickly turns about, dashes off around the end of ˈe fence and, running without hesitation, reaches the food in a smooth curve. Sometimes, however, especially when the points *D* and *F* are close

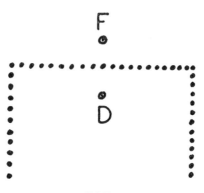

FIGURE 8

to each other, the solution is not so smooth; the dog may lose some time in barking, scratching, or jumping against the fence before he "conceives the bright idea" (as we would say) of going around.

It is interesting to compare the behavior of various animals put into the place of the dog. The problem is very easy for a chimpanzee or a four-year-old child (for whom a toy may be a more attractive lure than food). The problem, however, turns out to be surprisingly difficult for a hen who runs back and forth excitedly on her side of the fence and may spend considerable time before getting at the food if she gets there at all. But she may succeed, after much running, accidentally.

5. We should not build a big theory upon just one simple experiment which was only sketchily reported. Yet there can be no disadvantage in noticing obvious analogies provided that we are prepared to recheck and revalue them.

Going around an obstacle is what we do in solving any kind of problem; the experiment has a sort of symbolic value. The hen acted like people who solve their problem muddling through, trying again and again, and succeeding eventually by some lucky accident without much insight into the reasons for their success. The dog who scratched and jumped and barked before turning around solved his problem about as well as we did ours about the two containers. Imagining a scale that shows the waterline in our containers was a sort of almost useless scratching, showing only that what we seek lies deeper under the surface. We also tried to work forwards first, and came to the idea of turning round afterwards. The dog who, after brief inspection of the situation, turned round and dashed off gives, rightly or wrongly, the impression of superior insight.

No, we should not even blame the hen for her clumsiness. There is a certain difficulty in turning round, in going away from the goal, in proceeding without looking continually at the aim, in not following the direct path to the desired end. There is an obvious analogy between her difficulties and our difficulties.

PART XVI

The Vocabulary of Mathematics

1. New Names for Old
   by EDWARD KASNER *and* JAMES R. NEWMAN

Double Infinite Rapport
and Other Mathematical Jargon

THIS essay, written a few years ago, offers a twenty-minute tour of modern mathematics. The intention was to convey a notion of recent developments by reviewing the words that have been attached to new concepts. It will be seen that most of them are terms in ordinary usage, but given a special meaning. This is a fairly common practice in mathematics. One might make a similar tour of medical therapy, say, by explaining the names of the latest drugs. The names, though by now familiar, are essentially difficult: streptomycin, penicillin, sulfanilamide, and so on. Adrenocorticotropic hormone is so exceedingly polysyllabic that everyone, in the trade and out, now calls it ACTH. We suggested that mathematics, on the other hand, could be defined as the science that uses easy words— "group," "ring," "limit," "simple curve," "inside," and "outside"—for hard ideas. There are exceptions to this definition, depending to some extent on the concept-maker's temperament. The great British mathematician J. J. Sylvester, a prolific inventor of both concepts and names, was flowery of speech and poetic. He coined strange names such as "Hessian," "Jacobian," "discriminant," "umbral notation" (denoting quantities which he said were "mere shadows")—so many, in fact, that he was called the mathematical Adam.

Looking through the mathematics books which have accumulated on my shelves in the last decade, I now find that the easy word definition has become less suitable. The vocabulary of mathematics is getting more abstruse, though it does not follow that the ideas are getting more profound. One of the reasons for the harder words is that mathematics has moved further into the great world of affairs and has been serving strange gods. Communication and information theory, operational research, the calculus of games, are among the subjects in which mathematics has assumed major responsibilities, requiring, in each case, an appropriate jargon. One encounters, to give a small sample from some thoroughly respectable books and pamphlets, the "maximization of 'moral expectation'" (having to do with improving businessmen's profits and not the level of ethics), "correlogram analysis," "isoquants," "asymptotic minimax solutions of sequential point estimation problems," "confidence regions for linear regression," "subminimax solutions of compound statistical decision problems," the "Monte Carlo method," the "homotopy groups of Hurewicz," "vacuous games," "half-spaces," "pay-off functions," "hemibel thinking," "eigen values," "hole functions," "double infinite rapport," "pheno-typical inversion." Most of these words and expressions were unknown to mathematics twenty-five years ago; some are newly coined, some newly changed. It is safe to say that the vocabulary gives a somewhat inflated impression of the recent progress in mathematical thought. Mathematics has advanced, to be sure, but neither as fast nor as far as the grandiose terminology would suggest.

*   *   *

Edward Kasner, who was my collaborator in the writing of *Mathematics and the Imagination*, from which the following selection and two others in this book were taken, was a leading mathematician of the twentieth century. His specialty was higher geometry, to which he contributed a large number of original papers; he also was known for his skill as a teacher.

Born in New York City in 1878, Kasner was educated at the College of the City of New York, Columbia and Göttingen universities. In 1900 he joined the teaching staff at Columbia, in 1910 he was appointed professor, and in 1937 he received the Adrain chair in mathematics. His exceptional capacity as a lecturer made his courses among the most popular at the university. To a fertile imagination and natural mathematical gifts he joined a sense of humor and a wide appreciation of cultural values. He could make you see intricate mathematical relationships by his verbal images, by the graceful gestures of his delicate hands, and by the spidery, badly executed, but remarkably illuminating diagrams he liked so much to scrawl on the blackboard. It was his way to tease his students, to lead them disingenuously down the garden path. Then, by the absurdity of their position, he would force them to find the way out. By such methods, tinctured with irony and gentle malice, he succeeded both in explaining his subject and in whetting the student's appetite for more. I had the good fortune to attend several of his courses as a graduate student, and, like many others, I owe to him a true awakening of interest in mathematics and an appreciation of its rare excellence.

Some of his pupils attained distinction in mathematics; all remembered their instruction under him as an intellectual delight. And in this large assembly are included not only college and graduate school audiences, and those who attended his courses at the New School for Social Research, but the children of kindergarten age to whom he would often lecture on the mathematics of infinity, topology and other recondite matters.

Kasner's numerous honors included membership in the National Academy of Sciences. He never married. He died in New York, January 7, 1955, at the age of seventy-six.

For out of olde feldes, as men seith,
Cometh al this newe corne fro yeere to yere;
And out of olde bokes, in good feith,
Cometh al this newe science that men lere.

—CHAUCER

1 New Names for Old

By EDWARD KASNER
and JAMES R. NEWMAN

EVERY once in a while there is house cleaning in mathematics. Some old names are discarded, some dusted off and refurbished; new theories, new additions to the household are assigned a place and name. So what our title really means is new *words* in mathematics; not new names, but new words, new terms which have in part come to represent new concepts and a reappraisal of old ones in more or less recent mathematics. There are surely plenty of words already in mathematics as well as in other subjects. Indeed, there are so many words that it is even easier than it used to be to speak a great deal and say nothing. It is mostly through words strung together like beads in a necklace that half the population of the world has been induced to believe mad things and to sanctify mad deeds. Frank Vizetelly, the great lexicographer, estimated that there are 800,000 words in use in the English language. But mathematicians, generally quite modest, are not satisfied with these 800,000; let us give them a few more.

We can get along without new names until, as we advance in science, we acquire new ideas and new forms. A peculiar thing about mathematics is that it does not use so many long and hard names as the other sciences. Besides, it is more conservative than the other sciences in that it clings tenaciously to old words. The terms used by Euclid in his *Elements* are current in geometry today. But an Ionian physicist would find the terminology of modern physics, to put it colloquially, pure Greek. In chemistry, substances no more complicated than sugar, starch, or alcohol have names like these: Methylpropenylenedihydroxycinnamenylacrylic acid, or, 0-anhydrosulfaminobenzoine, or, protocatechuicaldehydemethylene. It would be inconvenient if we had to use such terms in everyday conversation. Who could imagine even the aristocrat of science at the breakfast table asking, "Please pass the O-anhydrosulfaminobenzoic acid," when all he wanted was sugar for his coffee? Biology also has some tantalizing tongue twisters. The purpose of these long words is not to frighten the exoteric, but to describe with scientific curtness what the literary man would take half a page to express.

In mathematics there are many easy words like "group," "family," "ring," "simple curve," "limit," etc. But these ordinary words are sometimes given a very peculiar and technical meaning. In fact, here is a booby-prize definition of mathematics: *Mathematics is the science which uses easy words for hard ideas.* In this it differs from any other science. There are 500,000 known species of insects and every one has a long Latin name. In mathematics we are more modest. We talk about "fields," "groups," "families," "spaces," although much more meaning is attached to these words than ordinary conversation implies. As its use becomes more and more technical, nobody can guess the mathematical meaning of a word any more than one could guess that a "drug store" is a place where they sell ice-cream sodas and umbrellas. No one could guess the meaning of the word "group" as it is used in mathematics. Yet it is so important that whole courses are given on the theory of "groups," and hundreds of books are written about it.

Because mathematicians get along with common words, many amusing ambiguities arise. For instance, the word "function" probably expresses the most important idea in the whole history of mathematics. Yet, most people hearing it would think of a "function" as meaning an evening social affair, while others, less socially minded, would think of their livers. The word "function" has at least a dozen meanings, but few people suspect the mathematical one. The mathematical meaning is expressed most simply by a *table.* Such a table gives the relation between two variable quantities when the value of one variable quantity is determined by the value of the other. Thus, one variable quantity may express the years from 1800 to 1938, and the other, the number of men in the United States wearing handle-bar mustaches; or one variable may express in decibels the amount of noise made by a political speaker, and the other, the blood pressure units of his listeners. You could probably never guess the meaning of the word "ring" as it has been used in mathematics. It was introduced into the newer algebra within the last thirty years. The theory of rings is much more recent than the theory of groups. It is now found in most of the new books on algebra, and has nothing to do with either matrimony or bells.

Other ordinary words used in mathematics in a peculiar sense are "domain," "integration," "differentiation." The uninitiated would not be able to guess what they represent; only mathematicians would know about them. The word "transcendental" in mathematics has not the meaning it has in philosophy. A mathematician would say: The number π, equal to 3.14159 . . ., is transcendental, because it is not the root of any algebraic equation with integer coefficients.

Transcendental is a very exalted name for a small number, but it was coined when it was thought that transcendental numbers were as rare as quintuplets. The work of Georg Cantor in the realm of the infinite has since proved that of all the numbers in mathematics, the transcendental ones are the most common, or, to use the word in a slightly different sense, the least transcendental. Immanuel Kant's "transcendental episte-mology" is what most educated people might think of when the word transcendental is used, but in that sense it has nothing to do with mathe-matics. Again, take the word "evolution," used in mathematics to denote the process most of us learned in elementary school, and promptly forgot, of extracting square roots, cube roots, etc. Spencer, in his philosophy, de-fines evolution as "an integration of matter, and a dissipation of motion from an indefinite, incoherent homogeneity to a definite, coherent heter-ogeneity," etc. But that, fortunately, has nothing to do with mathematical evolution either. Even in Tennessee, one may extract square roots without running afoul of the law.

As we see, mathematics uses simple words for complicated ideas. An example of a simple word used in a complicated way is the word "simple." "Simple curve" and "simple group" represent important ideas in higher mathematics.

FIGURE 1

The above is not a simple curve. A simple curve is a closed curve which does not cross itself and may look like Figure 2. There are many impor-tant theorems about such figures that make the word worth while. They are found in a queer branch of mathematics called "topology." A French

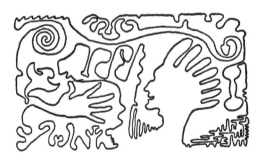

FIGURE 2

mathematician, Jordan, gave the fundamental theorem of this study: every simple curve has one inside and one outside. That is, every simple curve divides the plane into two regions, one inside the curve, and one out-side.

There are some groups in mathematics that are "simple" groups. The definition of "simple group" is really so hard that it cannot be given here. If we wanted to get a clear idea of what a simple group was, we should probably have to spend a long time looking into a great many books, and then, without an extensive mathematical background, we should probably miss the point. First of all, we should have to define the concept "group." Then we should have to give a definition of subgroups, and then of self-conjugate subgroups, and then we should be able to tell what a simple group is. A simple group is simply a group without any self-conjugate sub-groups—simple, is it not?

Mathematics is often erroneously referred to as the science of common sense. Actually, it may transcend common sense and go beyond either imagination or intuition. It has become a very strange and perhaps fright-ening subject from the ordinary point of view, but anyone who penetrates into it will find a veritable fairyland, a fairyland which is strange, but makes sense, if not common sense. From the ordinary point of view mathematics deals with strange things. We shall show you that occasion-ally it does deal with strange things, but mostly it deals with familiar things in a strange way. If you look at yourself in an ordinary mirror, regardless of your physical attributes, you may find yourself amusing, but not strange; a subway ride to Coney Island, and a glance at yourself in one of the distorting mirrors will convince you that from another point of view you may be strange as well as amusing. It is largely a matter of what you are accustomed to. A Russian peasant came to Moscow for the first time and went to see the sights. He went to the zoo and saw the giraffes. You may find a moral in his reaction as plainly as in the fables of La Fontaine. "Look," he said, "at what the Bolsheviks have done to our horses." That is what modern mathematics has done to simple geom-etry and to simple arithmetic.

There are other words and expressions, not so familiar, which have been invented even more recently. Take, for instance, the word "turbine." Of course, that is already used in engineering, but it is an entirely new word in geometry. The mathematical name applies to a certain diagram. (Geometry, whatever others may think, is the study of different shapes, many of them very beautiful, having harmony, grace and symmetry. Of course, there are also fat books written on abstract geometry, and abstract space in which neither a diagram nor a shape appears. This is a very im-portant branch of mathematics, but it is not the geometry studied by the Egyptians and the Greeks. Most of us, if we can play chess at all, are content to play it on a board with wooden chess pieces; but there are some who play the game blindfolded and without touching the board. It

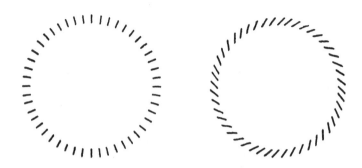

FIGURE 3—Turbines.

might be a fair analogy to say that abstract geometry is like blindfold chess—it is a game played without concrete objects.) Above you see a picture of a turbine, in fact, two of them.

A turbine consists of an infinite number of "elements" filled in continuously. An element is not merely a point; it is a point with an associated direction—like an iron filing. A turbine is composed of an infinite number of these elements, arranged in a peculiar way: the points must be arranged on a perfect circle, and the inclination of the iron filings must be at the same angle to the circle throughout. There are thus an infinite number of elements of equal inclination to the various tangents of the circle. In the special case where the angle between the direction of the element and the direction of the tangent is zero, what would happen? The turbine would be a circle. In other words, the theory of turbines is a generalization of the theory of the circle. If the angle is ninety degrees, the elements point toward the center of the circle. In that special case we have a normal turbine (see left-hand diagram).

There is a geometry of turbines, instead of a geometry of circles. It is a rather technical branch of mathematics which concerns itself with working out continuous groups of transformations connected with differential equations and differential geometry. The geometry connected with the turbine bears the rather odd name of "turns and slides."

*   *   *   *   *

The circle is one of the oldest figures in mathematics. The straight line is the simplest line, but the circle is the simplest nonstraight curve. It is often regarded as the limit of a polygon with a infinite number of sides. You can see for yourself that as a series of polygons is inscribed in a circle with each polygon having more sides than its predecessor, each polygon gets to look more and more like a circle.

The Greeks were already familiar with the idea that as a regular polygon increases in the number of its sides, it differs less and less from the circle in which it is inscribed. Indeed, it may well be that in the eyes of an omniscient creature, the circle would look like a polygon with an infinite number of straight sides. However, in the absence of complete omniscience, we shall continue to regard a circle as being a nonstraight

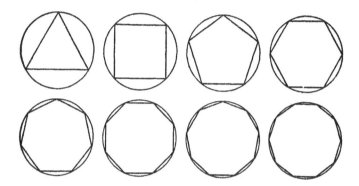

FIGURE 4—The circle as the limit of inscribed polygons.

curve. There are some interesting generalizations of the circle when it is viewed in this way. There is, for example, the concept denoted by the word "cycle," which was introduced by a French mathematician, Laguerre. A cycle is a circle with an arrow on it, like this:

FIGURE 5

If you took the same circle and put an arrow on it in the opposite direction, it would become a different cycle.

The Greeks were specialists in the art of posing problems which neither they nor succeeding generations of mathematicians have ever been able to solve. The three most famous of these problems are the squaring of the circle, the duplication of the cube, and the trisection of an angle. Many well-meaning, self-appointed, and self-anointed mathematicians, and a motley assortment of lunatics and cranks, knowing neither history nor mathematics, supply an abundant crop of "solutions" of these insoluble problems each year. However, some of the classical problems of antiquity have been solved. For example, the theory of cycles was used by Laguerre in solving the problem of Apollonius: given three fixed circles, to find a circle that touches them all. It turns out to be a matter of elementary high school geometry, although it involves ingenuity, and any brilliant

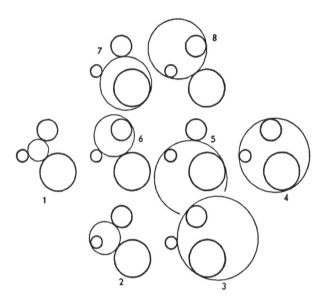

FIGURE 6(a)—The eight solutions of the problem of Apollonius. Each lightly drawn circle is in
contact with 3 heavily drawn ones.

high school student could work it out. It has eight answers, as shown in
Figure 6(a).

They can all be constructed with ruler and compass, and many methods
of solution have been found. Given three *circles*, there will be eight circles
touching all of them. Given three *cycles*, however, there will be only one
clockwise cycle that touches them all. (Two cycles are said to touch each
other only if their arrows agree in direction at the point of contact.) Thus,
by using the idea of cycles, we have one definite answer instead of eight.
Laguerre made the idea of cycles the basis of an elegant theory.

Another variation of the circle introduced by the eminent American
mathematician, C. J. Keyser, is obtained by taking a circle and removing
one point.[1] This creates a serious change in conception. Keyser calls it "a
patho-circle" (from pathological circle). He has used it in discussing the
logic of axioms.

We have made yet another change in the concept of circle, which introduces another word and a new diagram. Take a circle and instead of
leaving one point out, simply emphasize one point as the initial point.

[1] N.B. This is a diagram which the reader will have to imagine, for it is beyond the
capacity of any printer to make a circle with one point omitted. A point, having no
dimensions, will, like many of the persons on the Lord High Executioner's list, never
be missed. So the circle with one point missing is purely conceptual, not an idea
which can be pictured.

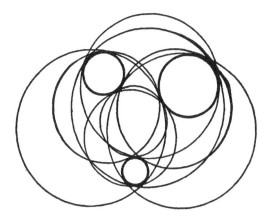

FIGURE 6(b)—The eight solutions of Apollonius merged into one diagram.

This is to be called a "clock." It has been used in the theory of polygenic functions. "Polygenic" is a word recently introduced into the theory of complex functions—about 1927. There was an important word, "monogenic," introduced in the nineteenth century by the famous French mathematician, Augustin Cauchy, and used in the classical theory of functions. It is used to denote functions that have a single derivative at a point, as in the differential calculus. But most functions, in the complex domain, have an infinite number of derivatives at a point. If a function is not monogenic, it can never be bigenic, or trigenic. Either the derivative has one value or an infinite number of values—either monogenic or polygenic, nothing intermediate. Monogenic means one rate of growth. Polygenic means many rates of growth. The complete derivative of a polygenic function is represented by a congruence (a double infinity) of clocks, all with different starting points, but with the same uniform rate of rotation.

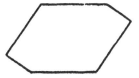

FIGURE 7—The parhexagon.

It would be useless to attempt to give a simplified explanation of these concepts. (The neophyte will have to bear with us over a few intervals like this for the sake of the more experienced mathematical reader.)

The going has been rather hard in the last paragraph, and if a few of the polygenic seas have swept you overboard, we shall throw you a hexagonal life preserver. We may consider a very simple word that has been introduced in elementary geometry to indicate a certain kind of hexagon. The word on which to fix your attention is "parhexagon." An ordinary hexagon has six arbitrary sides. A parhexagon is that kind of hexagon in which any side is both equal and parallel to the side opposite to it (as in Figure 7).

If the opposite sides of a quadrilateral are equal and parallel, it is called a parallelogram. By the same reasoning that we use for the word parhexagon, a parallelogram might have been called a parquadrilateral.

Here is an example of a theorem about the parhexagon: take any irregular hexagon, not necessarily a parhexagon, ABCDEF. Draw the diagonals AC, BD, CE, DF, EA, and FB, forming the six triangles, ABC, BCD, CDE, DEF, EFA, and FAB. Find the six centers of gravity, A', B', C', D', E', and F' of these triangles. (The center of gravity of a triangle is the point at which the triangle would balance if it were cut out of cardboard and supported only at that point; it coincides with the point of in-

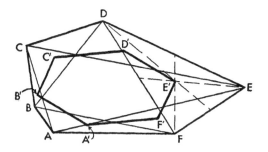

FIGURE 8—ABCDEF is an irregular hexagon. A'B'C'D'E'F' is a parhexagon.

tersection of the medians.) Draw A'B', B'C', C'D', D'E', E'F', and F'A'. Then the new inner hexagon A'B'C'D'E'F' will always be a parhexagon.

The word radical, favorite call to arms among Republicans, Democrats, Communists, Socialists, Nazis, Fascists, Trotskyites, etc., has a less hortatory and bellicose character in mathematics. For one thing, everybody knows its meaning: i.e., square root, cube root, fourth root, fifth root, etc. Combining a word previously defined with this one, we might say that the extraction of a root is the evolution of a radical. The square root of 9 is 3; the square root of 10 is greater than 3, and the most famous and the simplest of all square roots, the first incommensurable number discovered by the Greeks, the square root of 2, is 1.414. . . . There are also composite radicals—expressions like $\sqrt{7} + \sqrt[5]{10}$. The symbol for a radical is not the hammer and sickle, but a sign three or four centuries old, and the idea of the mathematical radical is even older than that. The concept of the "hyperradical," or "ultraradical," which means something higher than a radical, but lower than a transcendental, is of recent origin. It has a symbol which we shall see in a moment. First, we must say a few words about radicals in general. There are certain numbers and functions in mathematics which are not expressible in the language of radicals and which are generally not well understood. Many ideas for which there are no concrete or diagrammatic representations are difficult to explain. Most people find it impossible to think without words; it is necessary to give them a word and a symbol to pin their attention. Hyperradical or ultra-radical, for which hitherto there have been neither words, nor symbols, fall into this category.

We first meet these ultraradicals in trying to solve equations of the fifth degree. The Egyptians solved equations of the first degree perhaps 4000 years ago. That is, they found that the solution of the equation $ax + b = 0$, which is represented in geometry by a straight line, is $x = \dfrac{-b}{a}$. The quadratic equation $ax^2 + bx + c = 0$ was solved by the Hindus and the Arabs with the formula $x = \dfrac{-b \pm \sqrt{b^2 - 4ac}}{2a}$. The various conic sections, the circle, the ellipse, the parabola, and the hyperbola, are the geometric pictures of quadratic equations in two variables.

Then in the sixteenth century the Italians solved the equations of third and fourth degree, obtaining long formulas involving cube roots and square roots. So that by the year 1550, a few years before Shakespeare was born, the equation of the first, second, third, and fourth degrees had been solved. Then there was a delay of 250 years, because mathematicians were struggling with the equation of the fifth degree—the general quintic. Finally, at the beginning of the nineteenth century, Ruffini and Abel showed that equations of the fifth degree could not be solved with radicals. The general quintic is thus not like the general quadratic, cubic or bi-quadratic. Nevertheless, it presents a problem in algebra which theoretically can be solved by algebraic operations. Only, these operations are so hard that they cannot be expressed by the symbols for radicals. These new

FIGURE 9—A portrait of two ultraradicals.

higher things are named "ultraradicals," and they too have their special symbols (shown in Figure 9).

With such symbols combined with radicals, we can solve equations of the fifth degree. For example, the solution of $x^5 + x = a$ may be written $x = \sqrt[\wedge]{a}$ or $x = \sqrt[\smile]{a}$. The usefulness of the special symbol and name is apparent. Without them the solution of the quintic equation could not be compactly expressed.

*   *   *   *   *

We may now give a few ideas somewhat easier than those with which we have thus far occupied ourselves. These ideas were presented some time ago to a number of children in kindergarten. It was amazing how well they understood everything that was said to them. Indeed, it is a fair inference that kindergarten children can enjoy lectures on graduate mathematics as long as the mathematical concepts are clearly presented.

It was raining and the children were asked how many raindrops would fall on New York. The highest answer was 100. They had never counted higher than 100 and what they meant to imply when they used that number was merely something very, very big—as big as they could imagine. They were asked how many raindrops hit the roof, and how many hit New York, and how many single raindrops hit all of New York in 24 hours. They soon got a notion of the bigness of these numbers even though they did not know the symbols for them. They were certain in a little while that the number of raindrops was a great deal bigger than a hundred. They were asked to think of the number of grains of sand on the beach at Coney Island and decided that the number of grains of sand and the number of raindrops were about the same. But the important thing is that they realized that the number was *finite, not infinite*. In this respect they showed their distinct superiority over many scientists who to this day use the word infinite when they mean some big number, like a billion billion.

Counting, something such scientists evidently do not realize, is a precise operation.[2] It may be wonderful but there is nothing vague or mysterious about it. If you count something, the answer you get is either perfect or all wrong; there is no half way. It is very much like catching a train. You either catch it or you miss it, and if you miss it by a split second you might as well have come a week late. There is a famous quotation which illustrates this:

> Oh, the little more, and how much it is!
> And the little less, and what worlds away!

A big number is big, but it is definite and it is finite. Of course in poetry, the finite ends with about three thousand; any greater number is infinite. In many poems, the poet will talk to you about the infinite number of stars. But, if ever there was a hyperbole, this is it, for nobody, not even the poet, has ever seen more than three thousands stars on a clear night, without the aid of a telescope.

---

[2] No one would say that $1 + 1$ is "about equal to 2." It is just as silly to say that a billion billion is not a finite number, simply because it is big. Any number which can be named, or conceived of in terms of the integers is finite. *Infinite means something quite different.*

With the Hottentots, infinity begins at three.[3] Ask a Hottentot how many cows he owns, and if he has more than three he'll say "many." The number of raindrops falling on New York is also "many." It is a large finite number, but nowhere near infinity.

Now here is the name of a very large number: "Googol." [4] Most people would say, "A googol is so large that you cannot name it or talk about it; it is so large that it is infinite." Therefore, we shall talk about it, explain exactly what it is, and show that it belongs to the very same family as the number 1.

A googol is this number which one of the children in the kindergarten wrote on the blackboard:

10000000000000000000000000000000000000000000000000000000
0000000000000000000000000000000000000000000000000000000.

The definition of a googol is: 1 followed by a hundred zeros. It was decided, after careful mathematical researches in the kindergarten, that the number of raindrops falling on New York in 24 hours, or even in a year or in a century, is much less than a googol. Indeed, the googol is a number just larger than the largest numbers that are used in physics or astronomy. All those numbers require less than a hundred zeros. This information is, of course, available to everyone, but seems to be a great secret in many scientific quarters.

A very distinguished scientific publication recently came forth with the revelation that the number of snow crystals necessary to form the ice age was a billion to the billionth power. This is very startling and also very silly. A billion to the billionth power looks like this:

$$1000000000^{1000000000}$$

A more reasonable estimate and a somewhat smaller number would be 10^{30}. As a matter of fact, it has been estimated that if the entire universe, which you will concede is a trifle larger than the earth, were filled with protons and electrons, so that no vacant space remained, the total number of protons and electrons would be 10^{110} (i.e., 1 with 110 zeros after it). Unfortunately, as soon as people talk about large numbers, they run amuck. They seem to be under the impression that since zero equals nothing, they can add as many zeros to a number as they please with practically no serious consequences. We shall have to be a little more careful than that in talking about big numbers.

To return to Coney Island, the number of grains of sand on the beach is about 10^{20}, or more descriptively, 100000000000000000000. That is a large number, but not as large as the number mentioned by the divorcee

---

[3] Although, in all fairness, it must be pointed out that some of the tribes of the Belgian Congo can count to a million and beyond.
[4] Not even approximately a Russian author.

in a recent divorce suit who had telephoned that she loved the man "a million billion billion times and eight times around the world." It was the largest number that she could conceive of, and shows the kind of thing that may be hatched in a love nest.

Though people do a great deal of talking, the total output since the beginning of gabble to the present day, including all baby talk, love songs, and Congressional debates, totals about 10^{16} words. This is ten million billion. Contrary to popular belief, this is a larger number of words than is spoken at the average afternoon bridge.

A great deal of the veneration for the authority of the printed word would vanish if one were to calculate the number of words which have been printed since the Gutenberg Bible appeared. It is a number somewhat larger than 10^{16}. A recent popular historical novel alone accounts for the printing of several hundred billion words.

The largest number seen in finance (though new records are in the making) represents the amount of money in circulation in Germany at the peak of the inflation. It was less than a googol—merely

$$496,585,346,000,000,000,000.$$

A distinguished economist vouches for the accuracy of this figure. The number of marks in circulation was very nearly equal to the number of grains of sand on Coney Island beach.

The number of atoms of oxygen in the average thimble is a good deal larger. It would be represented by perhaps 10000000000000000000000000-000. The number of electrons, in size exceedingly smaller than the atoms, is much more enormous. The number of electrons which pass through the filament of an ordinary fifty-watt electric lamp in a minute equals the number of drops of water that flow over Niagara Falls in a century.

One may also calculate the number of electrons, not only in the average room, but over the whole earth, and out through the stars, the Milky Way, and all the nebulae. The reason for giving all these examples of very large numbers is to emphasize the fact that no matter how large the collection to be counted, a finite number will do the trick. Mathematics, to be sure, exhibits a great variety of infinite collections, but those encountered in nature, though sometimes very large, are all definitely finite. A celebrated scientist a few years ago stated in all seriousness that he believed that the number of pores (through which leaves breathe) of all the leaves, of all the trees in all the world, would certainly be infinite. Needless to say, he was not a mathematician. The number of electrons in a single leaf is much bigger than the number of pores of all the leaves of all the trees of all the world. And still the number of all the electrons in the entire universe can be found by means of the physics of Einstein. It is a good deal less than a googol—perhaps one with seventy-nine zeros, 10^{79}, as estimated by Eddington.

Words of wisdom are spoken by children at least as often as by scientists. The name "googol" was invented by a child (Dr. Kasner's nine-year-old nephew) who was asked to think up a name for a very big number, namely, 1 with a hundred zeros after it. He was very certain that this number was not infinite, and therefore equally certain that it had to have a name. At the same time that he suggested "googol" he gave a name for a still larger number: "Googolplex." A googolplex is much larger than a googol, but is still finite, as the inventor of the name was quick to point out. It was first suggested that a googolplex should be 1, followed by writing zeros until you got tired. This is a description of what would happen if one actually tried to write a googolplex, but different people get tired at different times and it would never do to have Carnera a better mathematician than Dr. Einstein, simply because he had more endurance. The googolplex then, is a specific finite number, with so many zeros after the 1 that the number of zeros is a googol. A googolplex is much bigger than a googol, much bigger even than a googol times a googol. A googol times a googol would be 1 with 200 zeros, whereas a googolplex is 1 with a googol of zeros. You will get some idea of the size of this very large but finite number from the fact that there would not be enough room to write it, if you went to the farthest star, touring all the nebulae and putting down zeros every inch of the way.

One might not believe that such a large number would ever really have any application; but one who felt that way would not be a mathematician. A number as large as the googolplex might be of real use in problems of combination. This would be the type of problem in which it might come up scientifically:

Consider this book which is made up of carbon and nitrogen and of other elements. The answer to the question, "How many atoms are there in this book?" would certainly be a finite number, even less than a googol. Now imagine that the book is held suspended by a string, the end of which you are holding. How long will it be necessary to wait before the book will jump up into your hand? Could it conceivably ever happen? One answer might be "No, it will never happen without some external force causing it to do so." But that is not correct. The right answer is that it will almost *certainly* happen *sometime* in less than a googolplex of years—perhaps tomorrow.

The explanation of this answer can be found in physical chemistry, statistical mechanics, the kinetic theory of gases, and the theory of probability. We cannot dispose of all these subjects in a few lines, but we will try. Molecules are always moving. Absolute rest of molecules would mean absolute zero degrees of temperature, and absolute zero degrees of tem-

perature is not only nonexistent, but impossible to obtain. All the molecules of the surrounding air bombard the book. At present the bombard-
ment from above and below is nearly the same and gravity keeps the book
down. It is necessary to wait for the favorable moment when there
happens to be an enormous number of molecules bombarding the book
from below and very few from above. Then gravity will be overcome and
the book will rise. It would be somewhat like the effect known in physics
as the Brownian movement, which describes the behavior of small particles
in a liquid as they dance about under the impact of molecules. It would
be analogous to the Brownian movement on a vast scale.

But the probability that this will happen in the near future or, for
that matter, on any specific occasion that we might mention, is between

$\dfrac{1}{\text{googol}}$ and $\dfrac{1}{\text{googolplex}}$. To be reasonably sure that the book will rise, we

should have to wait between a googol and a googolplex of years.

When working with electrons or with problems of combination like the
one of the book, we need larger numbers than are usually talked about. It
is for that reason that names like googol and googolplex, though they may
appear to be mere jokes, have a real value. The names help to fix in our
minds the fact that we are still dealing with finite numbers. To repeat, a
googol is 10^{100}; a googolplex is 10 to the googol power; which may be
written $10^{10^{100}} = 10^{\text{googol}}$.

We have seen that the number of years that one would have to wait to
see the miracle of the rising book would be less than a googolplex. In that
number of years the earth may well have become a frozen planet as dead
as the moon, or perhaps splintered to a number of meteors and comets.
The real miracle is not that the book will rise, but that with the aid of
mathematics, we can project ourselves into the future and predict with
accuracy *when* it will probably rise, i.e., some time between today and
the year googolplex.

PART XVII

Mathematics as an Art

1. Mathematics as an Art *by* JOHN WILLIAM NAVIN SULLIVAN

COMMENTARY ON

JOHN WILLIAM NAVIN SULLIVAN

JOHN WILLIAM NAVIN SULLIVAN was a sensitive, uncommonly perceptive interpreter of scientific thought, an activity to which he brought occasionally a mystic, invariably an aesthetic emphasis. It cannot be maintained, of course, that this point of view—Sullivan himself described it as "humanistic"—automatically assures clarity and exactitude of expression. The universe may be, as J. B. S. Haldane has said, "not only queerer than we suppose, but queerer than we can suppose," but it does not follow that scientific explanation therefore gains from being queer. The great popularizers, Jeans and Eddington, seemed in later years to be drawn to this conclusion, but one may argue that this marked a decline in the quality of their writings rather than an ascent to a higher plane of insight. Sullivan's outlook differs from theirs in one major respect. Like them, he was a "devotee and apostle of clarity" [1] in explaining what can be explained; unlike them, he did not feel compelled to refer the inexplicable to God.

Sullivan had a capable grasp of mathematics and physics; he was a conscientious and unpretentious expositor who neither shirked difficulties nor sought refuge from them in misty sentimentalities. His special power was to make others feel the beauty of scientific ideas: not alone the strangeness of the world but the strange and wonderful gifts of imagination whereby men make the world intelligible to themselves. Sullivan regarded science as an art, like music or painting. There are, he admitted, scientific works—star catalogues, for example—which are not art; but the theoretical structures of Gauss, Einstein or Maxwell are original, individual, "very personal" responses and expressions of exactly the same kind as the creative works of Beethoven or Dostoievski. A scientific theory is, in Sullivan's words, the manifestation of a certain kind of "awareness." It is an awareness possessing elements of freedom and constraint, of skepticism and faith; an awareness which is not merely passive but one which actually makes the world as we understand it. To be thus aware is to have a sense of one's inadequacy "in face of the fundamental problems of existence," yet to refuse to be crushed by them.

It is this awareness and this originality of response which Sullivan re-

[1] The quotation is from a memoir on Sullivan by the English historian of science, Charles Singer, introducing Sullivan's posthumous biography of Isaac Newton (*Isaac Newton, 1642–1729*, New York, 1938).

gards as "intimately connected with what is called mysticism." Mysticism, in Sullivan's sense, is the capacity to wonder and to dare. The creators in mathematics or music must, in addition to possessing certain positive qualities, suffer from an important disability: that of being unable simply to accept the world, as practical men can do. To minds of original temper the mystical experience must at least be possible, "although it may not, as a matter of fact, ever have occurred." The fact that it is possible means that "certain fatal inhibitions are absent," inhibitions which stifle curiosity and the ability to form independent ideas.[2] These are, I think, more sensible and compelling opinions than are usually heard when mysticism is discussed.

Sullivan was born in London on January 22, 1886, the son of an official of a Protestant Mission and of a mother who was a teacher and possessed, it is said, considerable musical talent.[3] The family was genteel but not prosperous and could spend little to educate their only son. After a few years in London schools, he got a job at the age of fourteen with a submarine cable company. The officials of the company had learned of the boy's unusual mathematical powers and permitted him to continue his education part-time at the North London Polytechnic School. It was there and at University College, London, where he attended a few courses, that Sullivan received his scientific training.

In 1910 he took a position with an electrical company in the United States, but he was temperamentally unfitted for the job and soon abandoned it. The one good result of his visit to this country was meeting the brilliantly imaginative mathematician and logician Alfred J. Lotka who strongly influenced the direction of his future work.[4] After returning to England in 1913, Sullivan held a few temporary posts, undertook small journalistic assignments, and then, on the outbreak of war, became an ambulance driver on the Serbian front. He was soon invalided home with "an exhausting" malady, and upon recovery went to work in the British Censorship Department. His bureau chief, the writer and editor J. Middleton Murry, encouraged Sullivan toward a literary career. By 1917 he was contributing to the *Athenaeum, Nature* and the *Times Literary Supplement*, and his reputation as a science writer was established.

The same year he published a wretched novel, *An Attempt at Life*; he was no better equipped for writing fiction than for the electrical business. Charles Singer portrays Sullivan as being without any "deep understanding

---

[2] J. W. N. Sullivan, *Contemporary Mind* (Essay on "Science and Art"), London, 1934.

[3] I have seen the father described as "a poor Irish sailor." The statement is untrue, the source of the confusion being Sullivan's own words in what was wrongly supposed to be his autobiography. The book, *But for the Grace of God* (London, 1932), tells a great deal about the author but is wholly unreliable in its biographical details.

[4] This and other biographical facts are from the memoir by Singer referred to in Note 1 above.

of his fellows," innocent of politics, devoid of practical interests, without comprehension of "social or spiritual motives." The judgment is severe but, so far as his writings reveal, perhaps essentially correct. Sullivan was a philosopher, a musician, a master of scientific thought and, above all, a dreamer. His boyhood, as he states, was spent in reveries; his later years were a mixture of withdrawnness and dazzling conversation (with a few chosen spirits), of mathematical physics, music, hard work, and deep— often dreamy—speculation.

Sullivan's literary output was quite large. It included *Aspects of Science* (1922); *Atoms and Electrons* (1924); *Three Men Discuss Relativity* (1926); *History of Mathematics in Europe from the Fall of Greek Science to the Rise of the Conception of Mathematical Rigour* (1925)—a delightful book, and the most systematic and documented of his writings; *Gallio, or the Tyranny of Science* (1926); *Beethoven* (1927), an admirable biography, many times reprinted; and a posthumous life of Isaac Newton (1938) which was more than ten years in preparation.

One of Sullivan's books, *How Things Behave—A Child's Introduction to Physics* (1932), was dedicated to his daughter, a child of his first marriage which was dissolved in 1922. He remarried in 1928, and had a son by his second wife. A notable event in his life was his meeting with Einstein in Berlin in 1924; their conversations "inspired and prompted" ideas reflected in much of his later writing. During the same year he passed a long holiday with Aldous Huxley in Florence, from which association he also derived stimulus and thoughts developed in future work.[5] But though these persons and others inspired and encouraged him, Sullivan was an independent spirit. It was characteristic of him to refuse to embrace either extreme position in the Free Will vs. Determinism controversy that sprang up in physics in the 1920s. The famous men lined up in both camps impressed but did not overawe him. He made his way within himself; he was his own man.

Sullivan died August 11, 1937, after years of suffering from locomotor ataxia. The last two years he was bedridden and had lost even the physical capacity for writing. Nevertheless, with the help of a devoted wife who acted also as his secretary, he finished the biography of Newton by dictation.

The selection which follows, "Mathematics as an Art," is typical of Sullivan's literary skill and of his approach to science. It is taken from his second *Aspects of Science*, published in 1925.

[5] Singer, *op. cit.*

> *Nature gets credit which should in truth be reserved for ourselves: the rose*
> *for its scent, the nightingale for its song; and the sun for its radiance. The*
> *poets are entirely mistaken. They should address their lyrics to themselves*
> *and should turn them into odes of self congratulation on the excellence of*
> *the human mind.* —ALFRED NORTH WHITEHEAD

1 Mathematics as an Art

By JOHN WILLIAM NAVIN SULLIVAN

THE prestige enjoyed by mathematicians in every civilized country is not altogether easy to understand. Anything which is valued by the generality of men is either useful or pleasant, or both. Farming is a valued occupation, and so is piano-playing, but why are the activities of the mathematician considered to be important? It might be said that mathematics is valued for its applications. Everybody knows that modern civilization depends, to an unprecedented extent, upon science, and a great deal of that science would be impossible in the absence of a highly developed mathematical technique. This is doubtless a weighty consideration; and it is true that even mathematics has benefited by the increased esteem in which science is held as a consequence of the magnificent murderous capacities it exhibited in the late war. But it is doubtful whether this consideration alone is adequate to explain the exalted position accorded to mathematics throughout a larger part of its history. On the other hand, it does not seem as if we could attach much importance to the claim made by many mathematicians that their science is a delightful art. Their claim is doubtless justified; but the fact that a few, a very few, unusual individuals obtain great pleasure from some incomprehensible pursuit is no reason why the ordinary man should admire them and support them. Chess professorships are not established, but there are probably more people who appreciate the "beauties" of chess than appreciate the beauties of mathematics. The present position accorded to mathematics by the non-mathematical public is due partly to the usefulness of mathematics and partly to the persistence, in a more or less vague form, of old and erroneous ideas respecting its real significance. It is only within quite recent times, indeed, that the correct status of mathematics has been discovered, although there are many and very important aspects of this wonderful activity which still remain mysterious.

It is probable that mathematics originated with Pythagoras. There is no clear evidence that that distinctive activity we call mathematical reasoning was fully recognised and practised by any one before Pythagoras. Certain arithmetical results had long been known, of course, but neither geometry nor algebra had been created. The geometrical formulas used by the ancient Egyptians for example, deal chiefly with land-surveying problems, and were evidently obtained empirically. They are usually wrong and are nowhere accompanied by proofs. It seems strange that this particular possibility of the mind should have been discovered so late, for it is completely independent of external circumstances. Even music, the most independent of what are usually classed as the arts, is more dependent on its *milieu* than is mathematics. Nevertheless, both music and mathematics, the two most "subjective" of human creations, have been singularly late and slow in their development. And just as it is impossible for us to understand what their rudimentary music meant to the Greeks, so it is impossible to enter into the difficulties of the pre-mathematical mind. The musical enthusiasms of Plato are just as remote from us as are the difficulties of that Chinese Emperor who could not be convinced by the abstract proof that the volume of a sphere varies as the cube of its radius. He had various sized spheres made, filled with water, and weighed. This was his conception of a proof. And this must have been typical of the ancient mind. They lacked a faculty, just as the Greeks lacked a harmonic sense.

It is not surprising, therefore, that when the mind first became aware of this unsuspected power it did not undersand its true nature. It appeared vastly more significant—or at least significant in a different way from what it really is. To the Pythagoreans, overwhelmed by the æsthetic charm of the theorems they discovered, number became the principle of all things. Number was supposed to be the very essence of the real; other things that could be predicted of the real were merely aspects of number. Thus the number one is what, in a certain aspect, we call reason, for reason is unchangeable, and the very essence of unchangeableness is expressed by the number one: the number two, on the other hand, is unlimited and indeterminate; "opinion" as contrasted with "reason," is an expression of the number two: again, the proper essence of marriage is expressed by the number five, since five is reached by combining three and two, that is, the first masculine with the first feminine number: the number four is the essence of justice, for four is the product of equals. To understand this outlook it is only necessary to enter into that condition of mind which takes any analogy to represent a real bond. Thus odd and even, male and female, light and darkness, straight and curved, all become expressions of some profound principle of opposition which informs the world. There are many mystical and semi-mystical writers of the present day who find themselves able to think in this manner; and it must be admitted that there is a not uncommon type of mind, otherwise orthodox, which is able to adopt this kind of reasoning without discomfort. Even Goethe, in his *Farbenlehre*, finds that a triangle has a mystic significance.

As long as the true logical status of mathematical propositions remained unknown it was possible for many mathematicians to surmise that they must have some profound relation to the structure of the universe. Mathematical propositions were supposed to be true quite independently of our minds, and from this fact the existence of God was deduced. This doctrine was, indeed, a refinement on the Pythagorean fantasies, and was held by many who did not believe in the mystic properties of numbers. But the mystical outlook on numbers continued to flourish for many centuries. Thus St. Augustine, speaking of the perfection of the number six, says:—

> Six is a number perfect in itself, and not because God created all things in six days; rather the inverse is true, that God created all things in six days because this number is perfect, and it would remain perfect, even if the work of the six days did not exist.

From speculations of this sort the Pythagorean doctrine developed, on the one hand, in a thoroughly respectable philosophic manner into the doctrine of necessary truths, and on the other descended to cabbalistic imbecilities. Even very good mathematicians became cabbalists. The famous Michael Stifel, one of the most celebrated algebraists of the sixteenth century, considered that by far the most important part of his work was his cabbalistic interpretation of the prophetic books of the Bible. That this method enjoyed a high prestige is sufficiently shown by the general belief accorded to his prophecy that the world would come to an end on October 3, 1533—with the result that a large number of people abandoned their occupations and wasted their substance, to find, when the date came and passed, that they were ruined. Such geometric figures as star-polygons, also, were supposed to be of profound significance; and even Kepler, after demonstrating their mathematical properties with perfect rigour, goes on to explain their use as amulets or conjurations. As another instance of the persistence of this way of regarding mathematical entities it may be mentioned that the early development of infinite series was positively hampered by the exaggerated significance attached to mathematical operations. Thus in the time of Leibnitz it was believed that the sum of an infinite number of zeros was equal to ½; and it was attempted to make this obvious idiocy plausible by saying that it was the mathematical analogue of the creation of the world out of nothing.

There is sufficient evidence, then, that there has existed a widespread tendency to attribute a mystic significance to mathematical entities. And there are many indications, even at the present day, that this tendency persists. It is probable, then, that the prestige enjoyed by the mathematician is not altogether unconnected with the prestige enjoyed by any master of the occult. The position accorded to the mathematician has been, to some extent, due to the superstitions of mankind, although doubtless it can be justified on rational grounds. For a long time, particularly in India and Arabia, men became mathematicians to become astronomers, and they became astronomers to become astrologers. The aim of their activities was superstition, not science. And even in Europe, and for some years after the beginning of the Renaissance, astrology and kindred subjects were important justifications of mathematical researches. We no longer believe in astrology or mystic hexagons and the like; but nobody who is acquainted with some of the imaginative but non-scientific people can help suspecting that Pythagoreanism is not yet dead.

When we come to consider the other justification of mathematics derived from the Pythagorean outlook—its justification on the ground that it provided the clearest and most indubitable examples of necessary truths —we find this outlook, so far from being extinct, still taught by eminent professors of logic. Yet the non-Euclidean geometries, now a century old, have made it quite untenable. The point of view is well expressed by Descartes in a famous passage from his Fifth Meditation:—

J'imagine un triangle, encore qu'il n'y ait peut-être en aucun lieu du monde hors de ma pensée une telle figure et qu'il n'y en ait jamais eu, il ne laisse pas néanmoins d'y avoir une certaine nature ou forme, ou essence déterminée de cette figure, laquelle est immuable et éternelle, que je n'ai point inventée et qui ne dépend en aucune façon de mon esprit; comme il paraît, de ce que l'on peut démontrer diverses propriétés de ce triangle, a savoir que ses trois angles sont égaux à deux droits, que le plus grand angle est soutenu par le plus grand côté, et autres semblables, lesquelles maintenant, soit que je le veuille ou non, je reconnais très clairement et très évidemment être en lui, encore que je n'y aie pensé auparavant en aucune façon, lorsque je me suis imaginé la première fois un triangle, et, pourtant, on ne peut pas dire que je les ai feintes ni inventées.

A triangle, therefore, according to Descartes, does not depend in any way upon one's mind. It has an eternal and immutable existence quite independent of our knowledge of it. Its properties are discovered by our minds, but do not in any way depend upon them. This way of regarding geometrical entities lasted for two thousand years. To the Platonists geometrical propositions, expressing eternal truths, are concerned with the world of Ideas, a world apart, separate from the sensible world. To the followers of St. Augustine these Platonic Ideas became the ideas of God; and to the followers of St. Thomas Aquinas they became aspects of the Divine Word. Throughout the whole of scholastic philosophy the necessary truth of geometrical propositions played a very important part; and, as we have said, there are certain philosophers of the present day who regard the axioms of Euclid's geometry as unescapable truths. If this outlook be justified, then the mathematical faculty gives us access, as it were, to an eternally existing, although not sensible, world. Before the discovery of mathematics this world was unknown to us, but it nevertheless existed, and Pythagoras no more invented mathematics than Columbus invented America. Is this a true description of the nature of mathematics? Is mathematics really a body of knowledge about an existing, but supersensible, world? Some of us will be reminded of the claims certain theorists have made for music. Some musicians have been so impressed by the extraordinary impression of *inevitability* given by certain musical works that they have declared that there must be a kind of heaven in which musical phrases already exist. The great musician discovers these phrases —he hears them, as it were. Inferior musicians hear them imperfectly; they give a confused and distorted rendering of the pure and celestial reality. The faculty for grasping celestial music is rare; the faculty for grasping celestial triangles, on the other hand, seems to be possessed by all men.

These notions, so far as geometry is concerned, rest upon the supposed necessity of Euclid's axioms. The fundamental postulates of Euclidean geometry were regarded, up to the early part of the nineteenth century, by practically every mathematician and philosopher, as necessities of thought. It was not only that Euclidean geometry was considered to be the geometry of existing space—it was the necessary geometry of any space. Yet it had quite early been realized that there was a fault in this apparently impeccable edifice. The well-known definition of parallel lines was not, it was felt, sufficiently obvious, and the Greek followers of Euclid made attempts to improve it. The Arabians also, when they acquired the Greek mathematics, found the parallel axiom unsatisfactory. No one doubted that this was a necessary truth, but they thought there should be some way of deducing it from the other and simpler axioms of Euclid. With the spread of mathematics in Europe came a whole host of attempted demonstrations of the parallel axiom. Some of these were miracles of ingenuity, but it could be shown in every case that they rested on assumptions which were equivalent to accepting the parallel axiom itself. One of the most noteworthy of these investigations was that of the Jesuit priest Girolamo Saccheri, whose treatise appeared early in the eighteenth century. Saccheri was an extremely able logician, too able to make unjustified assumptions. His method was to develop the consequences of denying Euclid's parallel axiom while retaining all the others. In this way he expected to develop a geometry which should be self-contradictory, since he had no doubt that the parallel axiom was a necessary truth. But although Saccheri struggled very hard he did not succeed in contradicting himself; what he actually did was to lay the foundations of the first non-Euclidean geometry. But even so, and although D'Alembert was expressing the opinion of all the mathematicians of his time in declaring the parallel axiom to be the "scandal" of geometry, no one seems seriously to have doubted it. It appears that the first mathematician to realize that the parallel axiom could be denied and yet a perfectly self-consistent geometry constructed was Gauss. But Gauss quite realized how staggering, how shocking, a thing he had done, and was afraid to publish his researches. It was reserved for a Russian, Lobachevsky, and a Hungarian, Bolyai, to publish the first non-Euclidean geometry. It at once became obvious that Euclid's axioms were not necessities of thought, but something quite different, and that there was no reason to suppose that triangles had any celestial existence whatever.

The further development of non-Euclidean geometry and its application to physical phenomena by Einstein have shown that Euclid's geometry is not only not a necessity of thought but is not even the most convenient geometry to apply to existing space. And with this there has come, of course, a profound change in the status we ascribe to mathematical entities, and a different estimate of the significance of the mathematician's activities. We can start from any set of axioms we please, provided they are consistent with themselves and one another, and work out the logical consequences of them. By doing so we create a branch of mathematics. The primary definitions and postulates are not given by experience, nor are they necessities of thought. The mathematician is entirely free, within the limits of his imagination, to construct what worlds he pleases. What he is to imagine is a matter for his own caprice; he is not thereby discovering the fundamental principles of the universe nor becoming acquainted with the ideas of God. If he can find, in experience, sets of entities which obey the same logical scheme as his mathematical entities, then he has applied his mathematics to the external world; he has created a branch of science. Why the external world should obey the laws of logic, why, in fact, science should be possible, is not at all an easy question to answer. There are even indications in modern physical theories which make some men of science doubt whether the universe will turn out to be finally rational. But, however that may be, there is certainly no more reason to suppose that natural phenomena must obey any particular geometry than there is to suppose that the music of the spheres, should we ever hear it, must be in the diatonic scale.

Since, then, mathematics is an entirely free activity, unconditioned by the external world, it is more just to call it an art than a science. It is as independent as music of the external world; and although, unlike music, it can be used to illuminate natural phenomena, it is just as "subjective," just as much of a product of the free creative imagination. And it is not at all difficult to discover that the mathematicians are impelled by the same incentives and experience the same satisfactions as other artists. The literature of mathematics is full of æsthetic terms, and the mathematician who said that he was less interested in results than in the beauty of the methods by which he found the results was not expressing an unusual sentiment.

But to say that mathematics is an art is not to say that it is a mere amusement. Art is not something which exists merely to satisfy an "æsthetic emotion." Art which is worthy of the name reveals to us some aspect of reality. This is possible because our consciousness and the external world are not two independent entities. Science has advanced sufficiently far for us to be able to say that the external world is, at least very largely, our own creation; and we understand much of what we have created by understanding the laws of our own being, the laws in accordance with which we must create. There is no reason to suppose that there is a heavenly storehouse of musical phrases, but it is true that the musician can reveal to us a reality which is profounder than that of common sense. "He who understands the meaning of my music," Beethoven is reported to have said, "shall be free from the miseries that afflict other men." We may not know what he meant, but it is evident that he regarded music as something that had meaning, something that revealed a reality which cannot normally be perceived. And it seems that the mathematician, in creating his art, is exhibiting that movement of our minds that has created the spatio-temporal material universe we know. Mathematics, as much as music or any other art, is one of the means by which we rise to a complete self-consciousness. The significance of mathematics resides precisely in the fact that it is an art; by informing us of the nature of our own minds it informs us of much that depends on our minds. It does not enable us to explore some remote region of the eternally existent; it helps to show us how far what exists depends upon the way in which we exist. We are the law-givers of the universe; it is even possible that we can experience nothing but what we have created, and that the greatest of our mathematical creations is the material universe itself.

We return thus to a sort of inverted Pythagorean outlook. Mathematics is of profound significance in the universe, not because it exhibits principles that we obey, but because it exhibits principles that we impose. It shows us the laws of our own being and the necessary conditions of experience. And is it not true that the other arts do something similar in those regions of experience which are not of the intellect alone? May it not be that the meaning Beethoven declared his music to possess is that, although man seems to live in an alien universe, yet it is true of the whole of experience as well as of that part of it which is the subject of science that what man finds is what he has created, and that the spirit of man is indeed free, eternally subject only to its own decrees? But however this may be it is certain that the real function of art is to increase our self-consciousness; to make us more aware of what we are, and therefore of what the universe in which we live really is. And since mathematics, in its own way, also performs this function, it is not only æsthetically charming but profoundly significant. It is an art, and a great art. It is on this, besides its usefulness in practical life, that its claim to esteem must be based.

Index